problems in quantum mechanics

editor D ter Haar

pion · london

third edition

problems in quantum mechanics

editor D ter Haar

Pion Limited, 207 Brondesbury Park, London NW2 5JN

First published 1960
Second edition 1964
Third revised and enlarged edition 1975
Reprinted January 1978

ISBN 0 85086 050 4

Printed in Great Britain

Contents

Preface to second edition

This is essentially an enlarged and revised second edition of a collection of problems which consisted of a text by Gol'dman and Krivchenkov augmented by a selection from a similar text by Kogan and Galitskii.

In preparing the present edition I have used the opportunity to revise some of the problems in the first edition, to change a few of the solutions, and to make the notation both uniform and conforming to English usage. Also, I have added a few problems from a collection by Irodov on atomic physics and a number of new problems which were mainly taken from Oxford University Examination papers. I should like to express my thanks to the Oxford University Press for permission to include these problems.

These problems can be used either in conjunction with any modern textbook, such as those by Schiff, Kramers, Landau and Lifshitz, Messiah, or Davydov, or as advanced reading for anybody who is familiar with the basic ideas of quantum mechanics from a more elementary textbook.

Oxford,
September 1963 D. ter Haar

Preface to third edition

In preparing the third edition I have dropped some of the problems, slightly rearranged the order of the problems, and added new problems to the old chapters, as well as added new sections on the density matrix and annihilation and creation operator problems and on relativistic wave equations. Otherwise the aim and scope of the book remain much as they were before, but to help readers I have added stars to more complicated problems.

Oxford,
September 1974 D. ter Haar

problems

1

One-dimensional motion

1. Determine the energy levels and the normalised wave functions of a particle in a "potential well". The potential energy V of the particle is:

$$V = \infty \quad \text{for} \quad x < 0 \text{ and for } x > a;$$

$$V = 0 \quad \text{for} \quad 0 < x < a.$$

2. Show that for particles in a "potential well" (see preceding problem) the following relations hold:

$$\bar{x} = \tfrac{1}{2}a,$$

$$\overline{(x-\bar{x})^2} = \frac{a^2}{12}\left(1 - \frac{6}{n^2\pi^2}\right).$$

Show also that for large values of n the above result agrees with the corresponding classical result.

3. Determine the momentum probability distribution function for particles in the nth energy state in a "potential well".

4. A particle in an infinitely deep rectangular potential well is in a state described by the wave function

$$\psi(x) = Ax(a-x),$$

where a is the well width and A a constant.

Find the probability distribution for the different energies of the particle and also the average value and the dispersion of the energy.

5. A particle is in the ground state in a potential well of length a. At time $t = 0$ the wall at $x = a$ is suddenly moved to $x = 2a$. Calculate the probability that, at time $t > 0$,

(a) the energy of the particle is the same as before $t = 0$; and

(b) the energy of the particle is less than before $t = 0$.

6*. A particle is enclosed in a one-dimensional rectangular potential well with infinitely high walls. Evaluate the average force exerted by the particle on the wall of the well.

7. Determine the energy levels and wave functions of a particle in an asymmetrical potential well (see fig. 1). Consider the case where $V_1 = V_2$.

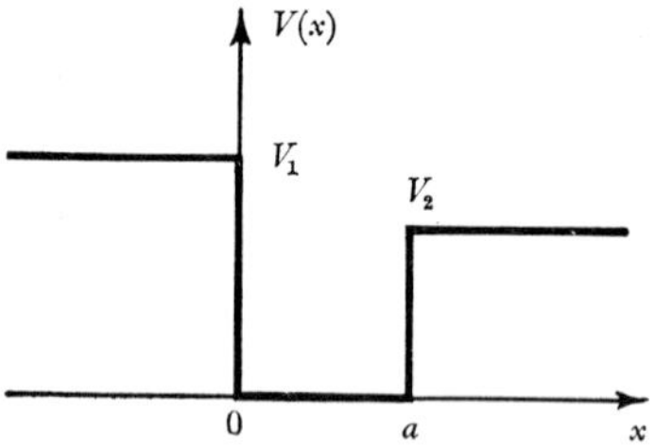

Fig. 1.

8. The Hamiltonian of an oscillator is equal to $\hat{H} = \hat{p}^2/2\mu + \mu\omega^2\hat{x}^2/2$, where $\hat{p}$ and $\hat{x}$ satisfy the commutation relationships $\hat{p}\hat{x} - \hat{x}\hat{p} = -i\hbar$. In order to eliminate $\hbar$, μ, and ω from the calculations, we introduce new variables $\hat{P}$ and $\hat{Q}$,†

$$\hat{P} = \frac{1}{\sqrt{(\mu\hbar\omega)}}\,\hat{p}, \quad \hat{Q} = \sqrt{\left(\frac{\mu\omega}{\hbar}\right)}\,\hat{x} \quad (P\hat{Q} - \hat{Q}P = -i),$$

and the energy E will be expressed in units $\hbar\omega$ ($E = \epsilon\hbar\omega$). The Schrödinger equation for the oscillator in the new variables will be of the form

$$\hat{H}'\psi = \tfrac{1}{2}(\hat{P}^2 + \hat{Q}^2)\psi = \epsilon\psi.$$

(a) Use the commutation relation $\hat{P}\hat{Q} - \hat{Q}\hat{P} = -i$, to show that

$$\tfrac{1}{2}(\hat{P}^2 + \hat{Q}^2)(\hat{Q} \pm i\hat{P})^n\psi = (\epsilon \mp n)(\hat{Q} \pm i\hat{P})^n\psi.$$

(b) Determine the normalised wave functions and the energy levels of the oscillator.

(c) Determine the commutator of the operator $\hat{a} = (1/\sqrt{2})(\hat{Q} + i\hat{P})$ and its Hermitean conjugate operator $\hat{a}^+ = (1/\sqrt{2})(\hat{Q} - i\hat{P})$. Express the wave function of the nth excited state in terms of the wave function of the ground state using the operator $\hat{a}$.

(d) Determine the matrix elements of the operators $\hat{P}$ and $\hat{Q}$ in the energy representation.

Hint. $\hat{P}^2 + \hat{Q}^2 - 1 = (\hat{P} + i\hat{Q})(\hat{P} - i\hat{Q})$.

9. Using the results of the preceding problem, show by direct multiplication of matrices that for an oscillator in the nth stationary state we have

$$\overline{(\Delta x)^2} = \overline{x^2} = \frac{\hbar}{\mu\omega}(n + \tfrac{1}{2}); \quad \overline{(\Delta p)^2} = \overline{p^2} = \mu\hbar\omega(n + \tfrac{1}{2}).$$

† Operators are indicated by a caret ^.

Show that, for any stationary state, the root-mean-square value of x is the same as it would have been for a classical oscillator with the same energy.

10. A particle moves in a potential $V(x) = \frac{1}{2}\mu\omega^2 x^2$. Determine the probability w to find the particle outside the classical limits, when it is in its ground state.

11. Find the energy levels of a particle moving in a potential of the following form:

$$V(x) = \infty \quad (x < 0); \quad V(x) = \frac{\mu\omega^2 x^2}{2} \quad (x > 0).$$

12. Write down the Schrödinger equation for an oscillator in the "p-representation" and determine the momentum probability distribution function.

13. Find the wave functions and energy levels of a particle in a potential $V(x) = V_0(a/x - x/a)^2$ $(x > 0)$ (see fig. 2) and show that the energy spectrum is the same as the oscillator spectrum.

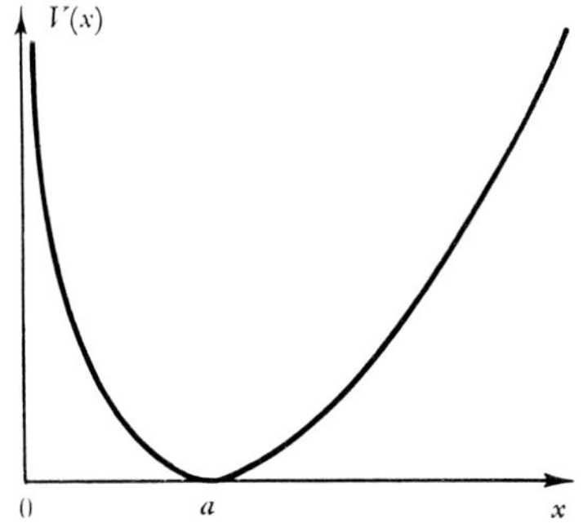

Fig. 2.

14*. Determine the energy levels for a particle in a potential $V = -V_0/\cosh^2(x/a)$ (see fig. 3).

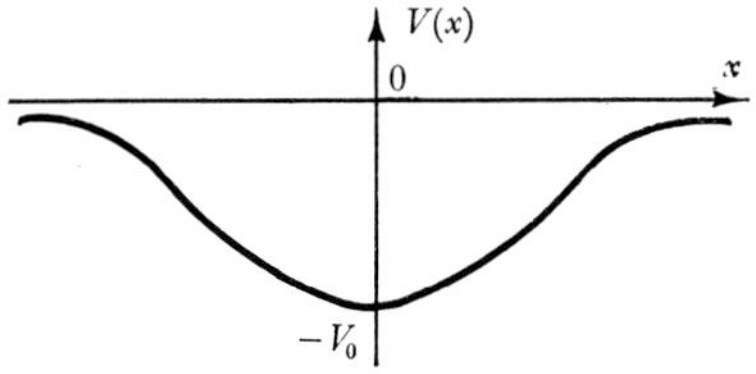

Fig. 3.

15*. Determine the energy levels and wave functions for a particle in the potential $V = V_0 \cot^2(\pi x/a)$ $(0 < x < a)$ (see fig. 4), and derive the normalisation constant of the ground state wave function.

Consider the limiting cases of small and large values of V_0.

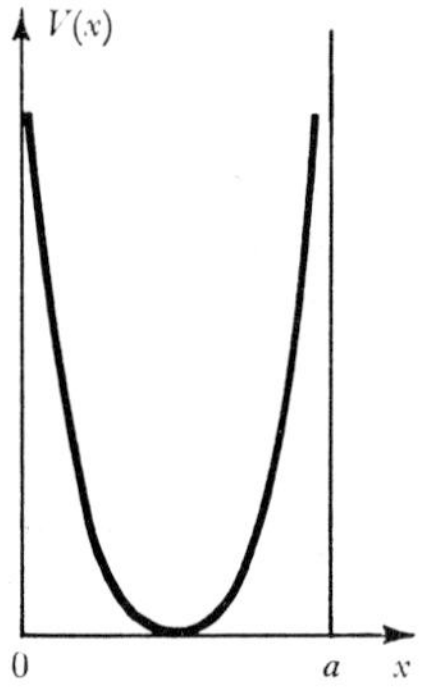

Fig. 4.

16. Determine the energy levels for a particle in a potential

$$V(x) = D(1-e^{-ax})^2.$$

17. An electron of mass μ moves in a one-dimensional potential

$$V(x) = -\frac{\hbar^2 P}{\mu}\,\delta(x^2-a^2),$$

where P is a positive dimensionless constant, $\delta(x)$ the Dirac delta-function, and a a constant length. Discuss the bound states for this potential as a function of P.

18*. A one-dimensional "hydrogen" atom is one in which an electron confined to the x-axis is acted upon by a force inversely proportional to the square of its distance from the origin. Find the energy eigenvalues and the eigenfunctions of this system.

19. Determine the wave functions of a charged particle in a uniform field $V(x) = -Fx$.

20*. Find the wave functions and energy levels of the stationary states of a particle of mass μ in a uniform gravitational field $\boldsymbol{g}$ for the case when the region of the motion of the particle is limited from below by a perfectly reflecting plane. (As a classical analogy of this system we can take a heavy solid ball, bobbing up and down on a metallic plate. We note that all calculations and results of this problem are clearly correct also for the case of the motion of a particle of charge $\boldsymbol{e}$ in a uniform electric field $\mathscr{E}$, in the presence of a reflecting plane, provided we replace in all equations $\boldsymbol{g}$ by $(e/m)\mathscr{E}$.) Take the limit to classical mechanics.

21*. Derive expressions for the wavefunction of a particle moving in a potential $V(x)$ using the semi-classical approximation. Give conditions for the applicability of the approximation and determine the quantization condition.

22. Use the semi-classical approximation to derive an expression for the number of discrete levels of a particle moving in a given potential.

23. Determine in the semi-classical approximation the energy spectrum of a particle in the following potentials:

(a) $V = \frac{1}{2}\mu\omega^2 x^2$ (oscillator);

(b) $V = V_0 \cot^2(\pi x/a)$ $(0 < x < a)$.

24. Use the semi-classical approximation to determine the bound energy levels for a particle of mass μ moving in a potential which equals $-V_0$ for $x = 0$, changes linearly with x until it vanishes at $x = \pm a$, and is zero for $|x| > a$ (see fig. 5). Determine also the total number of discrete energy levels.

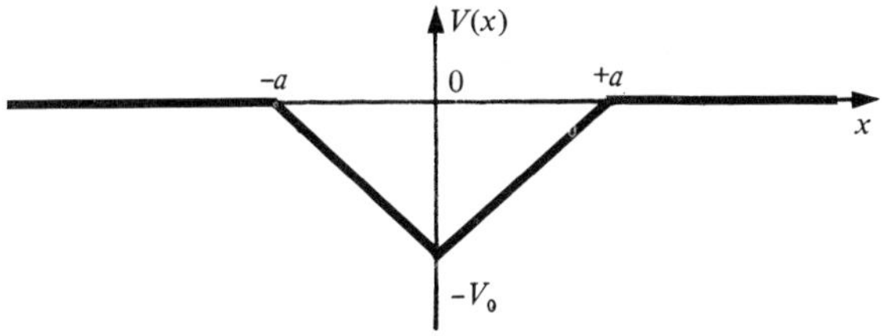

Fig. 5.

25. Determine in the semi-classical approximation the average value of the kinetic energy in a stationary state.

26. Use the result of the preceding question to find in the semi-classical approximation the average kinetic energy of a particle in the following potentials:

(a) $V = \frac{1}{2}\mu\omega^2 x^2$;

(b) $V = V_0 \cot^2(\pi x/a)$ $(0 < x < a)$ (see problem 23).

27. Determine the form of the energy spectrum of a particle in a potential $V(x) = ax^\nu$, using the semi-classical approximation and applying the virial theorem.

28. Obtain the semi-classical expression for the energy levels of a particle in a uniform gravitational field for the case where its motion is limited from below by a perfectly reflecting plane.

29. A particle oscillates in a one-dimensional potential field between two turning points $x = a$ and $x = b$. The former is due to a vertical potential wall, while the latter is of the more usual type with dV/dx finite. Apply the WKB method to find the quantisation condition for a stationary state in such a potential.

30*. Determine in the semi-classical approximation the form of the potential energy $V(x)$ for a given energy spectrum E_n. $V(x)$ may be assumed to be an even function $V(x) = V(-x)$, which increases monotonically for $x > 0$.

31*. Find the semi-classical solution of the Schrödinger equation in the momentum-representation.

Show that the same semi-classical function is obtained by going over from the "x-representation" to the "p-representation" starting from the usual semi-classical coordinate wave function.

32. Find the wave functions and energy levels of the stationary states of a plane rotator with moment of inertia I.

A rotator is a system of two rigidly connected particles rotating in a plane (or in space). The moment of inertia of a rotator is equal to $I = \mu a^2$, where μ is the reduced mass of the particles and a their distance apart.

33. If the wave function of a plane rotator at $t = 0$ is given by $\psi(\varphi, 0) = A \sin^2 \varphi$, where A is a normalising constant, what will be the wave function $\psi(\varphi, t)$ at time t?

34. A bead of mass μ is confined to a thin wire which forms a rigid circular loop of radius a. Find an expression for the tension in the wire when the system is in a stationary state, assuming the wire to be unstressed before the bead is placed on it.

35. Write down the Schrödinger equation in the "p-representation" for a particle moving in a periodic potential $V(x) = V_0 \cos bx$.

36. Write down the Schrödinger equation in the "p-representation" for a particle moving in a periodic potential $V(x) = V(x+b)$.

37*. Determine the allowed energy bands of a particle moving in the periodic potential given by fig. 6. Investigate the limiting case where $V_0 \to \infty$, and $b \to 0$ while $V_0 b =$ constant.

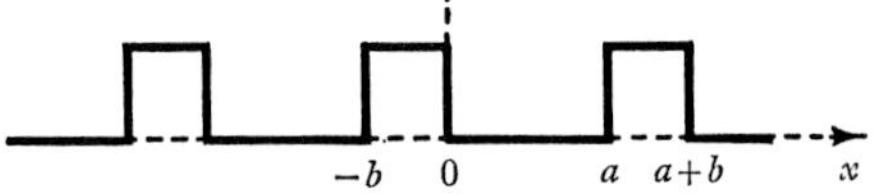

Fig. 6.

38*. A simple model of the electronic energy levels in a metal uses a one-dimensional potential of the form

$$V(x) = \frac{\hbar^2 P}{2\mu a} \sum_{n=-\infty}^{+\infty} \delta(x+na),$$

where μ is the electron mass, a the lattice constant, P a positive, dimensionless constant, and $\delta(x)$ the Dirac delta-function. Find expressions for the effective mass at the upper band edges.

39*. A particle moves in a periodic field $V(x)$:

$$V(x+a) = V(x).$$

Using a suitable semi-classical approximation obtain a transcendental equation to determine the allowed energy bands. Discuss this equation.

40. Show that for particles scattered by a complex potential, $V(x)(1+i\xi)$, the probability current density,

$$j = \frac{\hbar}{2\mu i}\left[\psi^* \frac{\partial\psi}{\partial x} - \psi\frac{\partial\psi^*}{\partial x}\right],$$

and the probability density, $\rho = \psi^*\psi$, satisfy the "continuity" equation

$$\frac{\partial j}{\partial x} + \frac{\partial \rho}{\partial t} = \frac{2V(x)\xi\rho}{\hbar}.$$

41. Use a variational principle to prove that any purely attractive one-dimensional potential has at least one bound state.

42. A particle of mass μ moves in a one-dimensional potential $\lambda V(x)$, where $V(x)$ satisfies the conditions

$$V(x) = 0, \quad x<0; \quad V(x) = 0, \quad x>a, \quad \lambda\int_0^a V(x)\,dx < 0.$$

Prove that, if λ is sufficiently small, there exists a bound state with an energy E which is approximately given by

$$E = -\frac{\mu\lambda^2}{2\hbar^2}\left[\int_0^a V(x)\,dx\right]^2.$$

Tunnel effect

1. In studying the emission of electrons from metals, it is necessary to take into account the fact that electrons with an energy sufficient to leave the metal may be reflected at the metal surface. Consider a one-dimensional model with a potential V which is equal to $-V_0$ for $x<0$ (inside the metal) and equal to zero for $x>0$ (outside the metal) (see fig. 7), and determine the reflection coefficient at the metal surface for an electron with energy $E>0$.

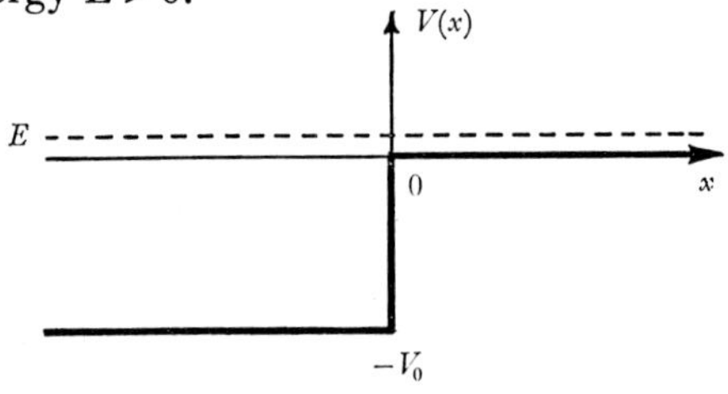

Fig. 7.

2*. In the preceding problem it was assumed that the potential changed discontinuously at the metal surface. In a real metal this change in potential takes place continuously over a region of the dimensions of the order of the interatomic distance in the metal. Approximate the potential near the metal surface by the function

$$V = -\frac{V_0}{e^{x/a}+1} \quad \text{(see fig. 8)}$$

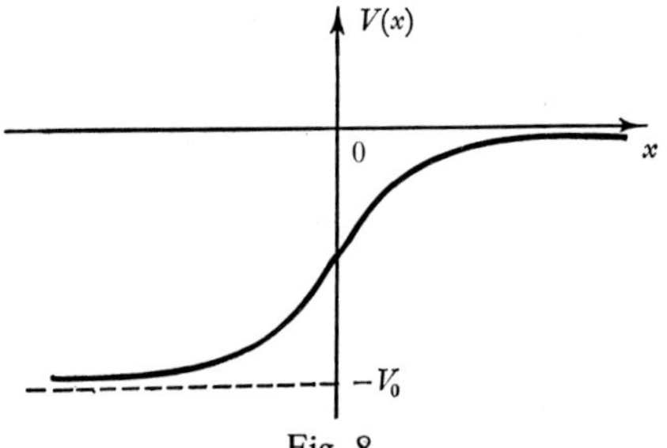

Fig. 8.

and determine the reflection coefficient of an electron with energy $E>0$.

3. Determine the coefficient of transmission of a particle through a rectangular barrier (see fig. 9).

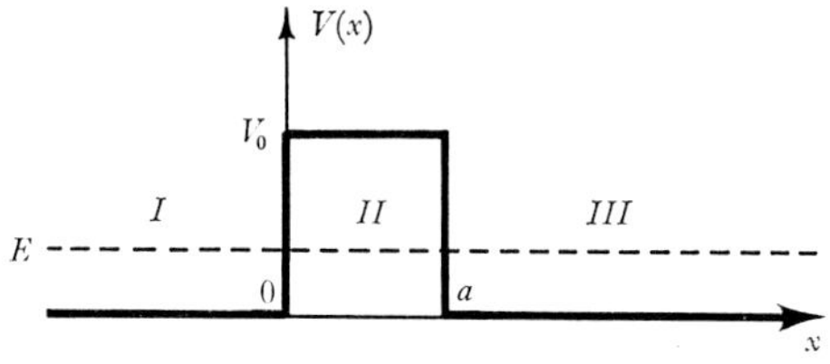

Fig. 9.

4. Determine the coefficient of reflection of a particle by a rectangular barrier in the case where $E > V_0$ (reflection above the barrier).

5. A particle is moving along the x-axis. Find the probability for transmission of the particle through a delta-function potential barrier at the origin.

6. Determine approximately the energy levels and wave functions of a particle in the symmetrical potential given by fig. 10 for the case where $E \ll V_0$ and the penetrability of the barrier is small $[(2\mu V_0/\hbar^2)\, b^2 \gg 1]$.

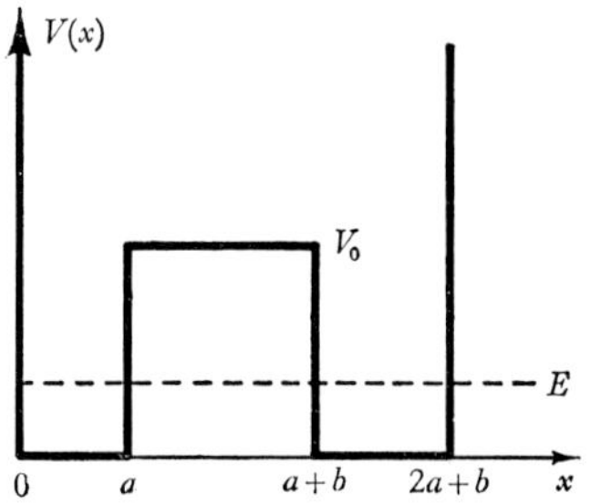

Fig. 10.

7*. Find the coefficient of transmission of a particle through a triangular barrier (see fig. 11). Consider the limiting cases of small and of large pentrability.

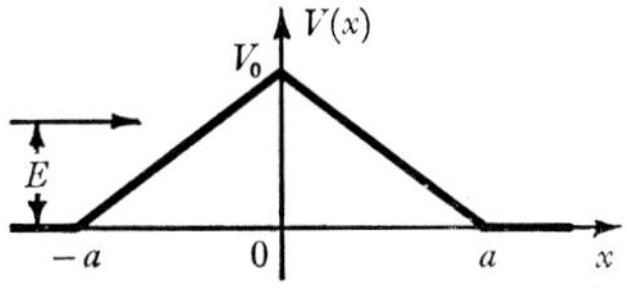

Fig. 11.

8*. Calculate the coefficient of transmission through a potential barrier $V(x) = V_0/\cosh^2(x/a)$ (see fig. 12) for particles moving with an energy $E < V_0$.

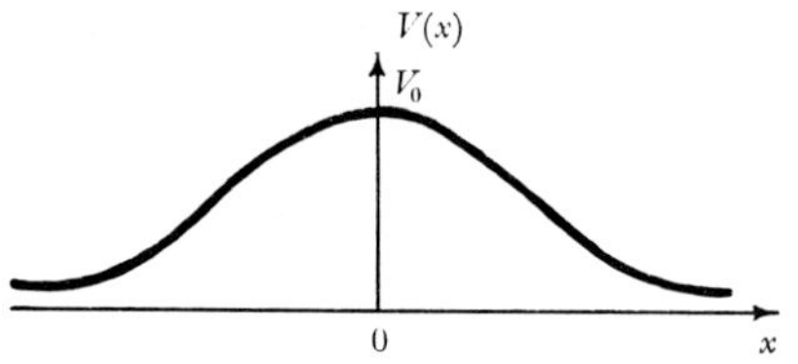

Fig. 12.

9. Evaluate in the semi-classical approximation the transmission coefficient for a parabolic potential barrier of the following form (see fig. 13):

$$V(x) = \begin{cases} V_0\left(1 - \dfrac{x^2}{a^2}\right) & \text{for} \quad -a \leqslant x \leqslant a, \\ 0 & \text{for} \quad |x| \geqslant a. \end{cases} \tag{1}$$

Give the criterion for the applicability of the result obtained.

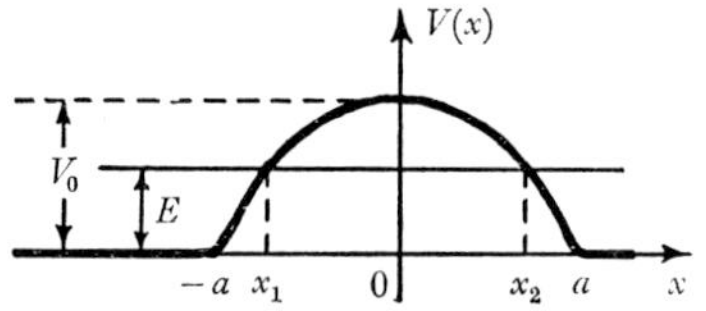

Fig. 13.

10. Calculate in the semi-classical approximation the coefficient of transmission of electrons through a metal surface under the action of a large electrical field strength F (fig. 14). Find the limits of applicability of the calculation.

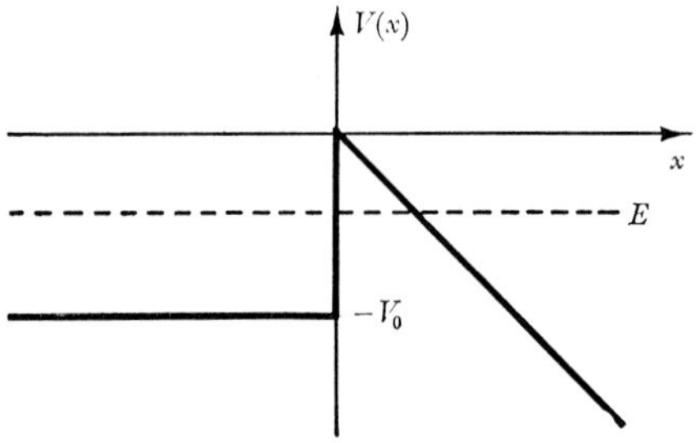

Fig. 14.

11*. The change of the potential near a metal surface is in reality a continuous one. For instance, the electrical image potential $V_{\text{e im}} = -e/4x$ will act at large distances from the surface. Determine the coefficient of transmission D of electrons through a metal surface under the action of an electrical field, taking into account the electrical image force (fig. 15).

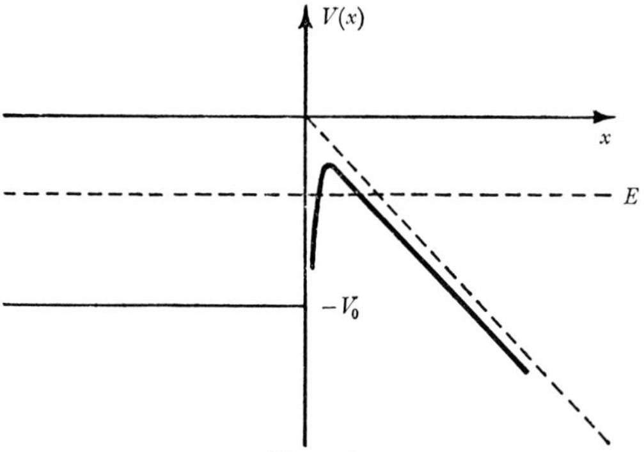

Fig. 15.

12*. A symmetrical potential $V(x)$ consists of two potential wells separated by a barrier (see fig. 16). Assuming that one may use a semi-classical argument, determine the energy levels of a particle in the potential $V(x)$. Compare the energy spectrum obtained with the energy spectrum of a single well.

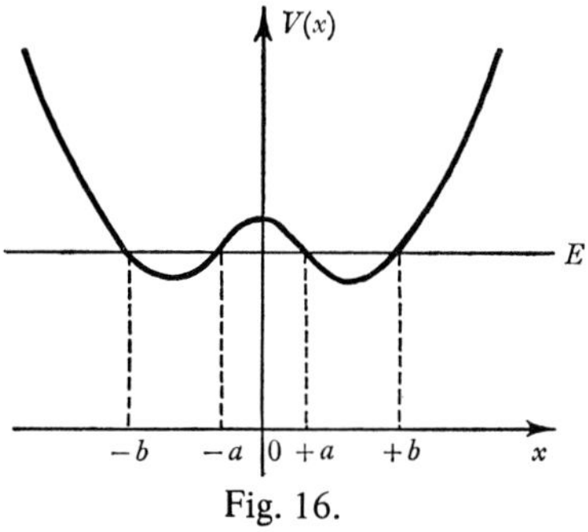

Fig. 16.

13. Assume that at $t = 0$ there exists an impenetrable partition between the two symmetrical potential wells (see preceding problem) and that a particle is in a stationary state in the well on the left.

Determine the time τ it takes after the partition is removed before the particle will be in the well on the right.

14*. The potential $V(x)$ consists of N identical potential wells separated by identical potential barriers (see fig. 17). Determine the energy levels in this potential, assuming that one can use the semi-classical approach.

Compare the energy spectrum obtained with the energy spectrum of a single well.

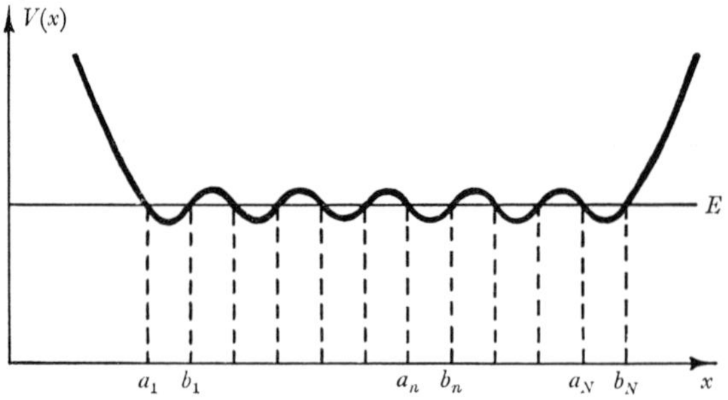

Fig. 17.

15*. Assuming that one may use the semi-classical approach, find the quasi-stationary levels of a particle in the symmetrical field given by fig. 18.

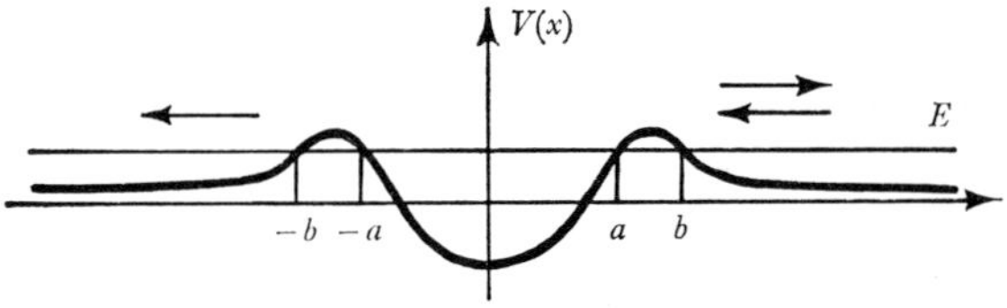

Fig. 18.

Find also the transmission coefficient $D(E)$ for a particle with energy $E < V_0$, where V_0 is the maximum value of the potential $V(x)$.

16. (i) Show generally that for any barrier the relation $R + D = 1$ is automatically satisfied, where R is the reflection coefficient and D the transmission coefficient.

(ii) A particle confined to one dimension encounters a symmetric potential barrier of finite extension with its centre at $x = 0$. For each energy $E = \hbar^2k^2/2\mu$ there is an even and an odd solution of the Schrödinger equation which to the right of the barrier have the form $A\cos(kx + \varphi)$ and $B\sin(kx + \varphi')$, respectively. Show that the reflection and transmission coefficients at this energy are, respectively, $\sin^2 2(\varphi - \varphi')$ and $\cos^2 2(\varphi - \varphi')$.

17. A particle is moving in a potential $V(x)$ which is zero for $|x| > a$ and finite for $|x| < a$. Check that the wavefunction $\psi(x)$ satisfies the integral equation

$$\psi(x) = e^{ikx} + \frac{\mu}{ik\hbar^2}\int_{-a}^{a} e^{ik|x-y|}\,V(y)\psi(y)dy\ .$$

Obtain an expression for the reflection coefficient to lowest order in V, if a beam of particles of mass μ and momentum $p = \hbar k$ is incident upon this potential $V(x)$.

3

Commutation relations; Heisenberg relations; spreading of wave packets; operators

1. Let $\hat{A}$, $\hat{B}$, and $\hat{C}$ be three operators. Express the commutator of the product $\hat{A}\hat{B}$ with $\hat{C}$ in terms of the commutators $[\hat{A}, \hat{C}]_-$ and $[\hat{B}, \hat{C}]_-$.

2. Show that for algebraic manipulations with commutators the distributive law holds, that is, that the commutator of a sum is the sum of commutators:

$$\Big[\sum_i \hat{A}_i, \sum_k \hat{B}_k\Big]_- = \sum_{i,k} [\hat{A}_i, \hat{B}_k]_-.$$

3. Prove that if $f(x)$ is a function of the coordinate x and $g(p)$ a function of the momentum p, we have

$$[\hat{p}, \hat{f}]_- = -i\hbar \frac{d\hat{f}}{dx} \quad \text{and} \quad [\hat{x}, \hat{g}]_- = i\hbar \frac{d\hat{g}}{dp} .$$

4. Prove the following relation:

$$\exp(\hat{L})\,\hat{a}\exp(-\hat{L})$$
$$= \hat{a} + [\hat{L}, \hat{a}]_- + \frac{1}{2!}[\hat{L}, [\hat{L}, \hat{a}]_-]_- + \frac{1}{3!}[\hat{L}, [\hat{L}, [\hat{L}, \hat{a}]_-]_-]_- + \dots .$$

5. Show that, if $[\hat{A}, \hat{B}]_- = K$ where K is a c-number and if λ is a c-number,

$$\exp[\lambda(\hat{A}+\hat{B})] = \exp(\lambda\hat{A})\exp(\lambda\hat{B})\exp(-\tfrac{1}{2}\lambda^2 K).$$

6. Prove that

$$[\hat{A}, e^{-\alpha\hat{B}}]_- = e^{-\alpha\hat{B}} \int_0^\alpha e^{\lambda\hat{B}}[\hat{A}, \hat{B}]_- e^{-\lambda\hat{B}}\, d\lambda \; ;$$

this relation is called Kubo's identity.

7. Show that if the two Hermitean operators $\hat{A}$ and $\hat{B}$ satisfy the commutation relation $\hat{A}\hat{B} - \hat{B}\hat{A} = i\hat{C}$, the following relation will hold:

$$\sqrt{[\overline{(\Delta\hat{A})^2}\,\overline{(\Delta\hat{B})^2}]} \geqslant \frac{|\bar{\hat{C}}|}{2}.$$

8. Find the uncertainty relation for the operators $\hat{q}$ and $F(\hat{p})$, if $\hat{q}$ and $\hat{p}$ satisfy the commutation relation $\hat{q}\hat{p} - \hat{p}\hat{q} = i\hbar$.

Hint. Express $F(\hat{p})$ in a Taylor series.

9. Use the Heisenberg relations to estimate the ground state energy of a harmonic oscillator.

10. Estimate the energy of an electron in the K-shell of an atom of atomic number Z both for the relativistic and the non-relativistic case.

11. Estimate the ground state energy of a two-electron atom with nuclear charge Z, using the Heisenberg relations.

12. The magnetic field produced by a free electron is partly due to its motion and partly due to the presence of its intrinsic magnetic moment.

It is known from electrodynamics that the magnetic field strength H_1 due to a moving charge is of the order of magnitude

$$H_1 \sim \frac{ev}{cr^2},$$

and the magnetic field strength H_2 due to a dipole moment μ

$$H_2 \sim \frac{\mu}{r^3}.$$

To determine the magnetic moment μ of a free electron from a measurement of the field strength produced by it, it is necessary that the following two conditions are satisfied:

$$H_2 \gg H_1 \tag{1}$$

and

$$\Delta r \ll r. \tag{2}$$

The meaning of the last condition is that the electron must be localised in a region Δr which is much smaller than the distance from that region to the point where the magnetic field is observed.

Is it possible to satisfy these two conditions simultaneously?

Hint. Take into account the Heisenberg relations and the value of the electronic magnetic moment $\mu = e\hbar/2mc$.

13. We consider a particle in a one-dimensional symmetrical potential well in which there is always, as is well known, at least one energy level (compare problem 41 of section 1).

If for a given depth of the well V_0 its width a is reduced until it satisfies the inequality

$$a^2 \ll \frac{\hbar^2}{\mu V_0},$$

then, at first sight, the spatial localisation of a particle bound in the well will become much more precise ($\Delta x \sim a$), and as the spread in momentum Δp in any case is limited to a value of the order $\sqrt{(\mu V_0)}$, the following inequality

$$\Delta p\, \Delta x \lesssim \sqrt{(\mu V_0)}\,.\,a \ll \hbar$$

would hold, violating the Heisenberg relations.

Show the error in the above argument and evaluate the product of the spread in coordinate and the spread in momentum of the particle.

14. The trajectory of a particle in a Wilson chamber is a chain of small droplets of linear dimensions of about 1 μ. Could one observe deviations from the classical motion for an electron with energy 1 keV?

15. For what value of the relative uncertainty in the angular momentum of the electron in the first Bohr orbit does its angular coordinate become completely undetermined?

16. Consider the operators $\hat{M}_z = -i\hbar\partial/\partial\varphi$ and $\hat{\varphi}$ for circular coordinates $-\pi < \varphi \leqslant \pi$. Discuss the Heisenberg relations as applied to $\hat{M}_z$ and $\hat{\varphi}$ (compare the derivation of the Heisenberg relations in problem 7 of section 3).

17. The wave function of a free particle at $t = 0$ is given by the expression

$$\psi(x, 0) = \varphi(x) \exp(ip_0 x/\hbar).$$

The function $\varphi(x)$ is real and appreciably different from zero only for values of x within the interval $-\delta < x < +\delta$. Determine in which region of x-values the wave function will be different from zero at time t.

18. Find the change of a wave function which is given at $t = 0$ for the following three cases (spreading of a wave packet):

(a) a free particle,

$$\psi(\boldsymbol{r}, 0) = \frac{1}{(\pi\delta^2)^{\frac{3}{4}}} \exp\left(\frac{i(\boldsymbol{p}_0 \cdot \boldsymbol{r})}{\hbar} - \frac{r^2}{2\delta^2}\right);$$

*(b) a particle moving in a uniform field,

$$\psi(\boldsymbol{r}, 0) = \frac{1}{(\pi\delta^2)^{\frac{3}{4}}} \exp\left(\frac{i(\boldsymbol{p}_0 \cdot \boldsymbol{r})}{\hbar} - \frac{r^2}{2\delta^2}\right);$$

(c) a particle moving in a potential $V = \frac{1}{2}\mu\omega^2 x^2$,

$$\psi(x, 0) = c \exp\left[\frac{ip_0 x}{\hbar} - \frac{\alpha^2(x - x_0)^2}{2}\right], \quad \alpha = \left(\frac{\mu\omega}{\hbar}\right)^{\frac{1}{2}}.$$

19. Show that the problem of how to determine the motion of an oscillator under the action of an external force $f(t)$ can be reduced to the simpler problem of determining the motion of a free oscillator, if we introduce the new variable $x_1 = x - \xi(t)$, where $\xi(t)$ satisfies the classical equation $\mu\ddot{\xi} = f(t) - \mu\omega^2 \xi$.

20*. Find the Green function for an oscillator whose eigenfrequency changes with time, and express it in terms of the solution of the classical equation of motion of an oscillator of varying frequency.

21*. Use the Green function obtained in the preceding problem to determine the time dependence of the probability density for particles moving in a potential $V(x) = \frac{1}{2}\mu\omega^2 x^2$ ($\omega = \text{const}$). The particle wave function at $t = 0$ is equal to

$$\psi(x, 0) = c\exp\left[-\frac{\alpha^2(x-x_0)^2}{2} + \frac{ip_0 x}{\hbar}\right].$$

22*. Find the Green function of an oscillator whose eigenfrequency changes in time and which is acted upon by a perturbing force $f(t)$.

23*. An oscillator is at $t = 0$ in its nth energy eigenstate. Determine the probability that it will make a transition to the mth eigenstate under the action of a perturbing force $f(t)$. Find the average value and the dispersion of its energy at time t.

24*. The point of suspension x_0 of an oscillator in its ground state starts to move at $t = 0$. The point of suspension moves according to the law $x_0 = x_0(t)$. At $t = T$ the point of suspension is fixed again.

Use the Green function (see problem 20 of section 3) to find the wave function of the system at any time $t > 0$, and also the probability of excitation of the nth level as a result of the process considered.

Consider the limiting cases of fast and slow processes ($\omega T \ll 1$ and $\omega T \gg 1$, where ω is the frequency of the oscillator).

25*. Discuss the motion of a particle governed by the Hamiltonian

$$\hat{H} = e^{-\gamma t}\frac{\hat{p}^2}{2\mu} - e^{\gamma t}F(t)x\,,$$

where γ is a positive constant.

Find the wavefunction for this system for the case where $F(t) = 0$ as $t \to \infty$, if at $t = 0$ the wavefunction is given by the formula

$$\psi_0 = A\exp\ ik_0 x - \frac{x^2}{2\sigma^2}\ , \qquad A = \text{constant}.$$

26. Inasmuch as the Schrödinger equation is a first-order differential equation with respect to time, $\psi(t)$ is uniquely determined by the value of $\psi(0)$. Write this connection in the form

$$\psi(t) = \hat{S}(t)\,\psi(0),$$

where $\hat{S}(t)$ is some operator.

(a) Show that the operator $\hat{S}(t)$ satisfies the equation

$$i\hbar\dot{\hat{S}}(t) = \hat{H}\hat{S}(t)$$

and is a unitary operator, that is, $\hat{S}^+ = \hat{S}^{-1}$.

(b) Show that in the case where $\hat{H}$ does not depend on time, $\hat{S}(t)$ is of the form

$$\hat{S}(t) = \exp(-i\hat{H}t/\hbar).$$

27. The average value of an operator $\hat{L}$ at time t follows from the following expression:

$$\bar{\hat{L}}(t) = \int \psi^*(t)\, \hat{L}\psi(t)\, d\tau.$$

(a) Show that the time dependence of the operator† $\hat{\mathscr{L}} = \hat{S}^{-1}(t)\,\hat{L}\hat{S}(t)$, with $\hat{S}(t)$ determined by the equation $\hat{S}(t)\,\psi(0) = \psi(t)$, satisfies the equation

$$\int \psi^*(0)\, \hat{\mathscr{L}}\psi(0)\, d\tau = \bar{\hat{L}}(t).$$

(b) Prove the following operator equation:

$$i\hbar\dot{\hat{\mathscr{L}}} = \hat{\mathscr{L}}\hat{\mathscr{H}} - \hat{\mathscr{H}}\hat{\mathscr{L}},$$

where

$$\hat{\mathscr{H}} = \hat{S}^{-1}\hat{H}\hat{S}.$$

(c) Show that if the operators $\hat{L}$ and $\hat{M}$ satisfy the commutation relation

$$\hat{L}\hat{M} - \hat{M}\hat{L} = i\hat{N},$$

the corresponding time-dependent operators satisfy the equation

$$\hat{\mathscr{L}}\hat{\mathscr{M}} - \hat{\mathscr{M}}\hat{\mathscr{L}} = i\hat{\mathscr{N}}.$$

28. Derive a law for the differentiation with respect to the time of the product of two operators.

29. Find the coordinate operator for the case of a free particle in the Heisenberg representation.

30. Find the coordinate and momentum operators in the Heisenberg representation for the case of a linear harmonic oscillator by solving the equations of motion for these operators.

31. In the Heisenberg picture the operators are time-dependent and satisfy the equation of motion (see problem 27 of section 3)

$$i\hbar\dot{\hat{\mathfrak{L}}} = [\hat{\mathfrak{L}}, \hat{\mathfrak{H}}]_-,$$

while the wavefunction is time-independent. On the other hand, in the Schrödinger picture the operators (unless explicitly dependent on the time) are time-independent, while the wavefunction satisfies the

† The operator $\hat{\mathscr{L}}$ is called the operator in the Heisenberg representation or in the Heisenberg picture, while $\hat{L}$ is the operator in the Schrödinger picture. See problem 31 of section 3 for the interaction picture.

Schrödinger equation

$$i\hbar \frac{\partial \psi}{\partial t} = \hat{H}\psi \,.$$

The operators and wavefunctions in the Heisenberg picture (indicated by a subscript H) are related to those in the Schrödinger picture (indicated by the subscript S) through the formulae

$$\hat{\mathcal{L}}_{\text{H}} = \hat{S}^{-1}(t)\hat{L}_{\text{S}}\hat{S}(t)\,, \quad \psi_{\text{S}} = \hat{S}(t)\psi_{\text{H}}\,, \quad \hat{S} = \exp(-i\hat{H}t/\hbar)\,.$$

It is often convenient to work in an intermediate picture, called the interaction picture. This is the case when the Hamiltonian is of the form

$$\hat{H} = \hat{H}_0 + \hat{H}_1\,.$$

In that case we can introduce operators and wavefunctions in the interaction picture (indicated by a subscript int) as follows

$$\hat{L}_{\text{int}} = \hat{S}_0^{-1}(t)\hat{L}_{\text{S}}\hat{S}_0(t)\,, \quad \psi_{\text{S}} = \hat{S}_0(t)\psi_{\text{int}}\,, \quad \hat{S}_0(t) = \exp(-i\hat{H}_0 t/\hbar)\,.$$

Find the equations of motion for $\hat{L}_{\text{int}}$ and ψ_{int}.

32. Prove for the one-dimensional case that if q and p are the generalized coordinate and momentum of a particle, we have

$$\overline{\dot{q}} = \overline{\frac{\partial H}{\partial p}}\,, \quad \overline{\dot{p}} = -\overline{\frac{\partial H}{\partial q}}\,,$$

where H is the Hamiltonian of the system, while the bar indicates a quantum-mechanical average. This is a particular example of the *Ehrenfest theorem*, which states that the classical equations of motion still hold, if we replace the physical quantities by quantum-mechanical averages.

33. Show that the average value of the time derivative of a physical quantity, which does not explicitly depend on the time, is equal to zero in a stationary state of the discrete spectrum.

34. Prove the virial theorem in quantum mechanics.

35. What is the physical meaning of the quantity p_0 in the expression for the wave function

$$\psi(x) = \varphi(x) \exp(ip_0 x/\hbar),$$

if $\varphi(x)$ is a real function?

36. Show that the average value of the momentum in a stationary state with a discrete energy eigenvalue is equal to zero.

37. Determine the time dependence of the coordinate operator $\hat{x}$ (in the coordinate representation) for (a) a free particle and (b) an oscillator.

38. Use the result of the preceding problem to determine the time dependence of the dispersion of the coordinate for the case of a free particle.

39. The wave function of a particle at $t = 0$ is of the form $\psi(x) = \varphi(x)\exp(ip_0 x/\hbar)$, where $\varphi(x)$ is real and normalised. Determine the value of the dispersion $\overline{(\Delta x)^2}$ at time t for the cases (a) and (b) of problem 37 of section 3. Show that in the case of the oscillator $\overline{(\Delta x)_t^2} = \overline{(\Delta x)_{t=0}^2}$, that is, the dispersion does not increase, if

$$\varphi(x) = c\exp(-\mu\omega x^2/\hbar)$$

(see problem 18c of section 3).

40. Determine the time dependence of the annihilation and creation operators for a harmonic oscillator (the $\hat{a}$ and $\hat{a}^+$ of problem 8 of section 1).

41. An oscillator is at $t = -\infty$ in its ground state. Determine the probability that at $t = +\infty$ the oscillator is in the nth excited state, if it is acted upon by a force $f(t)$, where $f(t)$ is an arbitrary even function of the time ($f = 0$ for $t \to \pm\infty$).

Evaluate the expression obtained for

(a) $f(t) = f_0 \exp(-t^2/\tau^2)$;

(b) $f(t) = f_0 \dfrac{1}{(t/\tau)^2 + 1}$.

42. Express the operator $\hat{\dfrac{1}{r}}$ in the "p-representation" and the operator $\hat{\dfrac{1}{p}}$ in the "r-representation".

43*. Express the operator $\hat{\dfrac{1}{p_x}}$ in the "x-representation" and the operator $\hat{\dfrac{1}{x}}$ in the "p_x-representation".

44. Find the matrices of the coordinate and the momentum in the energy representation for a particle in an infinite one-dimensional rectangular potential well.

45. Show that the average value of the square of a Hermitean operator is positive.

46. Does taking the complex conjugate correspond to: (1) a linear operator, (2) a Hermitean operator, (3) an operator which is its own complex conjugate?

47. If $\hat{T}$ is the translation operator, defined by the equation

$$\hat{T}f(x) = f(x+a)\,,$$

for any function $f(x)$, express $\hat{T}$ in terms of the momentum operator $\hat{p}$.

48. Assume λ to be a small quantity to find an expansion of the operator $(\hat{A} - \lambda\hat{B})^{-1}$ in powers of λ.

49. In a state of a quantum mechanical system described by a given wave function ψ_A, the mechanical quantity A has a well-defined value.

Can the quantity B also have a well-defined value in the case where the operators $\hat{A}$ and $\hat{B}$ (1) do not commute, (2) commute?

50. Show that if the Hamiltonian of a system is invariant under some coordinate transformation, the operator of this transformation will commute with the Hamiltonian.

51. Indicate which mechanical quantities or which combination of them are conserved when a system of N particles moves in the following external fields (consider energy, total angular momentum and its components, the components of the linear momentum, and parity):

(1) free motion;
(2) field of an infinite, homogeneous cylinder;
(3) field of an infinite, homogeneous plane;
(4) field of a homogeneous sphere;
(5) field of an infinite, homogeneous half-plane;
(6) field of two points;
(7) a uniform, variable field;
(8) field of a straight conductor with variable charge;
(9) field of a tri-axial ellipsoid;
(10) field of an infinite, homogeneous screw line;
(11) field of an infinite, homogeneous prism;
(12) field of a homogeneous cone;
(13) field of a cylindrical torus.

Ascertain also the commutability of the corresponding operators.

52. Prove the invariance of the non-relativistic Schrödinger equation under a Galilean transformation.

53. Prove that the function ψ, which at $t = 0$ describes a state with a well-defined value of the physical quantity A, will at $t > 0$ be an eigenfunction of the operator $\hat{A}(-t)$ corresponding to the same eigenvalue. The quantity $\hat{A}(\tau)$ is the operator corresponding to the quantity A in the Heisenberg representation at time τ.

54. The Hamiltonian $\hat{H}$ of a system performing a one-dimensional finite motion depends explicitly on the time. At every time t we assume that we know the eigenvalues $E_n(t)$ of the "instantaneous" Hamiltonian and the corresponding complete orthonormal set of eigenfunctions $\psi_n(t)$. (The term "instantaneous" is used here in the sense of "at a given time" which is different from its use in problems 73 and 74 of section 7, where it is synonymous with "sudden". The dependence of the ψ-function on the coordinate (q) is omitted here for the sake of simplicity.)

Give the wave function of the system in the representation whose base is the set of functions $\psi_n(t)$.

55. Let the Hamiltonian of the system, characterised in the preceding problem, be a slowly varying function of the time t. Assume that the system is at $t = 0$ in the mth quantum state and find its wave function at $t > 0$ in the first approximation of the adiabatic perturbation theory and discuss the conditions for the applicability of the result obtained.

56. The point of suspension of a linear oscillator in its ground state starts to move slowly at $t = 0$ and comes to rest again at $t = T$. Find the probability for the excitation of the oscillator in the adiabatic approximation (compare problem 24 of section 3) and ascertain the applicability of this approximation.

57. A particle is bound in an infinite rectangular potential well of width a. At $t = 0$ one of the walls of the well starts to move according to an arbitrarily given time dependence.

Reduce this problem to a wave equation with a Hamiltonian which depends explicitly on the time and discuss the particular case of the motion of the wall for which the variables in that equation can be separated.

58*. A particle is in the nth stationary state of an infinite rectangular potential well of width a. At $t = 0$ one of the walls of the well starts to move slowly according to a given time dependence. Find in the adiabatic approximation the probability that at $t > 0$ the particle will be found in the mth stationary state ($m \neq n$).

59. Evaluate the de Broglie wavelength of (i) an electron, (ii) a hydrogen atom, and (iii) a uranium atom, if the kinetic energy of each of them is 100 eV.

60. Find the kinetic energy of an electron and of a neutron with a de Broglie wavelength of 1 Å.

61. Under what circumstances will a particle of mass μ and velocity v ($\ll c$) show distinctly wave properties when it is scattered by a periodic structure of linear period d?

62. For what neutron energies may we expect especially strong diffraction effects when the neutrons are scattered by natural crystals with lattice constants between 2·5 and 6·0 Å?

63. Bearing in mind the wave properties of particles, show the limits of the applicability of the classical concepts for an electron of 10 eV energy and a proton with 1 MeV energy.

64. For the study of the structure of atomic nuclei several laboratories have constructed electron accelerators with energies up to 6 GeV. What is the electron de Broglie wavelength in that case, and what is the need for such high energies?

Angular momentum; spin

1. Consider the angular momentum operators $\hat{J}_x, \hat{J}_y$, and $\hat{J}_z$ which satisfy the commutation relations

$$[\hat{J}_x, \hat{J}_y]_- = i\hbar J_z \ , \quad [\hat{J}_y, \hat{J}_z]_- = i\hbar \hat{J}_x \ , \quad [\hat{J}_z, \hat{J}_x]_- = i\hbar \hat{J}_y \ ,$$

and introduce the non-Hermitean operators

$$\hat{J}_\pm = \hat{J}_x \pm i\hat{J}_y \ .$$

(i) Derive the commutation relations satisfied by $\hat{J}_+$, $\hat{J}_-$, and $\hat{J}_z$.

(ii) Prove the relations $\hat{J}_\mp \hat{J}_\pm = \hat{J}^2 - \hat{J}_z(\hat{J}_z \pm 1)$.

(iii) Let $\varphi_{a,\,b}$ be the joint eigenfunction of $\hat{J}^2$ and $\hat{J}_z$ corresponding to the eigenvalues a and b,

$$\hat{J}^2 \varphi_{a,\,b} = a\varphi_{a,\,b} \ , \quad \hat{J}_z \varphi_{a,\,b} = b\varphi_{a,\,b} \ ;$$

we shall use the Dirac notation

$$\varphi_{a,\,b} \equiv |a, b> \ ,$$

and choose the $\varphi_{a,\,b}$ to be an orthonormal set so that

$$\langle a', b' | a, b \rangle = \delta_{aa'} \delta_{bb'} \ .$$

Prove that

$$a \geqslant b^2 \ ,$$

and that $\hat{J}_+|a, b\rangle$ and $\hat{J}_-|a, b\rangle$ are also joint eigenfunctions of $\hat{J}^2$ and $\hat{J}_z$. Find the eigenvalues corresponding to $\hat{J}_+|a, b\rangle$ and $\hat{J}_-|a, b\rangle$.

Hence prove that there are states such that

$$\hat{J}_+|a, b_{\text{max}}\rangle = 0$$

and such that

$$\hat{J}_-|a, b_{\text{min}}\rangle = 0 \ .$$

If we put $b_{\text{max}}/\hbar = j$, prove that

$$a = j(j+1)\hbar^2 \ ,$$

and that j is either an integer or half-odd-integral.

(iv) If we now put $b = m\hbar$ and call the $|a, b\rangle$ state the $|j, m\rangle$ state in agreement with the usual notation for angular momentum eigenstates, find the matrix representation of J_x, J_y, and J_z in the $|j, m\rangle$ representation.

2. Express the operator of rotation over a finite angle φ_0 around the direction $\boldsymbol{n}$ in terms of the angular momentum operator (of a system of N particles).

3. Obtain expressions for the operators $\hat{l}_x, \hat{l}_y, \hat{l}_z$ in spherical coordinates starting from the fact that $\hat{l}_x, \hat{l}_y, \hat{l}_z$ are the operators of infinitesimal rotations.

4. Prove that under a rotation an operator $\hat{\Omega}$ changes to a new operator $\hat{\Omega}'$ such that

$$\hat{\Omega}' = \hat{R}\hat{\Omega}\hat{R}^{-1},$$

where $\hat{R}$ is the rotation operator (see problem 2 of section 4).

5. Give a simple interpretation of the commutability of the operators of the components of the linear momentum and the non-commutability of the operators of the components of the angular momentum, starting from the kinematic meaning of these operators, which are connected with infinitesimal translations and rotations.

6. Prove the following commutation relations:

(a) $[\hat{l}_i, \hat{x}_k]_- = ie_{ikl}\hat{x}_l$,

(b) $[\hat{l}_i, \hat{p}_k]_- = ie_{ikl}\hat{p}_l$,

where e_{ikl} is the antisymmetric unit tensor of third rank, the components of which change sign for any interchange of two of its indices, for instance, $e_{ikl} = -e_{ilk}$, and where $e_{123} = 1$ (1, 2, and 3 correspond to x, y, z).

7. Prove the following relations:

(a) $[\hat{\boldsymbol{l}}, (\hat{p}_x^2+\hat{p}_y^2+\hat{p}_z^2)]_- = 0$,

(b) $[\hat{\boldsymbol{l}}, (x^2+y^2+z^2)]_- = 0$.

8. Show that if $\psi_m^{(0)}$ is the eigenfunction of the $\hat{J}_z$ operator corresponding to an eigenvalue m, the function

$$\psi_m = \exp(-i\hat{J}_z\varphi)\exp(-i\hat{J}_y\vartheta)\psi_m^{(0)}$$

will be the eigenfunction of the operator:

$$\hat{J}_\xi = \hat{J}_x \sin\vartheta\cos\varphi + \hat{J}_y \sin\vartheta\sin\varphi + \hat{J}_z\cos\vartheta,$$

corresponding to the same eigenvalue, that is:

$$\hat{J}_\xi\psi_m = m\psi_m.$$

Hint. Use the relations (see problem 21 of section 3):

$$\exp(-i\hat{J}_y\vartheta)\hat{J}_z\exp(i\hat{J}_y\vartheta) = \hat{J}_z\cos\vartheta + \hat{J}_x\sin\vartheta,$$
$$\exp(-i\hat{J}_z\varphi)\hat{J}_x\exp(i\hat{J}_z\varphi) = \hat{J}_x\cos\varphi + \hat{J}_y\sin\varphi.$$

9. Show that in a state ψ with a well-defined value of $\hat{l}_z$ $(\hat{l}_z\psi = m\psi)$ the average values of $\hat{l}_x$ and $\hat{l}_y$ are equal to zero.

Hint. Find the average value in the state ψ for the left-hand side and for the right-hand side of the commutation relations

$$\hat{l}_y\hat{l}_z-\hat{l}_z\hat{l}_y = i\hat{l}_x, \qquad \hat{l}_z\hat{l}_x-\hat{l}_x\hat{l}_z = i\hat{l}_y.$$

10. Obtain an expression for the operator of the angular momentum relative to an arbitrary axis (z') in terms of the operators $\hat{l}_x, \hat{l}_y, \hat{l}_z$.

11. Show that in the state ψ for which $\hat{l}_z\psi = m\psi$ the average value of the angular momentum about an axis z' which makes an angle θ with the z-axis is equal to $m\cos\theta$.

This result can be visualised as follows. The angular momentum vector in the state ψ_m is evenly "spread out" over a cone with its axis along the z-axis, its slant height equal to $\sqrt{[l(l+1)]}$, and its height equal to m. The average value of its projection on the xy-plane is equal to zero, and its component along the z'-axis is, after averaging, equal to $m\cos\theta$.

12. Find the matrix giving the transformation of the components of the spin function of particles of spin 1 corresponding to an arbitrary rotation of the system of coordinates.

13. In a Stern–Gerlach type of experiment the deflection of a beam of atoms with total angular momentum J depends on the value of the component of the angular momentum in the direction of the magnetic field in the apparatus. If the particles in the beam have a well-defined value of the angular momentum relative to an axis which is not along the direction of the applied magnetic field, the beam is split into $2J+1$ components.

Determine the relative intensity of these components if $J = 1$, and if the component of the angular momentum along an axis which makes an angle θ with the direction of the applied magnetic field has the well-defined value $M(+1, 0, -1)$.

14. The three Hermitean operators $\hat{\sigma}_x$, $\hat{\sigma}_y$, and $\hat{\sigma}_z$ satisfy the relations

$$[\hat{\sigma}_x, \hat{\sigma}_y]_- = 2i\hat{\sigma}_z\ , [\hat{\sigma}_y, \hat{\sigma}_z]_- = 2i\hat{\sigma}_x\ , [\hat{\sigma}_z, \hat{\sigma}_x]_- = 2i\hat{\sigma}_y\ , \hat{\sigma}_x^2+\hat{\sigma}_y^2+\hat{\sigma}_z^2 = 3\ .$$

Find a representation of these so-called Pauli matrices in which $\hat{\sigma}_z$ is diagonal.

15. If $\hat{\sigma}$ is the vector operator with components $\hat{\sigma}_x$, $\hat{\sigma}_y$, and $\hat{\sigma}_z$, prove that

(i) if $\hat{A}$ and $\hat{B}$ commute with $\hat{\sigma}$, we have

$$(\hat{\sigma}\,.\,\hat{A})(\hat{\sigma}\,.\,\hat{B}) = (\hat{A}\,.\,\hat{B})+i(\hat{\sigma}\,.\,[\hat{A}\wedge\hat{B}])\ ;$$

and (ii) $\hat{\sigma}_x\hat{\sigma}_y\hat{\sigma}_z = i$.

16. Find the transformation operator for a spin-function (spinor for a spin-$\frac{1}{2}$ particle) corresponding to a rotation described by the Euler angles ϑ, ψ, and φ (see fig. 19).

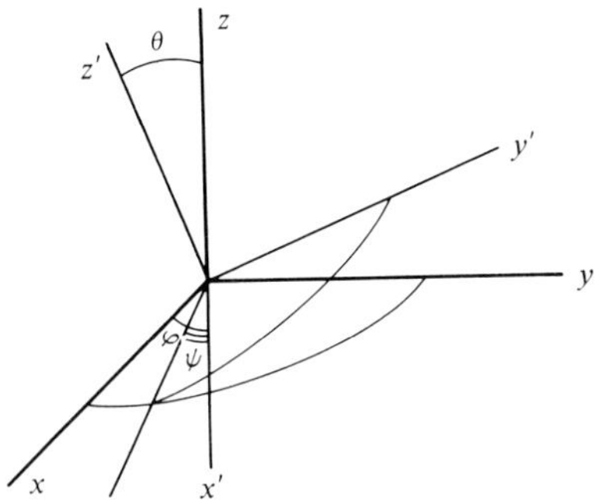

Fig. 19.

17. Find the transformation matrix for the components of a spinor, if the system of coordinates is rotated over an angle Φ with respect to an axis with direction cosines α, β, and γ.

18. Find the eigenfunctions of the operator $\alpha\hat{\sigma}_x+\beta\hat{\sigma}_y+\gamma\hat{\sigma}_z$, where $\alpha^2+\beta^2+\gamma^2=1$, and show that the expansion coefficients of an arbitrary function $\begin{pmatrix}\psi_1\\ \psi_2\end{pmatrix}$ in terms of these functions determine the probability that the value of the spin component in the direction characterised by the direction cosines α, β, and γ is equal to $+\frac{1}{2}$ or $-\frac{1}{2}$.

19. If the z-component of the electron spin is equal to $+\frac{1}{2}$, what is the probability that its component along a direction z' which makes an angle θ with the z-axis is equal to $+\frac{1}{2}$ or $-\frac{1}{2}$? Determine the average value of the component of the spin along this direction.

20. The most general form of the spin function of a particle of spin $\frac{1}{2}$ in the $|\frac{1}{2}, m\rangle$-representation is

$$\psi_1 = e^{i\alpha}\cos\delta, \qquad \psi_2 = e^{i\beta}\sin\delta.$$

This function describes a particle state in which the probability that the z-component of the spin is equal to $+\frac{1}{2}$ (or $-\frac{1}{2}$) is equal to $\cos^2\delta$ (or $\sin^2\delta$).

What would be the result of a measurement of the component of the spin along an arbitrary direction?

21. The spin function of a spin-$\frac{1}{2}$ particle has the following form in the $|\frac{1}{2}, m\rangle$-representation:

$$\begin{pmatrix}\psi_1\\ \psi_2\end{pmatrix} = \begin{pmatrix}e^{i\alpha}\cos\delta\\ e^{i\beta}\sin\delta\end{pmatrix}.$$

Is there a direction in space along which the spin component has the well-defined value of $+\frac{1}{2}$?

If such a direction exists, find the polar angles (θ, Φ) of its direction.

Hint. Find θ and Φ from requiring the second component of the spin function to be equal to zero.

22. Consider a system of non-interacting identical particles. Let their momentum be the same and their spin be equal to $\frac{1}{2}$. If these particles did not possess spin we could describe the system by a "pure case" ensemble. However, we do not know whether the spins of all the particles are parallel.

Is it possible to use an experiment of the Stern–Gerlach type to determine whether this beam of particles corresponds to a "pure case" or to a "mixture" ensemble?

23. Find that wave function of a system consisting of two spin-$\frac{1}{2}$ particles which is an eigenfunction of each of the two commuting operators, the square and the z-component of the total spin.

24. Show that in a system consisting of two spin-$\frac{1}{2}$ particles the total spin S is an integral of motion, provided the Hamiltonian is symmetric in the two spins.

25. Denote by $\hat{\sigma}_1$ and $\hat{\sigma}_2$ the spin operators of two particles and by $\boldsymbol{r}$ the radius vector connecting these particles. Show that any positive integral power of either of the operators

$$\hat{A} \equiv (\hat{\boldsymbol{\sigma}}_1 . \hat{\boldsymbol{\sigma}}_2) \quad \text{and} \quad \hat{S}_{12} = \frac{3(\hat{\boldsymbol{\sigma}}_1 . \boldsymbol{r})(\hat{\boldsymbol{\sigma}}_2 . \boldsymbol{r})}{r^2} - (\hat{\boldsymbol{\sigma}}_1 . \hat{\boldsymbol{\sigma}}_2),$$

and any product of such powers can be written as a linear combination of $\hat{A}$, $\hat{S}_{12}$, and the unit matrix.

26. Show that the operator $\hat{S}_{12}$ of the preceding question can be expressed as follows, in terms of the total spin operator $\hat{\boldsymbol{S}} = \frac{1}{2}(\hat{\boldsymbol{\sigma}}_1 + \hat{\boldsymbol{\sigma}}_2)$,

$$\hat{S}_{12} = \frac{6(\hat{\boldsymbol{S}} . \boldsymbol{r})^2}{r^2} - 2\hat{\boldsymbol{S}}^2$$

and that if the total spin of the two particles is equal to unity, $\hat{S}_{12}$ can be written in the form of the following 3×3 matrix:

$$\hat{S}_{12} = \frac{4\sqrt{\pi}}{\sqrt{5}} \begin{bmatrix} Y_{20} & -\sqrt{3}\,Y_{2,-1} & \sqrt{6}\,Y_{2,-2} \\ -\sqrt{3}\,Y_{21} & -2Y_{20} & \sqrt{3}\,Y_{2,-1} \\ \sqrt{6}\,Y_{22} & \sqrt{3}\,Y_{21} & Y_{20} \end{bmatrix}.$$

27. Consider two spin-$\frac{1}{2}$ particles interacting through a magnetic dipole–dipole interaction,

$$V = A\frac{(\hat{\boldsymbol{\sigma}}_1 . \hat{\boldsymbol{\sigma}}_2)\boldsymbol{r}^2 - (\hat{\boldsymbol{\sigma}}_1 . \boldsymbol{r})(\hat{\boldsymbol{\sigma}}_2 . \boldsymbol{r})}{r^5}.$$

If the two spins are at a fixed distance d apart and if at $t = 0$ one spin is parallel to $\boldsymbol{r}$ and the other one antiparallel to $\boldsymbol{r}$, calculate the time after which the parallel spin is antiparallel and the antiparallel spin parallel.

28. Find all the quartet and doublet spin wavefunctions for a system of three spin-$\frac{1}{2}$ particles.

29. Show that a totally symmetric spin wavefunction for a system of n spin-$\frac{1}{2}$ particles is an eigenfunction of the total spin operator $\hat{S}^2\left(\hat{S} = \sum_i \frac{1}{2}\hbar\hat{\sigma}_i\right)$ with quantum number $\frac{1}{2}n$ and that any other state of the system has a smaller total spin quantum number.

30. Let σ_i denote the spin variable of the ith electron. This variable can take on the two values $+1$ and -1. Show that if the operators

$$\hat{\sigma}_{lx} = \begin{pmatrix} 0 & 1 \\ 1 & 0 \end{pmatrix}_l, \quad \hat{\sigma}_{ly} = \begin{pmatrix} 0 & -i \\ i & 0 \end{pmatrix}_l, \quad \hat{\sigma}_{lz} = \begin{pmatrix} 1 & 0 \\ 0 & -1 \end{pmatrix}_l,$$

which refer to the lth electron, act upon a function $f(\sigma_1, \sigma_2, \ldots, \sigma_l, \ldots, \sigma_n)$ of the spin variables of n electrons, the result is the following one:

$$\hat{\sigma}_{lx} f = f(\sigma_1, \ldots, \sigma_{l-1}, -\sigma_l, \sigma_{l+1}, \ldots, \sigma_n),$$

$$\hat{\sigma}_{ly} f = -i\sigma_l f(\sigma_1, \ldots, \sigma_{l-1}, -\sigma_l, \sigma_{l+1}, \ldots, \sigma_n),$$

$$\hat{\sigma}_{lz} f = \sigma_l f(\sigma_1, \ldots, \sigma_{l-1}, \sigma_l, \sigma_{l+1}, \ldots, \sigma_n).$$

31. Show that the operator of the square of the total spin moment of n electrons can be written in the form

$$\hat{\boldsymbol{S}}^2 = n - \frac{n^2}{4} + \sum_{k<l} P_{kl},$$

where P_{kl} is the operator which interchanges the spin variables σ_k and σ_l, that is

$$P_{kl} f(\sigma_1, \ldots, \sigma_{k-1}, \sigma_k, \sigma_{k+1}, \ldots, \sigma_{l-1}, \sigma_l, \sigma_{l+1}, \ldots, \sigma_n)$$
$$= f(\sigma_1, \ldots, \sigma_{k-1}, \sigma_l, \sigma_{k+1}, \ldots, \sigma_{l-1}, \sigma_k, \sigma_{l+1}, \ldots, \sigma_n).$$

32. Obtain the result of problem 12 of section 4 by considering the transformation of a symmetric tensor of the second rank.

33. The angular momentum of a particle is equal to j, and its z-component has its largest possible value.

Determine the probability for different values of the angular momentum component along a direction which makes an angle θ with the z-axis.

34. How can one write down a wave function for a spin angular momentum for $s > \frac{1}{2}$, say for $s = 1$? (i) Do states exist which are oriented in an arbitrarily chosen direction? (ii) Is an arbitrary state oriented in some direction? Explain this situation.

35*. A system with total angular moment J is in the state with $J_z = M$. Determine the probability that a measurement (for instance, by a Stern–Gerlach experiment) of the angular momentum component along a direction z' which makes an angle ϑ with the z-axis leads to the value M'.

36*. A spin-$\frac{1}{2}$ particle moves in a central field of force. Find the wave function of this particle which is simultaneously an eigenfunction of the three commuting operators:

$$j_z = \hat{l}_z + \hat{s}_z, \quad \hat{l}^2, \quad \hat{j}^2.$$

37. A state of an electron is characterised by the quantum numbers l, j, and m. Using the wave functions obtained in the preceding question, determine the possible values of the components of the orbital and spin angular momentum and the probabilities that these values are realised. Find also the average values of these components.

38. We define the direction of the spin of a spin-$\frac{1}{2}$ particle as that direction along which the spin component has a well-defined value of $+\frac{1}{2}$. Let this direction be characterised by the polar angles Θ and Φ, and let the state of a particle be described by a wave function $\psi\,(l, j = l \pm \frac{1}{2}, m)$ (see problem 36 of section 4). It is clear that the spin direction of such a particle will, generally speaking, not be the same in all points of space. Find the relation between the angles Θ and Φ and the space coordinates of the particle.

39. A system consists of two particles, one with angular momentum $l_1 = 1$, and the other with angular momentum $l_2 = l$. The total angular momentum J can in that case take on the values $l+1$, l, and $l-1$. Express the eigenfunctions of the operators $\hat{J}^2$ and $\hat{J}_z$ in terms of the eigenfunctions of the square and the z-component of the angular momentum of the separate particles.

40. A system consists of two particles, one of spin $\frac{1}{2}$ and one of spin 0. Show that the orbital angular momentum is an integral of motion for any law of interaction between these particles.

41. Show that the normalised part of a $^3D_1$ state wave function which refers to the spin and angular dependence can be written in the following

form:

$$\frac{1}{4\sqrt{(2\pi)}}\hat{S}_{12}\begin{pmatrix}1\\0\\0\end{pmatrix} \quad (J=1,\ M=1,\ L=2,\ S=1),$$

$$\frac{1}{4\sqrt{(2\pi)}}\hat{S}_{12}\begin{pmatrix}0\\1\\0\end{pmatrix} \quad (J=1,\ M=0,\ L=2,\ S=1),$$

$$\frac{1}{4\sqrt{(2\pi)}}\hat{S}_{12}\begin{pmatrix}0\\0\\1\end{pmatrix} \quad (J=1,\ M=-1,\ L=2,\ S=1).$$

Hint. See problem 39 of section 4.

42. Prove the following equations:

(a) $[\hat{\boldsymbol{J}}^2, \hat{\boldsymbol{A}}]_- = i([\hat{\boldsymbol{A}}\wedge\hat{\boldsymbol{J}}]-[\hat{\boldsymbol{J}}\wedge\hat{\boldsymbol{A}}])$,

(b) $[\hat{\boldsymbol{J}}^2, [\hat{\boldsymbol{J}}^2, \hat{\boldsymbol{A}}]_-]_- = 2(\hat{\boldsymbol{J}}^2\hat{\boldsymbol{A}}+\hat{\boldsymbol{A}}\hat{\boldsymbol{J}}^2)-4\hat{\boldsymbol{J}}(\hat{\boldsymbol{J}}.\hat{\boldsymbol{A}})$,

(c) $(\hat{\boldsymbol{A}})_{nJM}^{n'JM'} = \dfrac{(\hat{\boldsymbol{J}}.\hat{\boldsymbol{A}})_{nJ}^{n'J}}{J(J+1)}(\boldsymbol{J})_{JM}^{JM'}.$

The arbitrary vector quantity $\hat{\boldsymbol{A}}$ satisfies the commutation rule

$$[\hat{J}_i, \hat{A}_k]_- = ie_{ikl}\hat{A}_l.$$

43. If $\hat{\boldsymbol{T}}$ is a vector operator whose commutator with the total angular momentum operator $\hat{\boldsymbol{J}}$ is governed by the relation

$$[(\boldsymbol{a}\,.\hat{\boldsymbol{J}}), \hat{\boldsymbol{T}}]_- = i\hbar[\hat{\boldsymbol{T}}\wedge\boldsymbol{a}],$$

where $\boldsymbol{a}$ is an arbitrary constant vector, show, by taking $\hat{\boldsymbol{T}}$ to be the electric dipole-moment operator that the quantum number m in the $|j, m\rangle$ representation (see problem 1 of section 4) changes by 0 or ±1 in an electric dipole transition.

Using the result (b) of the preceding problem show that

$$[(\lambda_j-\lambda_{j'})^2-2(\lambda_j+\lambda_{j'})]\langle j, m|\hat{\boldsymbol{T}}|j', m'\rangle = -4\langle j, m|\hat{\boldsymbol{J}}(\hat{\boldsymbol{J}}\,.\,\hat{\boldsymbol{T}})|j', m'\rangle,$$

where $\lambda_j = j(j+1)$.

Show also that $\hat{\boldsymbol{J}}$ commutes with $(\hat{\boldsymbol{J}}\,.\,\hat{\boldsymbol{T}})$ and hence that $\langle j, m|\hat{\boldsymbol{J}}(\hat{\boldsymbol{J}}\,.\,\hat{\boldsymbol{T}})|j', m'\rangle$ vanishes unless $j = j'$. Use these results

(i) in the case where $\hat{\boldsymbol{T}}$ is the electric dipole-moment operator, to show that j changes by 0 or ±1 in an electric dipole transition; and

(ii) in the case where $\hat{\boldsymbol{T}}$ is the electron spin $\hat{\boldsymbol{S}}$, and $j = j'$, to find the Landé g-factor in Russell–Saunders coupling (where the states are simultaneous eigenstates of $\hat{\boldsymbol{J}}^2$, $\hat{\boldsymbol{J}}_z$, $\hat{\boldsymbol{S}}^2$, and $\hat{\boldsymbol{L}}^2$.

44. Find the average value of the operator $\hat{\mu} = g_1 \hat{J}_1 + g_2 \hat{J}_2$ in the state characterised by the quantum numbers J, M_J, J_1, and J_2, if the total angular momentum $\hat{J}$ is equal to $\hat{J} = \hat{J}_1 + \hat{J}_2$.

Hint. Use the equations derived in problem 42 of section 4.

45. Find the magnetic moment (in nuclear magnetons) of the ^{15}N nucleus, for which one proton in a $p_{\frac{1}{2}}$ state is missing from a closed shell. The magnetic moment of a free proton is equal to $\mu_p = 2{\cdot}79$.

46. Evaluate the magnetic moment of the ^{17}O nucleus which contains one neutron in a $d_{\frac{5}{2}}$ state apart from closed shells. The magnetic moment of a free neutron is equal to $\mu_n = -1{\cdot}91$.

47. What would be the value of the magnetic moment of the deuteron if the deuteron were in the following states:

(a) $^3S_1$, (b) $^1P_1$, (c) $^3P_1$, (d) $^3D_1$?

48. Assuming the deuteron ground state to be a superposition of a $^3S_1$ and a $^3D_1$ state, determine the weight of the D-state, if

$$\mu_p = 2{\cdot}79, \quad \mu_n = -1{\cdot}91, \quad \mu_d = 0{\cdot}85.$$

49. Which of the following states can and which cannot occur: $^1P_0$, $^2P_{1/2}$, $^2P_1$, $^2P_{5/2}$, $^3P_0$, $^3P_{1/2}$, and $^3P_1$.

50. A nucleus A of spin 1 is excited in an even state. Energetically the emission of an α-particle is possible,

$$A \to B + \alpha.$$

The nucleus B produced in this reaction is stable, has zero spin, and is also in an even state. Use the principle of conservation of angular momentum and of parity to show that this reaction is forbidden.

51. Show that the orbital momentum L of the motion of two α-particles is always even ($L = 0, 2, 4, \ldots$).

52. Is it possible for the ^8_4Be nucleus in an excited state with spin 1 to decay into two α-particles?

53. Use the fact that the only bound state of the neutron–proton (n, p) system is even, that the total spin of that state is unity, and that the (n, n) and (n, p) interactions are identical to show that two neutrons cannot form a bound state.

54. Is it possible for a photon to decay spontaneously into (i) two photons or (ii) into three photons?

55. Show that for a system consisting of two identical particles with spin I the ratio of the number of states symmetrical in the two spins to the number of states antisymmetric in the two spins is equal to $(I+1)/I$.

56. A system consists of two spinless particles, each in a p-state. If $\hat{L} = \hat{l}^{(1)} + \hat{l}^{(2)}$, show that only one eigenvalue of $\hat{L}^2$ is permitted, if the wavefunction has to be antisymmetric in the two particles and find the expectation value of $(\hat{l}^{(1)} \cdot \hat{l}^{(2)})$ in the permitted states.

57. The quadrupole tensor of a system is defined by the equation

$$Q_{ik} = \sum_j e_j(3x_i^{(j)}x_k^{(j)} - \delta_{ik}r^{(j)2}),$$

where the summation is over all particles in the system and $r^{(j)}$ is their position vector.

If the system is in a state with a well-defined value of the total angular momentum, corresponding to a quantum number j, the quadrupole moment Q is defined as the value of Q_{zz} in the state with $m = j$.

Evaluate the average values of the Q_{ik} for a state characterized by a wavefunction of the form $\psi(r, \vartheta, \varphi) = f(r)Y_{l,m}(\vartheta, \varphi)$, using the relation

$$\cos\vartheta\, Y_{l,m} = \sqrt{\left[\frac{(l+m+1)(l-m+1)}{(2l+1)(2l+3)}\right]}Y_{l+1,m} + \sqrt{\left[\frac{(l+m)(l-m)}{(2l-1)(2l+1)}\right]}Y_{l-1,m}.$$

58*. Evaluate the average values of the Q_{ik} for a state of an electron moving in a central field of force, characterized by the quantum numbers j, l, and m_j.

59. Use the expressions for the matrix elements of vectors to show that the quadrupole moment of a nucleus is equal to

$$Q = I(2I-1)\sum_{i=1}^{Z}\sum_{n'}\{2(I+1)|(z_i)_{n',I+1}^{n,I}|^2 - 2I|(z_i)_{n',I-1}^{n,I}|^2\}.$$

The summation is over all Z protons, I is the nuclear spin, and n is the combination of all the other quantum numbers which characterise the state of the nucleus.

60. Evaluate the quadrupole moment of a nucleus with one proton of angular moment j outside a closed-shell spherically symmetric core, neglecting any deformation of the core.

5

Central field of force

1. A system consists of two particles of mass μ_1 and μ_2. Express the operators of the total orbital angular momentum $\hat{\boldsymbol{l}}_1+\hat{\boldsymbol{l}}_2$ and the total momentum $\hat{\boldsymbol{p}}_1+\hat{\boldsymbol{p}}_2$ in terms of the centre of mass coordinate

$$\boldsymbol{R} = \frac{\mu_1\boldsymbol{r}_1+\mu_2\boldsymbol{r}_2}{\mu_1+\mu_2}$$

and the relative coordinate $\boldsymbol{r} = \boldsymbol{r}_2-\boldsymbol{r}_1$. Show that if the potential energy of the interacting particles depends only on their distance apart, $V = V(|r_2-r_1|)$, the Hamiltonian can be put in the form

$$\hat{H} = -\frac{\hbar^2}{2(\mu_1+\mu_2)}\nabla_R^2-\frac{\hbar^2(\mu_1+\mu_2)}{2\mu_1\mu_2}\nabla_r^2+V(r),$$

where ∇_R^2 and ∇_r^2 are the Laplace operators referring to $\boldsymbol{R}$ and $\boldsymbol{r}$.

2. A particle moves in a central field. Write the equation for the radial part R_{nl} of the wave function in the form of a one-dimensional Schrödinger equation.

3. Show that in the case of a discrete spectrum in a central field the minimum value of the energy for a given value of the orbital quantum number l increases with increasing l.

4. Determine the wave functions and energy levels of a three-dimensional isotropic oscillator.

5. Solve the previous problem in Cartesian coordinates by separation of variables. Express the wave function for $n_r = 0$, $l = 1$ (see previous problem) as a linear combination of the wave functions obtained.

6. Assume that a nucleon in a light nucleus moves in an averaged potential of the form $V(r) = -V_0+\frac{1}{2}\mu\omega^2r^2$, to determine the number of particles of one kind (neutrons or protons) which can be accommodated in a closed shell. A shell is defined as the totality of all states with the same energy.

7. Calculate the theoretical radius of the closed shell nuclei ${}^4_2$He and ${}^{16}_8$O assuming the same potential as in the preceding question. The theoretical nuclear radius is defined as the distance from the centre of mass of the nucleus to the point where the "nuclear density"

$$\rho(\boldsymbol{r}) = \Sigma\psi_\nu^*(\boldsymbol{r})\psi_\nu(\boldsymbol{r})$$

(the summation is over all nucleons) decreases most steeply, that is,

$$\left(\frac{d^2\rho}{dr^2}\right)_{r=R} = 0.$$

8. We can write approximately $V(r) = -Ae^{-r/a}$ for the interaction between a proton and a neutron. Find the wave function of the ground state ($l = 0$) and determine the relation between the well depth A and the quantity a, which characterises the range of the force, by using the empirical value $E = -2{\cdot}2$ MeV for the deuteron energy.

9. Determine approximately the deuteron ground state energy for the potential $V(r) = -A\,e^{-r/a}$ ($A = 32$ MeV, $a = 2{\cdot}2$ fermi) using the Ritz variational principle. Use as trial wave function a function of the form $R = c\,e^{\alpha r/2a}$, depending on one parameter α. The value of c follows from the normalisation condition $\int_0^\infty R^2 r^2\,dr = 1$.

10. Determine the energy levels and wave functions of a particle in a spherical "potential well"

$$V(r) = 0 \quad (r < a); \quad V(r) = \infty \quad (r > a).$$

Consider the case where $l = 0$.

11. Determine the discrete energy spectrum of a particle with zero angular momentum in a spherical potential well

$$V(r) = \begin{cases} -V_0 & (r < a), \\ 0 & (r > a). \end{cases}$$

12. Find the p-state wavefunction which is proportional to $\cos\vartheta$ in the potential of the preceding problem.

Show that if $2\mu a^2 V_0 > \pi^2\hbar^2$, there are at least three bound p-states.

13. Apply perturbation theory to determine qualitatively the change in energy levels when we change the potential

$$V(r) = \begin{cases} -V_0 & (r < a) \\ 0 & (r > a), \end{cases}$$

to the potential given in fig. 20.

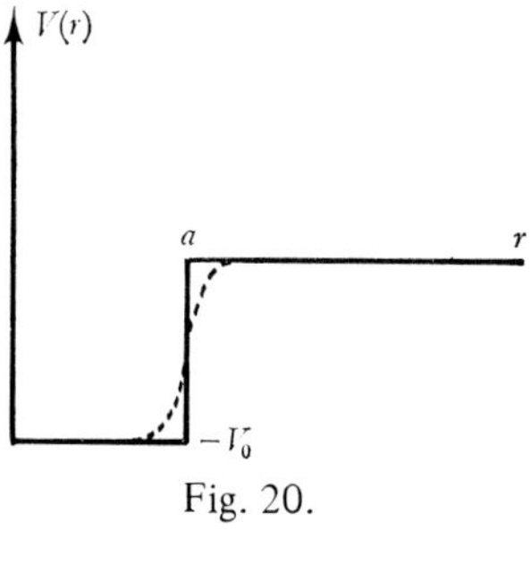

Fig. 20.

14*. The potential energy of an α-particle in a nucleus consists of two parts: the Coulomb repulsion and the short-range nuclear attractive force. The total potential energy is sketched in fig. 21. The emission

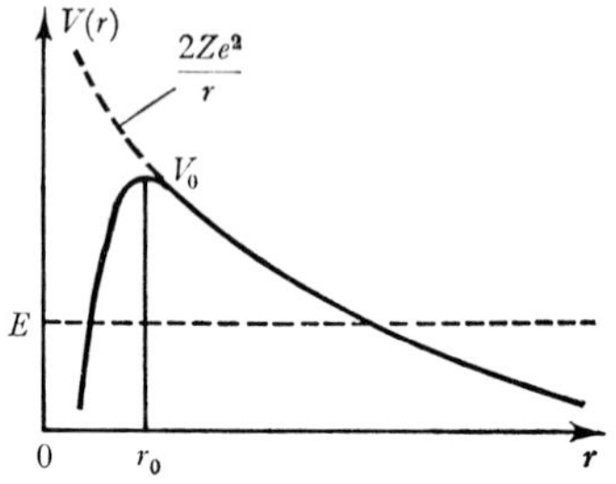

Fig. 21.

of an α-particle is a typical quantum effect, determined by a tunnel effect. Consider the transmission of a particle of zero angular momentum through a spherical potential barrier of the following simple form

$$V(r) = 0 \quad (r < r_1),$$
$$V(r) = V_0 \quad (r_1 < r < r_2),$$
$$V(r) = 0 \quad (r_2 < r).$$

Find a relation between the lifetime and the energy.

15*. Investigate the motion of a negative muon in the field of a nucleus of charge Ze (to be considered as a sphere of radius R, with the charge uniformly spread over its volume), assuming that the forces of interaction between the meson and the nucleus are purely electrostatic of character.

Find the wave functions and energy levels of the stationary states of the meson in the limiting cases of small and of very large Z.

16. Find the selection rules for transitions between stationary states of particles in a central field of force under the influence of a perturbation, whose operator is proportional to the gradient operator (∇).

17*. Indicate how one can calculate $\langle r^p \rangle$ $(p > -2l-3)$ for a particle in a Coulomb potential in a state characterized by the quantum numbers n, l, and m, and evaluate this average for $p = -4, -3, -2, -1, +1$, and $+2$.

18*. Find the radial part of the wave function of a particle in a central field using the semi-classical approximation.

19. Find the wavefunction of a free particle with angular momentum quantum number l (i) exactly, and (ii) using the semi-classical approximation.

20*. Apply the semi-classical method to the radial Schrödinger equation to derive the eigenvalues of the hydrogen atom.

21. Use the semi-classical approximation to derive an expression for the number of discrete levels of a particle moving in a given central field potential $V(r)$ (compare problem 22 of section 1).

Motion of particles in a magnetic field

1. Let the wave function of an electron at $t = 0$ be of the form

$$\Psi(x, y, z, 0) = \psi(x, y, 0)\,\varphi(z, 0).$$

The wave function at time t in a homogeneous magnetic field $\mathscr{H}$ which is along the z-axis will then also be of the form

$$\Psi(x, y, z, t) = \psi(x, y, t) \times \varphi(z, t),$$

since in the Schrödinger equation the z-coordinate is separable. Show that the function $\psi(x, y, T)$ regains its initial value, apart from a phase factor, if T is the period of the classical motion in the magnetic field.

2. Show that if a magnetic field is present the velocity component operators satisfy the following commutation relations:

$$\hat{v}_x \hat{v}_y - \hat{v}_y \hat{v}_x = \frac{ie\hbar}{\mu^2 c} \mathscr{H}_z;$$

$$\hat{v}_y \hat{v}_z - \hat{v}_z \hat{v}_y = \frac{ie\hbar}{\mu^2 c} \mathscr{H}_x;$$

$$\hat{v}_z \hat{v}_x - \hat{v}_x \hat{v}_z = \frac{ie\hbar}{\mu^2 c} \mathscr{H}_y.$$

3. Use the results of problem 2 of section 6 and problem 8 of section 1, to determine the energy of a charged particle moving in a constant, uniform magnetic field.

4. Determine the energy spectrum of a charged particle moving in a uniform magnetic and a uniform electric field which are at right-angles to each other.

5. Determine the wave functions of a charged particle in a uniform magnetic and a uniform electric field which are at right-angles to one another.

6. A charged particle moves in a homogeneous magnetic field and in a central field of the form $V(r) = \frac{1}{2}\mu\omega_0^2 r^2$. Determine its energy spectrum.

7. Determine the time dependence of the coordinate operators $\hat{x}$ and $\hat{y}$ of a charged particle in a uniform magnetic field (vector potential $A_x = -\frac{1}{2}\mathscr{H}y,\ A_y = \frac{1}{2}\mathscr{H}x,\ A_z = 0$).

8*. Determine the energy levels and wave functions of a charged particle in a uniform magnetic field. Use cylindrical coordinates. Write the vector potential in the form $A_\varphi = \frac{1}{2}\mathscr{H}\rho$, $A_\rho = A_z = 0$.

9. Find the current density components for a particle in a uniform magnetic field in the state characterised by the quantum numbers n, m, k_z (see preceding problem).

10. Find in cylindrical coordinates the energy levels of a charged particle in a uniform magnetic field using the semi-classical approximation.

11. Determine the classically accessible region of the radial motion of a particle in a magnetic field (see preceding problem).

12. Estimate the minimum "spreading out" in the radial direction of the orbit of a charged particle in a magnetic field.

13. Use classical mechanics to express the coordinate of the centre of the circle along which a charged particle in a uniform magnetic field moves in terms of the coordinates x and y and the generalised momenta p_x and p_y. Consider the coordinates and momenta in that expression to be operators and find the commutation relation for the coordinates of the "centre of the orbit" introduced in this way and the corresponding Heisenberg relation. Show that the sum of the squares of the coordinates of the "centre of the orbit" takes on the discrete values $(2\hbar c/|e|\mathscr{H})(n+\frac{1}{2})$, where $n = 0, 1, 2, \ldots$.

14. Show that in a variable uniform magnetic field the wave function of a particle with spin can be written as the product of a space function and a spin function.

15. A spin-$\frac{1}{2}$ particle is in a uniform magnetic field along the z-axis, the absolute magnitude of which varies arbitrarily in time: $\mathscr{H} = \mathscr{H}(t)$. The spin function at $t = 0$ is of the form

$$\begin{pmatrix} e^{-i\alpha}\cos\delta \\ e^{i\alpha}\sin\delta \end{pmatrix}.$$

Determine the average value of the x- and y-components of the spin and also the direction in space along which the spin at time t has a well-defined value.

16. In the half-space $x > 0$ there is a uniform magnetic field $\mathscr{H}_x = \mathscr{H}_y = 0$, $\mathscr{H}_z = \mathscr{H}$, while there is no field in the half-space $x < 0$. A beam of polarised neutrons of momentum $\boldsymbol{p}$ is incident upon the plane $x = 0$ from the region $x < 0$. Find the reflection coefficient of the neutrons at the dividing boundary.

17. A spin-$\frac{1}{2}$ particle is in a uniform magnetic field, the absolute magnitude of which is constant and which varies in time according to

the equations

$$\mathscr{H}_x = \mathscr{H} \sin\vartheta \cos\omega t, \quad \mathscr{H}_y = \mathscr{H} \sin\vartheta \sin\omega t, \quad \mathscr{H}_z = \mathscr{H} \cos\vartheta.$$

At $t = 0$, the spin component in the direction of the magnetic field is equal to $+\frac{1}{2}$. Determine the probability that at time t the particle has made a transition to the state where the spin component along the magnetic field direction is equal to $-\frac{1}{2}$.

18. At time $t = 0$ a spin-$\frac{1}{2}$ particle with spin in the $+x$-direction enters a region of space in which there is a uniform magnetic field H in the z-direction. Find the probability that at time t the spin is still in the $+x$-direction.

19. A spin-$\frac{1}{2}$ particle is placed in a magnetic field $(H, 0, 0)$. At time $t = 0$ the spin is in the $+z$-direction. Evaluate the time τ at which the spin will be in the $+y$-direction.

20. A spin-$\frac{1}{2}$ particle is placed in a time-dependent magnetic field $(He^{-\gamma t}, 0, 0)$, where H and γ (>0) are constants. If at $t = 0$ the spin is in the $+z$-direction find the probability that at time t the spin is in the $-z$-direction.

21. A spin-$\frac{1}{2}$ particle with a magnetic moment μ is in a non-uniform magnetic field of the form

$$\mathscr{H}_z = \mathscr{H}_0 + kz, \quad \mathscr{H}_y = -ky, \quad \mathscr{H}_x = 0 \quad (\operatorname{div} \mathscr{H} = 0).$$

(a) Find an expression for the time dependence of the x-, y-, z-coordinate operators.

(b) Determine the average values of the coordinates and the time dependence of their dispersion, assuming the particle wave function at $t = 0$ to be of the form

$$\psi = \varphi(x, y, z) \exp(ip_0 x/\hbar) \begin{pmatrix} \alpha \\ \beta \end{pmatrix}.$$

22. A spin-$\frac{1}{2}$ particle with magnetic moment $\mu = g\mu_B J$ (J is the total angular momentum in units $\hbar$, μ_B the Bohr magneton) is in a magnetic field $\mathscr{H}$ in the z-direction. For time $t \leqslant 0$ the spin is in the $+z$-direction.

At $t = 0$, the magnetic field is instantaneously rotated through 90° so that it points in x-direction.

(i) Find the wave function of the particle for all times $t > 0$.

(ii) Find the expectation values of $\hat{J}_x$, $\hat{J}_y$, and $\hat{J}_z$ for $t > 0$.

(iii) If the field is rotated *slowly* to the x-direction, taking a total time T, the expectation value of $\hat{J}_x$ is approximately equal to $\frac{1}{2}$ for all $t > T$. Estimate the shortest time T for which this is a correct description.

23. Find the energy levels of a system of two spin-$\frac{1}{2}$ particles in an external magnetic field $\mathscr{H}$, described by the Hamiltonian

$$\hat{H} = A[1-(\hat{\boldsymbol{\sigma}}_1.\hat{\boldsymbol{\sigma}}_2)]+\tfrac{1}{2}g\mu_B\mathscr{H}(\hat{\sigma}_{1z}+\hat{\sigma}_{2z}),$$

where A is a constant, g an effective Landé factor, μ_B the Bohr magneton, and $\hat{\boldsymbol{\sigma}}_1$ and $\hat{\boldsymbol{\sigma}}_2$ vectors whose components are the Pauli matrices referring to the two particles.

24. Positronium is a bound state of an electron and a positron. The effective Hamiltonian for the system in the 1S state in a magnetic field $\mathscr{H}$ can be written in the form

$$\hat{H} = \hat{H}_0+A(\hat{\boldsymbol{\sigma}}_1.\hat{\boldsymbol{\sigma}}_2)+\mu_B\mathscr{H}(\hat{\sigma}_{1z}-\hat{\sigma}_{2z}),$$

where the notation is the same as in the previous problem, where 1 and 2 refer respectively to the electron and the positron, and where $\hat{H}_0$ contains kinetic energies and central force potentials.

(i) Find the value of A if in zero magnetic field the $1\,^1S$ and the $1\,^3S$ states are separated by 2.10^5 Mc/s, with the singlet state the lowest.

(ii) Discuss the principles which restrict the $1\,^1S$ and $1\,^3S$ states to decay primarily by two and three quantum emission, respectively.

(iii) Find the energy eigenvalues and eigenfunctions for non-vanishing values of $\mathscr{H}$.

(iv) If the lifetime of the $1\,^1S$ state is 10^{-10} sec and of the $1\,^3S$ state is 10^{-7} sec, estimate the value of $\mathscr{H}$ which will cause the lifetime of the $1\,^3S$ state to be reduced to 10^{-8} sec.

25*. A neutral particle is in a spatially uniform magnetic field; the direction of this field, but not its absolute magnitude, varies with time.

Write down an equation for the spin function in the "ξ-representation" where the ξ-axis is along the direction of the magnetic field. Show that if the variation in the direction of the magnetic field is sufficiently slow, the probability for different values of the component of the spin along the magnetic field direction will not change in time.

26. Find the conditions under which the Hamiltonian of a charged particle in a magnetic field

$$\hat{H} = \frac{1}{2\mu}\left(\hat{\boldsymbol{p}}-\frac{e}{c}\hat{\boldsymbol{A}}\right)^2$$

is of the form

$$\hat{H}' = \frac{1}{2\mu}\hat{\boldsymbol{p}}^2-\frac{e}{\mu c}(\hat{\boldsymbol{p}}.\hat{\boldsymbol{A}})+\frac{e^2}{2\mu c^2}\hat{\boldsymbol{A}}^2.$$

27. A particle of mass μ and charge e is confined to move on a circle of radius a in the x–y-plane, but is otherwise free. A magnetic field $\mathscr{H}$ is applied in the z-direction. Find the ground state energy as a function of $\mathscr{H}$.

Atoms

1. Use the inequality

$$\int |\nabla\psi + Z\psi\nabla r|^2 \, d\tau \geqslant 0$$

to find the minimum energy of a one-electron atom and the corresponding wave function. Show that for the ground state of the atom the inequality $2\bar{T} \geqslant |\bar{V}|$ is satisfied.

2. An electron moving in the Coulomb field of a nucleus of charge Z is in its ground state. Show that the average electrostatic potential in the neighbourhood of the nucleus and the electron is given by the equation

$$\varphi = \frac{e(Z-1)}{r} + e\left(\frac{Z}{a} + \frac{1}{r}\right) e^{-2Zr/a}, \quad \left(a = \frac{\hbar^2}{\mu e^2}\right).$$

3. The wave function $\psi(\boldsymbol{r})$ describes the relative motion of two particles, a proton and an electron. Let the coordinates of the centre of mass of the hydrogen atom be accurately known and be equal to

$$X = 0, \quad Y = 0, \quad Z = 0.$$

Show that in this case the probability density for the proton is of the form

$$w(\boldsymbol{r}) = \left(\frac{m+M}{m}\right)^3 \left| \psi\left(\frac{m+M}{m} \boldsymbol{r}\right) \right| ,$$

where m and M are, respectively, the electron and the proton masses.

4. Find the momentum distribution of the hydrogen atom electron in the $1s$, $2s$, and $2p$ state.

5*. Find the correction to the hydrogen atom energy levels taking into account the relativistic change of mass with velocity (consider the terms of order v^2/c^2).

6. Show that the quadrupole moment of the hydrogen atom is equal to

$$Q_0 = -\frac{j-\frac{1}{2}}{j+1} \overline{r^2} .$$

7. Determine the total probability for excitation and ionisation of the tritium atom ^{3}H when its β-decay takes place. Calculate also the probability for exciting the nth level.

8. Determine the contribution to the magnetic field strength at the centre of a hydrogen atom due to the orbital motion of the electron. Evaluate this quantity for the $2p$ state.

9. What is the change in the expression for the magnetic moment of the hydrogen atom if the motion of the nucleus is taken into account?

10. Determine the separation of the hyperfine structure terms for an s-electron in the hydrogen atom.

11*. Determine the energy of the hyperfine structure in a one-electron atom with non-vanishing angular momentum.

12. Determine the shift of the atomic energy levels due to the motion of the nucleus. Evaluate this shift for helium in the triplet and in the singlet $(1s)(np)$ state, using the eigenfunctions in the form of hydrogen-like functions with an effective charge for the different electrons.

13. Give the possible values of the total angular momentum in a ^1S, ^3S, ^3P, ^2D, or ^4D state.

14. Which states (terms) can be realised for the following two-electron systems:

(a) $(ns)(n's)$, (b) $(ns)(n'p)$, (c) $(ns)(n'd)$, (d) $(np)(n'p)$?

15. Give the possible terms of the following configurations:

(a) $(np)^3$, (b) $(nd)^2$, (c) $(ns)(n'p)^4$.

16. Determine the lowest term of the following elements:

O $[(1s)^2(2s)^2(2p)^4]$,
Cl $[(1s)^2(2s)^2(2p)^6(3s)^2(3p)^5]$,
Fe $[(1s)^2(2s)^2(2p)^6(3s)^2(3p)^6(3d)^6(4s)^2]$,
Co $[(1s)^2(2s)^2(2p)^6(3s)^2(3p)^6(3d)^7(4s)^2]$,
As $[(1s)^2(2s)^2(2p)^6(3s)^2(3p)^6(3d)^{10}(4s)^2(4p)^3]$, and
La $[(1s)^2(2s)^2(2p)^6(3s)^2(3p)^6(3d)^{10}(4s)^2(4p)^6(4d)^{10}(5s)^2(5p)^6(5d)(6s)^2]$.

Hint. It is necessary to use the following empirically established rules (Hund rules):

(a) The lowest energy corresponds to the terms of the largest value of S for a given configuration of the electrons and to the largest value of L possible for this value of S.

(b) The ground state of the atom corresponds to $J = |L-S|$ if the unfilled shell is less than half full, and to $J = L+S$ if it is more than half full.

17. Determine the parity of the ground states of the following elements: K, Zn, B, C, N, O, Cl.

18. A system of N electrons is characterised by N triples of quantum numbers n, l, m_l. Determine the number of states corresponding to a given value M_S of the z-component of the total spin.

19. Find the number of states corresponding to the configuration $(nl)^x$.

20. Show that if $x \leqslant 2l+1$, the term with the largest value of L for the $(nl)^x$ configuration will be a singlet term with $L = xl - \frac{1}{4}x(x-2)$ if x is even, and a doublet with $L = xl - \frac{1}{4}(x-1)^2$ if x is odd.

21*. Construct the eigenfunctions corresponding to a p^3 configuration and characterised by the quantum numbers S, L, M_S, M_L, using the wave functions of the one-electron problem.

Hint. Consider the action of the operators $\hat{L}_x - i\hat{L}_y$ and $\hat{S}_x - i\hat{S}_y$ on the zeroth order antisymmetric function.

22*. Obtain the eigenfunctions for each of the two ^2D terms of the d^3 configuration.

23*. Two electrons move in a central field. Consider the electrostatic interaction between the electrons as a perturbation. Find the first-order energy shifts for the terms of the $(np)(n'p)$ configuration.

Hint. The sum of the roots of the secular equation is equal to the trace of the secular determinant.

24. Estimate the order of magnitude of the following quantities using the Thomas–Fermi model:

(a) average distance of an electron from the nucleus;
(b) average Coulomb interaction energy of two electrons in the atom;
(c) average kinetic energy of an electron;
(d) energy necessary to ionise the atom completely;
(e) average velocity of an electron in the atom;
(f) average angular momentum of an electron;
(g) average radial quantum number of an electron.

25. Use the Thomas–Fermi model to obtain an approximate expression for the energy of an atom in terms of the electronic density $\rho(\boldsymbol{r})$.†

26. Show that one obtains the Fermi–Thomas equation by minimising the total energy with respect to the density $\rho(\boldsymbol{r})$ subject to the normalisation condition

$$\int \rho(\boldsymbol{r})\, d^3\boldsymbol{r} = \text{constant}. \tag{1}$$

27. Show that if one adds to the constraint (1) of the preceding problem, the constraint of constant total angular momentum, one can prove that the rotation of the Fermi–Thomas atom is rigid.

28*. Use variational methods to find the best approximate expression for the electronic density in the Thomas–Fermi model, using trial functions of the form $\rho = A e^{-x}/x^3$, $x = \sqrt{(r/\lambda)}$, where A follows from the

† It is convenient to use in problems on the Thomas-Fermi model units in which $e = \hbar = \mu = 1$.

normalisation condition $\int \rho d\tau = N$ ($N = Z$ for neutral atoms) and where λ is a variational parameter. Find the energy of the atom (or ion).

Note. In determining the form of the trial functions, we take into account the fact that the exact solution must for small values of r be of the form $\rho \sim \text{const}/r^{\frac{3}{2}}$.

29. Prove the virial theorem for the Thomas–Fermi model.

30. Use the virial theorem to show that in the Thomas–Fermi model for the case of a neutral atom the electrostatic interaction energy of the atoms is one-seventh of the electrostatic interaction energy between the electrons and the nucleus.

31*. Evaluate in the Thomas–Fermi approximation the energy needed to ionise an atom (ion) completely.

32*. A diamagnetic atom is placed in an external magnetic field. Determine the magnitude of the induced magnetic field strength at the centre of the atom.

33*. Solve the preceding problem in the case of helium.

34*. The usual quantum rule of the semi-classical approximation is obtained for the case where the region in which the particle can move is bounded by two turning points, in the neighbourhood of which the semi-classical approximation cannot be applied. Another case is met with when one considers the motion of electrons in the Thomas–Fermi distribution which are in an s-state. (The same result applies to the motion of electrons in any central field of force which at small distances from the centre goes over into a Coulomb field; the Thomas–Fermi model is only taken to fix our ideas.) The region in which these electrons can move is bounded for small values of r by the point $r = 0$, which is not a turning point. On the other hand, it is wrong to demand that the semi-classical function is finite at $r = 0$, since in the neighbourhood of this point the semi-classical approximation cannot be applied.

Obtain the quantum rule for s-electrons of the Thomas–Fermi distribution.

35. Estimate the order of magnitude of the polarisability of a Thomas–Fermi atom, that is, the ratio of the dipole moment d of the Thomas–Fermi electron distribution, produced by the action of an applied electric field, to the value of the electric field strength $\mathscr{E}$.

36. Find the range over which the Landé g-factor can change for given values of L and S.

37. Show that the terms ${}^4D_{\frac{1}{2}}, {}^5F_1, {}^6G_{\frac{3}{2}}$ do not show a splitting which is linear in the field.

38*. Determine the shift in the term levels of a one-electron atom for the case of an intermediate field $(e\hbar/2\mu c)\mathscr{H} \sim |\Delta E_{jj}|$.

39*. Find the wave functions of an electron under the conditions of the preceding problem.

40*. Determine the splitting of the hydrogen energy levels in a strong magnetic field $[(e\hbar/2\mu c)\mathscr{H} > |E_{nlj} - E_{nlj'}|]$. To apply perturbation theory it is necessary to require that the energy of the atom in the magnetic field is small compared to the difference in energy between different multiplets, that is,

$$\frac{e\hbar\mathscr{H}}{2\mu c} < |E_{nlj} - E_{n'lj}| .$$

41. Determine the Zeeman splitting of the components of the hyperfine structure term ${}^2S_{\frac{1}{2}}$ $(j = \frac{1}{2}, l = 0)$ for the case of an intermediate magnetic field $[(e\hbar/2\mu c)\mathscr{H} \sim |\Delta E_{ff'}|]$. (The splitting produced by the field is of the same order of magnitude as the hyperfine structure splitting.)

42. Show that the sum total of the changes produced by an arbitrary magnetic field in the energies of all the states of a given value of M_J is equal to

$$\frac{e\hbar}{2\mu c}\mathscr{H} M_J \Sigma\left[1 + \frac{J(J+1) - L(L+1) + S(S+1)}{2J(J+1)}\right].$$

The summation is over J, satisfying the conditions

$$L + S \leqslant J \leqslant |L - S|, \quad J \geqslant M_J.$$

43. Show that the spin-orbit perturbation given by the equation $\hat{V}_{SL} = A(\hat{S}.\hat{L})$ is such that the average perturbation of all states corresponding to a given term (which is characterised by given L and S) is equal to zero.

44. Find the energy level splitting of an atom in a weak magnetic field when $(e\hbar/2\mu c)\mathscr{H} \ll |\Delta E_{JJ'}|$, where $\Delta E_{JJ'}$ is the distance between the levels of a multiplet.

45. Show that for a hydrogen atom in a uniform electric field
(a) the energy of a state with $l = n-1$, $m = n-1$ does not change to the first order in the field;
(b) the position of the centre of gravity of the displaced term is not changed;
(c) states which differ only in the sign of the z-component of the angular momentum have the same energy.

46*. Evaluate the shift of the hydrogen atom energy levels in a weak electric field (Stark effect small compared to the fine structure splitting).

47. Find the polarisability of the hydrogen atom in its ground state.

48. Find the magnetic moment of a hydrogen atom in a weak electric field.

49*. Evaluate the shift of the $n = 2$ hydrogen atom term in an electric field of intermediate intensity (Stark effect and fine structure splitting of the same order of magnitude).

50. A plane rotator with moment of inertia I and electrical dipole moment $\boldsymbol{d}$ is placed in a uniform electric field $\mathscr{E}$, in the plane of rotation. Consider $\mathscr{E}$ to be a perturbation and evaluate the first non-vanishing correction to the energy levels of the rotator.

51. Find the polarisability of a particle of mass μ and charge e in its ground state in an infinite square well.

52*. A three-dimensional rotator with moment of inertia I and electrical dipole moment $\boldsymbol{d}$ along the axis of the rotator is placed in a uniform electric field $\mathscr{E}$ which is considered to be a perturbation. Evaluate the first non-vanishing correction to the ground state energy of the rotator.

53. A hydrogen atom is in parallel magnetic and electric fields. Find the energy level splitting in the following cases:

(a) weak fields (Stark effect and Zeeman effect small compared to the fine structure splitting);

(b) intermediate fields, for the terms of principal quantum number $n = 2$.

54. A hydrogen atom is in an $n = 2$ state in mutually perpendicular electric and magnetic fields. Determine the energy level splitting assuming that the fields are strong (energy of the electron in the external electric and magnetic fields large compared to the fine structure splitting).

55. Find, by considering trial wavefunctions of the form $\psi = A \exp(-cr^2)$, an upper bound to the binding energy of a hydrogen atom.

56. Use a trial wavefunction of the same form as the one in the preceding problem to find an upper bound to the ground state energy of a one-dimensional harmonic oscillator.

57. Use the results from the preceding problem to find by a variational approach an upper bound to the energy of the first excited state of a one-dimensional harmonic oscillator.

58. Use the variational method to find the ground state energy of a two-electron system in the field of a nucleus of charge Z. Use as trial functions a product of hydrogen-like wave functions with effective charge Z'. Neglect relativistic corrections.

59. To a fair approximation we can use for the wave function of a helium atom the following expression:

$$\psi = \frac{Z'^3}{\pi a^3} e^{-Z'(r_1+r_2)/a} \quad (Z' = \tfrac{27}{16}) \quad \text{(see preceding problem).}$$

Show that the electrostatic potential around the atom is of the form:

$$\varphi(r) = 2e\left(\frac{1}{r} + \frac{Z'}{a}\right) e^{-2Z'r/a}.$$

60. Evaluate the diamagnetic susceptibility of helium using the approximate wave function of its ground state (see problem 58 of section 7).

61*. Use the variational method to determine the energy of the ground state of a lithium atom taking exchange into account. Take for the eigenfunctions of the electrons the following hydrogen-like functions: for the 1*s* electrons a function of the form $\psi_{100} = 2Z_1^{\frac{3}{2}} e^{-Z_1 r}$, and for the 2*s* electron a function of the form

$$\psi_{200} = cZ_2^{\frac{3}{2}} e^{-\frac{1}{2}Z_2 r}(1 - \gamma Z_2 r),$$

where Z_1 and Z_2 are variational parameters, where c follows from the normalisation of ψ_{200}, and γ from the fact that ψ_{100} and ψ_{200} are orthogonal on to one another. If we apply to this problem ordinary perturbation theory, we put $Z_1 = Z_2 = 3$. By introducing Z_1 and Z_2 as variational parameters, we include in our considerations the screening effect of the electrons.

62. Consider an atom acted upon by a perturbing potential u. Using perturbation theory we obtain for the wave function ψ in first approximation an expression of the following form:

$$\psi = \psi_0 + \sum_{n \neq 0} \frac{u_{n0}}{E_0 - E_n} \psi_n,$$

and for the energy

$$E = E_0 + u_{00} + \sum_{n \neq 0} \frac{(u_{n0})^2}{E_0 - E_n}.$$

To apply now the variational method we simplify as far as possible the form of ψ. Since

$$\sum_{n \neq 0}^{\infty} u_{n0} \psi_n = -u_{00} \psi_0 + \sum_{n=0}^{\infty} u_{n0} \psi_n = \psi_0 (u - u_{00}),$$

we use the following approximate form for ψ:

$$\psi \approx \psi_0 \left(1 + \frac{u - u_{00}}{E'}\right),$$

where E' in many cases can be considered to be equal to the average value of $E_0 - E_n$. Since the perturbed wave function is now approximately determined, we can apply the variational method to determine the energy.

Use the variational method to determine the energy of an atom acted upon by a perturbing potential u. Look for the minimum energy using a trial function of the form $\psi = \psi_0(1+\lambda u)$, where λ is a variational parameter.

63. Use the result of the preceding problem to find a formula for the polarisability of an atom. Give a numerical estimate of the polarisability of a hydrogen and of a helium atom, in their ground states.

64. Determine the shift of the energy levels due to the finite size of the nucleus. Assume the potential within the nucleus ($r < a$) to be constant (this corresponds physically to a distribution of the nuclear charge over the surface of a sphere of radius a).

65. Find the first-order shift in the energy of a hydrogenic ground state due to the finite extension of an arbitrary spherical nuclear charge distribution of mean square radius $\langle r^2 \rangle$ which is small compared to a_B^2 (a_B: Bohr radius) (compare problem 26 of section 7).

66*. Evaluate the value of $\psi^2(0)$ for a valence s-electron in an atom of large Z, using the semi-classical approximation.

67*. Evaluate in first-order perturbation theory the ground state energy of a two-electron atom or ion with nuclear charge Z, taking the interaction between the electrons as a perturbation.

Evaluate also the first ionisation potential of the atom (ion) considered.

68. Evaluate the first- and second-order corrections to the energy levels of a "slightly" anharmonic oscillator with potential energy $\frac{1}{2}\mu\omega^2 x^2 + \alpha x^3 + \beta x^4$.

69. Two identical spin-$\frac{1}{2}$ particles are subject to the same potential $V = \frac{1}{2}\mu\omega^2 x^2$ and interact through a potential $V' = V_0(\hbar/\mu\omega)^{1/2}\delta(x_1 - x_2)$. Calculate the first-order corrections to the energies of the lowest three levels of the unperturbed two-particle system, pointing out the splitting of any degenerate level which may occur.

70*. A linear harmonic oscillator is acted upon by a uniform electric field which is considered to be a perturbation and which depends as follows on the time:

$$\mathscr{E}(t) = A \frac{1}{\sqrt{(\pi)}\,\tau} e^{-(t/\tau)^2},$$

where A is a constant. (Since the action of a uniform field is equivalent to a shift of the point of suspension, this problem can be solved not only by perturbation theory, but also exactly, using the methods of problem 24 of section 3.)

Assuming that when the field is switched on (that is, at $t = -\infty$) the oscillator is in its ground state, evaluate to a first approximation the probability that it is excited at the end of the action of the field (that is, at $t = +\infty$).

71. Solve the preceding problem for a field which varies as follows:

$$\mathscr{E}(t) \sim \frac{1}{t^2+\tau^2},$$

and which corresponds to a given total classical imparted impulse P.

72. Solve the preceding problem for a field proportional to $e^{-t/\tau}$ corresponding to a given total classical imparted impulse P.

73*. A uniform periodic electric field acts upon a hydrogen atom which at $t = 0$ is in its ground state.

Determine the minimum frequency of the field necessary to ionise the atom and use perturbation theory to evaluate the probability for ionisation per unit time. For the sake of simplicity assume the electron in the final state to be free.

74*. Find the probability for the ejection of a K-electron from an atom accompanied by a dipole transition of the nucleus assuming a direct electrostatic interaction between the electron and the protons of the nucleus (internal conversion, neglecting retardation).

Use the wave functions of a K-electron in a hydrogen-like atom for the initial wave function, and assume the velocity of the electron in its final state to be much larger than atomic.

75. The nucleus of an atom in a stationary state ψ_0 experiences a sudden impact during a time τ which imparts to it a velocity v.

Assume that $\tau \ll T$ and $\tau \ll a/v$, where T and a are of the order of magnitude of the electronic periods and the dimensions of the electron shells, and express in general form the probability that the atom has made a transition to a state ψ_n as the result of such a "kick".

76. Use the result of the preceding problem to evaluate the total probability of excitation or ionisation of a hydrogen atom (initially in its ground state) as the result of a sudden "kick" during which the proton receives a momentum $\boldsymbol{p}$.

Discuss the conditions for the applicability of the result obtained.

77. How does the probability of the capture of a negative meson, which is in the K-orbit of a mesonic atom, depend on the charge Z of the nucleus?

78. Find the effective (average) potential φ acting on a charged meson passing through a non-excited hydrogen atom (whose polarisation may be neglected). Obtain the limiting expression for φ for large and small distances of the meson from the nucleus.

79. Let the potential energy $V(x, y, z)$ be a homogeneous function of the coordinates of degree of homogeneity ν,

$$V(\lambda x, \lambda y, \lambda z) = \lambda^{\nu} V(x, y, z).$$

Show that the average value of the kinetic energy in a state of the discrete spectrum is connected to the average value of the potential energy by the relation $2\bar{T} = \nu \bar{V}$ (virial theorem).

80. Investigate the influence of the finiteness of the nuclear mass M on the energy levels of an atom with n electrons, assuming that the energy levels are known for $M = \infty$.

81. Evaluate the "exchange" correction (that is, the correction determined by the Pauli principle) to the terms of a two-electron atom (or ion), assuming one of the two electrons to be in the 1s-state, neglecting the electrostatic interaction between the electrons, and taking into account the fact that the nuclear mass is finite.

82. Show that the average value of the dipole moment of a system of charged particles which is in a state of well-defined parity is equal to zero.

83. Show that if a system of N charged particles moves in a finite region of space, the following relation (the so-called "sum rule") holds:

$$\frac{2\mu}{\hbar^2 e^2} \sum_n (E_n - E_m) |d_{mn}|^2 = N,$$

where d_{mn} is the matrix element of some component of the dipole moment of the system and where the summation extends over all states of the system, while μ and e are the mass and charge of each particle.

84. Prove that for a system of N interacting particles of mass μ we have

$$\sum_n (E_n - E_m) |\langle n | \sum_j \exp\{i(\boldsymbol{q} \cdot \boldsymbol{r}_j)\} | m \rangle|^2 = N \frac{\hbar^2 q^2}{2\mu} .$$

85. Derive the sum rule

$$\sum_n (E_n - E_m) |p_{nm}|^2 = -\tfrac{1}{2}\hbar^2 V''_{mm},$$

where p_{nm} and V''_{mm} are the matrix elements of the linear momentum and of the second derivative of the potential energy of a one-dimensional system with Hamiltonian

$$\hat{H} = \frac{\hat{p}^2}{2\mu} + V(x).$$

86. Prove the sum rules

$$S_1 = \sum_{n'} f_{n,l}^{n',l+1} = \frac{(l+1)(2l+3)}{3(2l+1)}, \quad S_2 = \sum_{n'} f_{n,l}^{n',l-1} = \frac{-l(2l-1)}{3(2l+1)}, \qquad \text{(A)}$$

where

$$f_a^b = \frac{2\mu}{3\hbar^2}(E_b - E_a)[|x_{ab}|^2 + |y_{ab}|^2 + |z_{ab}|^2], \qquad \text{(B)}$$

with x_{ab}, y_{ab}, and z_{ab} the matrix elements of x, y, and z.

8

Molecules

1. Obtain the Schrödinger equation for a diatomic molecule assuming that the centre of mass of the molecules is approximately the same as the centre of mass of the nuclei. To describe the motion of the electrons use a moving system of coordinates fixed at the nuclei. Spin effects can be neglected.

2. Solve the preceding problem by considering the spin states of the electrons and describing these states in the moving system of coordinates ξ, η, ζ.

3. For small vibrations of the nuclei the wave functions of a molecule can be approximated by a product of three functions

$$\Phi_{el}(\xi_i, \eta_i, \zeta_i, \sigma_i, p), \quad f(\rho), \quad \Theta(\theta, \varphi).$$

The first function describes the motion of the electrons while the nuclei are fixed, while the second and third describe the vibrational and rotational states of the molecule. Find the equations which determine the vibrational and rotational parts of the wave functions of diatomic molecules.

4. Estimate for the case of diatomic molecules the relative order of magnitude of the following quantities: (a) the intervals between the electronic, vibrational, and rotational levels; (b) the internuclear distance and the zero point amplitude of the nuclear vibrations; (c) the characteristic periods and velocities of the electronic and nuclear motion.

5. Determine the possible terms of the diatomic molecules N_2, Br_2, LiH, HBr, and CN, which can be obtained by combining the two atoms in their ground state.

6. Find an equation which determines the electronic terms of a hydrogen atom interacting with a helium atom, assuming both atoms to be in their ground state.

7. Find the vibrational and rotational energy spectrum of a diatomic molecule, taking into account the fact that the nuclei move in a potential of the form

$$V(r) = -2D\left(\frac{1}{\rho} - \frac{1}{2\rho^2}\right), \quad \text{where} \quad \rho = \frac{r}{a}.$$

8. Express the effective potential of the preceding problem, $V(r)+(\hbar^2/2\mu r^2)K(K+1)$, near its minimum in the form of an oscillator potential and find the energy levels for small vibrations.

9. Determine the moment of inertia and the internuclear distance of the $^1H^{35}Cl$ molecule, if the difference in the frequency of two neighbouring lines of the rotational–vibrational (infra-red) band of $^1H^{35}Cl$ is equal to $\Delta\nu = 20{\cdot}9\ cm^{-1}$.

Evaluate the corresponding $\Delta\nu$ for the DCl spectrum.

10. Evaluate the ratio of the difference in energy of the first two rotational levels to the energy difference of the first two vibrational levels for the HF-molecule. The moment of inertia of HF is equal to $I = 1{\cdot}35\times10^{-40}\ g\,cm^2$ and the vibrational frequency is equal to $\Delta\nu_{vib} = 3987\ cm^{-1}$.

11. Determine the dissociation energy of the D_2-molecule, if the dissociation and zero point energy of the H_2 molecule are respectively equal to 4·46 eV and 0·26 eV.

12*. One often uses for the approximate form of the potential energy of a diatomic molecule a Morse potential $V = D(1-e^{-2\beta\xi})^2$; $\xi = (r-a)/a$. Determine the vibrational energy spectrum for $K = 0$.

13. Show that the operator of the square of the total angular momentum of a diatomic molecule can be written in the form

$$\hat{J}^2 = -\left[\frac{1}{\sin\theta}\frac{\partial}{\partial\theta}\left(\sin\theta\frac{\partial}{\partial\theta}\right)+\frac{1}{\sin^2\theta}\left(\frac{\partial}{\partial\varphi}-i\hat{M}_\zeta\cos\theta\right)^2\right]+\hat{M}_\xi^2 .$$

14. The ξ, η, ζ axes are axes of an orthogonal system of coordinates which is rigidly connected with a rotating rigid body. Find the operators $\hat{J}_\xi, \hat{J}_\eta, \hat{J}_\zeta$ of the ξ, η, ζ components of the angular momentum of the rigid body.

15. Show that the operators $\hat{J}_\xi, \hat{J}_\eta, \hat{J}_\zeta$ obey the following commutation relations

$$\hat{J}_\xi\hat{J}_\eta-\hat{J}_\eta\hat{J}_\xi = -i\hat{J}_\zeta,$$

$$\hat{J}_\eta\hat{J}_\zeta-\hat{J}_\zeta\hat{J}_\eta = -i\hat{J}_\xi,$$

$$\hat{J}_\zeta\hat{J}_\xi-\hat{J}_\xi\hat{J}_\zeta = -i\hat{J}_\eta,$$

which shows that the commutation relations for the angular momentum components in a moving system of coordinates differ from the commutation relations in a non-moving system only in the sign on the right-hand side of the relevant equations.

16. In classical mechanics we have the Euler equations:

$$A\frac{dp}{dt}+(C-B)\,qr = 0,$$

$$B\frac{dq}{dt}+(A-C)\,rp = 0,$$

$$C\frac{dr}{dt}+(B-A)\,pq = 0,$$

or

$$\frac{dJ_\xi}{dt}+\left(\frac{1}{B}-\frac{1}{C}\right)J_\eta J_\zeta = 0, \quad \text{etc.}$$

Show that in quantum mechanics the last relations are of the form

$$\frac{d\hat{J}_\xi}{dt}+\frac{1}{2}\left(\frac{1}{B}-\frac{1}{C}\right)(\hat{J}_\eta\hat{J}_\zeta+\hat{J}_\zeta\hat{J}_\eta) = 0, \quad \text{etc.}$$

17. Molecules with two or more axes of symmetry of third or higher order (for instance, CH_4) can be considered to be spherical tops. For such molecules the inertial ellipsoid goes over into a sphere: $A = B = C$. Determine the energy levels of a spherical top.

18. Molecules with an axis of symmetry of third or higher order (for instance, SO_2, NH_3, CH_3Cl) and molecules with lower symmetry or even without any symmetry, but of which two of their principal moments of inertia are equal can be considered to be symmetrical tops, $A = B \neq C$.

Determine the energy levels of a symmetrical top.

19. Give the Schrödinger equation of a symmetrical top.

20*. Find the eigenfunctions of the operator

$$\hat{J}^2 = -\left[\frac{1}{\sin\theta}\frac{\partial}{\partial\theta}\left(\sin\theta\frac{\partial}{\partial\theta}\right)+\frac{1}{\sin^2\theta}\left(\frac{\partial^2}{\partial\varphi^2}+\frac{\partial^2}{\partial\psi^2}\right)-2\frac{\cos\theta}{\sin^2\theta}\frac{\partial^2}{\partial\varphi\,\partial\psi}\right].$$

21. Evaluate the matrix elements of the Hamiltonian of an asymmetric top.

22. Determine the energy levels of an asymmetric top for $J = 1$.

23*. Find the wave functions of an asymmetric top for $J = 1$.

24. Use the properties of the Pauli matrices to show that even if spin–spin interactions are taken into account the $^2\Sigma$ terms of a diatomic molecule are not split.

25. Determine the multiplet splitting of a $^3\Sigma$ term related to a b-type bond.†

† An a- (or b-) type bond is one where the spin–orbit interaction energy (more precisely, the spin–axis interaction energy) is large (or small) compared to the energy differences between rotational levels.

26*. When the Schrödinger equation is approximately solved (see problem 3 of section 8) one does not take into account the operator

$$\frac{i\hbar^2}{2M\rho^2}\left\{\cot\theta(\hat{M}_\xi - i\hat{M}_\eta\hat{M}_\zeta - i\hat{M}_\zeta\hat{M}_\eta) + \frac{2}{\sin\theta}\hat{M}_\eta\frac{\partial}{\partial\varphi} + 2\hat{M}_\xi\frac{\partial}{\partial\theta}\right\} = \hat{w}$$

since its diagonal elements are equal to zero. A consideration of the off-diagonal elements related to the same electronic (n, Λ) and vibrational (v) states leads to an effect, the so-called rotational distortion of the spin. Considering the operator $\hat{w}$ to be a perturbation, determine the change in the doublet levels due to this perturbation.

27. Derive a connection between the value of the total nuclear spin angular momentum of the D_2 molecule in a Σ-state and the possible values of the quantum number K.

28. Determine the Zeeman splitting of a diatomic molecule term, if the term is a case *a* term. The magnetic field is assumed to be small, that is, the interaction energy of the spin with the external magnetic field is small compared to the difference in energy of consecutive rotational levels.

29. Determine the Zeeman splitting of a diatomic molecule term if the term is a case *b* term and the magnetic field is such that the interaction energy of the spin with the external magnetic field is small compared to the spin–axis interaction energy.

30. Solve the preceding problem for the case where the spin-axis interaction energy is small compared with the energy splitting produced by the external magnetic field.

31*. Determine the Zeeman splitting of a diatomic molecule doublet term (case *b*) in a magnetic field of such a magnitude that the interaction energy of the magnetic moment with the field is of the same order of magnitude as the spin–axis interaction energy.

32. Determine the splitting of a diatomic molecule term (case *a*) in an electric field, if the molecule possesses a constant dipole moment $\boldsymbol{p}$.

33. Solve the preceding problem for a case *b* term.

34. Consider an idealised model of an atom which is an electron of charge $-e$ and mass μ moving along a straight line, under the influence of an attractive force of magnitude k times its displacement from a fixed centre on this line which carries a compensating charge $+e$. Two such identical atoms are placed a large distance R apart, and their axes of vibration are parallel, and perpendicular to the line joining their centres.

Assuming that the interaction between the atoms is solely due to the electrostatic energies, find the dispersion energy of attraction in the ground state to the lowest power of R^{-1}.

35*. Determine the interaction energy between a hydrogen atom in its ground state and a proton for distances apart large compared to the Bohr radius.

36*. Use the variational method to estimate the dissociation energy of the hydrogen molecule ion H_2^+.

37*. Use perturbation theory to determine the interaction between two non-excited hydrogen atoms at a large distance R apart.

38. Estimate the interaction energy between a hydrogen atom in its ground state and another hydrogen atom in a p-state for distances apart large compared to the Bohr radius.

39*. Consider a system of atoms with spherical charge distributions. We showed in the preceding problem that two such atoms at large distances apart interact by a so-called dispersion force. The dispersion law is a quantum phenomenon and in contradistinction to the classical polarisation force it is additive. Show that the interaction energy of two such atoms is independent of the presence of other similar atoms, that is, show that the interaction energy of a system of atoms is additive, that is, equal to the sum of the interaction energies of different pairs of atoms.

40. Consider an axially symmetric molecule. If λ is the molecular orbital quantum number and if the oscillator strengths $f_{\alpha\beta}$ for transitions between states α and β are given by the expression

$$f_{\alpha\beta} = \frac{2\mu(E_\alpha - E_\beta)}{3e^2\hbar^2}|P_{\alpha\beta}|^2,$$

where the $P_{\alpha\beta}$ are the electrical dipole moment matrix elements, prove that the f satisfy the sum rules

$$\sum_{n'} f_{\lambda n,\lambda' n'} = \frac{1+(\lambda'-\lambda)\lambda}{3}, \quad \lambda' = \lambda,\ \lambda-1, \text{ or } \lambda+1. \tag{A}$$

9

Scattering

1*. Find the cross-section for the scattering of low velocity particles by a potential well (de Broglie wavelength large compared to the well dimensions).

2*. Determine the cross-section for the scattering of slow particles by the repulsive potential

$$V(r) = V_0 \quad (r < a), \qquad V(r) = 0 \quad (r > a).$$

3*. Determine the first three coefficients of the expansion in Legendre polynomials of the elastic scattering cross-section $d\sigma/d\Omega$ in terms of the phase shifts.

4*. Find the phase shifts in the field $V = A/r^2$. Determine the scattering cross-section at small angles.

5. Show that the scattering amplitude of a particle in an arbitrary external field is connected with its wave function ψ through the equation

$$f(\mathbf{k}) = -\frac{\mu}{2\pi\hbar^2}\int e^{-i\mathbf{kr}}\, V\psi\, d\tau.$$

6. Evaluate the differential scattering cross-section in a repulsive field $V = A/r^2$ both in the Born approximation and according to classical mechanics. Determine the limit of applicability of the formulae obtained.

7*. Find the discrete levels for a particle in an attractive potential $V(r) = -V_0 e^{-r/a}$ for $l = 0$. Determine the phase shift δ_0 for this potential and analyse the connection between δ_0 and the discrete spectrum.

8. Show that in a Coulomb field there exists a one-to-one relation between the poles of the scattering amplitude and the discrete energy levels.

Hint. Use the equation

$$\exp(2i\delta_l) = \frac{\Gamma(l+1+i/k)}{\Gamma(l+1-i/k)}.$$

9. Determine in Born approximation the differential and the total scattering cross-sections for the following potentials:

(a) $V(r) = V_0 \exp(-\alpha^2 r^2)$,

(b) $V(r) = V_0 \exp(-\alpha r)$.

10. Use the Born approximation to find the differential and the total elastic cross-section for scattering of fast electrons by: (a) a hydrogen atom; (b) a helium atom.

11. Evaluate the cross-section for elastic scattering by the potential

$$V(r) = A\frac{e^{-\kappa r}}{r}$$

in the Born approximation. Discuss the applicability of this approximation.

12. Evaluate in the Born approximation the cross-section for scattering by a "delta-function" potential.

13. Derive in the semi-classical approximation an expression for the lth phase shift for the case of a scattering potential $V(r)$.

14. If $f_{\text{com}}(\vartheta, \varphi)$ is the scattering amplitude in the centre-of-mass system, determine the scattering amplitude $f_{\text{lab}}(\vartheta', \varphi')$ in the laboratory system.

15. A particle of mass μ is scattered by a non-local potential so that its Schrödinger equation is

$$-\frac{\hbar^2}{2\mu}\nabla^2\psi(r) + \int V(r, r')\psi(r')d^3r' = E\psi(r)\,,$$

with

$$V(r, r') = -\frac{\hbar^2}{2\mu}\lambda u(r)u(r')\,.$$

(i) Show that only s-waves are affected by the interaction.

(ii) Obtain a formula for the scattering amplitude.

Hint. Write the Schrödinger equation in its integral equation form; compare problem 5 of section 9.

(iii) Find the s-wave scattering phase shift for the case where

$$u(r) = \frac{e^{-r/b}}{r}\,.$$

16*. A spherical rotator (that is, a particle moving on the surface of a sphere of given radius a) is hit by a particle which interacts with the rotator by Coulomb forces. Use perturbation theory to evaluate the differential cross-section for the inelastic scattering over an angle θ with the simultaneous excitation of the lth level of the rotator.

17*. Determine the total cross-section for elastic scattering of fast particles by an impenetrable sphere of radius a (de Broglie wavelength $\lambda\!\!\!^{-} \ll a$).

18. Show the following properties of the scattering length, that is, the scattering amplitude for $E \to 0$ with the opposite sign:

(a) in a repulsive potential the scattering length is positive;

(b) in an attractive potential, the scattering length is negative, if there are no discrete levels;

(c) the scattering length tends to infinity if on deepening the potential well a new level appears.

Hint. If we go over from the Schrödinger equation to the equivalent integral equation ($E = 0$)

$$\psi = 1 - \frac{\mu}{2\pi\hbar^2} \int \frac{V(\mathbf{r}')\,\psi(\mathbf{r}')}{|\mathbf{r}-\mathbf{r}'|}\, d\tau',$$

we find for the asymptotic behaviour of ψ

$$\psi = 1 - \frac{a}{r},$$

where the scattering length

$$a = \frac{\mu}{2\pi\hbar^2} \int \psi V\, d\tau$$

is expressed in terms of the potential energy and $\psi_{E=0}$.

19*. Show that in the general case of inelastic scattering, the following formula, which connects the total scattering cross-section $\sigma = \sigma_{\text{el}} + \sigma_{\text{inel}}$ and the elastic scattering amplitude for $\vartheta = 0$, holds:

$$\sigma = \frac{4\pi}{k} \operatorname{Im} f(0).$$

Hint. Use the expansion of these quantities in terms corresponding to different values of the orbital angular momentum:

$$f(\vartheta) = \frac{1}{2ik} \sum_{l=0}^{\infty} (2l+1)(\eta_l - 1) P_l(\cos\vartheta),$$

$$\sigma_{\text{el}} = \frac{\pi}{k^2} \sum_{l=0}^{\infty} (2l+1)|1-\eta_l|^2,$$

$$\sigma_{\text{inel}} = \frac{\pi}{k^2} \sum_{l=0}^{\infty} (2l+1)(1-|\eta_l|^2).$$

20. Consider as a limiting case a so-called "black" nucleus. Let the nuclear radius R be large compared to the neutron de Broglie wavelength. Assume that all neutrons hitting the nucleus are absorbed. Determine the total scattering and the total absorption cross-sections.

21*. Prove that the phase shift δ_l for the scattering by a potential $V(r)$ can be found from the variational principle

$$\delta F_l = 0\,,$$

where

$$F_l = \frac{\int_0^\infty dr\,\widetilde{V}(r)\chi_l(r)\left\{\chi_l(r) - \int_0^\infty dr'\chi_l(r')\widetilde{V}(r')G(r,r')\right\}}{\left\{\int_0^\infty dr\,\widetilde{V}(r)\chi_l(r)j_l(kr)\right\}^2}\,,$$

where $\widetilde{V}(r) = (2\mu/\hbar^2)V(r)$, χ_l is the radial part of the wavefunction (see, e.g., problem 1 of section 9), j_l is a spherical Bessel function, and $G(r, r')$ is the Green function of the radial Schrödinger equation.

22. Consider a collision of two identical particles with interaction energy $V(r)$. Find the effective scattering cross-section for slow identical particles in the case of a short-range force.

23. Evaluate the cross-section for elastic scattering of an electron by an electron or of an α-particle by an α-particle.

24. The scattering of neutrons by protons depends on the total spin of the neutron and proton. At small energies the cross-section for the triplet state ($S = 1$) is equal to $\sigma^{\text{tr}} = 4\pi|f_3|^2 \approx 2.10^{-24}$ cm^2 and for the singlet state ($S = 0$) $\sigma^{\text{si}} = 4\pi|f_1|^2 \approx 78.10^{-24}$ cm^2.

Introduce the operator

$$\hat{f} = \frac{f_1+3f_3}{4} + \frac{f_3-f_1}{4}(\hat{\boldsymbol{\sigma}}_n\,.\,\hat{\boldsymbol{\sigma}}_p).$$

It is easily seen that its eigenvalues are f_3 and f_1 for the triplet and singlet state respectively. To determine the scattering cross-section for arbitrarily polarised neutrons we must average the operator $\hat{f}^2$,

$$\sigma = 4\pi\overline{\hat{f}^2}.$$

Let the spin state of the incoming neutrons be described by the function $\begin{pmatrix} e^{-i\alpha}\cos\beta \\ e^{i\alpha}\sin\beta \end{pmatrix}$ (the direction of the neutron spin in polar angles is $\Theta = 2\beta$, $\Phi = 2\alpha + \pi/2$, see problem 21 of section 4) and that of the proton by the function $\begin{pmatrix} 1 \\ 0 \end{pmatrix}$ (proton spin parallel to the z-axis). Determine for this case the cross-section for the scattering of neutrons by protons.

25. Find the probability that the spin of a slow neutron changes its direction when it is scattered by a proton if the neutron spin before the collision was in the direction of the z-axis and the proton spin in the opposite direction.

26. Slow neutrons are scattered by molecular hydrogen. If their energies are appreciably less than thermal ($\lambda\!\!\!^{-} > 10^{-8}$ cm, that is, the wavelength of the incoming neutrons is large compared to the distance apart of the protons in the molecule) their scattering amplitude is equal to the sum of the amplitudes due to each of the protons.

Thus we have

$$\hat{f} = \frac{f_1+3f_3}{2} + \frac{f_3+f_1}{4}(\hat{\boldsymbol{\sigma}}_n \cdot \hat{\boldsymbol{\sigma}}_{p_1} + \hat{\boldsymbol{\sigma}}_{p_2}).$$

The spins of the two protons in a hydrogen molecule can be parallel (orthohydrogen) or antiparallel (parahydrogen).

Determine the cross-section for the scattering of neutrons by para- or orthohydrogen.

27. Derive a relation between the cross-section for the scattering of unpolarised neutrons by a nucleus of spin I and the scattering length a_+ and a_- referring to the scattering of neutrons with their spin parallel or antiparallel to the nuclear spin.

28. The interaction energy between a neutron of spin $\frac{1}{2}\hbar$ and an atom in a ferromagnet can be approximated by

$$V(\boldsymbol{r}) = A\delta(\boldsymbol{r}) + (\boldsymbol{n}\cdot\boldsymbol{\sigma})f(r),$$

with $\boldsymbol{n}$ a unit vector in the direction of magnetisation of the atom and

$$f(r) = B, \quad r \leqslant a; \quad f(r) = 0, \quad r > a.$$

Calculate the scattering cross-section for the two possible spin orientations of a beam of slow neutrons ($\lambda\!\!\!^{-} \gg a$), if $a^3 B \ll A$.

Calculate the polarisation $P = (I_+ - I_-)/(I_+ + I_-)$ of a beam of slow neutrons which has passed through a block of saturated iron of thickness L and with N atoms per unit volume. Here I_+ and I_- refer to the transmitted intensities of beams with spins respectively parallel and antiparallel to the direction of magnetisation.

29*. Consider the elastic scattering of spin-$\frac{1}{2}$ particles by scalar (spin-zero) particles. Find the differential cross-section for scattering involving a spin reversal. Consider the scattering of S and P waves.

30. Consider the nuclear reaction

$$A + a \rightarrow C \rightarrow B + b.$$

What will be the angular distribution of the products of this nuclear reaction in the centre of mass system, or, what is the same, in the system in which the compound nucleus is at rest, for the following three cases:

(a) the spin of the compound nucleus is equal to zero;
(b) the orbital angular momentum of the relative motion of the products of the reaction is equal to zero;
(c) the orbital angular momentum of the relative motion of the reacting particles is equal to zero (the spin of the compound nucleus different from zero)?

31*. Find the eigenfunctions of the operators of the square and of the z-component of the isotopic spin, I^2 and I_z, for the nucleon–meson system.

32*. In the centre of mass system the scattering of mesons by nucleons is the problem of scattering of particles by a stationary scattering centre. Far from the scattering centre the incoming wave with well-defined values of the z-components of spin S_z and of isotopic spin τ_z can be written in the form

$$e^{ikz}\begin{pmatrix}1\\0\end{pmatrix}\delta(\pi-\pi_i)\,\delta(n-\tau_z).$$

The quantity π_i can take on the values:

$$\pi_i = 1 \text{ for a } \pi^+ \text{ meson,} \qquad \delta(\pi-1) = \varphi_+;$$
$$\pi_i = 0 \text{ for a } \pi^0 \text{ meson,} \qquad \delta(\pi) = \varphi_0;$$
$$\pi_i = -1 \text{ for a } \pi^- \text{ meson,} \qquad \delta(\pi+1) = \varphi_-;$$

while τ_z takes on the values:

$$\tau_z = \tfrac{1}{2} \text{ for a proton,} \qquad \delta(n-\tfrac{1}{2}) = \psi_p;$$
$$\tau_z = -\tfrac{1}{2} \text{ for a neutron,} \qquad \delta(n+\tfrac{1}{2}) = \psi_n.$$

Expand the incoming wave in terms of the eigenfunctions of the operators J^2, J_z, I^2, I_z.

Hint. In the meson–nucleon system the parity $(-1)^l$, the total angular momentum J, and the isotopic spin I are integrals of motion.

Since the total spin in such a system is equal to $\frac{1}{2}$, the orbital angular momentum will also be conserved. If we expand the incoming wave in terms of the eigenfunctions of operators, the absolute magnitude of which is conserved, we can sum over l and I instead of over J and I.

33. Show that the elastic scattering process

$$\pi^+ + p \to \pi^+ + p$$

can take place only in the isotopic spin state with $I = \frac{3}{2}$, while

$$\pi^+ + n \to \pi^+ + n$$

can take place in both of the isotopic spin states with $I = \frac{3}{2}$ and $I = \frac{1}{2}$.

34*. Express the scattering amplitude of pions, scattered by nucleons, in terms of phase shifts for the following three cases:

$$\pi^+ + p \to \pi^+ + p,$$

$$\pi^- + p \to \pi^- + p,$$

$$\pi^- + p \to \pi^0 + n.$$

35*. Show that the scattering amplitudes for all possible meson–nucleon reactions can be expressed in terms of those considered in the preceding problem assuming isotopic invariance.

Express these quantities in terms of the scattering amplitudes for the states with isotopic spin $\frac{3}{2}$ and $\frac{1}{2}$.

36*. Express the total cross-sections for the following reactions in terms of phase shifts:

$$\pi^+ + p \to \pi^+ + p,$$

$$\pi^- + p \to \pi^- + p,$$

$$\pi^- + p \to \pi^0 + n.$$

37*. In the region of low energies when the meson wavelength is large compared with the range of the meson–nucleon forces, the main contribution to the scattering comes from S- and P-waves only.

Find the differential scattering cross-section in terms of phase shifts for the following reactions:

$$\pi^+ + p \to \pi^+ + p,$$

$$\pi^- + p \to \pi^- + p,$$

$$\pi^- + p \to \pi^0 + n.$$

38*. A pion beam is scattered by a target of unpolarised protons, that is, the number of protons with $S_z = \frac{1}{2}$ in the target is equal to the number of protons with $S_z = -\frac{1}{2}$. It turns out that the protons which were originally unpolarised are polarised in the scattering process. Determine the magnitude of the proton polarisation considering only S- and P-waves.

39. Use the principle of detailed balancing to connect the cross-section for the radiative capture of a neutron by a proton with that for the photo-dissociation of the deuteron.

40. A system consisting of a negative and a positive pion is in a state with a well-defined orbital angular momentum l.

Find the selection rule for the process

$$\pi^- + \pi^+ \rightarrow \pi^0 + \pi^0$$

with respect to whether l is even or odd.

41. Show that the process

$$\pi^- + d \rightarrow n + n$$

is forbidden for a scalar pion (that is, a pion with spin 0 and even parity) when it is captured in the S-state of "meso-deuterium".

42. Show that it follows from the experimental fact that the process

$$\pi^- + d \rightarrow n + n + \pi^0$$

has a very small probability that the negative and the neutral pion possess the same internal parity, assuming that the negative pion capture leads to the S-state of "meso-deuterium".

43. Show that the production of a pseudoscalar neutral pion during a $p+p$ collision is forbidden, if the neutral pion emerges in a P-state. Assume that the reaction takes place near the threshold so that the effective relative angular momentum in the final state is equal to zero.

44. Which are the eigenfunctions and eigenvalues of the isotopic spin for a system of two nucleons?

45. Show that

$$\frac{d\sigma(p+d \rightarrow d+n+\pi^+)}{d\sigma(p+d \rightarrow d+p+\pi^0)} = 2,$$

where the $d\sigma$ are the differential cross-sections of the corresponding reactions taken at the same relative energy, angle of separation, and relative orientation of the spins.

46. Show that the differential cross-sections for the production of pions when accompanying the formation of a deuteron when two nucleons collide

$$p + p \rightarrow d + \pi^+ \tag{1}$$

and

$$n + p \rightarrow d + \pi^0 \tag{2}$$

are connected by the relation

$$\frac{d\sigma(p+p \rightarrow d+\pi^+)}{d\sigma(n+p \rightarrow d+\pi^0)} = 2.$$

47. If the decay of the Λ°-hyperon ($T = 0$) obeys the selection rule $\Delta T = \frac{1}{2}$, obtain the branching ratio for the two decay modes $\Lambda^\circ \to \pi^- + P$ and $\Lambda^\circ \to \pi^0 + N$.

48. Show that the differential cross-section for the processes

$$n + p \to p + p + \pi^- \tag{1}$$

and

$$n + p \to n + n + \pi^+, \tag{2}$$

taken at the same relative energies, angles of separation, and relative orientation of the spins are the same.

49. Prove that a free electron can neither absorb nor emit a photon.

Creation and annihilation operators; density matrix

1. Let $\varphi_n(j)$ be a complete orthonormal set of single-particle functions which are arguments of both the spatial coordinates r_j and the spin coordinates s_j of the jth particle, together denoted by the argument j. Consider a system of N identical particles, let $P1, P2, ..., Pj, ..., PN$ indicate a permutation of $1, ..., j, ..., N$, and let ϵ_P be a function which is +1 or −1 for even or odd permutations, if we are considering a system of N fermions, and which is always +1, if we are considering a system of N bosons. Using the Dirac bra and ket notation, prove that the ket set

$$|i_1, i_2, ..., i_N\rangle = \frac{1}{N!}\sum_P \epsilon_P \varphi_{i_1}(P1)\varphi_{i_2}(P2) \ldots \varphi_{i_N}(PN)\,, \tag{1a}$$

where the summation is over all $N!$ possible permutations of $1, ..., N$, together with the bra set

$$\langle i_1, i_2, ..., i_N| = \frac{1}{N!}\sum_P \epsilon_P \varphi^*_{i_1}(P1)\varphi^*_{i_2}(P2) \ldots \varphi^*_{i_N}(PN)\,, \tag{1b}$$

where φ^* is the Hermitean conjugate of φ, if we use spinors for particles with non-zero spin, satisfies the orthonormality relation

$$\langle i'_1, i'_2, ..., i'_N|i_1, i_2, ..., i_N\rangle = \frac{1}{N!}\sum_P \epsilon_P\, \delta(i_1 - i'_{P1}) \ldots \delta(i_N - i'_{PN})\,, \tag{2}$$

where we have assumed the quantum number index i to be a continuous parameter. If it is a discrete parameter, the Dirac delta-functions in (2) must be replaced by Kronecker deltas.

Prove also that the set (1) satisfies a closure relation.

2. If $|\Psi\rangle$ is a properly symmetrised wavefunction of the N-particle system and if $\hat{\Omega}$ is an operator acting on functions in the Hilbert space spanned by the set (1) of the preceding problem, express $|\Psi\rangle$ and $\hat{\Omega}$ in terms of this set.

3. Consider the Hilbert space which is the (direct) product space of the Hilbert spaces corresponding to 0-, 1-, ..., N-, ... particle systems. The complete orthonormal set spanning this Hilbert space will be the set

$$|0\rangle, |i\rangle, |i_1, i_2\rangle, ..., |i_1, i_2, ..., i_N\rangle, ..., \tag{1}$$

where $|0\rangle$ is the vacuum state and the other states are defined in problem

1 of section 10. The creation operator $\hat{a}^{+}(i)$, producing an eigenvector corresponding to $N+1$ particles from one corresponding to N particles is defined by the equation

$$\hat{a}^{+}(i)\,|i_1, ..., i_N\rangle = \sqrt{(N+1)}\,|i, i_1, ..., i_N\rangle\,. \qquad (2)$$

Express the $|i_1, ..., i_N\rangle$ in terms of the $\hat{a}^{+}(i)$ and the vacuum state.

4. If $[\hat{A}, \hat{B}]_{\pm} \equiv \hat{A}\hat{B} \pm \hat{B}\hat{A}$ are the commutator and anticommutator of $\hat{A}$ and $\hat{B}$, and if the $\hat{a}^{+}(i)$ are the operators defined in the preceding problem, prove that

$$[\hat{a}^{+}(i), \hat{a}^{+}(j)]_{-} = 0, \qquad \text{for a system of bosons,}$$

and

$$[\hat{a}^{+}(i), \hat{a}^{+}(j)]_{+} = 0, \qquad \text{for a system of fermions.}$$

5. If we introduce an operator $\hat{a}(i)$—the Hermitean conjugate of $\hat{a}^{+}(i)$—by the equation

$$\langle i_1, ..., i_N|\,\hat{a}(i) = \sqrt{(N+1)}\,\langle i, i_1, i_2, ..., i_N|\,, \qquad (1)$$

prove (i) that

$$[\hat{a}(i), \hat{a}(j)]_{-} = 0, \qquad \text{for a system of bosons,}$$

and

$$[\hat{a}(i), \hat{a}(j)]_{+} = 0, \qquad \text{for a system of fermions;}$$

and (ii) that

$$\hat{a}(i)\,|i_1, ..., i_N\rangle = \frac{1}{\sqrt{N}}\{\delta(i-i_1)\,|i_2, ..., i_N\rangle + ... + (\pm 1)^{N-1}\delta(i-i_N)\,|i_1, ..., i_{N-1}\rangle\}, \qquad (2)$$

where the upper (lower) sign refers to a system of bosons (fermions). Equation (2) shows that one can interpret the $\hat{a}(i)$ as annihilation operators.

6. Prove that the annihilation and creation operators $\hat{a}(i)$ and $\hat{a}^{+}(i)$ satisfy the relations

$$[\hat{a}(i), \hat{a}^{+}(j)]_{-} = \delta(i-j), \qquad \text{for a system of bosons,}$$

and

$$[\hat{a}(i), \hat{a}^{+}(j)]_{+} = \delta(i-j), \qquad \text{for a system of fermions.}$$

7. If the operator $\hat{\Omega}$ is of the form

$$\hat{\Omega} = \sum_{k} \hat{\Omega}_{k}^{(1)} + \tfrac{1}{2} \sum_{k,l} \hat{\Omega}_{kl}^{(2)}\,, \qquad (1)$$

where the $\hat{\Omega}_k^{(1)}$ and $\hat{\Omega}_{kl}^{(2)}$ are, respectively, single- and two-particle operators which differ only in the particles on which they operate, express $\hat{\Omega}$ in terms of the annihilation and creation operators.

8. Consider a system contained in a finite volume v, so that we can take for the original complete orthonormal set a plane-wave set,

$$\varphi_i \to \frac{1}{\sqrt{v}} e^{i(\boldsymbol{k}\,.\,\boldsymbol{r})}, \tag{1}$$

where the $\boldsymbol{k}$ now form a complete set. In that case, one usually writes $\hat{a}_{\boldsymbol{k}}^{+}, \hat{a}_{\boldsymbol{k}}$ rather than $\hat{a}^{+}(\boldsymbol{k}), \hat{a}(\boldsymbol{k})$.

If the $\hat{\Omega}_j^{(1)}$ and $\hat{\Omega}_{ij}^{(2)}$ of the preceding problem are of the form

$$\hat{\Omega}_j^{(1)} = -\frac{\hbar^2}{2\mu}\nabla_j^2 + V(\boldsymbol{r}_j), \quad \hat{\Omega}_{ij}^{(2)} = U(r_{ij}), \tag{2}$$

where $\boldsymbol{r}_j$ is the spatial coordinate of the jth particle and $r_{ij} = |\boldsymbol{r}_i - \boldsymbol{r}_j|$, express $\hat{\Omega}$ in terms of the $\hat{a}_{\boldsymbol{k}}$ and the $\hat{a}_{\boldsymbol{k}}^{+}$.

9. Give an expression for $\hat{a}^{+}(i)\hat{a}(i)\,|i_1, ..., i_N\rangle$.

What is the physical meaning of the operator

$$\hat{n}(i) = \hat{a}^{+}(i)\hat{a}(i)?$$

10. Prove that, if $\hat{a}_{\boldsymbol{k}}^{+}$ creates a particle of momentum $\hbar\boldsymbol{k}$ and energy $\epsilon_{\boldsymbol{k}}$, $\hat{a}_{\boldsymbol{k}}^{+}$ takes one from any eigenstate of the free-particle Hamiltonian operator

$$\hat{H} = \sum_{\boldsymbol{k}} \epsilon_{\boldsymbol{k}} \hat{a}_{\boldsymbol{k}}^{+} \hat{a}_{\boldsymbol{k}}$$

to another differing in energy by $\epsilon_{\boldsymbol{k}}$.

11. We define the Schrödinger field operators by the equations

$$\hat{\Psi}^{+}(\boldsymbol{r}) = \frac{1}{\sqrt{v}} \sum_{\boldsymbol{k}} \hat{a}_{\boldsymbol{k}}^{+} e^{i(\boldsymbol{k}\,.\,\boldsymbol{r})}, \quad \hat{\Psi}(\boldsymbol{r}) = \frac{1}{\sqrt{v}} \sum_{\boldsymbol{k}} \hat{a}_{\boldsymbol{k}} e^{-i(\boldsymbol{k}\,.\,\boldsymbol{r})},$$

where v is the volume of the system and the $\hat{a}_{\boldsymbol{k}}$ and $\hat{a}_{\boldsymbol{k}}^{+}$ were defined in problem 8 of section 10.

Prove that

$$[\hat{\Psi}(\boldsymbol{r}), \hat{\Psi}^{+}(\boldsymbol{r}')]_{\pm} = \delta(\boldsymbol{r}-\boldsymbol{r}'), \quad [\hat{\Psi}(\boldsymbol{r}), \hat{\Psi}(\boldsymbol{r}')]_{\pm} = [\hat{\Psi}^{+}(\boldsymbol{r}), \hat{\Psi}^{+}(\boldsymbol{r}')]_{\pm} = 0,$$

where the upper (lower) signs refer to fermion (boson) systems.

What is the physical meaning of these operators?

12. If the Hamiltonian of a system of bosons has the form

$$\hat{H} = \sum_i \left\{-\frac{\hbar^2}{2\mu}\nabla_i^2 + V(\boldsymbol{r}_i)\right\} + \tfrac{1}{2}\sum_{i,j} U(\boldsymbol{r}_i, \boldsymbol{r}_j),$$

express H in terms of the operators $\hat{\Psi}^{+}(\boldsymbol{r})$ and $\hat{\Psi}(\boldsymbol{r})$ of the preceding problem.

13. Consider the Hamiltonian

$$\hat{H} = \hat{H}_0 + \hat{H}_1, \quad \hat{H}_0 = \epsilon \hat{a}^{+}\hat{a}, \quad \hat{H}_1 = \lambda\epsilon(\hat{a}^{+} + \hat{a}) ,$$

where $\hat{a}$ and $\hat{a}^{+}$ are boson annihilation and creation operators. Find, to second order in λ, the shift in the energy levels (i) treating $\hat{H}_1$ as a perturbation, and (ii) exactly.

14*. Consider a system of two interacting one-dimensional harmonic oscillators with Hamiltonian:

$$\hat{H} = -\frac{\hbar^2}{2\mu}\frac{\partial^2}{\partial x_1^2} - \frac{\hbar^2}{2\mu}\frac{\partial^2}{\partial x_2^2} + \tfrac{1}{2}\mu\omega^2(x_1^2 + x_2^2 + 2\lambda x_1 x_2) ,$$

where λ is real and $\lambda < 1$. Find the energy levels of this system, by expressing $\hat{H}$ in terms of suitable creation and annihilation operators.

15. Let $\hat{a}_{\boldsymbol{k}}$ and $\hat{a}_{-\boldsymbol{k}}$ be fermion annihilation operators corresponding to the single-particle states $|\boldsymbol{k}\rangle$ and $|-\boldsymbol{k}\rangle$. If new annihilation operators are defined by the equations

$$\hat{b}_{\boldsymbol{k}} = u_{\boldsymbol{k}}\hat{a}_{\boldsymbol{k}} + v_{\boldsymbol{k}}\hat{a}^{+}_{-\boldsymbol{k}}, \quad \hat{b}_{-\boldsymbol{k}} = u_{-\boldsymbol{k}}\hat{a}_{-\boldsymbol{k}} + v_{-\boldsymbol{k}}\hat{a}^{+}_{\boldsymbol{k}}, \tag{1}$$

with real coefficients $u_{\boldsymbol{k}}, v_{\boldsymbol{k}}, u_{-\boldsymbol{k}}, v_{-\boldsymbol{k}}$, find the conditions that the $\hat{b}_{\boldsymbol{k}}$, $\hat{b}_{-\boldsymbol{k}}$ are again fermion operators.

If the state $|0\rangle$ is characterised by the conditions

$$\hat{b}_{\boldsymbol{k}}|0\rangle = \hat{b}_{-\boldsymbol{k}}|0\rangle = 0 ,$$

evaluate the expectation value of $\hat{a}^{+}_{\boldsymbol{k}}\hat{a}_{\boldsymbol{k}}$ in the state $|0\rangle$.

16. What would be the answers to the preceding problem, if the $\hat{a}_{\boldsymbol{k}}$, $\hat{a}_{-\boldsymbol{k}}$ were boson operators?

17*. Consider a system of bosons described by the Hamiltonian

$$\hat{H} = -\sum_{j=1}^{N}\frac{\hbar^2}{2\mu}\nabla_j^2 + \sum_{i,j=1}^{N} V(r_{ij}) \tag{1}$$

(compare problem 8 of section 10). Prove that, if the interaction is weak, the low-lying eigenvalues of the system are approximately given by the expression

$$E = E_0 + \sum_{\boldsymbol{k}\neq 0}\epsilon_{\boldsymbol{k}} n_{\boldsymbol{k}} , \tag{2}$$

where the $n_{\boldsymbol{k}}$ are the eigenvalues of some quasi-particle occupation number operator. Find expressions for E_0 and the $\epsilon_{\boldsymbol{k}}$.

18*. Indicate how one can reduce the boson Hamiltonian

$$\hat{H} = \sum_{i=1}^{N} L_{ii}\hat{a}_i^{+}\hat{a}_i + \sum_{i\neq j} L_{ij}\hat{a}_i^{+}\hat{a}_j, \quad L_{ij} = L_{ji}^{*} , \tag{1}$$

to the diagonal form

$$\hat{H} = \sum_{k=1}^{N} E_k \hat{b}_k^+ \hat{b}_k , \tag{2}$$

and how one can find the quasi-particle energies E_k.

19*. Indicate how one can reduce the boson Hamiltonian

$$\hat{H} = \sum_{i,j} L_{ij} \hat{a}_i^+ \hat{a}_j + \tfrac{1}{2} \sum_{i,j} \{M_{ij} \hat{a}_i^+ \hat{a}_j^+ + M_{ij}^* \hat{a}_i \hat{a}_j\} , \tag{1}$$

with

$$L_{ij} = L_{ji}^* , \quad M_{ij} = M_{ji} , \tag{2}$$

to the diagonal form

$$\hat{H} = E_0 + \sum_k E_k \hat{b}_k^+ \hat{b}_k . \tag{3}$$

20. Often one is dealing with a system which is part of a larger system. Let x and q denote, respectively, the coordinates of the smaller system and the coordinates of the remainder of the larger system. The larger system will be described by a normalised wavefunction $\Psi(q, x)$ which can not necessarily be written as a product of functions depending on x and q only. Let $\hat{A}$ be an operator acting on the x only, let $\hat{H}$ be the Hamiltonian describing the smaller system, and let the *density matrix* or *density operator* $\hat{\rho}$ be defined in the coordinate representation by the equation

$$\langle x | \hat{\rho} | x' \rangle = \int \Psi^* (q, x') \Psi(q, x) dq , \tag{1}$$

where the integration is over all the degrees of freedom of the remainder of the larger system.

(i) Express the expectation value of $\hat{A}$ in terms of $\hat{\rho}$ for the case where the larger system is described by the wavefunction $\Psi(q, x)$.

(ii) What is the normalisation condition for $\hat{\rho}$?

(iii) Find the equation of motion for $\hat{\rho}$.

21. If the wavefunction $\Psi(q, x)$ of the preceding problem can be written in the form

$$\Psi(q, x) = \Phi(q) \chi(x) , \tag{1}$$

we are dealing with a so-called pure case.

Prove that the necessary and sufficient condition for a pure case is that $\hat{\rho}$ is idempotent, that is, that

$$\hat{\rho}^2 = \hat{\rho} . \tag{2}$$

22. The density matrix is especially useful, if the coordinates x of the smaller system are discrete variables. An instance of such a system is a spin-$\frac{1}{2}$ particle, if one is interested only in its spin variable—the q might in that case be the spatial coordinates of the particle.

(i) Show that the density matrix $\hat{\rho}$ describing a spin-$\frac{1}{2}$ particle can be written in the form

$$\hat{\rho} = \tfrac{1}{2}\{\hat{1} + (\boldsymbol{P} \,.\, \hat{\sigma})\}, \tag{1}$$

where $\hat{\sigma}_x$, $\hat{\sigma}_y$, and $\hat{\sigma}_z$ are the Pauli matrices (see problem 14 of section 4) and $\hat{1}$ is the unit 2×2 matrix.

(ii) What is the physical meaning of the vector $\boldsymbol{P}$ in equation (1)?

(iii) If the system is placed in a constant magnetic field $\mathcal{H}$, find the equation of motion for $\boldsymbol{P}$.

23. By analogy with the polarisation of spin-$\frac{1}{2}$ particles (see preceding problem) the polarisation of a light quantum can be described by a two-component wavefunction $\begin{pmatrix} a \\ b \end{pmatrix}$, where $|a|^2$ and $|b|^2$ are the probabilities that the photon is polarised in one or the other of two mutually perpendicular directions or that the photon is right- or left-handed circularly polarised.

If we want to determine the polarisation of a photon we could, for instance, use a filter, which we shall call a detector, although strictly speaking it is not a detector but a device to prepare for a measurement. Such a filter would correspond to a pure state, described by a wavefunction

$$\psi^{\text{det}} = c_1^{\text{det}} \psi_1 + c_2^{\text{det}} \psi_2 , \tag{1}$$

where

$$\psi_1 \equiv \begin{pmatrix} 1 \\ 0 \end{pmatrix}, \quad \psi_2 \equiv \begin{pmatrix} 0 \\ 1 \end{pmatrix}, \tag{2}$$

are the wavefunctions corresponding to the two polarisation states. This pure state corresponds to a detector density matrix $\hat{\rho}^{\text{det}}$ given by its matrix elements

$$\rho_{ij}^{\text{det}} = c_i^{\text{det}} (c_j^{\text{det}})^* , \tag{3}$$

or, in matrix form

$$\hat{\rho}^{\text{det}} = \begin{Vmatrix} |c_1^{\text{det}}|^2 & c_1^{\text{det}} (c_2^{\text{det}})^* \\ (c_1^{\text{det}})^* c_2^{\text{det}} & |c_2^{\text{det}}|^2 \end{Vmatrix}. \tag{4}$$

Find an expression for the probability of a response of a detector described by $\hat{\rho}^{\text{det}}$ to a photon in a state described by a density matrix $\hat{\rho}$.

24. The result of an annihilation process of an electron and a positron producing a pair of photons, A and B, can be described by a 4×4 density matrix given by the equation

$$\hat{\rho}_{\text{AB}} = \tfrac{1}{4}\{\hat{1}_{\text{A}} \hat{1}_{\text{B}} - (\hat{\omega}_{\text{A}} \,.\, \hat{\omega}_{\text{B}})\}, \tag{1}$$

where $\hat{1}_A$, $\hat{\omega}_A$ (1_B, $\hat{\omega}_B$) are 2×2 matrices operating on the polarisation wavefunction of photon A (B), and where $\hat{\omega}_{Ax}$ ($\hat{\omega}_{Bx}$), $\hat{\omega}_{Ay}$ ($\hat{\omega}_{By}$), and $\hat{\omega}_{Az}$ ($\hat{\omega}_{Bz}$) have the same form as the Pauli matrices $\hat{\sigma}_x$, $\hat{\sigma}_y$, and $\hat{\sigma}_z$.

Evaluate the polarisation of the two photons and the correlation between the two polarisations.

25. A 2×2 scattering matrix $\hat{S}$ for the scattering of spin-$\frac{1}{2}$ particles by a spin-zero target is defined such that

$$\psi_f = \hat{S}\psi_i , \tag{1}$$

where ψ_i and ψ_f are, respectively, the initial wavefunction and the wavefunction after scattering. If

$$\hat{S} = g\hat{1} + h(\boldsymbol{n} \cdot \hat{\sigma}) , \tag{2}$$

where $|g|^2 + |h|^2 = 1$ while $\boldsymbol{n}$ is a unit vector, find the polarisation after scattering, if the polarisation before the scattering were zero.

11

Relativistic wave equations†

1. If one makes the usual substitution

$$p_\mu \to -i\hbar\partial_\mu \,, \qquad p_\mu \equiv \boldsymbol{p} \,, \frac{iE}{c} \,, \tag{1}$$

into the classical relativistic Hamiltonian equation

$$\sum_\mu \left(p_\mu - \frac{e}{c}A_\mu\right)^2 + m^2c^2 = 0 \,, \qquad A_\mu \equiv \boldsymbol{A} \,, i\phi \,, \tag{2}$$

where $\boldsymbol{A}$ and ϕ are, respectively, the vector and scalar potential of the electromagnetic field, one obtains the Klein-Gordon equation for a particle of mass m and charge e in an electromagnetic field

$$\sum_\mu \left(\partial_\mu - \frac{ie}{\hbar c}A_\mu\right)^2 \psi - \frac{m^2c^2}{\hbar^2}\psi = 0 \,. \tag{3}$$

In the case of a free particle the Klein-Gordon equation is

$$\sum_\mu \partial_\mu^2 \psi - \frac{m^2c^2}{\hbar^2}\psi = 0 \,,$$

$$\nabla^2\psi - \frac{1}{c^2}\frac{\partial^2\psi}{\partial t^2} - \frac{m^2c^2}{\hbar^2}\psi = 0 \,. \tag{4}$$

Prove that, if we introduce charge and current densities, ρ and $\boldsymbol{j}$, through the equations

$$j_\mu \equiv \boldsymbol{j} \,, ic\rho \,, \tag{5}$$

$$j_\mu = \frac{e\hbar}{2mi}(\psi^*\partial_\mu\psi - \psi\partial_\mu\psi^*) \,, \tag{6}$$

they satisfy in the case of a free particle, whose wavefunction satisfies equation (4), the continuity equation

$$\sum_\mu \partial_\mu j_\mu = 0 \,. \tag{7}$$

† In this section we use a fourth coordinate $x_4 = ict$ and Greek indices such as μ for 1, 2, 3, 4, while Latin indices such as k run from 1 to 3. We do not use the repeated index convention. We use m for the mass in this section, as μ is used as an index, and we shall use the abbreviated notation ∂_μ for $\partial/\partial x_\mu$.

Note that we have introduced a charge density rather than a probability density. This is connected with the fact that ρ/e is not positive definite, which introduces complications we do not want to consider.

Generalise j_μ such that it will still satisfy equation (7) for the case of a charged particle moving in an electromagnetic field, so that ψ satisfies equation (3).

2. Show that the Klein-Gordon equation for a free particle and the corresponding j_μ lead in the non-relativistic limit to the Schrödinger equation and the corresponding expressions for $\boldsymbol{j}$ and ρ.

3. Find the free-particle solutions of the Klein-Gordon equation corresponding to well-defined momentum values, and find the corresponding charge density.

4. We saw in the preceding problem that there is an additional degree of freedom in the case of the Klein-Gordon equation, which suggests that we should work with a two-component wavefunction. At the same time it would be convenient, if one could get a wave equation which has only first derivatives with respect to the time in it, so that giving the wavefunction at $t = 0$ determines it at later times. We therefore write

$$\psi = \varphi + \chi, \qquad i\hbar\frac{\partial\psi}{\partial t} = mc^2(\varphi - \chi). \tag{1}$$

(i) Find the first-order differential equations for φ and χ which are equivalent to the free-particle Klein-Gordon equation.

(ii) Express these equations in spinor form, using the 2×2 matrices $\hat{\tau}$ and $\hat{1}$, where $\hat{1}$ is the unit 2×2 matrix and $\hat{\tau}_1$, $\hat{\tau}_2$, and $\hat{\tau}_3$ are 2×2 matrices which have the same form as the Pauli matrices $\hat{\sigma}_x$, $\hat{\sigma}_y$, and $\hat{\sigma}_z$ (see problem 14 of section 4).

(iii) Express the charge density in terms of the spinor $\Psi = \begin{pmatrix}\varphi\\ \chi\end{pmatrix}$.

5. Find the free-particle solutions of the Klein–Gordon equation corresponding to well-defined values of the momentum in terms of the functions φ and χ of the preceding problem. Find the non-relativistic limit of these solutions.

6. Prove that the orbital angular momentum operator $[\boldsymbol{r} \wedge \boldsymbol{p}]$ commutes with the Klein–Gordon Hamiltonian in the form given in the solution to problem 4 of section 11. This is a justification for the statement that the Klein–Gordon equation describes a relativistic zero-spin particle.

7. Find the energy levels of a π-mesic atom, that is, a system of a nucleus of charge Ze, which one may assume to be infinitely heavy, and a π^--meson.

8. We saw in problem 1 of section 11 that, if we start from the Klein–Gordon equation in the form of equation (3) or (4) of that problem, the probability density is not necessarily positive definite. This was connected with the fact that giving the wavefunction ψ at time $t = 0$ did not fix ψ at a later time. Dirac suggested that one should start from a multi-component wavefunction, or spinor, Ψ:

$$\Psi \equiv \begin{pmatrix} \psi_1 \\ \psi_2 \\ \cdot \\ \cdot \\ \cdot \\ \cdot \end{pmatrix}, \tag{1}$$

and require (a) that the charge density be given by the expression

$$\rho = e \sum_j \psi_j^* \psi_j , \tag{2}$$

and (b) that the ψ_j satisfy first-order differential equations, which we can write in the form

$$\frac{1}{c}\frac{\partial \psi_j}{\partial t} + \sum_i \sum_{k=1}^{3} \alpha_{ji}^{(k)} \frac{\partial \psi_i}{\partial x_k} + \frac{imc}{\hbar} \sum_i \beta_{ji} \psi_i = 0 , \tag{3}$$

or in matrix form

$$\frac{1}{c}\frac{\partial \Psi}{\partial t} + (\hat{\alpha} \cdot \nabla)\Psi + \frac{imc}{\hbar}\hat{\beta}\Psi = 0 , \tag{4}$$

where $\hat{\alpha}$ is a vector with components $\hat{\alpha}_1, \hat{\alpha}_2, \hat{\alpha}_3$.

Find the conditions to be satisfied by the matrices $\hat{\alpha}$ and $\hat{\beta}$ in order that ρ satisfies an equation of continuity and that each of the ψ_i satisfies the wave equation (4) of problem 1 of section 11.

9. Prove that if we introduce the matrices

$$\hat{\gamma}_\mu \equiv \hat{\gamma}, \hat{\gamma}_4 , \tag{1}$$

with

$$\hat{\gamma} = -i\hat{\beta}\hat{\alpha} , \quad \hat{\gamma}_4 = \hat{\beta} , \tag{2}$$

the Dirac equation (4) of the preceding problem can be written in the form

$$\left\{ \sum_\mu \hat{\gamma}_\mu \hat{p}_\mu - imc \right\} \Psi = 0 , \quad \hat{p}_\mu \equiv -i\hbar \partial_\mu . \tag{3}$$

What is the expression for the four-vector j_μ in terms of the $\hat{\gamma}_\mu$?

10. Simplify the following products of the matrices $\hat{\gamma}_\mu$ defined in the preceding problem:

$$\sum_\mu \hat{\gamma}_\mu \hat{\gamma}_\mu , \quad \sum_\mu \hat{\gamma}_\mu \hat{\gamma}_\nu \hat{\gamma}_\mu , \quad \sum_\mu \hat{\gamma}_\mu \hat{\gamma}_\lambda \hat{\gamma}_\nu \hat{\gamma}_\mu , \quad \sum_\mu \hat{\gamma}_\mu \hat{\gamma}_\lambda \hat{\gamma}_\nu \hat{\gamma}_\rho \hat{\gamma}_\mu .$$

11. One can construct from the $\hat{\gamma}_\mu$ the following sixteen elements:

$$\hat{1},\ \hat{\gamma}_\mu,\ \hat{\gamma}_\mu\hat{\gamma}_\nu\ (\mu \neq \nu),\ \hat{\gamma}_\mu\hat{\gamma}_\nu\hat{\gamma}_\rho\ (\mu \neq \nu \neq \rho \neq \mu),\ \hat{\gamma}_1\hat{\gamma}_2\hat{\gamma}_3\hat{\gamma}_4 \equiv \hat{\gamma}_5\ .$$

If these elements are denoted by $\hat{\Gamma}_j$ ($j = 1, 2, \ldots, 16$), prove that

(i) $\hat{\Gamma}_j^2 = \hat{1}$,

(ii) $\hat{\Gamma}_j\hat{\Gamma}_k = \lambda\hat{\Gamma}_l$, where $\lambda = \pm 1$ or $\pm i$,

(iii) $\mathrm{Tr}\,\hat{\Gamma}_j = 0$, unless $\hat{\Gamma}_j = \hat{1}$.

12. Find the solutions of the Dirac equation of a free particle (equation (4) of problem 8 of section 11) corresponding to states with well-defined values of p_μ.

13. Find the non-relativistic limit of the solutions of the Dirac equation of a free particle found in the preceding problem and also the non-relativistic limit of the charge and current densities.

14. An electron with momentum $\boldsymbol{p}_1$ $(= 0, 0, p_1)$ is incident from the $-z$-direction with its spin in the $+z$-direction onto a potential step, described by the potentials

$$\boldsymbol{A} = 0\ ; \quad \phi = 0\ , \quad z \leqslant 0\ , \quad \phi = V_0/e\ , \quad z > 0\ , \tag{1}$$

where V_0 is a constant which satisfies the inequalities

$$0 < V_0 < E - mc^2\ .$$

Find expressions for the reflection and transmission coefficients, considering positive energy states only.

15. Find in the Heisenberg representation the position and velocity of a free particle described by the Dirac equation.

16. Determine how a Dirac wavefunction transforms under a Lorentz transformation

$$x'_\mu = \sum_\nu a_{\mu\nu}x_\nu\ , \tag{1}$$

where

$$\sum_\mu a_{\mu\nu}a_{\mu\nu'} = \delta_{\nu\nu'}\ . \tag{2}$$

Discuss in some detail the following special cases: (a) a proper Lorentz transformation from one inertial frame to another one moving relative to the first one with a velocity v in the direction of the x-axis; (b) a rotation around the z-axis; (c) an inversion; and (d) time reversal.

17. Show that the wavefunction given by equation (15) of problem 12 of section 11, that is, the wavefunction of a free particle with a well-defined momentum, can be obtained by a Lorentz transformation from the rest frame.

18. Discuss the transformation properties of the following bi-linear quantities:

$$G_j \equiv \overline{\Psi}\hat{\Gamma}_j\Psi ,$$

where $\overline{\Psi}$ is the adjoint of Ψ (see equation (8) of problem 9 of section 11) and the $\hat{\Gamma}_j$ are defined in problem 11 of section 11.

19. The Dirac equation for a charged particle, moving in an electromagnetic field is obtained from equation (3) of problem 9 of section 11 in the usual way, by replacing $\hat{p}_\mu$ by $\hat{p}_\mu - (e/c)A_\mu$. Hence we have

$$\left\{\sum_\mu \hat{\gamma}_\mu(\hat{p}_\mu - (e/c)A_\mu) - imc\right\}\Psi = 0 . \tag{1}$$

If we now consider the Dirac equation for a particle with the same mass, but opposite charge, we have

$$\left\{\sum_\mu \hat{\gamma}_\mu(\hat{p}_\mu + (e/c)A_\mu) - imc\right\}\Psi_c = 0 , \tag{2}$$

where we have indicated the *charge conjugate* wavefunction by Ψ_c.

Show how one can obtain Ψ_c from Ψ.

20. Discuss the conservation of angular momentum for a particle satisfying the Dirac equation for a free particle, or of a particle moving in a central field of force.

21. Consider the non-relativistic limit of the motion of a charged spin-$\frac{1}{2}$ particle moving in a weak electromagnetic field and find the magneto-gyric ratio for such a particle.

22. Reduce the Dirac equation for the motion of an electron in an electrostatic potential $\phi \equiv V(r)/e$ to the equation for the two large components for the positive energy case and discuss the extra terms which appear as compared to the Schrödinger equation.

23. Find the correction to the hydrogen atom energy levels due to the spin-orbit interaction.

24. Find the energy levels of an electron in a uniform magnetic field.

25. (i) Show that for a free zero-rest-mass spin-$\frac{1}{2}$ particle the wave equation can be written in the form

$$c(\hat{\sigma} \cdot \hat{\boldsymbol{p}})\Psi = i\hbar\frac{\partial\Psi}{\partial t} , \tag{1}$$

where the components of $\hat{\sigma}$ are the Pauli matrices and $\hat{\boldsymbol{p}}$ is the linear momentum operator.

(ii) Discuss the conservation of angular momentum in this case.

(iii) Show that the spin of the particle in a positive (negative) energy state is parallel (antiparallel) to its momentum.

solutions

1

One-dimensional motion

1. $$E_n = \frac{\pi^2 \hbar^2}{2\mu a^2} n^2, \quad \psi_n(x) = \sqrt{\left(\frac{2}{a}\right)} \sin \frac{\pi n}{a} x.$$

3. $$|a(p)|^2 = \frac{4n^2 \pi a}{\hbar} \frac{1}{\left(\frac{p^2 a^2}{\hbar^2} - n^2 \pi^2\right)^2} \begin{cases} \cos^2 \frac{pa}{2\hbar} & \text{for odd } n \\ \sin^2 \frac{pa}{2\hbar} & \text{for even } n. \end{cases}$$

4. To find the probability distribution we want, we expand the wave function of the (non-stationary) state under consideration in terms of the wave functions ψ_n of the stationary states of the particle in the well,

$$\psi(x) = \sum_{n=1}^{\infty} c_n \psi_n(x), \tag{1}$$

where

$$\psi_n(x) = \sqrt{\left(\frac{2}{a}\right)} \sin \frac{\pi n}{a} x \quad (n = 1, 2, 3, \ldots). \tag{2}$$

The probability to find the particle in the nth state is equal to:

$$w_n = |c_n|^2 = \left| \int_0^a \psi(x) \psi_n^*(x) \, dx \right|^2, \tag{3}$$

and

$$\sum_{n=1}^{\infty} w_n = \int_0^a |\psi(x)|^2 \, dx.$$

To obtain the normalised probabilities w_n we normalise the original wave function. To do this we must take

$$A^2 = \left[\int_0^a x^2 (a-x)^2 \, dx \right]^{-1} = 30 a^{-5}.$$

Substituting ψ and ψ_n into equation (3), we have:

$$w_n = \frac{30}{a^5} \cdot \frac{2}{a} \left[\int_0^a x(a-x) \sin \frac{\pi n x}{a} \, dx \right]^2.$$

Evaluating the integral we have finally:

$$w_n = \frac{240}{(\pi n)^6} [1 - (-1)^n]^2. \tag{4}$$

The probability $w_n = c_n^2$ which we have obtained is different from zero only for $n = 1, 3, 5, \ldots$. These values of n correspond according to equation (2) to even (with respect to the middle of the well $x = a/2$) functions $\psi_n(x)$. Thus in the superposition (1) only even states are represented (as should be the case, since the original wave function is even with respect to $x = a/2$).

Let us verify the normalisation of the probabilities w_n.

We use the equation

$$\sum_{k=1}^{\infty} \frac{1}{(2k-1)^{2m}} = \frac{(2^{2m}-1)\,\pi^{2m}}{2.(2m)!} |B_{2m}|, \tag{5}$$

where B_{2m} are the so-called Bernouilli numbers, and we have (in our case $2m = 6$, $B_{2m} = B_6 = 1/42$):

$$\sum_{n=1}^{\infty} w_n = \frac{240}{\pi^6}.2^2 \sum_{n=1,3,5,\ldots}^{\infty} \frac{1}{n^6} = \frac{960}{\pi^6} \sum_{k=1}^{\infty} \frac{1}{(2k-1)^6} = \frac{960}{\pi^6} \cdot \frac{(2^6-1)\,\pi^6}{2.6!} \cdot \frac{1}{42} = 1,$$

as was to be proved.

According to equation (4) the probability to find the particle in an even state corresponding to a quantum number n decreases fast with increasing n. The probability to find it in the ground state ($n = 1$) is equal to

$$w_1 = \frac{240}{\pi^6}.2^2 \approx 0{\cdot}999,$$

so that the total probability to find the particle in all the excited states ($n = 3, 5, \ldots$) is equal to 0·001. This is connected with the fact that the original wave function is everywhere very close to the wave function of the ground state of the particle in the well

$$\psi_1 = \sqrt{\left(\frac{2}{a}\right)} \sin \frac{\pi x}{a}.$$

One can thus say that the state under consideration is "almost stationary".†

The energy levels of the particle in the well potential considered are given by the equation:

$$E_n = \frac{\pi^2 \hbar^2}{2\mu a^2} n^2 \quad (n = 1, 2, 3, \ldots). \tag{6}$$

† This nomenclature is only used for this particular problem and should not be confused with the conventional term "quasi-stationary state", which has a completely different meaning.

The average energy of the particle in the non-stationary state considered is, according to equations (4) and (6), equal to

$$\bar{E} = \sum_{n=1}^{\infty} E_n w_n = \frac{\pi^2 \hbar^2}{2\mu a^2} \cdot \frac{240 . 2^2}{\pi^6} \sum_{n=1,3,5,\dots}^{\infty} \frac{1}{n^4}.$$

Using equation (5), we find ($2m = 4, B_4 = -1/30$):

$$\sum_{n=1,3,5,\dots}^{\infty} \frac{1}{n^4} = \frac{\pi^4}{96},$$

so that

$$\bar{E} = \frac{\pi^2 \hbar^2}{2\mu a^2} \cdot \frac{10}{\pi^2} \approx 1{\cdot}014 E_1, \tag{7}$$

where E_1 is the ground state energy.

The same result (7) is, finally, obtained by taking the average of the Hamiltonian $\hat{H} = \hat{T} + U$ over the (normalised) wave function,

$$\bar{E} = \int \psi^* \hat{H} \psi \, dx = -\frac{\hbar^2}{2\mu} \int_0^a \psi \frac{d^2\psi}{dx^2} dx = \frac{\hbar^2 a^3}{6\mu} A^2 = \frac{5\hbar^2}{\mu a^2}. \tag{7'}$$

The average of the square of the energy of the particle in the state under consideration is equal to

$$\overline{E^2} = \sum_{n=1}^{\infty} E_n^2 w_n = \left(\frac{\pi^2 \hbar^2}{2\mu a^2}\right)^2 \cdot \frac{960}{\pi^6} \sum_{n=1,3,5,\dots}^{\infty} \frac{1}{n^2} = 30 \left(\frac{\hbar^2}{\mu a^2}\right)^2. \tag{8}$$

According to equations (7′) and (8) the energy dispersion is equal to

$$\Delta E = \sqrt{[\overline{(E-\bar{E})^2}]} = \sqrt{[\overline{E^2} - (\bar{E})^2]} = \sqrt{(5)} \frac{\hbar^2}{\mu a^2}, \tag{9}$$

which is comparable with $\bar{E} \approx E_1$.

This relatively appreciable value of the energy dispersion is clearly explained by the appreciable contribution of the excited state to the quantity $\overline{E^2}$ (by virtue of the fact that $E_n^2 \sim n^4$).

5. (i) $\frac{1}{2}$; (ii) $\frac{32}{9\pi^2}$.

6*. To avoid computational difficulties, we consider first of all a symmetric well of very great, but finite depth V_0 (see fig. 22) and later on take the limit $V_0 \to \infty$.

We evaluate the average force $\bar{f}$ exerted, for instance, on the right-hand wall of the well by the particle.

The operator of the force exerted on the particle by the external field (the wall of the well) is equal to $[-dV(x)/dx]$, where $V(x)$ is the operator of the potential energy of the particle. According to Newton's third law, the force exerted by the particle on the wall is equal, but opposite in sign, to that force, or

$$f(x) = \frac{dV(x)}{dx}, \tag{1}$$

and its quantum mechanical average (which we require) is equal to:

$$\bar{f} = \int_{-\infty}^{\infty} f(x)\,[\psi(x)]^2\,dx, \tag{2}$$

where $\psi(x)$ is the (real) wave function and $f(x)$ is taken for the (right-hand) wall under consideration.

Since $V(x)$ is everywhere constant except at $x = a$ where $V(x)$ jumps suddenly by an amount V_0, equation (1) may be written in the form:

$$f(x) = V_0\,\delta(x-a), \tag{3}$$

where δ is the delta-function.† Substituting expression (3) into equation (2), we get:

$$\bar{f} = V_0[\psi(a)]^2. \tag{4}$$

To find $\psi(a)$ we consider the solution of the well-known problem of a rectangular symmetrical well of finite depth. The wave function is of the form (see problem 7 of section 1, with $V_1 = V_2 = V_0$):

$$\left.\begin{aligned} \psi(x) &= c_1 \exp(\kappa x) \quad (x \leqslant 0), \\ \text{where}\quad \kappa &= \sqrt{\left[\frac{2\mu}{\hbar^2}(V_0 - E)\right]}, \end{aligned}\right\} \tag{5}$$

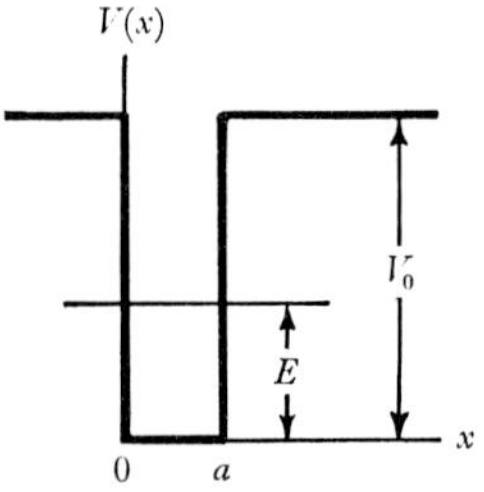

Fig. 22.

† Indeed, from equation (1) we get ($\epsilon \to 0$):

$$\int_{a-\epsilon}^{a+\epsilon} f(x)\,dx = V(a+\epsilon) - V(a-\epsilon) = V_0;$$

the same follows clearly also from equation (3).

$$\left.\begin{aligned} \psi(x) &= c\sin(kx+\delta) \quad (0\leqslant x\leqslant a), \\ \text{where} \qquad k &= \sqrt{\left(\frac{2\mu E}{\hbar^2}\right)}, \\ \sin(ka+\delta) &= -\sin\delta = -\frac{\hbar k}{\sqrt{(2\mu V_0)}}, \end{aligned}\right\} \tag{6}$$

$$\psi(x) = c_2\exp(-\kappa x) \quad (x\geqslant a). \tag{5'}$$

According to equation (6) the value of $\psi(a)$ in which we are interested is equal to:

$$\psi(a) = c\sin(ka+\delta) = -c\frac{\hbar k}{\sqrt{(2\mu V_0)}}. \tag{7}$$

As we use the wave function to evaluate the average force by (2), it must be normalised. From this requirement, we can immediately determine the coefficient c if we take into account the fact that the normalising integral $\int_{-\infty}^{\infty}[\psi(x)]^2\,dx = 1$ is appreciable only in the interval $0\leqslant x\leqslant a$. Indeed, outside that interval the wave function decreases exponentially according to equations (5) and (5′) over a very small distance of the order of magnitude $1/\kappa \approx \hbar/\sqrt{(2\mu V_0)}$ (we remind ourselves that we are interested in the limit $V_0\to\infty$; for the same reason we can neglect E compared to V_0) so that the contribution of the regions $x\leqslant 0$ and $x\geqslant a$ to the total probability $\int_{-\infty}^{\infty}\psi^2\,dx$ is negligibly small.

We have thus

$$1 = \int_{-\infty}^{\infty}[\psi(x)]^2\,dx \approx \int_0^a[\psi(x)]^2\,dx = c^2\int_0^a \sin^2(kx+\delta)\,dx$$

$$= c^2\left[\frac{a}{2}-\frac{1}{2k}\sin ka\,.\cos(ka+2\delta)\right]. \tag{8}$$

Let us now take the limit $V_0\to\infty$.

According to equation (6) we have for $V_0\to\infty$: $\delta\to n\pi$, $(ka+\delta)\to n_1\pi$, and consequently $ka\to(n_1-n)\pi$ where n and n_1 are integers, so that the last term in equation (8) tends to zero.

Indeed,

$$\frac{1}{2k}\sin ka\cos(ka+2\pi n) = \frac{1}{4k}\sin 2ka \to \frac{1}{4k}\sin[2\pi(n_1-n)] = 0.$$

We have thus for a well with $V_0 = \infty$:

$$c = \sqrt{\left(\frac{2}{a}\right)}. \tag{9}$$

Using equations (9), (7), (6), and (4), we find finally

$$\bar{f} = \frac{2E}{a}. \qquad (10)$$

As V_0 has disappeared from the result, this is also valid for the limiting case where $V_0 \to \infty$.

The force $\bar{f}$ is obtained clearly with the correct sign ($\bar{f} > 0$ on the right-hand wall of the well). (The average force acting on both walls is clearly equal to zero. Indeed, the operator of this total force is

$$V_0\{\delta(x-a) - \delta(x)\},$$

so that we find, taking the symmetry of the problem into account:

$$\bar{f}_{\text{tot}} = V_0\{[\psi(a)]^2 - [\psi(0)]^2\} = 0.$$

This result is correct for finite motion in any arbitrary form of the potential energy.)

We note that the result (10) which does not contain $\hbar$ explicitly retains its form exactly in classical mechanics. Indeed, in that case $\bar{f}_{\text{cl}}$ (time average!) is equal to the product of the momentum imparted to the wall at each collision ($2\mu v$) and the number of collisions with the wall considered per unit time $v/2a$ (v is the particle velocity), or

$$\bar{f}_{\text{cl}} = 2\mu v . \frac{v}{2a} = \frac{2E}{a}.$$

7.

$$\frac{\hbar^2}{2\mu}\frac{d^2\psi}{dx^2} + (E - V_1)\psi = 0 \quad (x < 0),$$

$$\frac{\hbar^2}{2\mu}\frac{d^2\psi}{dx^2} + E\psi = 0 \qquad (0 < x < a),$$

$$\frac{\hbar^2}{2\mu}\frac{d^2\psi}{dx^2} + (E - V_2)\psi = 0 \quad (x > a).$$

If we introduce the following notation

$$\kappa_1 = \frac{\sqrt{[2\mu(V_1 - E)]}}{\hbar}, \quad \kappa = \frac{\sqrt{(-2\mu E)}}{\hbar}, \quad \kappa_2 = \frac{\sqrt{[2\mu(V_2 - E)]}}{\hbar},$$

we find that the general solution in each of the three regions has the following form:

$$\psi = A_1 \exp(-\kappa_1 x) + B_1 \exp(\kappa_1 x) \quad (x < 0),$$

$$\psi = A \exp(-\kappa x) + B \exp(\kappa x) \qquad (0 < x < a),$$

$$\psi = A_2 \exp(-\kappa_2 x) + B_2 \exp(\kappa_2 x) \quad (x > a).$$

Let us consider the discrete spectrum $E < V_2$; κ_1 and κ_2 are then real. If we put in the interval $0 < x < a$ $\kappa = ik$, where k is real, we can write

$$\psi = \sin(kx+\delta) \quad (0 < x < a).$$

Since the wave function must be bounded, we have $A_1 = 0$, $B_2 = 0$.

The condition that ψ and $d\psi/dx$ must be continuous can be expressed by requiring the logarithmic derivative $(1/\psi)(d\psi/dx)$ to be continuous:

$$\kappa_1 = k\cot\delta,$$
$$-\kappa_2 = k\cot(ka+\delta).$$

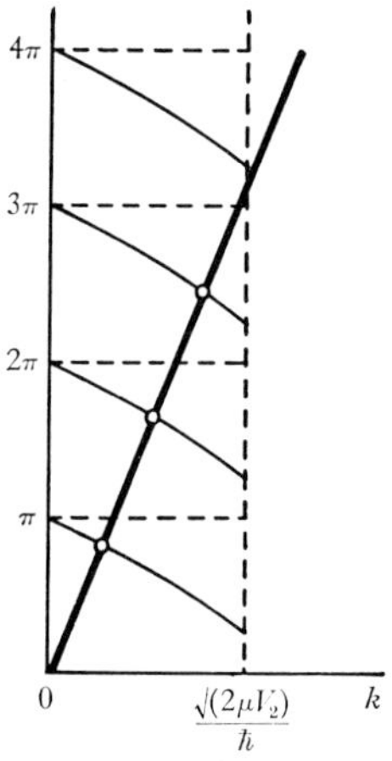

Fig. 23.

Expressing κ_1 and κ_2 in terms of k, we can rewrite these last conditions as follows:

$$\sqrt{\left(\frac{2\mu V_1}{\hbar^2 k^2}-1\right)} = \cot\delta,$$
$$-\sqrt{\left(\frac{2\mu V_2}{\hbar^2 k^2}-1\right)} = \cot(ka+\delta).$$

Since the cotangent is periodic with period π, the values of δ and of $ka+\delta$ can be written as follows:

$$\delta = \arcsin\frac{\hbar k}{\sqrt{(2\mu V_1)}}+n_1\pi,$$

$$ka+\delta = -\arcsin\frac{\hbar k}{\sqrt{(2\mu V_2)}}+n_2\pi,$$

where the value of arc sin lies between 0 and $\frac{1}{2}\pi$. Eliminating δ we find a transcendental equation to determine the discrete energy levels

$$ka = n\pi - \arcsin\frac{\hbar k}{\sqrt{(2\mu V_1)}} - \arcsin\frac{\hbar k}{\sqrt{(2\mu V_2)}}, \quad k = \frac{\sqrt{(2\mu E)}}{\hbar} > 0.$$

The values of k satisfying this equation must be obtained graphically, by finding the intersection of the line $y = ak$ and the curve

$$y = n\pi - \arcsin\frac{\hbar k}{\sqrt{(2\mu V_1)}} - \arcsin\frac{\hbar k}{\sqrt{(2\mu V_2)}} \quad \text{(see fig. 23).}$$

Let us consider a symmetrical potential well $V_1 = V_2 = V$. It is easily seen that in this case there is always at least one level, whatever the values of V and a. If $\beta = [\sqrt{(2\mu V)}/\hbar]\,a \ll 1$, one can without difficulty find the value of the only discrete energy level. Expanding

$$\pi - 2\arcsin\sqrt{(E/V)}$$

in a power series in β, we find $E = V - (\mu a^2/2\hbar^2)\,V^2$. In general, the number of discrete levels will be equal to N, where N follows from the relation

$$N > \frac{\sqrt{(2\mu V a)}}{\pi\hbar} > N - 1.$$

8. (b) The eigenvalues of the operator $\hat{H}'$ are always positive. If ψ_0 is the wave function corresponding to the state of lowest energy, we have from (a)

$$(\hat{Q} + i\hat{P})\,\psi_0 = 0 \quad \text{or} \quad \left(\frac{\partial}{\partial Q} + Q\right)\psi_0 = 0.$$

Hence, we find that

$$\psi_0 = c_0\, e^{-\frac{1}{2}Q^2}, \quad \epsilon_0 = \tfrac{1}{2}.$$

In this way we find that the energy of the oscillator is equal to

$$\epsilon_n = n + \tfrac{1}{2} \quad (n \geqslant 0),$$

and the corresponding wave function has the form

$$\psi_n = A\left(\frac{\partial}{\partial Q} - Q\right)^n e^{-\frac{1}{2}Q^2}.$$

The normalising constant A_n follows from the condition

$$\int_{-\infty}^{+\infty} |\psi_n(Q)|^2\, dQ = 1.$$

For the ground state the normalising constant A_0 is equal to $1/\sqrt[4]{\pi}$, and thus:

$$\psi_0 = \frac{1}{\sqrt[4]{\pi}} e^{-\frac{1}{2}Q^2}.$$

We can express as follows the wave function of the nth state in terms of the wave function of the $(n-1)$th state,

$$\psi_n(Q) = c_n(\hat{Q} - i\hat{P})\,\psi_{n-1}(Q),$$

where c_n follows from the condition

$$c_n^2 \int [(Q - iP)\,\psi_{n-1}(Q)]^2\, dQ = 1.$$

Replacing P by $-i(\partial/\partial Q)$ and integrating by parts we find

$$c_n^2 \int \psi_{n-1}(P^2 + Q^2 + 1)\,\psi_{n-1}\, dQ = c_n^2 . 2n = 1,$$

so that $c_n = 1/\sqrt{(2n)}$ and

$$\begin{aligned}\psi_n &= c_n\left(Q - \frac{\partial}{\partial Q}\right)\psi_{n-1} \\ &= c_n . c_{n-1} \ldots c_1\left(Q - \frac{\partial}{\partial Q}\right)^n \psi_0 = A_n\left(Q - \frac{\partial}{\partial Q}\right)^n \psi_0.\end{aligned}$$

Finally we get

$$\psi_n = \frac{1}{\sqrt{(2^n\, n!)}}\, \frac{1}{\sqrt[4]{\pi}} \left(Q - \frac{\partial}{\partial Q}\right)^n e^{-\frac{1}{2}Q^2}.$$

The polynomial of degree n

$$H_n(Q) = e^{\frac{1}{2}Q^2}\left(Q - \frac{\partial}{\partial Q}\right)^n e^{-\frac{1}{2}Q^2}$$

is called a Hermite polynomial.

(c) $$\hat{a}\hat{a}^+ - \hat{a}^+\hat{a} = 1; \quad \psi_n = \frac{1}{\sqrt{n!}}(\hat{a}^+)^n\,\psi_0;$$

(d) $$(\hat{P} + i\hat{Q})^n_{n-1} . [(\hat{P} - i\hat{Q})^n_{n-1}]^* = 2n.$$

The wave functions ψ_n we have taken to be real so that the matrix elements of $\hat{Q}$ and $i\hat{P} = \partial/\partial Q$ are also real,

$$(\hat{Q})^n_{n-1} = \sqrt{\left(\frac{n}{2}\right)},$$

$$(\hat{P})^n_{n-1} = i\sqrt{\left(\frac{n}{2}\right)},$$

or, going back to the original variables, we have

$$(\hat{x})^n_{n-1} = (\hat{x})^{n-1}_n = \sqrt{\left(\frac{n\hbar}{2\mu\omega}\right)},$$

$$(\hat{p})^n_{n-1} = -(\hat{p})^{n-1}_n = i\sqrt{\left(\frac{n\mu\hbar\omega}{2}\right)}.$$

10.
$$w = \int_1^\infty e^{-y^2}\,dy \Big/ \int_0^\infty e^{-y^2}\,dy \approx 0{\cdot}16.$$

11. The wave function must vanish at $x = 0$. For $x > 0$ it satisfies the differential equation of an ordinary oscillator. One sees easily that the wave functions of the oscillator for odd values of $n = 2k+1$ vanish at $x = 0$ and are, in the region $x \geqslant 0$, solutions of the problem under consideration. Hence

$$E_k = \hbar\omega(2k+\tfrac{3}{2}) \quad (k = 0, 1, 2, \ldots).$$

12.
$$\left(\frac{p^2}{2\mu} - \frac{\mu\omega^2\hbar^2}{2}\frac{\partial^2}{\partial p^2}\right) a_n(p) = E_n\, a_n(p),$$

$$|a_n(p)|^2 = \frac{1}{2^n . n!\sqrt{(\pi\mu\omega\hbar)}} \exp(-p^2/\mu\omega\hbar)\, H_n^2\left(\frac{p}{\sqrt{(\mu\omega\hbar)}}\right).$$

13. An investigation of the behaviour of the solutions of the Schrödinger equation

$$-\frac{\hbar^2}{2\mu}\frac{d^2\psi}{dx^2} - \left[E - V_0\left(\frac{a}{x} - \frac{x}{a}\right)^2\right]\psi = 0$$

as $x \to \infty$ shows that ψ behaves asymptotically as $\exp(-\frac{1}{2}\xi)$ where ξ is a new independent variable

$$\xi = \frac{\sqrt{(2\mu V_0)}}{\hbar a}x^2.$$

As $x \to 0$ ψ is proportional to $\xi^{\nu/2}$, with

$$\nu = \frac{1}{2}\left[\sqrt{\left(\frac{8\mu V_0 a^2}{\hbar^2} + 1\right)} + 1\right].$$

We make the substitution

$$\psi = \exp(-\tfrac{1}{2}\xi)\,\xi^{\frac{1}{2}\nu}\,u(\xi)$$

and find for $u(\xi)$ the following equation:

$$\xi u'' + (\nu + \tfrac{1}{2} - \xi)\,u' - \left[\frac{\nu}{2} + \frac{1}{4} - \frac{\mu a(E + 2V_0)}{2\hbar\sqrt{(2\mu V_0)}}\right] u = 0. \qquad (1)$$

Equation (1) is the equation for the confluent hypergeometric function and its general solution is of the form

$$u(\xi) = c_1 F(\alpha, \nu+\tfrac{1}{2}, \xi) + c_2 F(\alpha-\nu+\tfrac{1}{2}, \tfrac{3}{2}-\nu, \xi)\,.\,\xi^{\frac{1}{2}-\nu},$$

where we have written α for the expression within square brackets in equation (1).

From the requirement that $\psi(0)$ must be finite, it follows that $c_2 = 0$. Apart from this we must require that the wave function should decrease as $x \to \infty$, that is, that the function $u(\xi)$ reduces to a polynomial. This can be attained by putting α equal to $-n$ ($n = 0, 1, 2, \ldots$) so that we find the energy levels

$$E_n = \frac{2\hbar}{a}\sqrt{\left(\frac{2V_0}{\mu}\right)}\left\{n+\tfrac{1}{2}+\frac{1}{4}\left[\sqrt{\left(\frac{8\mu V_0 a^2}{\hbar^2}+1\right)}-\sqrt{\left(\frac{8\mu V_0 a^2}{\hbar^2}\right)}\right]\right\}.$$

The energy spectrum is thus the same as for an oscillator of angular frequency $\omega = \sqrt{(8V_0/\mu a^2)}$ provided the zero of the energy scale is suitably chosen. It is of interest to note that the zero-point energy of a particle in the potential $V_0(a/x - x/a)^2$ is always larger than the zero-point energy of the corresponding oscillator. The wave functions are of the form

$$\psi_n = c_n x^\nu \exp\left[-\sqrt{\left(\frac{\mu V_0}{2\hbar^2 a^2}\right)}x^2\right] F\left[-n, \nu+\tfrac{1}{2}, \sqrt{\left(\frac{2\mu V_0}{\hbar^2 a^2}\right)}x^2\right],$$

where $\nu = \frac{1}{4}[\sqrt{(8\mu V_0 a^2/\hbar^2+1)}+1]$, while the constant c_n must be found from the normalisation condition.

14*. In the Schrödinger equation

$$-\frac{\hbar^2}{2\mu}\frac{d^2\psi}{dx^2}-\left[E+\frac{V_0}{\cosh^2(x/a)}\right]\psi = 0$$

we make the substitution

$$\psi = \left(\cosh\frac{x}{a}\right)^{-2\lambda} u, \quad \lambda = \frac{1}{4}\left[\sqrt{\left(\frac{8\mu V_0 a^2}{\hbar^2}+1\right)}-1\right].$$

The equation for u takes the following form

$$\frac{d^2 u}{dx^2}-\frac{4\lambda}{a}\tanh\frac{x}{a}\frac{du}{dx}+\frac{4}{a^2}(\lambda^2-\kappa^2)u = 0,$$

where

$$\kappa = \sqrt{\left(-\frac{\mu E a^2}{2\hbar^2}\right)}$$

(we consider the discrete spectrum, $E < 0$).

If we introduce a new independent variable

$$z = -\sinh^2 \frac{x}{a},$$

then the equation for u reduces to the hypergeometric equation

$$z(1-z)\frac{d^2u}{dz^2} + [\tfrac{1}{2} - (1-2\lambda)z]\frac{du}{dz} - (\lambda^2 - \kappa^2)u = 0. \qquad (1)$$

The parameters α, β, γ, which occur in the general form of the hypergeometric equation

$$z(1-z)\frac{d^2u}{dz^2} + [\gamma - (\alpha+\beta+1)z]\frac{du}{dz} - \alpha\beta u = 0,$$

have in our case the following values:

$$\gamma = \tfrac{1}{2}, \quad \alpha = \kappa - \lambda, \quad \beta = -\kappa - \lambda.$$

The two solutions of equation (1) which lead respectively to the even and the odd wave functions ψ are of the form

$$u_1 = F(-\lambda+\kappa, -\lambda-\kappa, \tfrac{1}{2}; z), \qquad (2)$$

$$u_2 = \sqrt{(z)}\, F(-\lambda+\kappa+\tfrac{1}{2}, -\lambda-\kappa+\tfrac{1}{2}, \tfrac{3}{2}; z). \qquad (3)$$

These solutions lead to finite values of the wave functions at

$$x = 0 \quad (z = 0).$$

In order that the wave function

$$\psi = \left(\cosh \frac{x}{a}\right)^{-2\lambda} u$$

tends to zero as $x \to \pm\infty$ $(z \to -\infty)$, the hypergeometric functions in equations (2) and (3) should reduce to polynomials. This condition means, for instance, for u_1 that either $\lambda - \kappa$ or $\lambda + \kappa$ must be a non-negative integer. The second case can, however, be discarded since in that case the wave function increases exponentially as $x \to \pm\infty$. We find thus $\lambda - \kappa = k$ $(k = 0, 1, 2, \ldots)$ and for the energy levels

$$E_k = -\frac{\hbar^2}{2\mu a^2}\left[\frac{1}{2}\sqrt{\left(\frac{8\mu V_0 a^2}{\hbar^2} + 1\right)} - 2k - \tfrac{1}{2}\right]^2.$$

We find similarly for expression (3) that the condition that the wave function be finite as $x \to \pm\infty$ is satisfied provided

$$\lambda - \kappa - \tfrac{1}{2} = l \quad (l = 0, 1, 2, \ldots)$$

and hence

$$E_l = -\frac{\hbar^2}{2\mu a^2}\left[\frac{1}{2}\sqrt{\left(\frac{8\mu V_0 a^2}{\hbar^2}+1\right)}-(2l+1)-\tfrac{1}{2}\right]^2.$$

Combining these expressions we find

$$E_n = -\frac{\hbar^2}{2\mu a^2}\left[\frac{1}{2}\sqrt{\left(\frac{8\mu V_0 a^2}{\hbar^2}+1\right)}-(n+\tfrac{1}{2})\right]^2 \quad (n = 0, 1, 2, \ldots).$$

The number of discrete levels is equal to the largest integer satisfying the inequality

$$N < \frac{1}{2}\sqrt{\left(\frac{8\mu V_0 a^2}{\hbar^2}+1\right)}-\tfrac{1}{2}.$$

We note that the energy spectrum which we have obtained coincides for a suitable choice of parameters with the spectrum of a Morse potential (see problem 12 of section 8).

15*. We make in the wave equation

$$-\frac{\hbar^2}{2\mu}\frac{d^2\psi}{dx^2}-\left(E-V_0\cot^2\frac{\pi}{a}x\right)\psi = 0$$

the following substitution

$$\psi = \left(\sin\frac{\pi}{a}x\right)^{-2\lambda} u.$$

Putting

$$\lambda = \frac{1}{4}\left[\sqrt{\left(\frac{8\mu V_0 a^2}{\pi^2\hbar^2}+1\right)}-1\right],$$

$$\nu = \sqrt{\left[\frac{\mu a^2}{2\hbar^2\pi^2}(E+V_0)\right]},$$

we get the following equation for u

$$\frac{d^2u}{dx^2}-4\frac{\pi}{a}\lambda\cot\frac{\pi x}{a}\frac{du}{dx}+\frac{4\pi^2}{a^2}(\nu^2-\lambda^2)u = 0.$$

By introducing as the independent variable

$$z = \cos^2\frac{\pi x}{a}$$

the last equation is reduced to the hypergeometric equation

$$z(1-z)\frac{d^2u}{dz^2}+[\tfrac{1}{2}-(1-2\lambda)z]\frac{du}{dz}+(\nu^2-\lambda^2)u = 0. \qquad (1)$$

Comparing this equation with the general form of the hypergeometric equation

$$z(1-z)\frac{d^2u}{dz^2}+[\gamma-(\alpha+\beta+1)z]\frac{du}{dz}-\alpha\beta u = 0,$$

we find for the parameters:

$$\gamma = \tfrac{1}{2}, \quad \alpha = -\nu-\lambda, \quad \beta = \nu-\lambda.$$

Equation (1) has two solutions. One of these solutions is different from zero and finite at $z = 0$ (which corresponds to $x = \frac{1}{2}a$),

$$u_1 = F(-\nu-\lambda, \nu-\lambda, \tfrac{1}{2};\, z).$$

The other solution

$$u_2 = \sqrt{(z)}\,F(-\nu-\lambda+\tfrac{1}{2}, \nu-\lambda+\tfrac{1}{2}, \tfrac{3}{2};\, z)$$

vanishes at $z = 0$ ($x = \frac{1}{2}a$). To determine the behaviour of the solutions at $z = 1$ (which corresponds to $x = 0$, or $x = a$), we use the relation $F(\alpha, \beta, \gamma;\, z) = (1-z)^{-\alpha} F[\alpha, \gamma-\beta, \gamma;\, z/(z-1)]$. We find for u_1 and u_2

$$u_1 = (1-z)^{\nu+\lambda} F\left(-\nu-\lambda,\ -\nu+\lambda+\tfrac{1}{2}, \tfrac{1}{2};\frac{z}{z-1}\right), \tag{2}$$

$$u_2 = \sqrt{(z)}\,(1-z)^{\nu+\lambda-\frac{1}{2}} F\left(-\nu-\lambda+\tfrac{1}{2},\ -\nu+\lambda+1, \tfrac{3}{2};\frac{z}{z-1}\right). \tag{3}$$

In order that the wave function ψ vanishes at $x = 0$ and $x = a$ it is necessary that the power series in $z/(z-1)$ is a polynomial.

The hypergeometric series in expression (2) for u_1 breaks off if either $\nu+\lambda$ or $\nu-\lambda-\frac{1}{2}$ is a non-negative integer.

However, the condition $\psi = 0$ at $x = 0$ and $x = a$ is satisfied only in the second case,

$$\nu-\lambda-\tfrac{1}{2} = k \quad (k = 0, 1, \ldots).$$

The energy levels are now given by the equation

$$E_k = [(2k+1)^2+4(2k+1)\lambda-2\lambda]\frac{\pi^2\hbar^2}{2\mu a^2}. \tag{4}$$

A similar discussion of equation (3) shows that the energy levels satisfy the condition

$$\nu-\lambda = l \quad (l = 1, 2, 3, \ldots). \tag{5}$$

We can combine expressions (4) and (5) for the energy levels as follows:

$$E_n = (n^2+4n\lambda-2\lambda)\frac{\pi^2\hbar^2}{2\mu a^2} \quad (n = 1, 2, 3, \ldots).$$

For odd values of n we get the wave functions

$$\psi_n = c_n \left(\sin\frac{\pi x}{a}\right)^{-2\lambda} F\left(-\frac{n}{2}-2\lambda, \frac{n}{2}, \tfrac{1}{2}; \cos^2\frac{\pi x}{a}\right),$$

while even n corresponds to

$$\psi_n = c_n \left(\sin\frac{\pi x}{a}\right)^{-2\lambda} \cos\frac{\pi x}{a} F\left(-\frac{n}{2}-2\lambda+\tfrac{1}{2}, \frac{n}{2}+\tfrac{1}{2}, \tfrac{3}{2}; \cos^2\frac{\pi x}{a}\right).$$

The normalised wave function of the ground state is given by the equation

$$\psi = \sqrt{\left[\frac{\pi\Gamma(2\lambda+1)}{a\Gamma(\frac{1}{2})\,\Gamma(2\lambda+\frac{3}{2})}\right]} \left(\sin\frac{\pi x}{a}\right)^{2\lambda+1}, \quad \Gamma(\tfrac{1}{2}) = \sqrt{\pi}.$$

Let us consider the limiting case where $V_0 \to 0$. The problem goes then over into the problem of a particle in a potential well (see problem 1 of section 1). The quantity λ tends to zero and, as we should have expected, we find for the energy levels

$$E_n = \frac{\pi^2\hbar^2}{2\mu a^2} n^2.$$

In the opposite case where $\lambda \gg 1$ we find for the low-lying levels ($n \ll \lambda$)

$$E_n = \hbar\omega(n+\tfrac{1}{2}) \quad (n = 1, 2, \ldots),$$

where

$$\omega = \frac{\pi}{a}\sqrt{\left(\frac{2V_0}{\mu}\right)}.$$

This result could also have been obtained by expanding the potential energy near the point $x = a/2$ and retaining only the quadratic terms.

16. The Schrödinger equation is

$$\psi'' + \frac{2\mu}{\hbar^2}[E - D + 2D\exp(-ax) - D\exp(-2ax)]\psi = 0.$$

Introducing a new variable ξ as follows

$$\xi = \frac{2\sqrt{(2\mu D)}}{a\hbar}\exp(-ax),$$

and putting

$$s = \frac{\sqrt{[-2\mu(E-D)]}}{a\hbar}, \quad n = \frac{\sqrt{(2\mu D)}}{a\hbar} - (s+\tfrac{1}{2}), \tag{1}$$

the Schrödinger equation becomes (primes now indicate differentiation with respect to ξ)

$$\psi'' + \frac{\psi'}{\xi} + \left(-\tfrac{1}{4} + \frac{n+s+\frac{1}{2}}{\xi} - \frac{s^2}{\xi^2}\right)\psi = 0.$$

Making the substitution

$$\psi = \exp(-\tfrac{1}{2}\xi)\,\xi^{S}\,w(\xi),$$

we get for w the equation for a confluent hypergeometric function, and from the requirement that ψ remains finite as $\xi \to \infty$, we find that n must be a non-negative integer, whence we get for E from (1)

$$E = \sqrt{\left(\frac{2D}{\mu}\right)}a\hbar(n+\tfrac{1}{2}) - \frac{a^2\hbar^2}{2\mu}(n+\tfrac{1}{2})^2,$$

where n must be positive and less than $\sqrt{(2\mu D)}/a\hbar - \frac{1}{2}$ as s is positive by definition.

17. Because of the normalisation of the delta-function we can write

$$\delta(x^2 - a^2) = \frac{1}{2a}[\delta(x-a) + \delta(x+a)].$$

For a bound state $E < 0$ and putting $E = -\hbar^2\kappa^2/2\mu$, we can write the eigenfunctions corresponding to bound states in the form

$$\psi = A\exp(\kappa x), \quad x < -a; \quad \psi = B\exp(\kappa x) + C\exp(-\kappa x), \quad -a < x < a;$$

$$\psi = D\exp(-\kappa x), \quad a < x.$$

By an argument similar to the one leading to equation (8) in problem 35 of section 1 we see that the wave function must satisfy the conditions

$$\psi'(-a+) - \psi'(-a-) = -\frac{P}{a}\psi(-a),$$

$$\psi'(a+) - \psi'(a-) = -\frac{P}{a}\psi(a),$$

as well as the requirement of continuity of ψ at $x = \pm a$.

We see that the ensuing equations can be satisfied by either an even or an odd function corresponding to $A = D$, $B = C$ or $A = -D$, $B = -C$. For the energy eigenvalues we get easily the following equations:

$$\text{even parity: } \coth\kappa a = 2\kappa a/P; \quad (1)$$

$$\text{odd parity: } \tanh\kappa a = 2\kappa a/P. \quad (2)$$

One sees from (1) and (2): (i) there is always one and only one solution of (1) which as $P \to 0$ corresponds to $\kappa \to 0$ and thus to $E \to 0$; the fact that there is always at least one solution is a special case of the general theorem proved in problem 41 of section 1; (ii) there is a solution of (2) only if $P > 2$. As $P \to \infty$, the two solutions approach to one another and $\kappa \to \infty$ in that limit.

18*. The problem here is to take the singularity at the origin properly into account. The solution of the Schrödinger equation for $x > 0$ and $x < 0$ is straightforward and leads in the usual way to the associated Laguerre polynomials. The solutions can be divided into even and odd wave functions. The odd solutions vanish necessarily at the origin and do not give rise to any difficulties, but the even solutions have to be treated with more care.

The Schrödinger equation

$$\frac{\hbar^2}{2\mu}\psi'' + \left(E + \frac{e^2}{|x|}\right)\psi = 0 \tag{1}$$

can be written in the form (primes now indicate differentiation with respect to z)

$$\psi'' - \tfrac{1}{4}\psi + \frac{\alpha}{|z|}\psi = 0, \tag{2}$$

where

$$E = -\hbar^2/2\mu a_B^2\,\alpha^2, \quad a_B = \hbar^2/\mu e^2, \quad z = 2x/\alpha a_B. \tag{3}$$

The solution of (2) which is everywhere well behaved and finite occurs only if α is an integer ($\alpha = N$). In that case the energy levels are exactly the same as for the three-dimensional hydrogen atom, except that $N = 0$ is an allowed value: the ground state of the one-dimensional hydrogen atom is $-\infty$, and one sees from (2) that the corresponding wave function is $\exp(-\frac{1}{2}|z|)$, an even function with a discontinuity of slope at the origin. The other wave functions are given by the equations

$$\psi = A\exp(-\tfrac{1}{2}z)\,zF(1-N, 2;\ z), \quad z = 2x/Na_0, \quad x > 0; \tag{4}$$

$$\psi = B\exp(\tfrac{1}{2}z)\,zF(1-N, 2;\ -z), \quad z = 2x/Na_0, \quad x < 0. \tag{5}$$

The quantities A and B are normalisation constants and $F(\alpha, \beta;\ x)$ is the confluent hypergeometric function (which for $\alpha = N$ reduces to an associated Laguerre polynomial). There are two possibilities: either we choose $A = B$ when we have an even eigenfunction with a discontinuity of slope at the origin, or we choose the odd eigenfunction with $A = -B$: the energy levels with $N > 0$ are thus twofold degenerate.

We refer to a paper by Loudon (*Am. J. Phys.* **27**, 649 (1959)) for a detailed discussion including a discussion of the possibility of degenerate levels for a one-dimensional potential and of various ways of approximating the one-dimensional Coulomb potential by suitably truncated potentials.

19. In the case under consideration there is only a continuous spectrum and the eigenfunctions are non-degenerate.

Let us in the Schrödinger equation

$$-\frac{\hbar^2}{2\mu}\frac{d^2\psi}{dx^2}-[E+Fx]\psi = 0,$$

go over from the coordinate to the momentum representation; we get

$$\frac{p^2}{2\mu}a(p)-Ea(p) = i\hbar F\frac{d}{dp}a(p).$$

The solution of this equation corresponding to an eigenvalue E,

$$a_E(p) = c\exp\left[-\frac{i}{\hbar F}\left(\frac{p^3}{6\mu}-Ep\right)\right],$$

is a wave function in the momentum representation. We normalise the function $a(p)$ to $\delta(E-E')$,

$$\int a_E^*(p)\,a_{E'}(p)\,dp = \delta(E-E'),$$

or

$$cc^*\int\exp\left[-\frac{ip}{\hbar F}(E-E')\right]dp = cc^*\,2\pi\hbar F\delta(E-E'),$$

whence

$$c = \frac{1}{\sqrt{(2\pi\hbar F)}}.$$

The wave function in coordinate representation is

$$\psi(x) = \frac{\alpha}{2\pi\sqrt{F}}\int_{-\infty}^{+\infty}\exp\left(i\frac{u^3}{3}-iuq\right)du = \frac{\alpha}{\pi\sqrt{F}}\int_0^\infty\cos\left(\frac{u^3}{3}-uq\right)du,$$

$$q = \left(x+\frac{E}{F}\right)\alpha, \quad \alpha = \left(\frac{2\mu F}{\hbar^2}\right)^{\frac{1}{3}}.$$

This integral can be expressed in terms of the Airy function $\Phi(q)$:

$$\Phi(q) = \frac{1}{\sqrt{\pi}}\int_0^\infty\cos\left(\frac{u^3}{3}+uq\right)du, \quad \psi(x) = \frac{\alpha}{\sqrt{(\pi F)}}\Phi(-q).$$

20*. The potential energy of the particle under consideration is of the form

$$V(z) = \begin{cases}\infty, & z\leqslant 0,\\ \mu g z, & z\geqslant 0,\end{cases}$$

where the direction $-\boldsymbol{g}$ is, as usual, taken along the positive z-axis and where $z = 0$ is the reflecting plane.

This potential energy and also the energy levels and wave functions which we shall obtain are depicted in fig. 24. We restrict ourselves to

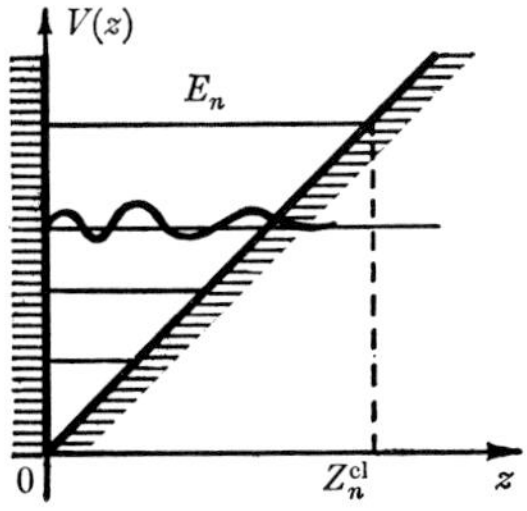

Fig. 24.

an investigation of the motion of the particle in the z-direction since the motion in the xy-plane is free and is of no interest at the moment. The time-independent Schrödinger equation for the particle wave function $\psi(z)$ in the region $z \geqslant 0$ is then:

$$-\frac{\hbar^2}{2\mu}\frac{d^2\psi}{dz^2}+\mu g z\psi = E\psi, \tag{1}$$

where E is the particle energy. This equation must clearly be solved with the boundary conditions

$$\psi \to 0 \quad \text{as} \quad z \to \infty \tag{2}$$

and

$$\psi = 0 \quad \text{for} \quad z = 0. \tag{3}$$

To simplify equation (1) we introduce a new independent variable ζ which is connected with z by the relation

$$\frac{2\mu^2 g}{\hbar^2}z-\frac{2\mu E}{\hbar^2} = c\zeta, \tag{4}$$

where c is for the moment an arbitrary constant.

Substituting this variable into equation (1) transforms it to

$$\frac{d^2\psi}{d\zeta^2}-c^3\left(\frac{\hbar^2}{2\mu^2 g}\right)^2\zeta\psi(\zeta) = 0.$$

It is now clear that if we take $c = (2\mu^2 g/\hbar^2)^{\frac{2}{3}}$, that is, write equation (4) in the form

$$\zeta = \left(\frac{2\mu^2 g}{\hbar^2}\right)^{\frac{1}{3}}\left(z-\frac{E}{\mu g}\right), \tag{4'}$$

we are led to the very simple form

$$\psi''(\zeta)-\zeta\psi(\zeta) = 0.$$

The solution of this equation which is finite for all values of ζ, that is, also for all values of z, is

$$\psi = A\Phi(\zeta) \equiv A\frac{1}{\sqrt{\pi}}\int_0^{\infty}\cos\left(u\zeta+\tfrac{1}{3}u^3\right)du, \tag{5}$$

where $\Phi(\zeta)$ is the Airy function and A a normalisation factor to be determined in a moment.

To begin with, it is clear that the energy levels of a particle which is in the potential well shown in fig. 24 form a discrete spectrum and that their number is infinite. To find these levels we use the boundary condition (3) which, by means of equations (5) and (4′), can be written in the form

$$\Phi\left(-\frac{2^{\frac{1}{3}}}{\mu^{\frac{1}{3}}g^{\frac{2}{3}}\hbar^{\frac{2}{3}}}E\right) = 0. \tag{3′}$$

Putting the argument of equation (3′) equal to the roots of the Airy function which we shall denote by $-\alpha_n$ $(n = 1, 2, \ldots)$

$$0 < \alpha_1 < \alpha_2 < \ldots < \alpha_n < \ldots,$$

we find for the solutions E_n of this transcendental equation

$$-\left(\frac{2}{\mu g^2\hbar^2}\right)^{\frac{1}{3}} E_n = -\alpha_n.$$

Hence the required energy spectrum of the particle has the form

$$E_n = \left(\frac{\mu g^2\hbar^2}{2}\right)^{\frac{1}{3}} . \alpha_n \quad (n = 1, 2, \ldots). \tag{6}$$

In particular, the energy of the ground state ($n = 1$, $\alpha_1 \approx 2{\cdot}34$) is equal to

$$E_1 \approx 2{\cdot}34\left(\frac{\mu g^2\hbar^2}{2}\right)^{\frac{1}{3}}. \tag{6′}$$

As it should be for the case of a one-dimensional bounded motion, the energy levels we have found are non-degenerate. The wave functions of the corresponding stationary states are, according to equations (5), (4′), and (6), of the form

$$\psi_n(z) = A_n\Phi(\zeta_n) = A_n\Phi\left(\frac{z}{a} - \alpha_n\right), \tag{7}$$

where a denotes a characteristic length

$$a = \left(\frac{\hbar^2}{2\mu^2 g}\right)^{\frac{1}{3}}. \tag{8}$$

The normalising factor A depends on n and a.

This function is oscillating in the classical region, and the number of nodes of the function ψ_n is equal to n (including the node at the edge of the region for $z = 0$), while the distance between consecutive nodes which is equal to†

$$\pi\bar{\lambda} = \frac{\pi\hbar}{\sqrt{\{2\mu[E-V(z)]\}}} \tag{9}$$

decreases towards $z = 0$ (that is, with increasing kinetic energy, see fig. 24).

The constant factor in equation (7) is determined from the normalisation of ψ. If we normalise with respect to ζ this constant is clearly equal to‡

$$A_n = \frac{1}{\sqrt{\left\{\int_{-\alpha_n}^{\infty} [\Phi(\zeta)]^2 \, d\zeta\right\}}}, \tag{10}$$

and if we normalise with respect to z, we have [see equations (4′) and (8)]

$$A_n = \frac{1}{\sqrt{a}} \cdot \frac{1}{\sqrt{\left\{\int_{-\alpha_n}^{\infty} [\Phi(\zeta)]^2 \, d\zeta\right\}}}. \tag{10′}$$

We investigate now the asymptotic behaviour of the wave functions $\psi_n(z)$ at sufficiently large distances from the classical turning point§

$$z_n^{\text{cl}} = \frac{E_n}{\mu g}, \tag{11}$$

and also the form of the energy spectrum E_n for $n \gg 1$, that is, in the semi-classical case.

According to equations (6) and (8) the lengths z_n^{cl} and a are connected by the relation

$$z_n^{\text{cl}} = a \,.\, \alpha_n. \tag{11′}$$

We use equations (8), (11), and (11′) to write equation (4′) in the form

$$\zeta = \frac{z - z_n^{\text{cl}}}{a} = \frac{z}{a} - \alpha_n$$

and consider asymptotic expressions for the functions $\psi_n(z)$.

† It is clear that the quantity given by equation (9) has this meaning only for $n \gg 1$. In this case the bar denotes an averaging over a section Δz which contains a large number of wavelengths $2\pi\bar{\lambda}$, but which is small compared to the distance over which $V(z)$ varies appreciably.

‡ As the Airy function is real for real values of its argument, we have

$$|\Phi(\zeta)|^2 = [\Phi(\zeta)]^2.$$

§ z_n^{cl} is clearly the greatest height which a (classical) particle of weight μg and given energy E_n can attain.

(1) The classically inaccessible region and values of z not too near z_n^{cl}:

$$z - z_n^{\text{cl}} \gg a, \qquad \text{i.e.,} \qquad \zeta \gg 1.\dagger$$

Using the asymptotic expression for the Airy function

$$\Phi(\zeta) \approx \frac{1}{2\zeta^{\frac{1}{4}}} \exp\left(-\tfrac{2}{3}\zeta^{\frac{3}{2}}\right),$$

we find

$$\psi_n(z) \approx \frac{A_n}{2}\left(\frac{a}{z - z_n^{\text{cl}}}\right)^{\frac{1}{4}} \exp\left[-\frac{2}{3}\left(\frac{z - z_n^{\text{cl}}}{a}\right)^{\frac{3}{2}}\right] \tag{12}$$

(exponential decrease of the probability $|\psi_n|^2$ in the classically inaccessible region).

(2) The region of the classical motion and values of z not too near z_n^{cl}:

$$0 < z < z_n^{\text{cl}} \quad \text{and} \quad |z - z_n^{\text{cl}}| \gg a,$$

that is,

$$-\alpha_n < \zeta < 0 \quad \text{and} \quad |\zeta| \gg 1.$$

It is clear that to satisfy the second condition at least the following inequality must obtain: $\alpha_n \gg 1$. As we shall see below, α_n is proportional to $n^{\frac{2}{3}}$ which means that the interval of ζ-values considered here exists only for very highly excited (semi-classical) energy levels, that is, for $n \gg 1$.

The required asymptotic expansion of $\Phi(\zeta)$ has the form

$$\Phi(\zeta) \approx \frac{1}{|\zeta|^{\frac{1}{4}}} \sin\left(\tfrac{2}{3}|\zeta|^{\frac{3}{2}} + \frac{\pi}{4}\right), \tag{13}$$

so that

$$\psi_n(z) \approx A_n\left(\frac{a}{|z - z_n^{\text{cl}}|}\right)^{\frac{1}{4}} \sin\left[\frac{2}{3}\left(\frac{|z - z_n^{\text{cl}}|}{a}\right)^{\frac{3}{2}} + \frac{\pi}{4}\right]. \tag{14}$$

The asymptotic expressions (12) and (14) give us in this way the semi-classical wave functions. By means of (14) one can find explicit expressions for the roots of the Airy function $\zeta_n \equiv -\alpha_n$, and hence also for the energy levels (6) in the semi-classical case. To do this we must clearly equate the argument of the sine to an integral multiple of π:

$$\tfrac{2}{3}|\zeta|^{\frac{3}{2}} + \frac{\pi}{4} = (n+1)\pi \quad (n = 0, 1, 2, \ldots),$$

so that

$$\alpha_n = |\zeta_n| = \left[\frac{3}{2}\left(n\pi + \frac{3\pi}{4}\right)\right]^{\frac{2}{3}}, \tag{15}$$

† It can easily be shown that the inequality $|\zeta| \gg 1$ is equivalent to $|d\lambda/dz| \ll 1$ by virtue of equations (9) and (4′).

and [see equation (6)]

$$E_n = \tfrac{1}{2}(9\pi^2 \mu g^2 \hbar^2)^{\frac{1}{3}} (n + \tfrac{3}{4})^{\frac{2}{3}}. \tag{16}$$

The last result can, of course, also be obtained without knowing the wave functions by using the Bohr–Sommerfeld quantisation rule (see problem 28 of section 1).

Simple expressions may also be obtained for the normalisation constant A_n in the semi-classical case.

Indeed, since for $n \gg 1$ the wave function (7) oscillates rapidly in the classical region $0 < z < z_n^{\text{cl}}$ [see equation (14)] and outside this region decreases exponentially, we can in the integral of equation (10) replace $\sin^2$ by its average value $\frac{1}{2}$ and limit the integration over ζ by $\zeta = 0$. Taking into account that $n \gg 1$ we get thus:

$$A_n \approx (\alpha_n)^{-\frac{1}{4}} \approx \left(\frac{2}{3\pi n}\right)^{\frac{1}{6}}$$

and [see equations (10) and (10′)]:

$$A_n \approx \frac{1}{\sqrt{a}} \left(\frac{2}{3\pi n}\right)^{\frac{1}{6}}. \tag{17}$$

Let us now write down an expression for the average distance $\overline{\pi\lambda}$ [see equation (9)] between consecutive nodes of the semi-classical wave function. Using equations (4′) and (8), we find†:

$$\overline{\pi\lambda} = \frac{\pi a}{\sqrt{|\zeta_n|}} = \frac{\pi a}{\sqrt{|(z/a) - \alpha_n|}}. \tag{18}$$

Since in the semi-classical region $|\zeta_n| \gg 1$, we have $\overline{\pi\lambda} \ll a$.

Near the reflecting plane, that is, for $z/a \ll \alpha_n$, this "semi-wavelength" is a minimum and as

$$\alpha_n \approx \left(\frac{3\pi}{2} n\right)^{\frac{2}{3}},$$

it is equal to

$$(\overline{\pi\lambda})_{\min} \approx \left(\frac{2\pi^2}{3}\right)^{\frac{1}{3}} \frac{a}{n^{\frac{1}{3}}}. \tag{18'}$$

Finally, the distance between two successive levels ΔE_n follows from equation (16) which leads to

$$\Delta \ln E_n = \frac{\Delta E_n}{E_n} = \frac{2}{3}\frac{\Delta n}{n},$$

† The same result could have been obtained by using equation (13) and the obvious relation

$$\Delta(\tfrac{2}{3} |\zeta|^{\frac{3}{2}}) \approx |\zeta|^{\frac{1}{2}} \Delta\zeta = \pi,$$

taking into account the fact that $\overline{\pi\lambda} \equiv \Delta z = a\Delta\zeta$.

and by putting $\Delta n = 1$ to

$$\Delta E_n = \frac{2}{3}\frac{E_n}{n}. \tag{19}$$

Let us now show the limiting transition to classical mechanics. This limit can be obtained by averaging the quantum mechanical probability density $|\psi_n(z)|^2$ over an interval Δz which is small compared to the dimensions of the classical region of motion, but contains a large number of wavelengths, so that it satisfies the inequalities:

$$2\pi\bar{\lambda} \ll \Delta z \ll z_n^{\mathrm{cl}}.$$

According to equations (14) and (17) we have:

$$|\psi_n(z)|^2 = \frac{1}{a}\left(\frac{2}{3\pi n}\right)^{\frac{1}{3}} \sqrt{\left(\frac{a}{|z-z_n^{\mathrm{cl}}|}\right)} \sin^2\left[\frac{2}{3}\left(\frac{|z-z_n^{\mathrm{cl}}|}{a}\right)^{\frac{3}{2}} + \frac{\pi}{4}\right].$$

Averaging this expression over an interval Δz we can replace the square of the sine, which is oscillating many times in that interval, by its average value $\frac{1}{2}$, which leads after multiplication by Δz to

$$\overline{|\psi_n(z)|^2}\,\Delta z = \left(\frac{2}{3\pi n}\right)^{\frac{1}{3}} \frac{1}{2\sqrt{(a|z-z_n^{\mathrm{cl}}|)}}\Delta z. \tag{20}$$

In the classical limit ($\hbar \to 0, n \to \infty$) this probability must go over into $\Delta t/\frac{1}{2}T$ (since it is normalised) which is the ratio of the period spent in the interval Δz to the semi-period of the classical motion. This means that in the limit we must have

$$\overline{|\psi(z)|^2}\,\Delta z = \frac{\Delta t}{\frac{1}{2}T}.$$

Eliminating from equation (20) $a^{\frac{1}{2}} n^{\frac{1}{3}} \sim (n\hbar)^{\frac{1}{3}}$ by using equation (16) and omitting everywhere the index n we get

$$\frac{\Delta z}{\Delta t} = \frac{2}{\frac{1}{2}T}\sqrt{\left(\frac{EH}{\mu g}\right)}, \tag{21}$$

where $H = |z - z^{\mathrm{cl}}|$. This ratio must clearly be equal to the classical velocity of the particle

$$\left|\frac{dz}{dt}\right| \equiv v_{\mathrm{cl}} = \sqrt{(2gH)}.$$

Since

$$\tfrac{1}{2}T = \int_0^{H_{\max}} \frac{dH}{v} = \int_0^{E/\mu g} \frac{dH}{\sqrt{(2gH)}} = \left(\sqrt{\frac{2E}{\mu g^2}}\right), \tag{22}$$

the equation $\Delta z/\Delta t = \sqrt{(2gH)}$ is, indeed, satisfied.

In conclusion it is useful to estimate the order of magnitude of the quantities which we meet with when we apply the results obtained here to macroscopic particles, for instance, a ball of mass $\mu = 1$ g, freely falling from a height $H = 100$ cm in the earth's gravitational field ($g \approx 10^3$ cm sec^{-2}) on a solid slab.

The characteristic "zero point" energy $\epsilon \sim (\mu g^2 \hbar^2)^{\frac{1}{3}}$ [see equation (6′)] is of the order of magnitude

$$\epsilon \sim (1.10^6.10^{-54})^{\frac{1}{3}} = 10^{-16} \text{ erg},$$

and the energy of the ball $E = \mu g H \approx 1.10^3.100 = 10^5$ erg, so that the number of the state (the number of nodes of the wave function) n and the distance between adjacent energy levels ΔE_n are given by [cf. equations (16) and (19)]:

$$n \sim \left(\frac{E}{\epsilon}\right)^{\frac{3}{2}} \sim 10^{32},$$

$$\Delta E_n \sim \frac{E}{n} \sim 10^{-27} \text{ erg}.$$

The characteristic length $a \sim (\hbar^2/\mu^2 g)^{\frac{1}{3}}$ [see equation (8)] is equal to:

$$a \sim \left(\frac{10^{-54}}{1.10^3}\right)^{\frac{1}{3}} = 10^{-19} \text{ cm}$$

and the (minimum) wavelength of the ball is equal to

$$\lambda\!\!\!^{-} \sim \frac{a}{n^{\frac{1}{3}}} \sim 10^{-30} \text{ cm}$$

as it should be, since $\lambda\!\!\!^{-}.n \sim H$.

The magnitude of these quantities gives us an idea with what enormous precision the motion of macroscopic bodies follows the laws of classical mechanics.

21. We write the wavefunction $\psi(x)$ in the form

$$\psi(x) = \exp\frac{iS(x)}{\hbar}, \tag{1}$$

and substitute it into the time-independent Schrödinger equation. The equation for S then takes the form

$$\frac{1}{2\mu}\left(\frac{dS}{dx}\right)^2 - \frac{i\hbar}{2\mu}\frac{d^2S}{dx^2} + V(x) = E,$$

or

$$\left(\frac{dS}{dx}\right)^2 - i\hbar\frac{d^2S}{dx^2} = p^2(x), \tag{2}$$

where

$$p(x) = \sqrt{\{[E - V(x)]/2\mu\}}, \tag{3}$$

which is the classical momentum of the particle at the point x.

If

$$\left(\frac{dS}{dx}\right)^2 \gg \hbar \frac{d^2S}{dx^2}, \tag{4}$$

equation (2) becomes approximately

$$\left(\frac{dS}{dx}\right)^2 = p^2(x), \tag{5}$$

which is the classical Hamilton–Jacobi equation. Using equation (3) we can write inequality (4) in the form

$$\lambda \gg \frac{\lambda}{2\pi}\frac{d\lambda}{dx},$$

where $\lambda = h/p$ is the "local" de Broglie wavelength. This means that the approximation (5) is good, provided the de Broglie wavelength does not change much over a wavelength.

Using (3) we can also write the inequality in the form

$$p^3 \gg \mu\hbar\left|\frac{dU}{dx}\right|. \tag{6}$$

This means that the approximation is good, provided we are dealing with the motion of a particle with a large momentum in a potential with a small gradient.

If the inequality (6) is satisfied we can use a series expansion for $S(x)$:

$$S(x) = S_0(x) - i\hbar S_1(x) + (i\hbar)^2 S_2(x) \ldots . \tag{7}$$

This is the semi-classical or Wentzel–Kramers–Brillouin (WKB) approach.

Substituting (7) into (2) we get the following set of equations:

$$\left(\frac{dS_0}{dx}\right)^2 = p^2(x), \tag{8a}$$

$$\frac{dS_1}{dx}\frac{dS_0}{dx} + \frac{1}{2}\frac{d^2S_0}{dx^2} = 0, \tag{8b}$$

. . . .

The first equation leads to

$$\frac{dS_0}{dx} = \pm p(x), \tag{9}$$

and the second one to

$$S_1 = -\ln C\sqrt{p}, \tag{10}$$

where C is a constant.

Let us first consider the region where $E > V(x)$—the so-called classically allowed region—and write $k(x) = p(x)/\hbar$. From equations (1), (7), (9), and (10) we find

$$\psi(x) = \frac{A}{\sqrt{p}}\sin\left[\left[\int_a^x k(x')dx'\right] + \alpha,\right. \tag{11}$$

where A, a, and α are constants.

The points where $E = V(x)$ are the classical turning points. Of course, near the turning points inequality (6) is violated. If near the turning point $x = x_t$ we write

$$p^2 = 2\mu[E - V(x)] \approx 2\mu\left|\frac{dU}{dx}\right||x - x_t|, \tag{12}$$

and substitute that into condition (6), we find after some manipulations that the semi-classical approximation can be used for points x satisfying the condition

$$|x - x_t| \gg \frac{\hbar}{2p} = \frac{\lambda}{4\pi}. \tag{13}$$

The region where $E < V(x)$ is the classically unattainable region. If in that region we write $k(x) = i\kappa(x)$, so that

$$\kappa(x) = \frac{1}{\hbar}\sqrt{\{2\mu[V(x) - E]\}}, \tag{14}$$

the wavefunction has the form

$$\psi(x) = \frac{B}{\sqrt{|p|}}\exp\left[-\int_a^x \kappa(x')dx'\right] + \frac{C}{\sqrt{|p|}}\exp\left[\int_a^x \kappa(x')dx'\right]. \tag{15}$$

If x_1 and x_2 are the turning points (see fig. 24a) and a_1, b_1 and a_2, b_2 the

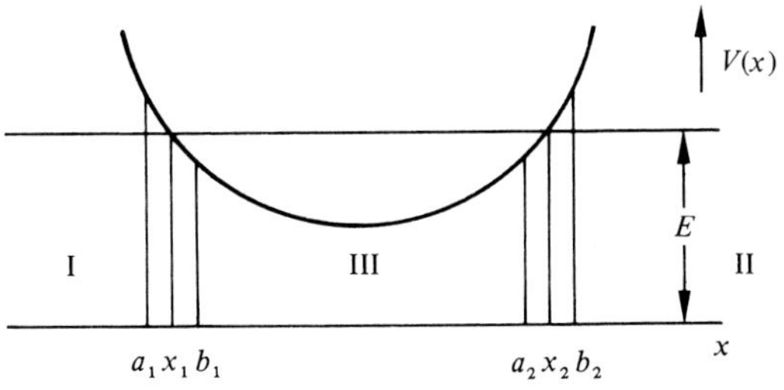

Fig. 24a.

regions where condition (13) is violated, we have in regions I ($x < a_1$) and II ($x > b_2$) the following solutions:

$$\psi_{\mathrm{I}}(x) = \frac{C}{\sqrt{|p|}} \exp\left[-\int_x^{x_1} \kappa(x')dx'\right], \quad x < a_1, \tag{16}$$

$$\psi_{\mathrm{II}}(x) = \frac{B}{\sqrt{|p|}} \exp\left[-\int_{x_2}^{x} \kappa(x')dx'\right], \quad x > b_2, \tag{17}$$

while in the region III ($b_1 < x < a_2$) the solution is given by (11) with $a \equiv x_1$.

In the region $a_1 < x < b_1$ we can use for the potential the approximation [cf.(12)]

$$V(x) = E-(x-x_1)F, \quad F = \left|\frac{dU}{dx}\right|_{x=x_1}. \tag{18}$$

The Schrödinger equation then reduces to the Schrödinger equation considered in problem 20 of section 1 and the connection between the wavefunction in regions I, II, and III follows as in that problem. The result is that at x_1 we must have

$$C = \tfrac{1}{2}A \quad \text{and} \quad \alpha = \tfrac{1}{4}\pi. \tag{19}$$

Similarly for the connection at x_2 we find

$$B = (-1)^{n+1}\tfrac{1}{2}A \quad \text{and} \quad \alpha = -\int_{x_1}^{x_2} k(x')dx' - n\pi + \tfrac{1}{4}\pi. \tag{20}$$

Comparing equations (19) and (20) we find

$$2\int_{x_1}^{x_2} p(x)dx = (n+\tfrac{1}{2})h, \tag{21}$$

the required quantization condition.

We note that we have earlier seen that the semi-classical approximation is good, provided we are several wavelengths away from the turning points [see condition (13)]; this means that in the interval x_1,x_2 there must be many wavelengths, that is, that the quantum number n must be large.

22. The maximum momentum of a particle at point x is equal to $\sqrt{[-2\mu V(x)]}$, and the number of eigenstates corresponding to an interval dx and a momentum interval dp is equal to $dxdp/h$. The total number of discrete levels, N, will thus be given by the expression

$$N = \frac{\sqrt{(2\mu)}}{h}\int [-V(x)]^{1/2} dx, \tag{1}$$

where the integration is over those values of x for which $V(x) < 0$.

23. (a) $$E_n = (n+\tfrac{1}{2})\hbar\omega \quad (n = 0, 1, 2, \ldots),$$

(b) $$E_n = \frac{\pi^2\hbar^2}{2\mu a^2}\left[\sqrt{\left(\frac{2\mu a^2 V_0}{\pi^2\hbar^2}\right)} + (n+\tfrac{1}{2})\right]^2 - V_0$$
$$(n = 0, 1, 2, \ldots).$$

24. $E_n = -V_0 + V_0\left\{\frac{3h}{8a\sqrt{(2\mu V_0)}}(n+\tfrac{1}{2})\right\}^{2/3}$; the total number of discrete energy levels is one more than the largest integer n for which E_n is still negative. One can easily check that this number differs by $\frac{1}{2}$ from the number given in the solution to problem 22 of section 1. As the semi-classical approximation is valid only for large n, this difference can be neglected.

25. The average value of the kinetic energy in the stationary state ψ_n (we assume the wave function to be real) is given by the equation

$$\overline{T} = -\frac{\hbar^2}{2\mu}\int \psi_n \frac{d^2\psi_n}{dx^2}\,dx = \frac{\hbar^2}{2\mu}\int\left(\frac{d\psi_n}{dx}\right)^2 dx.$$

In the semi-classical approximation the wave function has in the classically accessible region $(a<x<b)$ the form

$$\psi_n = \frac{A_n}{\sqrt{p}}\cos\left(\frac{1}{\hbar}\int_a^x p\,dx - \frac{\pi}{4}\right), \quad p = \sqrt{[2\mu(E_n - V)]},$$

so that

$$\frac{d\psi_n}{dx} = -\frac{\sqrt{p}}{\hbar}A_n \sin\left(\frac{1}{\hbar}\int_a^x p\,dx - \frac{\pi}{4}\right) - \frac{1}{2}\frac{A_n}{p^{\frac{3}{2}}}\frac{dp}{dx}\cos\left(\frac{1}{\hbar}\int_a^x p\,dx - \frac{\pi}{4}\right).$$

If we substitute this expression into the integral which determines $\overline{T}$, the limits of integration must be the limits of the classically accessible region, since outside that region ψ_n decreases exponentially. Taking for the square of the fast oscillating trigonometric functions their average value $\frac{1}{2}$ and neglecting the integral containing the oscillating product

$$\sin\left(\frac{1}{\hbar}\int_x^a p\,dx - \frac{\pi}{4}\right).\cos\left(\frac{1}{\hbar}\int_a^x p\,dx - \frac{\pi}{4}\right) = \tfrac{1}{2}\sin\left(\frac{2}{\hbar}\int_a^x p\,dx - \frac{\pi}{2}\right),$$

we get

$$\overline{T} = \frac{A_n^2}{4\mu}\int_a^b\left[p + \frac{\hbar^2}{4p^3}\left(\frac{dp}{dx}\right)^2\right]dx.$$

The condition of applicability of the semi-classical approximation, $|d\lambda/dx| = (\hbar/p^2)|dp/dx| \ll 1$, means that the second term under the integral sign is small compared to the first term so that we find by using the quantum condition

$$\overline{T} = \frac{A_n^2}{4\mu}\int_a^b p\,dx = \frac{A_n^2}{4\mu}\pi\hbar(n+\tfrac{1}{2}).$$

The constant A_n follows from the normalisation condition

$$\int \psi_n^2\,dn \approx A_n^2\int_a^b \frac{1}{p}\cos^2\left(\frac{1}{\hbar}\int_a^x p\,dx - \frac{\pi}{4}\right)dx \approx \frac{A_n^2}{2}\int_a^b \frac{dx}{p} = 1.$$

On the other hand, differentiating the quantum condition

$$\int_a^b p\,dx = \int_a^b \sqrt{[2\mu(E_n - V)]}\,dx = \pi\hbar(n+\tfrac{1}{2})$$

with respect to n we get

$$\mu\frac{dE_n}{dn}\int_a^b \frac{dx}{\sqrt{[2\mu(E_n - V)]}} = \mu\frac{dE_n}{dn}\int_a^b \frac{dx}{p} = \pi\hbar,$$

so that

$$A_n^2 = \frac{2\mu}{\pi\hbar}\frac{dE_n}{dn}.$$

The expression for the average kinetic energy then becomes

$$\overline{T} = \frac{1}{2}\frac{dE_n}{dn}(n+\tfrac{1}{2}).$$

26. (a) $\overline{T} = \tfrac{1}{2}\hbar\omega(n+\tfrac{1}{2});$

(b) $$\overline{T} = \frac{\pi^2\hbar^2}{2\mu a^2}\left[\sqrt{\left(\frac{2\mu a^2 V_0}{\pi^2\hbar^2}\right)} + (n+\tfrac{1}{2})\right](n+\tfrac{1}{2}).$$

27. From the virial theorem it follows that

$$2\overline{T} = \nu\overline{V},$$

so that

$$E = \frac{2+\nu}{\nu}\overline{T}.$$

Substituting into that relation the average value of the kinetic energy

$$\overline{T} = \frac{1}{2}\frac{dE_n}{dn}(n+\tfrac{1}{2})$$

(see preceding problem) we get the equation

$$E = \frac{2+\nu}{2\nu}\frac{dE}{dn}(n+\tfrac{1}{2}),$$

the solution of which is of the form

$$E = A(n+\tfrac{1}{2})^{2\nu/(2+\nu)}. \tag{1}$$

For the case where ν is an even integer, the constant A can be found from the quantum condition

$$I(E) = 2\int_0^b \sqrt{[2\mu(E-ax^\nu)]}\,dx = n\hbar(n+\tfrac{1}{2}), \tag{2}$$

where $b^\nu = E/a$.

Differentiating (2) with respect to E, introducing a new variable $u = ax^\nu/E$, and using (1) we find

$$A = \left[\sqrt{\left(\frac{\pi}{2\mu}\right)}\,\nu\hbar a^{1/\nu}\,\frac{\Gamma(\frac{3}{2}+1/\nu)}{\Gamma(1/\nu)}\right]^{2\nu/(2+\nu)}.$$

If $\nu = 2$, $A = \hbar\sqrt{(2a/\mu)}$ and (1) reduces to the normal expression $E = \hbar\sqrt{(2a/\mu)}\,(n+\frac{1}{2})$.

28. The potential energy of the particle is of the form (fig. 25; see also problem 20 of section 1):

$$V(z) = \begin{cases} \infty & (z \leqslant 0), \\ \mu g z & (z \geqslant 0). \end{cases}$$

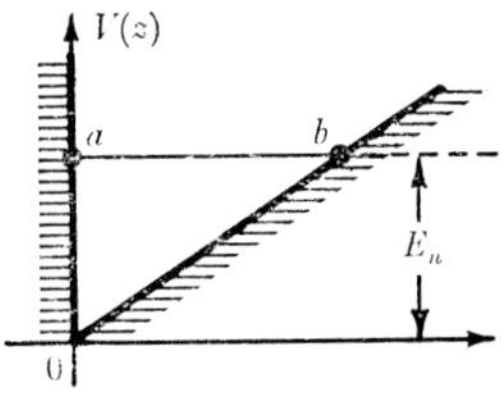

Fig. 25.

One of the turning points is $z = 0$ and the other one follows from the equation $V(z) = \mu g z = E$ and has thus the coordinate:

$$z = \frac{E}{\mu g}.$$

As far as the quantisation rule is concerned, its right-hand side must be slightly modified compared with the usual expression $\pi(n+\frac{1}{2})$. This is connected with the fact that in the usual derivation of the Bohr–Sommerfeld quantisation rule one assumes that the potential energy near both classical turning points $z = a$ and $z = b$ can be expanded in a power series in $(z-a)$ and $(z-b)$ beginning with a linear term. As a consequence, in the semi-classical expressions for the wave function the phase starts in both turning points at $\pi/4$:

$$\frac{c}{\sqrt{[p(z)]}} \sin\left[\frac{1}{\hbar}\int_a^z p(z)\,dz + \frac{\pi}{4}\right]$$

and

$$\frac{c'}{\sqrt{[p(z)]}} \sin\left[\frac{1}{\hbar}\int_z^b p(z)\,dz + \frac{\pi}{4}\right] \quad (a \leqslant z \leqslant b). \tag{1}$$

The requirement that these two expressions are the same in the interval $a \leqslant z \leqslant b$ leads to the condition

$$\frac{1}{\hbar}\int_a^b p\,dz + \frac{\pi}{2} = (n+1)\pi \quad (n = 0, 1, 2, \ldots) \tag{2}$$

(the total of the two phases in equations (1) is put equal to an integral multiple of π).

In the case under consideration one of the turning points ($z = a = 0$) corresponds to a "vertical potential wall" so that the first of the two phases of equation (1) does not start with a term $\pi/4$ (in contradistinction to an "inclined" potential wall the function behaves here semi-classically right up to the turning point $z = a$ itself where it must tend to zero, whence this result follows); in the second of the phases of equation (1) the term $\pi/4$ enters as usual. If we now add the two phases we get thus instead of equation (2) the condition

$$\frac{1}{\hbar}\int_a^b p\,dz + \frac{\pi}{4} = (n+1)\pi,$$

from which we obtain the quantisation rule we are looking for,

$$\frac{1}{\hbar}\int_a^b p\,dz = (n+\tfrac{3}{4})\pi \quad (n = 0, 1, 2, \ldots). \tag{3}$$

Substituting

$$p = \sqrt{\{2\mu[E - \mu(z)]\}} = \sqrt{[2\mu(E - \mu g z)]}, \quad a = 0, \quad b = E/\mu g,$$

we get after elementary calculations

$$E \equiv E_n = \tfrac{1}{2}(9\pi^2 \mu g^2 \hbar^2)^{\frac{1}{3}} (n+\tfrac{3}{4})^{\frac{2}{3}}, \tag{4}$$

which agrees exactly with the semi-classical result obtained in problem 20 of section 1, as should be the case.

If we borrow from equation (22) of that problem the value of the classical frequency of the bouncing ball

$$\omega = \frac{2\pi}{T} = \pi g \sqrt{\left(\frac{\mu}{2E}\right)}$$

and evaluate from equation (4) the distance between successive levels ($\Delta n = 1$),

$$\Delta E_n = \frac{dE_n}{dn},$$

we arrive at the relation

$$\Delta E = \hbar\omega,$$

which expresses the approximate equidistance of the semi-classical energy levels.

29.
$$\int_a^b \sqrt{\{2\mu[E-V(x)]\}}\, dx = (n+\tfrac{3}{4})\pi\hbar$$

(compare the discussion in problem 28 of section 1).

30*. We start from the quantisation rule

$$\int_{x_1}^{x_2} \sqrt{\{2\mu[E-V(x)]\}}\, dx = \pi\hbar(n+\tfrac{1}{2}),$$

which determines the spectrum, or more precisely $n(E)$, if the potential energy $V(x)$ is given. Inasfar as $V(x)$ is an even function we have

$$2\int_0^a \sqrt{\{2\mu[E-V(x)]\}}\, dx = \pi\hbar(n+\tfrac{1}{2}), \tag{1}$$

where $x_2 = -x_1 = a$, $E = V(a)$.

The integral equation (1) can be solved as follows.† Differentiating (1) with respect to E, and introducing V as variable instead of x, we get

$$\frac{\pi\hbar}{\sqrt{(2\mu)}}\frac{dn}{dE} = \int_0^a \frac{dx}{\sqrt{(E-V)}} = \int_0^E \frac{dx}{dV}\frac{dV}{\sqrt{(E-V)}}. \tag{2}$$

Multiplying (2) by $(\alpha-E)^{-\frac{1}{2}}$ and integrating over E from 0 to α, where α is a parameter to be determined, we get

$$\int_0^\alpha \frac{\pi\hbar}{\sqrt{(2\mu)}}\frac{dx}{dE}\frac{dE}{\sqrt{(\alpha-E)}} = \int_0^\alpha dE \int_0^E \frac{dx}{dV}\frac{dV}{\sqrt{[(\alpha-E)(E-V)]}}$$
$$= \int_0^\alpha \frac{dx}{dV}dV \int_V^\alpha \frac{dE}{\sqrt{[(\alpha-E)(E-V)]}} = \pi x(\alpha).$$

Putting $\alpha = V$, we get

$$x(V) = \frac{\hbar}{\sqrt{(2\mu)}}\int_{E_0}^V \frac{dE}{\frac{dE}{dn}\sqrt{(V-E)}},$$

† Compare Landau and Lifshitz, *Mechanics*, Pergamon Press, Oxford, 1960, §12.

where $x(V)$ is the inverse function of $V(x)$ and dE/dn is considered to be a function of E, while E_0 is the zero of the energy scale.

31*. (1) The Schrödinger equation in the momentum representation is of the form:

$$\left\{\frac{p^2}{2\mu}+\hat{V}\left(i\hbar\frac{\partial}{\partial p}\right)\right\}G(p)=EG(p),$$

where $\hat{V}[i\hbar(\partial/\partial p)] = V(\hat{x})$ is the operator of the potential energy of the particle.

We must look for a solution of this equation in the form:

$$G(p)=\exp\left[-(i/\hbar)\,S(p)\right], \tag{1}$$

where we expand $S(p)$ in powers of $\hbar$:

$$S(p)=S_0(p)+\hbar S_1(p)+\hbar^2S_2(p)+\ldots. \tag{2}$$

To find an expansion in powers of $\hbar$ of the expression which we obtained by letting the operator $\hat{V}[i\hbar(\partial/\partial p)]$ act on the wave function $G(p)$, we assume that the potential energy can be expanded in a power series in x. Thus,

$$\hat{V}\left(i\hbar\frac{\partial}{\partial p}\right)=\sum_{n=0}^{\infty}\frac{V^{(n)}(0)}{n!}\left(i\hbar\frac{\partial}{\partial p}\right)^n. \tag{3}$$

Let us consider the action of the different terms of this series on the function (1):

$$i\hbar\frac{\partial}{\partial p}\exp\left(-\frac{i}{\hbar}S\right)=\exp\left(-\frac{i}{\hbar}S\right)\frac{\partial S}{\partial p},$$

$$\left(i\hbar\frac{\partial}{\partial p}\right)^2\exp\left(-\frac{i}{\hbar}S\right)=\exp\left(-\frac{i}{\hbar}S\right)\left[\left(\frac{\partial S}{\partial p}\right)^2+i\hbar\frac{\partial^2 S}{\partial p^2}\right],$$

$$\left(i\hbar\frac{\partial}{\partial p}\right)^3\exp\left(-\frac{i}{\hbar}S\right)=\exp\left(-\frac{i}{\hbar}S\right)\left[\left(\frac{\partial S}{\partial p}\right)^3+i\hbar(1+2)\frac{\partial^2 S}{\partial p^2}\frac{\partial S}{\partial p}+O(\hbar^2)\right].$$

In general, as can be easily shown by induction:

$$\left(i\hbar\frac{\partial}{\partial p}\right)^n\exp\left(-\frac{i}{\hbar}S\right)$$

$$=\exp\left(-\frac{i}{\hbar}S\right)\left[\left(\frac{\partial S}{\partial p}\right)^n+i\hbar\left(\sum_{k=1}^{n-1}k\right)\frac{\partial^2 S}{\partial p^2}\left(\frac{\partial S}{\partial p}\right)^{n-2}+O(\hbar^2)\right]$$

$$=\exp\left(-\frac{i}{\hbar}S\right)\left[\left(\frac{\partial S}{\partial p}\right)^n+i\hbar\frac{n(n-1)}{2}\frac{\partial^2 S}{\partial p^2}\left(\frac{\partial S}{\partial p}\right)^{n-2}+O(\hbar^2)\right]. \tag{4}$$

According to equations (3) and (4) we find

$$\hat{V}\left(i\hbar\frac{\partial}{\partial p}\right)\exp\left(-\frac{i}{\hbar}S\right) = \exp\left(-\frac{i}{\hbar}S\right)\left[V\left(\frac{\partial S}{\partial p}\right)+\frac{i\hbar}{2}\frac{\partial^2 S}{\partial p^2}V''\left(\frac{\partial S}{\partial p}\right)+\dots\right].$$

The equation for the action function $S(p)$ has thus the following form:

$$\frac{p^2}{2\mu}+V\left(\frac{\partial S}{\partial p}\right)+\tfrac{1}{2}i\hbar\frac{\partial^2 S}{\partial p^2}V''\left(\frac{\partial S}{\partial p}\right)+\dots = E.$$

Substituting now the expansion (2) for S and taking together terms of the same order in $\hbar$ we get the following equations to determine S_0 and S_1:

$$\frac{p^2}{2\mu}+V\left(\frac{\partial S_0}{\partial p}\right) = E, \tag{5}$$

$$\frac{\partial S_1}{\partial p}V'\left(\frac{\partial S_0}{\partial p}\right)+\frac{i}{2}\frac{\partial^2 S_0}{\partial p^2}V''\left(\frac{\partial S_0}{\partial p}\right) = 0. \tag{6}$$

Let $x(p)$ denote the dependence of the coordinate on the momentum when the particle moves in the field $V(x)$. We find then by solving equation (5) with respect to $\partial S_0/\partial p$,

$$\frac{\partial S_0}{\partial p} = x(p),$$

so that

$$S_0 = \int x(p)\,dp.$$

From equation (6) we get:

$$\frac{\partial S_1}{\partial p} = -\frac{i}{2}\frac{\partial^2 S_0}{\partial p^2}\frac{V''(\partial S_0/\partial p)}{V'(\partial S_0/\partial p)}.$$

One sees easily that the right-hand side is the derivative of

$$-\frac{i}{2}\ln\left|V'\left(\frac{\partial S_0}{\partial p}\right)\right|,$$

with respect to the momentum, so that

$$S_1 = -\frac{i}{2}\ln\left|V'\left(\frac{\partial S_0}{\partial p}\right)\right|.$$

The semi-classical wave function has thus finally the following form in the momentum representation:

$$G(p) = \frac{c}{\sqrt{|V'(\partial S_0/\partial p)|}}\exp\left[-\frac{i}{\hbar}\int x(p)\,dp\right]. \tag{7}$$

The meaning of the factor in the denominator is clear when we note that $|V'(\partial S_0/\partial p)|$ is the force, $|dp/dt|$, expressed in terms of the momentum p. This means, as should be the case in the semi-classical case, that the probability to find the momentum between p and $p+dp$ is proportional to the corresponding interval of time dt:

$$|G|^2\,dp = \frac{|c|^2}{|dp/dt|}\,dp = |c|^2\,dt.$$

(2) We now turn to the second part of the problem, that is, to show that the wave function $G(p)$ which we have found can be obtained as the Fourier transform of the semi-classical wave function in the coordinate representation.

The semi-classical function in the x-representation has the form:

$$\psi(x) = \frac{c_1}{\sqrt{|p(x)|}}\exp\left[\frac{i}{\hbar}\int p(x)\,dx\right],$$

where $p(x)$ is the momentum as a function of the coordinate

$$p(x) = \sqrt{\{2\mu[E-V(x)]\}}.$$

The corresponding function in the momentum representation is†

$$\tilde{G}(p) = \frac{1}{\sqrt{(2\pi\hbar)}}\int dx\,\psi(x)\exp\left(-\frac{i}{\hbar}px\right)$$

$$= \frac{c_1}{\sqrt{(2\pi\hbar)}}\int\frac{dx}{\sqrt{|p(x)|}}\exp\left\{\frac{i}{\hbar}\left[\int p(x)\,dx - px\right]\right\}. \tag{8}$$

Using the fact that the action $S = \int p\,dx$ is large compared to $\hbar$ we can evaluate the integral in equation (8) by expanding the exponent in a series near its maximum (saddle-point method).

To find the maximum we put the first derivative of the expression $\varphi(x) \equiv \int p(x)\,dx - px$ equal to zero:

$$\frac{d\varphi}{dx} = p(x) - p = \sqrt{\{2\mu[E-V(x)]\}} - p = 0. \tag{9}$$

The solution of this equation is $x(p)$, that is, the coordinate as function of the momentum in the field V, or the function which we considered sub (1).

† We denote this function by $\tilde{G}$ to distinguish it for the moment from the semi-classical function in the momentum representation which we found sub (1); our problem is to prove the identity of these two functions.

For the second derivative we find:

$$\frac{d^2\varphi}{dx^2} = -\frac{\mu V'}{\sqrt{[2\mu(E-V)]}} = \mu\frac{(dp/dt)(x)}{p(x)}.$$

To find the value of φ'' at the maximum we must take $x(p)$ for x, and we thus find $p[x(p)] = p$, so that

$$\left(\frac{d^2\varphi}{dx^2}\right)_{\max} = \mu\frac{\frac{dp}{dt}(p)}{p}.$$

The expansion of $\varphi(x)$ near its maximum is thus of the form:

$$\varphi(x) = \int^{x(p)} \sqrt{[2\mu(E-V)]}\,dx - px(p) + \mu\frac{\frac{dp}{dt}(p)}{2p}[x-x(p)]^2.$$

If we substitute this expression into $\tilde{G}(p)$ and take out from under the integral sign the slowly varying factor $1/\sqrt{|p(x)|}$ at the maximum $x = x(p)$ we get:

$$\tilde{G}(p) = \frac{c_1}{\sqrt{(|p|2\pi\hbar)}}\exp\left\{\frac{i}{\hbar}\left[\int^{x(p)}\sqrt{\{2\mu(E-V)\}}\,dx - px(p)\right]\right\}$$

$$\times\int dx\exp\left\{\frac{i\mu}{\hbar}\frac{\frac{dp}{dt}(p)}{p}[x-x(p)]^2\right\}$$

$$= \frac{c_1}{\sqrt{(2\pi\hbar|p|)}}\exp\left\{\frac{i}{\hbar}\left[\int^{x(p)}\sqrt{\{2\mu(E-V)\}}\,dx - px(p)\right]\right\}\sqrt{\left(\frac{\hbar\pi}{2i\mu}\frac{|p|}{|dp/dt|}\right)}$$

$$= \frac{c_1}{\sqrt{(4\mu i)}}\frac{1}{\sqrt{|dp/dt|}}\exp\left\{\frac{i}{\hbar}\left[\int^{x(p)}\sqrt{\{2\mu(E-V)\}}\,dx - px(p)\right]\right\}.$$

We prove now that the expression within square brackets may be written as $-\int x(p)\,dp$.

To do this we take its derivative with respect to p:

$$\frac{d}{dp}\left(\int^{x(p)}\sqrt{\{2\mu[E-V(x)]\}}\,dx - px(p)\right)$$

$$= \frac{dx}{dp}\sqrt{(2\mu\{E-V[x(p)]\})} - p\frac{dx}{dp} - x(p).$$

According to equation (9) the first two terms cancel each other. Hence,

$$\frac{d}{dp}(\ldots) = -x(p),$$

and thus

$$\int^{x(p)} \sqrt{[2\mu(E-V)]}\,dx - px(p) = -\int x(p)\,dp.$$

The function $\tilde{G}$ is thus of the form

$$\tilde{G}(p) = \frac{c_1}{\sqrt{(4\mu i)}}\frac{1}{\sqrt{|dp/dt|}}\exp\left[-\frac{i}{\hbar}\int x(p)\,dp\right]$$

and is therefore the same, apart from a multiplying constant, as the function $G(p)$ of equation (7):

$$\tilde{G}(p) = \text{const}\,G(p),$$

as had to be proved.

32. The Hamiltonian of a free plane rotator reduces to the operator of the kinetic energy which has the form $\hat{M}_z^2/2I$, where $\hat{M}_z$ is the operator of the component of the angular momentum of the rotator along the axis of rotation, the z-axis.

The Schrödinger equation for the stationary states of the rotator takes the form (we replace $\hat{M}_z$ by $(\hbar/i)(\partial/\partial\varphi)$ where φ is the angle of rotation of the rotator around the z-axis):

$$-\frac{\hbar^2}{2I}\frac{d^2\psi}{d\varphi^2} = E\psi.$$

The general solution of this equation which satisfies the condition of uniqueness (which in this case is the requirement of periodicity in φ) is obtained by putting $\sqrt{(2EI/\hbar^2)}$ equal to an integer m ($m = 0, \pm 1, \pm 2, \ldots$) and has the form

$$\psi_m(\varphi) = c_1 \sin m\varphi + c_2 \cos m\varphi,$$

where c_1 and c_2 are arbitrary constants, and the corresponding energy levels are according to the considerations of a moment ago equal to

$$E_m = \frac{\hbar^2 m^2}{2I}.$$

As could be expected, each level for $m \neq 0$ is twofold degenerate with respect to the direction of rotation. The normalised wave function of the stationary state corresponding at the same time to a well-defined value of the angular momentum ($M_z = m\hbar$) is clearly equal to

$$\psi_m(\varphi) = \frac{1}{\sqrt{(2\pi)}}\exp(im\varphi).$$

33. $$\psi(\varphi, t) = \tfrac{1}{2}A[1 - \cos 2\varphi \exp(-2i\hbar t/I)],$$

where I is the moment of inertia of the rotator.

34. The tension T in the wire is related to the energy E of the systems by the usual thermodynamic relation $T = -\partial E/\partial a$. The energy levels of the particles are given by the equation (compare problem 32 of section 1)

$$E_n = \frac{n^2\hbar^2}{2\mu a^2},$$

and we have thus for the tension

$$T = \frac{n^2\hbar^2}{\mu a^3}.$$

35. The Hamiltonian operator has the following form for the given potential:

$$\hat{H} = \frac{1}{2\mu}p^2 + \tfrac{1}{2}V_0 \exp\left(b\hbar\frac{\partial}{\partial p}\right) + \tfrac{1}{2}V_0 \exp\left(-b\hbar\frac{\partial}{\partial p}\right).$$

Since

$$\exp\left(b\hbar\frac{\partial}{\partial p}\right) a(p) = a(p+b\hbar),$$

the Schrödinger equation can be written as a finite-difference equation

$$\frac{1}{2\mu}p^2 a(p) + \tfrac{1}{2}V_0 a(p+b\hbar) + \tfrac{1}{2}V_0 a(p-b\hbar) = Ea(p).$$

36. $$\frac{\hbar^2}{2\mu}k^2 a(k) + \sum_{-\infty}^{+\infty} V_n a\left(k+\frac{2\pi n}{b}\right) = Ea(k)$$

$$\left[k = \frac{p}{\hbar},\ V(x) = \sum_{-\infty}^{+\infty} V_n \exp\left(-\frac{2\pi n i x}{b}\right),\ V_n = V_{-n}^*\right].$$

37*. The wave function in the region of the well $(0 < x < a)$ is of the form

$$\psi = c_1 \exp(i\kappa_1 x) + c_2 \exp(-i\kappa_1 x), \quad \kappa_1 = \frac{\sqrt{(2\mu E)}}{\hbar},$$

and in the region of the barrier $-b < x < 0$

$$\psi = c_3 \exp(i\kappa_2 x) + c_4 \exp(-i\kappa_2 x), \quad \kappa_2 = \frac{\sqrt{[2\mu(E-V_0)]}}{\hbar}.$$

Since $\psi(x) = \text{const.}\psi(x+l)$ (the constant has an absolute magnitude of unity and is put equal to $\exp(-ikl)$, $l = a+b$; for a discussion of the reason for this choice see the answer to problem 35 of section 1) in the region of the next barrier $(a < x < a+b)$ we have

$$\psi = \exp(ikl)\{c_3 \exp[i\kappa_2(x-l)] + c_4 \exp[-i\kappa_2(x-l)]\}.$$

The requirement that the wave function and its derivatives are continuous at $x = 0$ and $x = a$ leads to the following four equations:

$$c_1 + c_2 = c_3 + c_4,$$

$$c_1 \exp(i\kappa_1 a) + c_2 \exp(-i\kappa_1 a) = \exp(ikl)\,[c_3 \exp(-i\kappa_2 b) + c_4 \exp(i\kappa_2 b)],$$

$$\kappa_1(c_1 - c_2) = \kappa_2(c_3 - c_4),$$

$$\kappa_1[c_1 \exp(i\kappa_1 a) - c_2 \exp(-i\kappa_1 a)] = \kappa_2[c_3 \exp(-i\kappa_2 b) - c_4 \exp(i\kappa_2 b)] \exp(ikl).$$

This set of equations has a non-trivial solution only if

$$\cos kl = \cos\kappa_1 a . \cos\kappa_2 b - \frac{\kappa_1^2 + \kappa_2^2}{2\kappa_1 \kappa_2} \sin\kappa_1 a . \sin\kappa_2 b. \tag{1}$$

We investigate two cases:

(a) $E < V_0$, κ_2 is purely imaginary.

Introducing the notation $\kappa_2 = i\kappa$, we can write equation (1) in the form

$$\cos kl = \cos\kappa_1 a . \cosh\kappa b + \frac{\kappa^2 - \kappa_1^2}{2\kappa_1 \kappa} \sin\kappa_1 a . \sinh\kappa b. \tag{2}$$

The allowed energy bands are thus determined from the relation

$$-1 \leqslant \cos\kappa_1 a . \cosh\kappa b + \frac{\kappa^2 - \kappa_1^2}{2\kappa_1 \kappa} \sin\kappa_1 a . \sinh\kappa b \leqslant 1.$$

To discuss the general regularity in the position of the allowed bands we consider the limiting case

$$\frac{\sqrt{(2\mu V_0)}}{\hbar} b \ll 1, \quad a \gg b, \quad E \ll V_0.$$

We write $(\mu V_0/\hbar^2)\, ab = \gamma$. Equation (2) is thus approximately of the form

$$\cos ka = \gamma \frac{\sin\kappa_1 a}{\kappa_1 a} + \cos\kappa_1 a. \tag{3}$$

In fig. 26 we have plotted the function $(\gamma \sin\kappa_1 a/\kappa_1 a) + \cos\kappa_1 a$ as a function of $\kappa_1 a$. The allowed energy bands are indicated on the $\kappa_1 a$-axis by heavy lines.

To the right of each point $\kappa_1 a = n\pi$ there is a forbidden energy band. From the figure it is clear that the forbidden energy bands get narrower with increasing number of the band. One can easily estimate the width of the forbidden bands. The left-hand side of expression (3) takes on the values $(-1)^n$, when

$$\cos(\kappa_1 a - \varphi) = (-1)^n \cos\varphi, \quad \tan\varphi = \frac{\gamma}{\kappa_1 a},$$

which is possible for $\kappa_1 a = n\pi$ and for $\kappa_1 a = n\pi + 2\varphi$. It follows thus that the width of the forbidden energy bands is equal to 2φ. For large values of n

$$2\varphi \approx \frac{\gamma}{n\pi}.$$

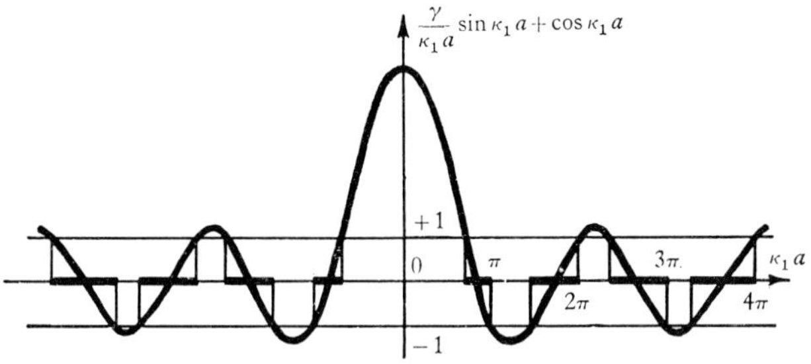

Fig. 26.

(b) $E > V_0$. In that case the energy bands are determined from the relation

$$-1 \leqslant \cos \kappa_1 a . \cos \kappa_2 b - \frac{\kappa_1^2 + \kappa_2^2}{2\kappa_1 \kappa_2} \sin \kappa_1 a . \sin \kappa_2 b \leqslant +1.$$

38*. Let us consider the general problem of the motion of a particle in a periodic potential. The Schrödinger equation can be written in the form

$$\psi'' + f(x)\psi = 0, \tag{1}$$

where $f(x+na) = f(x)$.

Let ψ_1 and ψ_2 be two linearly independent solutions of (1). It then follows easily that

$$\psi_1 \psi_2'' - \psi_2 \psi_1'' = 0,$$

or that the *Wronskian* $W \;(= \psi_1 \psi_2' - \psi_2 \psi_1')$ is independent of x.

As $f(x)$ is a periodic function $\psi_1(x+a)$ and $\psi_2(x+a)$ must also be solutions of (1) and can thus be expressed as linear combinations of $\psi_1(x)$ and $\psi_2(x)$:

$$\left.\begin{aligned} \psi_1(x+a) &= c_1 \psi_1(x) + c_2 \psi_2(x), \\ \psi_2(x+a) &= c_3 \psi_1(x) + c_4 \psi_2(x). \end{aligned}\right\} \tag{2}$$

From (2) it follows that the Wronskian at $x+a$ is equal to the Wronskian at x multiplied by $c_1 c_4 - c_2 c_3$ which must thus be equal to unity.

We now look for solutions $\psi_1(x)$ and $\psi_2(x)$ such that

$$\psi_1(x+a) = \lambda_1 \psi_1(x), \quad \psi_2(x+a) = \lambda_2 \psi_2(x),$$

or, from (2), we must find solutions of the equations

$$\begin{aligned} c_1\psi_1(x)+c_2\psi_2(x) &= \lambda\psi_1(x), \\ c_3\psi_1(x)+c_4\psi_2(x) &= \lambda\psi_2(x), \end{aligned} \tag{3}$$

which leads to an eigenvalue equation for λ:

$$\lambda^2-(c_1+c_4)\lambda+1 = 0, \tag{4}$$

where we have used the fact that $c_1 c_4 = c_2 c_3$. In order that $\psi_1(x)$ and $\psi_2(x)$ are physically acceptable solutions the absolute magnitude of λ must be equal to unity. This occurs when $\frac{1}{2}(c_2+c_4)<1$ and in that case λ_1 and λ_2 are each other's conjugate complex so that we can write

$$\begin{aligned} \psi_1(x+a) &= \exp(ika)\,\psi_1(x), \\ \psi_2(x+a) &= \exp(-ika)\,\psi_2(x), \end{aligned} \tag{5}$$

or, equivalently,

$$\begin{aligned} \psi_1(x) &= \exp(ikx)\,u_1(x), \\ \psi_2(x) &= \exp(-ikx)\,u_2(x), \end{aligned} \tag{6}$$

where $u_1(x)$ and $u_2(x)$ are periodic functions. To see the significance of (5) and (6) it is useful to consider the case where f is constant.† From a consideration of that case it follows that for certain energy values (5) holds while for other energy values we do not have physically acceptable solutions.

Consider now the potential energy

$$V(x) = \frac{\hbar^2 P}{2\mu a}\sum_{n=-\infty}^{+\infty}\delta(x+na). \tag{7}$$

As V contains singularities we have instead of the usual requirement that $\psi'(x)$ be continuous a boundary condition which we obtain by integrating the Schrödinger equation over a short interval including the singularity (from $-\epsilon$ to $+\epsilon$, where ϵ is positive and tends to zero); the result is

$$\psi'(0+)-\psi'(0-) = \frac{P}{a}\psi(0). \tag{8}$$

Using (6) we can write the solutions in the regions $0<x<a$ and $-a<x<0$ as follows:

$$\psi_1(x) = A\exp(i\kappa x)+B\exp(-i\kappa x), \quad \kappa = \sqrt{(2\mu E)}/\hbar, \quad 0<x<a,$$

$$\psi_1(x) = \exp(-ika)\{A\exp[i\kappa(x+a)]+B\exp[-i\kappa(x+a)]\}, \quad -a<x<0,$$

† For a detailed discussion we refer to H. Jones, *The Theory of Brillouin Zones and Electronic States in Crystals*. Amsterdam: North Holland Publishing Company, 1960. See also H. A. Kramers, *Physica* **2**, 483 (1935).

and from (8) and the requirement that ψ be continuous we have

$$A+B = \exp(-ika)[A\exp(i\kappa a)+B\exp(-i\kappa a)],$$

$$i\kappa[A-B]-i\kappa[A\exp(-ika+i\kappa a)-B\exp(-ika-i\kappa a)] = (P/a)(A+B),$$

from which follows the equation (compare (3) in problem 37 of section 1):

$$\cos ka = \cos\kappa a+\frac{P}{2}\frac{\sin\kappa a}{\kappa a}. \tag{9}$$

In fig. 26 we had plotted the right-hand side of (9), which we shall denote by $f(\kappa a)$, as function of κa. The width of the energy gaps was evaluated in problem 16 of section 1.

The band edges occur when $\cos ka = +1$ or -1, that is, for $k = 0, \pi/a, 2\pi a, \ldots$, or $k = k_n = n\pi/a$. To find the effective mass at the band edge, we must expand E (or κ) in powers of $k-k_n$ at the lower edge or in powers of k_n-k at the upper edge.

At the upper edges

$$\kappa_n a = n\pi,$$

or

$$E_n = \frac{\hbar^2}{2\mu}\left(\frac{\pi n}{a}\right)^2,$$

and writing

$$\kappa = \frac{n\pi}{a}-\kappa', \quad k = \frac{n\pi}{a}-k',$$

we find

$$\kappa' = \frac{k'^2 n\pi a}{P},$$

whence follows for the effective mass μ^* defined by the relation

$$E = E_n+\frac{\hbar^2 k'^2}{2\mu^*}$$

that

$$\mu^* = -\frac{\mu a P}{2\pi^3 n^3}.$$

The situation is much more complicated at the lower band edges, as the values of κ corresponding to those edges are the solutions of a transcendental equation.†

† See H. Jones, loc. cit.

39*. As the potential is periodic it follows that there must be solutions of the Schrödinger equation for which

$$\psi(x+a) = z\psi(x).$$

If $|z| \neq 1$ the wave function increases unrestrictedly in one direction or the other. The condition from which the allowed energy bands are determined is thus $|z| = 1$ or $z = \exp(i\varphi)$.

Let us now consider one period of the potential (see fig. 27). It can be divided into three regions: $x_1 < x < x_2$, $x_2 < x < x_3$, $x_3 < x$. In the first region the solution can be written in the form:

$$\psi_1 = \frac{c_1}{\sqrt{p}} \exp\left(\frac{i}{\hbar}\int_{x_1}^{x} p\, dx\right) + \frac{c_2}{\sqrt{p}} \exp\left(-\frac{i}{\hbar}\int_{x_1}^{x} p\, dx\right)$$

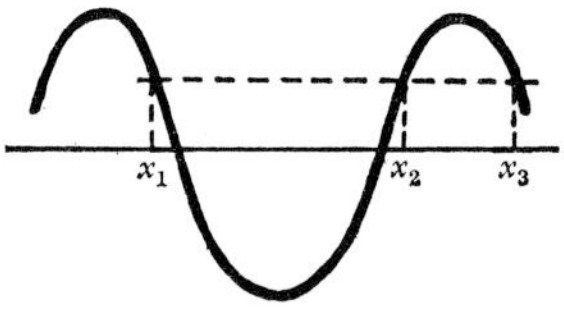

Fig. 27.

or, differently:

$$\psi_1 = \frac{c_1'}{\sqrt{p}} \exp\left[-i\left(\frac{1}{\hbar}\int_{x}^{x_2} p\, dx - \frac{\pi}{4}\right)\right] + \frac{c_2'}{\sqrt{p}} \exp\left[i\left(\frac{1}{\hbar}\int_{x}^{x_2} p\, dx - \frac{\pi}{4}\right)\right],$$

where

$$c_1' = c_1 \exp\left[i\left(\alpha - \frac{\pi}{4}\right)\right], \quad c_2' = c_2 \exp\left[-i\left(\alpha - \frac{\pi}{4}\right)\right], \quad \alpha = \frac{1}{\hbar}\int_{x_1}^{x_2} p\, dx. \quad (1)$$

Using the boundary conditions of the semi-classical case we obtain a solution in the second region:

$$\psi_{\rm II} = \frac{c_2' - c_1'}{i\sqrt{|p|}} \exp\left(\frac{1}{\hbar}\int_{x_2}^{x} |p|\, dx\right) + \frac{1}{2}\frac{c_2' + c_1'}{\sqrt{|p|}} \exp\left(-\frac{1}{\hbar}\int_{x_2}^{x} |p|\, dx\right)$$

or

$$\psi_{\rm II} = \frac{c_1'' \exp\left(-\frac{1}{\hbar}\int_{x}^{x_3} |p|\, dx\right)}{i\sqrt{|p|}} + \frac{1}{2}\frac{c_2''}{\sqrt{|p|}} \exp\left(\frac{1}{\hbar}\int_{x}^{x_3} |p|\, dx\right),$$

where

$$c_1'' = (c_2' - c_1') \exp(\beta), \quad c_2'' = (c_2' + c_1') \exp(-\beta), \quad \beta = \frac{1}{\hbar}\int_{x_2}^{x_3} |p|\, dx.$$

Using again the boundary conditions, we get for $x > x_3$:

$$\psi_{\mathrm{III}} = -i\frac{(c_1''-\frac{1}{4}c_2'')\exp(-\frac{1}{4}\pi i)}{\sqrt{p}}\exp\left(\frac{i}{\hbar}\int_{x_3}^{x} p\,dx\right) - i\frac{(c_1''+\frac{1}{4}c_2'')\exp(\frac{1}{4}\pi i)}{\sqrt{p}}\exp\left(-\frac{i}{\hbar}\int_{x_2}^{x} p\,dx\right).$$

The condition that the solution is quasi-periodic is of the form:

$$-i(c_1''-\tfrac{1}{4}c_2'')\exp(-\tfrac{1}{4}\pi i) = zc_1,$$

$$-i(c_1''+\tfrac{1}{4}c_2'')\exp(\tfrac{1}{4}\pi i) = zc_2.$$

Expressing c_1'' and c_2'' in terms of c_1 and c_2 we obtain the following set of equations:

$$c_1\exp(i\alpha)[\exp(-\beta)+4\exp(\beta)]+ic_2\exp(-i\alpha)[\exp(-\beta)-4\exp(\beta)] = 4zc_1,$$

$$c_1\exp(i\alpha)[\exp(-\beta)-4\exp(\beta)]+ic_2\exp(-i\alpha)[\exp(-\beta)+4\exp(\beta)] = 4izc_2.$$

The value of z is determined from the condition that this set can be solved (that is, by putting its determinant equal to zero):

$$z^2-2z[\exp(\beta)+\tfrac{1}{4}\exp(-\beta)]\cos\alpha+1 = 0,$$

so that

$$z_{1,2} = [\exp(\beta)+\tfrac{1}{4}\exp(-\beta)]\cos\alpha \pm \sqrt{\{[\exp(\beta)+\tfrac{1}{4}\exp(-\beta)]^2\cos^2\alpha-1\}}. \tag{2}$$

If the expression under the square root is positive, then z is real and different from unity. In the opposite case, z is complex and its modulus is equal to unity,

$$z_{1,2} = [\exp(\beta)+\tfrac{1}{4}\exp(-\beta)]\cos\alpha \pm i\sqrt{\{1-[\exp(\beta)+\tfrac{1}{4}\exp(-\beta)]^2\cos^2\alpha\}},$$

$$|z_{1,2}|^2 = [\exp(\beta)+\tfrac{1}{4}\exp(-\beta)]^2\cos^2\alpha + \{1-[\exp(\beta)+\tfrac{1}{4}\exp(-\beta)]^2\cos^2\alpha\} = 1.$$

Thus the equation which determines the allowed bands follows from the condition that the expression under the square root in equation (2) is negative,

$$[\exp(\beta)+\tfrac{1}{4}\exp(-\beta)]^2\cos^2\alpha \leqslant 1. \tag{3}$$

The condition that the semi-classical approximation can be applied is that the expression $\int p\,dx$ is large compared to $\hbar$, that is,

$$\alpha \gg 1, \quad \beta \gg 1.$$

If, in accordance with this, we neglect the term with the negative exponent we can write equation (3) in the form

$$\exp(2\beta)\cos^2\alpha \leqslant 1$$

or

$$\cos^2\alpha \leqslant \exp(-2\beta). \tag{3'}$$

To a first approximation, we have

$$\cos^2\alpha = 0, \quad \text{or} \quad \alpha = n\pi + \tfrac{1}{2}\pi \; (\equiv \alpha_n).$$

From (1) we thus get

$$\int_{x_1}^{x_2} p\,dx = \pi\hbar(n+\tfrac{1}{2}).$$

This is the Bohr–Sommerfeld quantisation rule.

From the solution of this equation we get a set of levels in a well which is a hole in a periodic potential with impenetrable walls:

$$E_0, E_1, E_2, \ldots, E_n, \ldots.$$

We now expand the left-hand side of equation (3′) near the value α_n corresponding to the nth energy level

$$\cos^2\alpha \approx \sin^2\alpha_n(\Delta\alpha_n)^2 \approx (\Delta\alpha_n)^2.$$

From the definition of α we have

$$\Delta\alpha = \frac{\Delta E}{\hbar}\frac{\partial}{\partial E}\int_{x_1}^{x_2} p\,dx = \frac{\Delta E}{\hbar}\int_{x_1}^{x_2}\frac{\partial p}{\partial E}\,dx = \frac{\Delta E}{\hbar}\int_{x_1}^{x_2}\frac{dx}{v} = \frac{\Delta E}{\hbar}\frac{T}{2},$$

where T is the period of motion of the particle in the corresponding well-level.

If we note that $\exp(-2\beta)$ is the probability of transmission through the barrier, $W(E)$, we can write equation (3′) in the form

$$(\Delta E_n)^2 \leqslant \frac{4\hbar^2}{T_n^2} W(E_n),$$

so that we get for the band-width:

$$\Delta E_n = \frac{2\hbar}{T_n}\sqrt{[W(E_n)]} \ll \hbar\omega_n.$$

41. If we choose as a (normalised) trial function

$$\psi = \sqrt[4]{(2\alpha/\pi)}\exp(-\alpha x^2), \tag{1}$$

we find that the ground state energy E_0 must satisfy the inequality

$$E_0 \leqslant \int \psi^* \left[-\frac{\hbar^2}{2\mu}\frac{d^2}{dx^2} + V(x)\right]\psi\, dx, \tag{2}$$

where $V(x) < 0$.

Substituting (1) into (2) we obtain $E \leqslant E_\alpha$ with

$$E_\alpha = \frac{\hbar^2 \alpha}{2\mu} + \sqrt{\left(\frac{2\alpha}{\pi}\right)}\int \exp(-2\alpha x^2)\, V(x)\, dx. \tag{3}$$

The minimum value of E_α, $E_\alpha^{\min}$, is obtained by finding α from the condition

$$\frac{\partial E_\alpha}{\partial \alpha} = 0 = \frac{\hbar^2}{2\mu} + \sqrt{\left(\frac{1}{2\alpha\pi}\right)}\int \exp(-2\alpha x^2)\, V(x)\, dx$$
$$- \sqrt{\left(\frac{2\alpha}{\pi}\right)}\int 2x^2 \exp(-2\alpha x^2)\, V(x)\, dx. \tag{4}$$

Combining (3) and (4) we find

$$E_\alpha^{\min} = \sqrt{\left(\frac{4\alpha}{2\pi}\right)}\int \exp(-2\alpha x^2)(1+4\alpha x^2)\, V(x)\, dx < 0,$$

from which follows that $E_0 < 0$, as was to be shown.

42. If λ is small, we should expect (i) only one bound state and (ii) a wave function corresponding to the ground state which is varying only very slowly. In the region where the potential vanishes this wave function is of the form

$$\psi = A\exp(\alpha x), \quad x<0; \quad \psi = B\exp(-\alpha x), \quad a<x, \tag{1}$$

where

$$E = -\frac{\hbar^2}{2\mu}\alpha^2. \tag{2}$$

We may expect αa to be small compared to unity so that continuity of the wave function will (approximately) mean that we have

$$A = B = \sqrt{\alpha} \quad \text{and} \quad \psi = \sqrt{\alpha} \quad \text{for} \quad 0<x<a. \tag{3}$$

To find the value of α we take into account that $\alpha a \ll 1$ so that to a first approximation the potential may be replaced by a δ-function potential V': $V' = -b\delta(x)$, where

$$b = -\lambda \int_0^a V(x)\, dx. \tag{4}$$

We can then find α from the condition (compare problems 38 and 17 of section 1) that

$$\psi'(0+)-\psi'(0-) = \frac{2\mu}{\hbar^2} b\psi(0), \tag{5}$$

and using (1), (3), and (4) we get

$$\alpha = -\frac{\mu b}{\hbar^2} = \frac{\lambda\mu}{\hbar^2}\int_0^a V(x)\,dx,$$

and thus

$$E = -\frac{\lambda^2\mu}{2\hbar^2}\left[\int_0^a V(x)\,dx\right]^2.$$

Tunnel effect

1. In the region of the metal ($x < 0$) the general form of the wave function corresponding to the eigenvalue E is the following:

$$\psi_{\mathrm{I}} = b\,e^{i\kappa x} + c\,e^{-i\kappa x}, \quad \kappa = \sqrt{[2\mu(E+V_0)]}/\hbar.$$

In the region $x > 0$ the eigenfunction has the form of a travelling wave emerging from the metal

$$\psi_{\mathrm{II}} = a\,e^{ikx}, \quad \text{where} \quad k = \sqrt{(2\mu E)}/\hbar.$$

On the metal boundary the wave functions ψ_{I} and ψ_{II} and their derivatives must be continuous:

$$\psi_{\mathrm{II}}(0) = \psi_{\mathrm{I}}(0), \quad \text{or} \quad a = b + c,$$
$$\psi'_{\mathrm{II}}(0) = \psi'_{\mathrm{I}}(0), \quad \text{or} \quad ak = (b-c)\,\kappa.$$

The ratio of the current density of the reflected wave to the current density of the incoming wave gives us the reflection coefficient

$$R_0 = \left(\frac{\kappa - k}{\kappa + k}\right)^2 = \left(\frac{\sqrt{(E+V_0)} - \sqrt{E}}{\sqrt{(E+V_0)} + \sqrt{E}}\right)^2 = \frac{V_0^2}{[\sqrt{(E+V_0)} + \sqrt{E}]^4}.$$

If the electron energy is zero the reflection coefficient R_0 is equal to unity, while with increasing energy R decreases rapidly; for $E \gg V_0$

$$R_0 \approx \frac{V_0^2}{16E^2}.$$

In the other limiting case $E \ll V_0$

$$R_0 \approx 1 - 4\sqrt{(E/V_0)}.$$

For normal metals $V_0 \sim 10$ eV. The reflection coefficient for 0·1 eV electrons is then

$$R_0 = 0{\cdot}67.$$

2*. In the Schrödinger equation

$$-\frac{\hbar^2}{2\mu}\psi'' - \left(E + \frac{V_0}{e^{x/a}+1}\right)\psi = 0 \tag{1}$$

we make the substitution

$$\xi = -e^{-x/a}, \quad \psi = \xi^{-ika}\,u(\xi), \quad \text{where} \quad k = \sqrt{(2\mu E)}/\hbar.$$

For the function $u(\xi)$ we get the hypergeometric equation

$$\xi(1-\xi)u''+(1-2ika)(1-\xi)u'-\kappa_0^2 a^2 u = 0 \quad (\kappa_0 = \sqrt{(2\mu V_0)}/\hbar).$$

The solution of equation (1) which for $x\to\infty$ ($\xi\to 0$) is finite and behaves asymptotically as a travelling wave $c\,e^{ikx}$ has the form

$$\psi = c\,e^{ikx}F\{i(\kappa-k)a,\ -i(\kappa+k)a, 1-2ika,\ -e^{-x/a}\}$$

$$\{\kappa = \sqrt{[2\mu(E+V_0)]}/\hbar\}.$$

To find the reflection coefficient it is necessary to determine the form of the wave function inside the metal ($x\to-\infty$):

$$\psi \approx c\frac{\Gamma(1-2ika)\,\Gamma(-2ika)}{\Gamma[-i(\kappa+k)a]\,\Gamma[1-i(k+\kappa)a]}e^{i\kappa x} + c\frac{\Gamma(1-2ika)\,\Gamma(2ika)}{\Gamma[i(\kappa-k)a]\,\Gamma[1+i(\kappa-k)a]}e^{-i\kappa x}.$$

Hence we find for the reflection coefficient

$$R_a = \left|\frac{\Gamma(2ika)\,\Gamma[-i(\kappa+k)a]\,\Gamma[1-i(\kappa+k)a]}{\Gamma(-2ika)\,\Gamma[i(\kappa-k)a]\,\Gamma[1+i(\kappa+k)a]}\right|^2 = \frac{\sinh^2\pi a(\kappa-k)}{\sinh^2\pi a(\kappa+k)}.$$

To evaluate R_a one must use the following relations:

$$\Gamma(z+1) = z\Gamma(z),$$

$$\Gamma(z)\,\Gamma(1-z) = \pi/\sin\pi z,$$

$$\Gamma^*(ix) = \Gamma(-ix)$$

(x real). As $a\to 0$ the equation for the reflection coefficient goes over into the expression for R_0 in the case of a rectangular potential wall (see preceding problem).

It is easily verified that $R_a < R_0$, that is, that the reflection coefficient in the case of a smooth change of the potential is less than in the case of a sudden change. For $a = 1$ Å, $V_0 = 10$ eV, $E = 0{\cdot}1$ eV we find $R_a = 0{\cdot}23$.

3. We consider a flux of particles with energy $E < V_0$ moving from left to right. In the region III the wave function is an outgoing wave

$$\psi_{\mathrm{III}} = C\,e^{ikx}, \quad k = \sqrt{(2\mu E)}/\hbar.$$

In region I we have both an incoming and a reflected wave

$$\psi_{\mathrm{I}} = e^{ikx} + A\,e^{-ikx}.$$

In region II the general solution of the Schrödinger equation

$$-\frac{\hbar^2}{2\mu}\psi''_{\mathrm{II}}-(E-V_0)\psi_{\mathrm{II}}=0$$

has the form

$$\psi_{\mathrm{II}}=B_1\,e^{\kappa x}+B_2\,e^{-\kappa x},\quad \kappa=\sqrt{[2\mu(V_0-E)]}/\hbar.$$

The coefficients A, B_1, B_2, C are determined by the condition of continuity of the wave function and its first derivative.

At $x=0$ this condition leads to the relations:

$$1+A=B_1+B_2,$$

$$ik(1-A)=\kappa(B_1-B_2).$$

Similarly we have at $x=a$

$$B_1\,e^{\kappa a}+B_2\,e^{-\kappa a}=C\,e^{ika},$$

$$\kappa(B_1\,e^{\kappa a}-B_2\,e^{-\kappa a})=ikC\,e^{ika}.$$

From these equations we find

$$A=C\frac{k^2+\kappa^2}{4ik\kappa}e^{ika}(e^{\kappa a}-e^{-\kappa a})=C\frac{2\mu V_0}{\hbar^2}\frac{e^{\kappa a}-e^{-\kappa a}}{4ik\kappa}e^{ika},$$

$$B_1=C\,.\tfrac{1}{2}(1+ik/\kappa)\,e^{ika-\kappa a},$$

$$B_2=C\,.\tfrac{1}{2}(1-ik/\kappa)\,e^{ika+\kappa a},$$

$$C=-\frac{4ik}{\kappa}\frac{e^{-ika}}{e^{\kappa a}(1-ik/\kappa)^2-e^{-\kappa a}(1+ik/\kappa)^2}.$$

Since we have taken for the expression for the incoming wave in ψ_{I} the form e^{ikx}, the transmission coefficient is $D=CC^*$. Evaluating this we have

$$D=\frac{4k^2\kappa^2}{(k^2+\kappa^2)^2\sinh^2\kappa a+4k^2\kappa^2}.$$

We note that the transmission coefficient tends to zero when we go over to classical mechanics, that is, as $\hbar\to 0$.

If $\kappa a\gg 1$ (that is, $V_0-E\gg\hbar^2/2\mu a^2$), the expression for the transmission coefficient simplifies

$$D\approx 16\frac{E}{V_0}\left(1-\frac{E}{V_0}\right)\exp\left[-\frac{2\sqrt{[2\mu(V_0-E)]}}{\hbar}a\right].$$

We consider two concrete cases:

(a) An electron with $E=1$ eV passes through a potential barrier with $V_0=2$ eV and $a=1$ Å. We find for D the value 0·777.

(b) Now let a proton with the same energy fall on the same potential barrier. In this case one can show that the transmission coefficient becomes vanishingly small: $D=3{\cdot}6\,.\,10^{-19}$.

4. $$R = \frac{(k^2-\kappa^2)^2 \sin^2 \kappa a}{4k^2\kappa^2+(k^2-\kappa^2)^2\sin^2\kappa a} \quad \left(k = \frac{\sqrt{(2\mu E)}}{\hbar}, \quad \kappa = \frac{\sqrt{[2\mu(E-V_0)]}}{\hbar}\right).$$

5. We have $V = a\delta(x)$ and we must match the solutions for $x < 0$ and $x > 0$:

$$\psi_{\mathrm{I}} = e^{ikx} + A\,e^{-ikx}, \quad k = \frac{\sqrt{(2\mu E)}}{\hbar}, \quad x < 0;$$

$$\psi_{\mathrm{II}} = B\,e^{ikx}, \quad x > 0.$$

The boundary conditions are (compare problem 38 of section 1)

$$\psi_{\mathrm{I}}(0) = \psi_{\mathrm{II}}(0) \quad \text{and} \quad \psi'_{\mathrm{II}}(0) - \psi'_{\mathrm{I}}(0) = \frac{2\mu a}{\hbar^2}\psi(0).$$

We thus find

$$A = \frac{-iC}{1+iC}, \quad B = \frac{1}{1+iC}, \quad C = \frac{\mu a}{k\hbar^2},$$

and for the transmission coefficient we find

$$D = |B|^2 = \frac{k^2\hbar^4}{k^2\hbar^4 + \mu^2 a^2}.$$

6. The wave function has the following form:

$$\psi_{\mathrm{I}} = A \sin kx, \qquad 0 < x < a,$$

$$\psi_{\mathrm{II}} = B_1 e^{\kappa x} + B_2 e^{-\kappa x}, \qquad a < x < a+b,$$

$$\psi_{\mathrm{III}} = C \sin k(2a+b-x), \qquad a+b < x < 2a+b,$$

where

$$k = \sqrt{(2\mu E)}/\hbar, \quad \kappa = \sqrt{[2\mu(V_0 - E)]}/\hbar.$$

The requirement that both the wave function and its derivative are continuous leads to the following relations:

$$A \sin ka = B_1 e^{\kappa a} + B_2 e^{-\kappa a},$$

$$Ak \cos ka = \kappa(B_1 e^{\kappa a} - B_2 e^{-\kappa a}),$$

$$B_1 e^{\kappa(a+b)} + B_2 e^{-\kappa(a+b)} = C \sin ka,$$

$$\kappa(B_1 e^{\kappa(a+b)} - B_2 e^{-\kappa(a+b)}) = -Ck \cos ka.$$

Eliminating B_1 and B_2 from these equations, we find:

$$\left(\frac{\kappa}{k}\tan ka + 1\right) A\,e^{\kappa b} = \left(\frac{\kappa}{k}\tan ka - 1\right) C,$$

$$\left(\frac{\kappa}{k}\tan ka - 1\right) A\,e^{-\kappa b} = \left(\frac{\kappa}{k}\tan ka + 1\right) C.$$

From the requirement

$$\begin{vmatrix} \left(\frac{\kappa}{k}\tan ka+1\right)e^{\kappa b} & -\left(\frac{\kappa}{k}\tan ka-1\right) \\ \left(\frac{\kappa}{k}\tan ka-1\right)e^{-\kappa b} & -\left(\frac{\kappa}{k}\tan ka+1\right) \end{vmatrix} = 0$$

we get

$$\left(\frac{\kappa}{k}\tan ka+1\right)e^{\kappa b} = \pm\left(\frac{\kappa}{k}\tan ka-1\right).$$

This equation determines the energy levels.

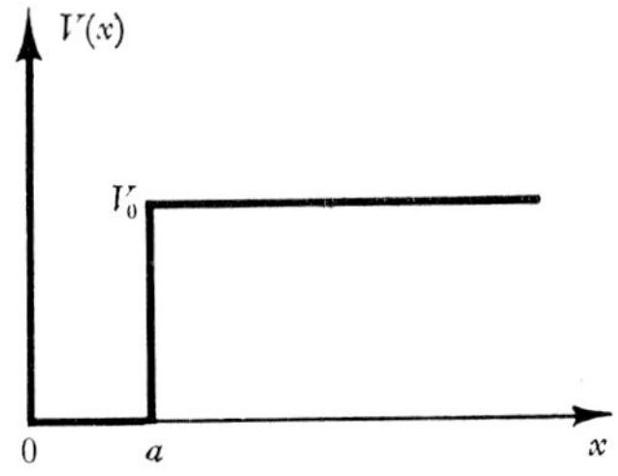

Fig. 28.

Using the inequality

$$\kappa b \gg 1,$$

the last equation can approximately be written in the form:

$$\tan ka = -\frac{k}{\kappa} \mp 2\frac{k}{\kappa}e^{-\kappa b}.$$

The right-hand side of this equation is a small quantity. In zeroth approximation ($k \ll \kappa$) we get

$$k_0 = \frac{n\pi}{a}, \quad E_n^{(0)} = \frac{n^2\pi^2\hbar^2}{2\mu a^2}$$

which is the value of the energy of a particle in an infinite potential well (see problem 1 of section 1). In the next approximation we have:

$$k = \frac{n\pi}{a} - \frac{k_0}{a\kappa_0} \mp 2\frac{k_0}{a\kappa_0}e^{-\kappa_0 b} \quad (n = 1, 2, 3, \ldots),$$

$$E_n = E_n^{(0)} - \frac{2E_n^{(0)}}{a\kappa_0} \mp 4\frac{E_n^{(0)}}{a\kappa_0}e^{-\kappa_0 b}, \quad \kappa_0 = \frac{\sqrt{[2\mu(V_0 - E_n^{(0)})]}}{\hbar}.$$

The first two terms $E_n^{(1)} = E_n^{(0)} - 2E_n^{(0)}/a\kappa_0$ do not depend on b and give the approximate values of the energy levels of a particle in the potential well depicted in fig. 28 ($b \to \infty$).

In the zeroth approximation the levels are twofold degenerate; this corresponds to the possibility of finding the particle either in region I or in region III. If we consider the case of finite b, we find that the possibility of transmission through the barrier leads to a splitting of the levels. This splitting is exponentially small. Let us find in the present approximation the coefficient A, B_1, B_2, and C. The lowest level

$$E_n^- = E_n^{(1)} - 4\frac{E_n^{(0)}}{a\kappa_0} e^{-\kappa_0 b}$$

corresponds to the following coefficients:

$$B_1 = (-1)^{n-1}\frac{k_0}{\kappa_0} e^{-\kappa_0(b+a)} A,$$
$$C = A,$$
$$B_2 = (-1)^{n-1}\frac{k_0}{\kappa_0} e^{\kappa_0 a} A.$$

The upper level

$$E_n^+ = E_n^{(1)} + 4\frac{E_n^{(0)}}{a\kappa_0} e^{-\kappa_0 b}$$

corresponds to the coefficients

$$B_1 = -(-1)^{n-1}\frac{k_0}{\kappa_0} e^{-\kappa_0(a+b)} A,$$
$$C = -A,$$
$$B_2 = (-1)^{n-1}\frac{k_0}{\kappa_0} e^{\kappa_0 a} A.$$

The value of A which follows from the normalisation condition is equal to $1/\sqrt{a}$ (in evaluating the normalising integral the contribution from region II can be neglected).

We find thus for the lower level the following wave function:

$$\psi_{\mathrm{I}} = \frac{1}{\sqrt{a}} \sin kx,$$
$$\psi_{\mathrm{II}} = (-1)^{n-1}\frac{1}{\sqrt{a}}\frac{k_0}{\kappa_0}[e^{-\kappa_0(x-a)} + e^{-\kappa_0(a+b-x)}],$$
$$\psi_{\mathrm{III}} = \frac{1}{\sqrt{a}} \sin k(2a+b-x).$$

For the upper level we find similarly:

$$\psi_{\mathrm{I}} = \frac{1}{\sqrt{a}} \sin kx,$$
$$\psi_{\mathrm{II}} = (-1)^{n-1}\frac{1}{\sqrt{a}}\frac{k_0}{\kappa_0}[e^{-\kappa_0(x-a)} - e^{-\kappa_0(a+b-x)}],$$
$$\psi_{\mathrm{III}} = -\frac{1}{\sqrt{a}} \sin k(2a+b-x).$$

In fig. 29 we have drawn the wave functions for $n = 1$ and $n = 4$.

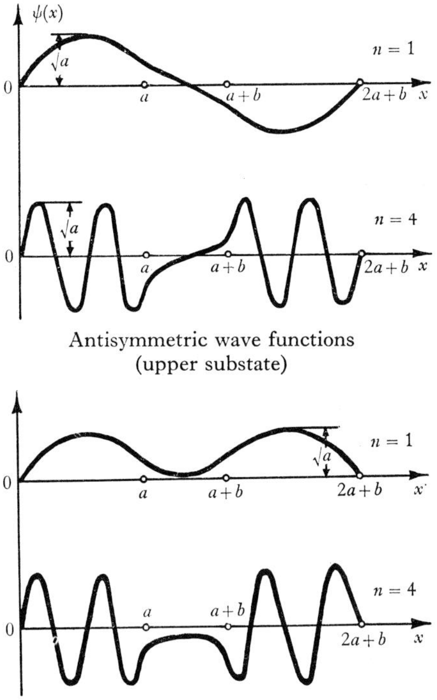

Antisymmetric wave functions
(upper substate)

Symmetric wave functions
(lower substate)

Fig. 29.

7*. The potential energy $V(x)$ is equal to $V_0(1+x/a)$ for $-a \leqslant x \leqslant 0$, to $V_0(1-x/a)$ for $0 \leqslant x \leqslant a$, and to zero for $|x| \geqslant a$. If we use the notation (E is the particle energy)

$$k_0^2 = \frac{2\mu E}{\hbar^2}, \quad \kappa_0^2 = \frac{2\mu V_0}{\hbar^2}, \tag{1}$$

we can write the Schrödinger equation in these regions as follows:

$$\frac{d^2\psi}{dx^2} + k_0^2\psi = 0 \quad (|x| \geqslant a), \tag{2}$$

$$\frac{d^2\psi}{dx^2} - \left(\kappa_0^2 - k_0^2 + \kappa_0^2\frac{x}{a}\right)\psi = 0 \quad (-a \leqslant x \leqslant 0), \tag{3}$$

$$\frac{d^2\psi}{dx^2} - \left(\kappa_0^2 - k_0^2 - \kappa_0^2\frac{x}{a}\right)\psi = 0 \quad (0 \leqslant x \leqslant a). \tag{4}$$

Making in equations (3) and (4) respectively the following change of variable,

$$\zeta = \left(\frac{a}{\kappa_0^2}\right)^{2/3}\left(\kappa_0^2 - k_0^2 + \kappa_0^2\frac{x}{a}\right), \tag{5}$$

$$\eta = \left(\frac{a}{\kappa_0^2}\right)^{2/3}\left(\kappa_0^2 - k_0^2 - \kappa_0^2\frac{x}{a}\right), \tag{6}$$

these equations become

$$\frac{d^2\psi}{d\eta^2} - \eta\psi = 0. \tag{4'}$$

$$\frac{d^2\psi}{d\zeta^2} - \zeta\psi = 0, \tag{3'}$$

The general solution of equations of the form (3′) or (4′) is a linear combination of two Airy functions v and u.

Assuming that the particle falls onto the barrier from the left (that is, there is only an outgoing wave for $x \geqslant a$) and taking the coefficient of the incoming wave equal to unity, we have thus the following wave function:

$$\left.\begin{array}{ll} \psi = e^{ik_0x} + A\,e^{-ik_0x} & (x \leqslant -a), \\ \psi = Bu(\zeta) + Cv(\zeta) & (-a \leqslant x \leqslant 0), \\ \psi = Du(\eta) + Ev(\eta) & (0 \leqslant x \leqslant a), \\ \psi = F\,e^{ik_0x} & (x \geqslant a). \end{array}\right\} \tag{7}$$

Connecting $\psi(x)$ and $d\psi/dx$ at the points $x = -a$, $x = 0$, and $x = a$, we get the following six equations for the six unknown constants A, B, C, D, E, F:

$$\left.\begin{array}{l} e^{-ik_0a} + A\,e^{ik_0a} = Bu(-\lambda) + Cv(-\lambda), \\ i\sqrt{(\lambda)}\,(e^{-ik_0a} - A\,e^{ik_0a}) = Bu'(-\lambda) + Cv'(-\lambda), \\ Bu(\mu) + Cv(\mu) = Du(\mu) + Ev(\mu), \\ Bu'(\mu) + Cv'(\mu) = -Du'(\mu) - Ev'(\mu), \\ Du(-\lambda) + Ev(-\lambda) = F\,e^{ik_0a}, \\ Du'(-\lambda) + Ev'(-\lambda) = -i\sqrt{(\lambda)}\,F\,e^{ik_0a}. \end{array}\right\} \tag{8}$$

We have used here the notation†

$$\lambda \equiv k_0^2\,a^{2/3}\,\kappa_0^{-4/3} = (\kappa_0 a)^{2/3}\left(\frac{k_0}{\kappa_0}\right)^2, \tag{9}$$

$$\mu \equiv a^{2/3}\,\kappa_0^{-4/3}(\kappa_0^2 - k_0^2) = (\kappa_0 a)^{2/3}\left(1 - \frac{k_0^2}{\kappa_0^2}\right). \tag{10}$$

† One should not confuse the mass μ and the dimensionless quantity μ defined by (10).

Since the velocity outside the barrier is the same to the left and to the right, the required transmission coefficient γ which is defined as the ratio of the current densities for $x > a$ and $x < -a$ is clearly equal to $|F|^2$. Solving equation (8) and assuming the Airy functions u and v to be normalised to satisfy

$$u'(t)\,v(t) - u(t)\,v'(t) = 1,$$

we find

$$\gamma = |F|^2 = \frac{\lambda}{[v(\mu)\,u'(-\lambda) - u(\mu)\,v'(-\lambda)]^2 + \lambda[v(\mu)\,u(-\lambda) - u(\mu)\,v(-\lambda)]^2}$$
$$\times \frac{1}{[v'(\mu)\,u'(-\lambda) - u'(\mu)\,v'(-\lambda)]^2 + \lambda[v'(\mu)\,u(-\lambda) - u'(\mu)\,v(-\lambda)]^2}. \tag{11}$$

Equations (11), (9), (10), and (1) together solve our problem. The functions $u(t)$ and $v(t)$ which are real for real values of t and their derivatives $u'(t)$ and $v'(t)$ have been tabulated.

We shall now find the limiting expressions for the penetrability of the barrier in the semi-classical case.

It is well known that one may consider the motion to be semi-classical provided the particle wavelength

$$\lambda\!\!\!^{-} = \frac{\hbar}{\sqrt{\{2\mu[E - V(x)]\}}}$$

changes little over a distance $\Delta x \sim \lambda\!\!\!^{-}$.

In the regions $x < -a$ and $x > a$ the motion is free and thus also semi-classical ($\lambda\!\!\!^{-} = 1/k_0 = \text{const}$, so that $d\lambda\!\!\!^{-}/dx = 0$). In the regions $-a \leqslant x \leqslant 0$ and $0 \leqslant x \leqslant a$ we have respectively from equations (1), (5), and (6):

$$\lambda\!\!\!^{-} = \left(\frac{a}{\kappa_0^2}\right)^{\frac{1}{3}} \frac{1}{i\sqrt{\zeta}}, \quad \frac{d\lambda\!\!\!^{-}}{dx} = \frac{d\lambda\!\!\!^{-}}{d\zeta}\frac{d\zeta}{dx} = -\frac{1}{2i\zeta^{\frac{3}{2}}},$$

and

$$\lambda\!\!\!^{-} = \left(\frac{a}{\kappa_0^2}\right)^{\frac{1}{3}} \frac{1}{i\sqrt{\eta}}, \quad \frac{d\lambda\!\!\!^{-}}{dx} = \frac{d\lambda\!\!\!^{-}}{d\eta}\frac{d\eta}{dx} = \frac{1}{2i\eta^{\frac{3}{2}}}.$$

The condition of semi-classicism $|d\lambda\!\!\!^{-}/dx| \ll 1$ leads thus to the requirement

$$|\zeta| \gg 1 \quad \text{and} \quad |\eta| \gg 1. \tag{12}$$

We consider a range of energy $E < V_0$ (that is, $k_0 < \kappa_0$, $\mu > 0$).

We denote the barrier width for an energy E by $2l$ and we have clearly:

$$l = a\frac{V_0 - E}{V_0} = a\left(1 - \frac{k_0^2}{\kappa_0^2}\right), \quad a - l = a\frac{k_0^2}{\kappa_0^2}. \tag{13}$$

Condition (12) is, of course, violated near the classical turning points $x = \pm l$ (where $\zeta = 0$, $\eta = 0$). One can, however, formulate a condition of "integral" semi-classicism (semi-classicism in the important region) which ensures a "semi-classical" penetrability of the barrier. For this one must obviously demand that the sections of non-semi-classical behaviour Δx (corresponding to the sections $\Delta\zeta \sim 1$, $\Delta\eta \sim 1$) are small compared to l and $(a-l)$.

We have

$$(\Delta x)_{\text{non-semicl}} = \frac{(\Delta\zeta)_{\text{non-semicl}}}{|d\zeta/dx|} \sim \left(\frac{a}{\kappa_0^2}\right)^{\frac{1}{3}}.$$

From equations (13), (9), and (10) it follows that the condition $(\Delta x)_{\text{non-semicl}} \ll l$ leads to the requirement that $\mu \gg 1$ and the condition $(\Delta x)_{\text{non-semicl}} \ll (a-l)$ to $\lambda \gg 1$.

We note that to satisfy the inequality $\lambda \gg 1$ it is in any case necessary that $k_0 a \equiv a/\lambda\!\!\!^{-}_0 \gg 1$, which can be seen from the definition (9) of λ if we write it in the form $\lambda = (k_0 a)^{\frac{2}{3}} (k_0/\kappa_0)^{\frac{4}{3}}$. We can convince ourselves in exactly the same way that for $E > V_0$ (that is, $k_0 > \kappa_0$, $\mu < 0$) the condition for semi-classicism is the inequality $|\mu| \gg 1$, and if that condition is fulfilled we have automatically $\lambda \gg 1$. It can be seen from equation (10) that the inequality $|\mu| \gg 1$ for $k_0 \gtrsim \kappa_0$ requires in any case that the above-mentioned "trivial" condition for semi-classicism $k_0 a \equiv a/\lambda\!\!\!^{-}_0 \gg 1$ is satisfied.

Thus, both for $E > V_0$ and for $E < V_0$ the semi-classical transmission coefficient γ can be obtained from equation (11) for $|\mu| \gg 1, \lambda \gg 1$, that is, by using asymptotic expansions for $u(t)$ and $v(t)$.

These asymptotic expansions have the following form (we take everywhere only the leading terms; $t > 0$ and $t \gg 1$, $x \equiv \frac{2}{3} t^{\frac{3}{2}}$):

$$\left.\begin{aligned} u(t) &\approx t^{-\frac{1}{4}} e^{x}, \quad u'(t) \approx t^{\frac{1}{4}} e^{x}; \\ v(t) &\approx \tfrac{1}{2} t^{-\frac{1}{4}} e^{-x}, \quad v'(t) \approx -\tfrac{1}{2} t^{\frac{1}{4}} e^{-x}; \end{aligned}\right\} \tag{14}$$

$$\left.\begin{aligned} u(-t) &\approx t^{-\frac{1}{4}} \cos\left(x + \frac{\pi}{4}\right), \\ u'(-t) &\approx t^{\frac{1}{4}} \sin\left(x + \frac{\pi}{4}\right); \\ v(-t) &\approx t^{-\frac{1}{4}} \sin\left(x + \frac{\pi}{4}\right), \\ v'(-t) &\approx -t^{\frac{1}{4}} \cos\left(x + \frac{\pi}{4}\right). \end{aligned}\right\} \tag{15}$$

We evaluate the semi-classical barrier penetrability for two cases:

I. $E < V_0, \quad \mu \gg 1, \quad \lambda \gg 1.$

II. $E > V_0, \quad \mu| \gg 1$ (and thus $\lambda \gg 1$).

In case I replacing $u(\mu), \ldots$ by equation (14) and $u(-\lambda) \ldots$ by equation (15) and dropping in the sums the exponentially small terms $\sim e^{-x}$ (note that taking these terms into account has in general no sense since we have neglected even exponentially large terms $\sim (1/x)\,e^{x}$ with respect to terms $\sim e^{x}$) we find easily

$$\gamma \approx \exp\left(-\tfrac{8}{3}\mu^{\frac{3}{2}}\right) = \exp\left[-\frac{8}{3}\,\frac{a\sqrt{(2\mu)}}{\hbar V_0}(V_0 - E)^{\frac{3}{2}}\right]. \tag{16}$$

As should be true in the semi-classical case $\gamma \ll 1$ (and $\gamma \to 0$ as $\hbar \to 0$).

Exactly the same result as equation (16) can, finally, be obtained also by using the well-known semi-classical formula

$$\gamma = \exp\left(-2\int_{-l}^{l}\sqrt{\left\{\frac{2\mu}{\hbar^2}[V(x) - E]\right\}}\,dx\right).$$

In case II replacing $u(\mu) = u(-|\mu|), \ldots$ and $u(-\lambda), \ldots$ by equation (15) we get in our present approximation the obvious result $\gamma \approx 1$.

We note in conclusion two rather trivial limiting cases.

If for fixed a and E, $V_0 \to 0$, we find $\lambda \to \infty$, $\mu \to -\infty$, so that according to case II $\gamma \to 1$.

If for fixed a and V_0, $E \to 0$, we find $\lambda \to 0$ so that always

$$\mu \to (\kappa_0 a)^{\frac{2}{3}}\,\gamma \to 0$$

[see equation (11)] as in the case of a barrier of any other shape.

8*. For the case $E < V_0$ the wave function can be obtained from the expressions for the wave functions of problem 14 of section 1 by changing the sign of E and V_0.

The general form of the wave function corresponding to an energy E is

$$\psi = c_1\left(\cosh\frac{x}{a}\right)^{-2\lambda} F\left(-\lambda + \frac{ika}{2}, -\lambda - \frac{ika}{2}, \tfrac{1}{2}; -\sinh^2\frac{x}{a}\right)$$

$$+ c_2\left(\cosh\frac{x}{a}\right)^{-2\lambda}\sinh\frac{x}{a}\,F\left(-\lambda + \frac{ika}{2} + \tfrac{1}{2}, -\lambda - \frac{ika}{2} + \tfrac{1}{2}, \tfrac{3}{2}; \sinh^2\frac{x}{a}\right), \tag{1}$$

where

$$\lambda = \tfrac{1}{4}[\sqrt{(1 - 8\mu V_0 a^2/\hbar^2)} - 1], \quad k = \sqrt{(2\mu E)}/\hbar.$$

The coefficients c_1 and c_2 are determined from the condition that as $x \to +\infty$ the wave function has the asymptotic form

$$\psi \sim e^{ikx}.$$

To find the asymptotic form of expression (1) we use the relation

$$F(\alpha,\beta,\gamma;\,z) = \frac{\Gamma(\gamma)\,\Gamma(\beta-\alpha)}{\Gamma(\beta)\,\Gamma(\gamma-\alpha)}(-z)^{-\alpha}\,F\left(\alpha,\alpha+1-\gamma,\alpha+1-\beta;\frac{1}{z}\right)$$

$$+\frac{\Gamma(\gamma)\,\Gamma(\alpha-\beta)}{\Gamma(\alpha)\,\Gamma(\gamma-\beta)}(-z)^{-\beta}\,F\left(\beta,\beta+1-\gamma,\beta+1-\alpha;\frac{1}{z}\right).$$

We find thus:

$$\psi_{x\to-\infty} \sim (-1)^{2\lambda}\,[(c_1 A_1 - c_2 A_2)(-\tfrac{1}{2})^{-ika}\,e^{ikx}$$

$$+(c_1 B_1 - c_2 B_2)(-\tfrac{1}{2})^{ika}\,e^{-ikx}], \qquad (2)$$

$$\psi_{x\to+\infty} \sim [(c_1 A_1 + c_2 A_2)(\tfrac{1}{2})^{-ika}\,e^{-ikx} + (c_1 B_1 + c_2 B_2)(\tfrac{1}{2})^{ika}\,e^{ikx}], \qquad (3)$$

where for the sake of simplicity we have introduced the notation

$$A_1 = \frac{\Gamma(\frac{1}{2})\,\Gamma(-ika)}{\Gamma(-\lambda-ika/2)\,\Gamma(\lambda+\frac{1}{2}-ika/2)},$$

$$A_2 = \frac{\Gamma(\frac{3}{2})\,\Gamma(-ika)}{\Gamma(-\lambda+\frac{1}{2}-ika/2)\,\Gamma(\lambda+1-ika/2)},$$

$$B_1 = \frac{\Gamma(\frac{1}{2})\,\Gamma(ika)}{\Gamma(-\lambda+ika/2)\,\Gamma(\lambda+\frac{1}{2}+ika/2)},$$

$$B_2 = \frac{\Gamma(\frac{3}{2})\,\Gamma(ika)}{\Gamma(-\lambda+\frac{1}{2}+ika/2)\,\Gamma(\lambda+1+ika/2)}.$$

The difference in sign of the coefficient c_2 in equations (2) and (3) is explained by the fact that $\sinh(x/a)$ is an odd function and the second term of expression (1) changes sign when we change from positive to negative values of x.

The requirement that at $+\infty$ there is only an outgoing wave leads to the following relation between c_1 and c_2:

$$c_1 A_1 + c_2 A_2 = 0.$$

The transmission coefficient will then be of the form:

$$D = \frac{|c_1 B_1 + c_2 B_2|^2}{|c_1 A_1 - c_2 A_2|^2}.$$

Substituting in this expression the values of the coefficients A_1, A_2, B_1, and B_2 and performing a simple transformation we find finally:

$$D = \frac{\sinh^2 \pi ka}{\sinh^2 \pi ka + \cos^2\{\frac{1}{2}\pi\sqrt{[1-(8\mu V_0 a^2)/\hbar^2]}\}}, \quad \text{for} \quad \frac{8\mu V_0 a^2}{\hbar^2} < 1$$

and

$$D = \frac{\sinh^2 \pi ka}{\sinh^2 \pi ka + \cosh^2 \{\frac{1}{2}\pi \sqrt{[(8\mu V_0 a^2)/\hbar^2 - 1]}\}}, \quad \text{for} \quad \frac{8\mu V_0 a^2}{\hbar^2} > 1.$$

9. If we denote the mass and the energy of the particle by μ and E ($E < V_0$) we have for the transmission (penetrability) coefficient in the semi-classical approximation the following well-known expression:

$$D \approx \exp\left(-2 \int_{x_1}^{x_2} \sqrt{\left\{\frac{2\mu}{\hbar^2}[V(x) - E]\right\}} dx\right), \tag{1}$$

where x_1 and x_2 are the coordinates of the classical turning points, which are determined from the equation $V(x_{1,2}) = E$,

$$x_2 = -x_1 = a\sqrt{\left(1 - \frac{E}{V_0}\right)}. \tag{2}$$

Substituting the potential $V(x)$ and equation (2) into equation (1), changing our variables in the integral by putting $x/x_2 = t$, and using the fact that

$$\int_{-1}^{1} \sqrt{(1 - t^2)}\, dt = \frac{\pi}{2},$$

we find finally

$$D \approx \exp\left[-\pi \sqrt{\left(\frac{2\mu}{V_0}\right)} \frac{a(V_0 - E)}{\hbar}\right]. \tag{3}$$

The criterion for the applicability of this formula is that the exponent is a large quantity,

$$\pi \sqrt{\left(\frac{2\mu}{V_0}\right)} \frac{a(V_0 - E)}{\hbar} \gg 1. \tag{4}$$

One can easily verify that if for the barrier under consideration condition (4) is satisfied we can at the same time validly put the factor which multiplies the exponential of expression (1) equal to unity.

Indeed, this factor is equal to unity in the case where the field $V(x)$ satisfies the condition of semi-classicism over all sections of the barrier, except those in the immediate neighbourhood of the turning points. This condition is of the form

$$\left|\frac{d\lambda}{dx}\right| \equiv \left|\frac{d}{dx}\left(\frac{\hbar}{\sqrt{\{2\mu[E - V(x)]\}}}\right)\right| \ll 1, \tag{5}$$

where λ is the de Broglie wavelength.

Inequality (5) can be transformed to

$$\left|\frac{t}{(t^2-1)^{\frac{3}{2}}}\right| \ll \sqrt{\left(\frac{2\mu}{V_0}\right)}\frac{a(V_0-E)}{\hbar}, \tag{5'}$$

with $t = x/x_2$.

From equation (4) it follows directly that the condition for semi-classicism (5) or (5′) is violated only near $t = \pm 1$ (that is, $x = \pm x_2$), while the width of the region of non-semi-classicism is of the order of magnitude

$$|\Delta x| \equiv |x - x_{1,2}| \sim \frac{a^{\frac{1}{3}}\hbar^{\frac{2}{3}}}{\mu^{\frac{1}{3}}(V_0-E)^{\frac{1}{3}}V_0^{\frac{1}{3}}}. \tag{6}$$

For the required "integral" semi-classicism of the field $V_0(x)$ we must clearly have:

$$|\Delta x| \ll a. \tag{7}$$

Using equations (6) and (4) we can easily check that condition (7) is satisfied, and thus that the factor in front of the exponential in equations (1) and (3) is equal to unity.

10. The potential energy of the electron is of the form depicted in fig. 14. The transmission coefficient is

$$D \approx \exp\left[-\frac{2}{\hbar}\int_0^{x_0}\sqrt{[2\mu(|E|-Fx)]}\,dx\right]$$

where the points $x = 0$, $x = x_0 = |E|/F$ limit the region which the particle according to classical mechanics cannot reach. Evaluating the integral in the exponent, we find

$$D \approx \exp\left[-\frac{4}{3}\frac{\sqrt{(2\mu)}}{\hbar F}|E|^{\frac{3}{2}}\right]. \tag{1}$$

To ascertain the limits of the applicability of this result, we note that semi-classical considerations cannot be used near the classical turning points x_0 over a region $x - x_0 \lesssim (\hbar^2/2\mu F)^{\frac{1}{3}}$. Equation (1) is applied when that region is less than the width of the barrier, $x_0 = |E|/F$,

$$\left(\frac{\hbar^2}{2\mu F}\right)^{\frac{1}{3}} \ll \frac{|E|}{F} \quad \text{or} \quad \frac{\sqrt{(2\mu)}}{\hbar F}|E|^{\frac{3}{2}} \gg 1.$$

This requirement is thus equivalent to the requirement that the transmission coefficient is small, $D \ll 1$.

The transmission coefficient D decreases steeply with increasing $|E$ and increases with increasing F (see table 1).

TABLE 1
Transmission coefficient

F	10^6	5×10^6	10^7	2×10^7	3×10^7	5×10^7	10^8 V/cm
			Without the electrical image force				
$E=-2$ V	10^{-84}	$1{\cdot}3\times10^{-17}$	$3{\cdot}5\times10^{-9}$	6×10^{-5}	$1{\cdot}5\times10^{-3}$	0·02	0·14
$E=-3$ V	10^{-154}	$1{\cdot}3\times10^{-31}$	$3{\cdot}5\times10^{-16}$	19×10^{-8}	7×10^{-6}	8×10^{-4}	0·029
$E=-5$ V	10^{-332}	4×10^{-67}	6×10^{-34}	$2{\cdot}5\times10^{-17}$	10^{-11}	$2{\cdot}5\times10^{-7}$	5×10^{-4}
			With the electrical image force				
$E=-2$ V	10^{-80}	8×10^{-15}	$1{\cdot}3\times10^{-6}$	0·013	1†	1†	1†
$E=-3$ V	10^{-150}	5×10^{-28}	7×10^{-14}	$2{\cdot}3\times10^{-6}$	7×10^{-4}	0·07	1†
$E=-5$ V	10^{-328}	8×10^{-65}	10^{-31}	2×10^{-15}	6×10^{-10}	10^{-5}	0·01

† When the transmission coefficient is equal to unity, the electron can leave the metal even according to classical mechanics.

11*. The total potential energy is equal to:

$$V = -Fx - \frac{e^2}{4x}.$$

One must note that this expression is incorrect for small values of x (of the order of magnitude of atomic distances). However, to evaluate the transmission coefficient, the exact shape of the potential in that region is inessential.

The transmission coefficient is

$$D \approx \exp\left\{-\frac{2}{\hbar}\int_{x_1}^{x_2} \sqrt{[2\mu(|E| - Fx - e^2/4x)]}\, dx\right\} = \exp\left(-\frac{2}{\hbar}\int_{x_1}^{x_2} p\, dx\right).$$

The turning points x_1 and x_2 follow from the condition that the classical particle momentum must vanish at x_1 and x_2,

$$p = \sqrt{[2\mu(|E| - Fx - e^2/4x)]} = 0,$$

$$x_{1,2} = \frac{|E| \pm \sqrt{(E^2 - e^2 F)}}{2F};$$

the integral

$$\int_{x_1}^{x_2} p\, dx = \sqrt{(2\mu)} \int_{x_1}^{x_2} \sqrt{(|E| - Fx - e^2/4x)}\, dx$$

is a complete elliptic integral. Taking as the independent variable

$(F/|E|)x = \xi$ this integral is reduced to a function of one parameter:

$$\int_{x_1}^{x_2} p\,dx = \tfrac{2}{3}\sqrt{(2\mu)}\frac{|E|^{\frac{3}{2}}}{F}\varphi(y),$$

$$y = \frac{\sqrt{(e^2 F)}}{2|E|}.$$

Here

$$\varphi(y) = \frac{3}{2}\int_{\xi_1}^{\xi_2}\sqrt{\left(1-\frac{y^2}{\xi}-\xi\right)}\,d\xi,$$

and the limits of integration ξ_1 and ξ_2 are the roots of the expression under the square root sign. Introducing the notation

$$k_0 = \tfrac{4}{3}\sqrt{(2\mu)}\frac{|E|^{\frac{3}{2}}}{\hbar F},$$

we get $D = \exp[-k_0\varphi(y)]$.

We note that the transmission coefficient becomes $D = \exp(-k_0)$ if we neglect the electrical image force ($y = 0$; see problem 10). The values of $\varphi(y)$ are given in table 2. The influence of the electrical image force on the coefficient of transmission through a barrier can be seen from table 1.

TABLE 2

y	0	0·2	0·3	0·4	0·5	0·6	0·7	0·8	0·9	1·0
$\varphi(y)$	1·000	0·951	0·904	0·849	0·781	0·696	0·603	0·494	0·345	0·000

12*. For the wave function in the region $x < -b$ we have

$$\psi = \frac{c}{\sqrt{|p|}}\exp\left(-\frac{1}{\hbar}\int_x^{-b}|p|\,dx\right)$$

(the solution must tend to zero at infinity). In the region $-b<x<-a$ we have

$$\psi = \frac{c}{\sqrt{p}}\exp\left(-\frac{i\pi}{4}+\frac{i}{\hbar}\int_{-b}^{x}p\,dx\right)+\frac{c}{\sqrt{p}}\exp\left(\frac{i\pi}{4}-\frac{i}{\hbar}\int_{-b}^{x}p\,dx\right)$$

$$= \frac{c}{\sqrt{p}}\exp\left(-\frac{i\pi}{4}+\frac{i}{\hbar}\int_{-b}^{-a}p\,dx-\frac{i}{\hbar}\int_{x}^{-a}p\,dx\right)$$

$$+\frac{c}{\sqrt{p}}\exp\left(\frac{i\pi}{4}-\frac{i}{\hbar}\int_{-b}^{-a}p\,dx+\frac{i}{\hbar}\int_{x}^{-a}p\,dx\right).$$

In the region $-a<x<+a$ we have

$$\psi = \frac{c}{\sqrt{|p|}}\exp\left(-\frac{i\pi}{4}+\frac{i}{\hbar}\int_{-b}^{-a}p\,dx\right)\left[\tfrac{1}{2}\exp\left(-\frac{i\pi}{4}-\frac{1}{\hbar}\int_{-a}^{x}|p|\,dx\right)\right.$$
$$\left.+\exp\left(\frac{i\pi}{4}+\frac{1}{\hbar}\int_{-a}^{x}|p|\,dx\right)\right]$$
$$+\frac{c}{\sqrt{|p|}}\exp\left(\frac{i\pi}{4}-\frac{i}{\hbar}\int_{-b}^{-a}p\,dx\right)\left[\tfrac{1}{2}\exp\left(\frac{i\pi}{4}-\frac{1}{\hbar}\int_{-a}^{x}|p|\,dx\right)\right.$$
$$\left.+\exp\left(-\frac{i\pi}{4}+\frac{1}{\hbar}\int_{-a}^{x}|p|\,dx\right)\right]$$
$$=\frac{c}{\sqrt{|p|}}\sin\left(\frac{1}{\hbar}\int_{-b}^{-a}p\,dx\right)\exp\left(-\frac{1}{\hbar}\int_{-a}^{+a}|p|\,dx+\frac{1}{\hbar}\int_{x}^{+a}|p|\,dx\right)$$
$$+\frac{2c}{\sqrt{|p|}}\cos\left(\frac{1}{\hbar}\int_{-b}^{-a}p\,dx\right)\exp\left(\frac{1}{\hbar}\int_{-a}^{+a}|p|\,dx-\frac{1}{\hbar}\int_{x}^{+a}|p|\,dx\right).$$

For $+a<x<+b$ we find similarly

$$\psi = \frac{c}{\sqrt{p}}\sin\left(\frac{1}{\hbar}\int_{-b}^{-a}p\,dx\right)\exp\left(-\frac{1}{\hbar}\int_{-a}^{+a}|p|\,dx\right)$$
$$\times\left[\tfrac{1}{2}\exp\left(-\frac{i\pi}{4}-\frac{i}{\hbar}\int_{+a}^{x}p\,dx\right)+\tfrac{1}{2}\exp\left(\frac{i\pi}{4}+\frac{i}{\hbar}\int_{+a}^{x}p\,dx\right)\right]$$
$$+\frac{2c}{\sqrt{p}}\cos\left(\frac{1}{\hbar}\int_{-b}^{-a}p\,dx\right)\exp\left(\frac{1}{\hbar}\int_{-a}^{+a}|p|\,dx\right)$$
$$\times\left[\exp\left(-\frac{i\pi}{4}+\frac{i}{\hbar}\int_{+a}^{x}p\,dx\right)+\exp\left(\frac{i\pi}{4}-\frac{i}{\hbar}\int_{a}^{x}p\,dx\right)\right]$$
$$=\frac{c}{\sqrt{p}}\left[\tfrac{1}{2}\sin\left(\frac{1}{\hbar}\int_{-b}^{-a}p\,dx\right)\exp\left(-\frac{1}{\hbar}\int_{-a}^{+a}|p|\,dx-\frac{i\pi}{4}\right)\right.$$
$$\left.+2\cos\left(\frac{1}{\hbar}\int_{-a}^{+a}p\,dx\right)\exp\left(\frac{1}{\hbar}\int_{-a}^{+a}|p|\,dx+\frac{i\pi}{4}\right)\right]$$
$$\times\exp\left(-\frac{i}{\hbar}\int_{+a}^{+b}p\,dx+\frac{i}{\hbar}\int_{x}^{b}p\,dx\right)$$
$$+\frac{c}{\sqrt{p}}\left[\tfrac{1}{2}\sin\left(\frac{1}{\hbar}\int_{-b}^{-a}p\,dx\right)\exp\left(-\frac{1}{\hbar}\int_{-a}^{+a}|p|\,dx+\frac{i\pi}{4}\right)\right.$$
$$\left.+2\cos\left(\frac{1}{\hbar}\int_{-a}^{+a}p\,dx\right)\exp\left(\frac{1}{\hbar}\int_{-a}^{+a}|p|\,dx-\frac{i\pi}{4}\right)\right]$$
$$\times\exp\left(\frac{i}{\hbar}\int_{+a}^{+b}p\,dx-\frac{i}{\hbar}\int_{x}^{+b}p\,dx\right).$$

Finally, extending this solution to the region $x > +b$ we have:

$$\psi = \frac{c}{\sqrt{|p|}}\left[\tfrac{1}{2}\sin\left(\frac{1}{\hbar}\int_a^b p\,dx\right)\cos\left(\frac{1}{\hbar}\int_a^b p\,dx\right)\exp\left(-\frac{1}{\hbar}\int_{-a}^{+a}|p|\,dx\right)\right.$$

$$\left.+2\cos\left(\frac{1}{\hbar}\int_a^b p\,dx\right)\sin\left(\frac{1}{\hbar}\int_a^b p\,dx\right)\exp\left(\frac{1}{\hbar}\int_{-a}^{+a}|p|\,dx\right)\right]$$

$$\times\exp\left(-\frac{1}{\hbar}\int_b^x|p|\,dx\right)$$

$$+\frac{c}{\sqrt{|p|}}\left[-\sin^2\left(\frac{1}{\hbar}\int_a^b p\,dx\right)\exp\left(-\frac{1}{\hbar}\int_{-a}^{+a}|p|\,dx\right)\right.$$

$$\left.+4\cos^2\left(\frac{1}{\hbar}\int_a^b p\,dx\right)\exp\left(\frac{1}{\hbar}\int_{-a}^{+a}|p|\,dx\right)\right]$$

$$\times\exp\left(\frac{1}{\hbar}\int_b^x|p|\,dx\right).$$

In order that this solution tends to zero as $x\to+\infty$ it is necessary that the coefficient

$$\exp\left(\frac{1}{\hbar}\int_b^x|p|\,dx\right)$$

tends to zero, that is,

$$-\sin^2\left(\frac{1}{\hbar}\int_a^b p\,dx\right)\exp\left(-\frac{1}{\hbar}\int_{-a}^{+a}|p|\,dx\right)$$

$$+4\cos^2\left(\frac{1}{\hbar}\int_a^b p\,dx\right)\exp\left(\frac{1}{\hbar}\int_{-a}^{+a}|p|\,dx\right)=0,$$

or

$$\cot\left(\frac{1}{\hbar}\int_a^b p\,dx\right)=\pm\tfrac{1}{2}\exp\left(-\frac{1}{\hbar}\int_{-a}^{+a}|p|\,dx\right).$$

Assuming the transparency of the barrier to be small, we get the following condition to determine the energy levels:

$$\frac{1}{\hbar}\int_a^b p\,dx=\pi(n+\tfrac{1}{2})\pm\tfrac{1}{2}\exp\left(-\frac{1}{\hbar}\int_{-a}^{+a}|p|\,dx\right).$$

We shall denote by $E_n^{(0)}$ the energy levels of a separate potential well

$$\frac{1}{\hbar}\int_a^b\sqrt{[2\mu(E_n^{(0)}-V)]}\,dx=\pi(n+\tfrac{1}{2}).$$

The energy levels in a double well, $E_n = E_n^{(0)} + \Delta E_n$, are found from the quantisation rule found a moment ago by expanding $\sqrt{[2\mu(E_n - V)]}$ in terms of ΔE_n and breaking off at the term linear in ΔE_n:

$$\Delta E_n \frac{\mu}{\hbar} \int_a^b \frac{dx}{\sqrt{[2\mu(E_n^{(0)} - V)]}} = \pm \tfrac{1}{2} \exp\left(-\frac{1}{\hbar}\int_{-a}^{+a} |p|\, dx\right)$$

or

$$\Delta E_n = \pm \frac{\hbar\omega}{2\pi} \exp\left(-\frac{1}{\hbar}\int_{-a}^{+a} |p|\, dx\right),$$

where ω is the angular frequency of the classical motion in a separate well:

$$\frac{2\pi}{\omega} = 2\mu \int_a^b \frac{dx}{p}.$$

The splitting of the level E_n is equal to $2|\Delta E_n|$.

13.
$$\tau = \frac{\pi^2}{\omega} \exp\left(\frac{1}{\hbar}\int_{-a}^{+a} |p|\, dx\right).$$

14*. We write the wave function in the region of the nth potential barrier $b_n < x < a_{n+1}$ in the form

$$\begin{aligned}\psi &= \frac{C_n}{\sqrt{|p|}} \exp\left(-\frac{1}{\hbar}\int_{b_n}^{x} |p|\, dx\right) + \frac{D_n}{\sqrt{|p|}} \exp\left(\frac{1}{\hbar}\int_{b_n}^{x} |p|\, dx\right)\\ &= \frac{C_n}{\sqrt{|p|}} \exp\left(-\frac{1}{\hbar}\int_{b_n}^{a_{n+1}} |p|\, dx + \frac{1}{\hbar}\int_{x}^{a_{n+1}} |p|\, dx\right)\\ &\quad + \frac{D_n}{\sqrt{|p|}} \exp\left(\frac{1}{\hbar}\int_{b_n}^{a_{n+1}} |p|\, dx - \frac{1}{\hbar}\int_{x}^{a_{n+1}} |p|\, dx\right);\end{aligned}$$

if we extend this function into the region of the $(n+1)$th potential barrier $b_{n+1} < x < a_{n+2}$, we get

$$\begin{aligned}\psi = \frac{1}{\sqrt{|p|}} &\exp\left(-\frac{1}{\hbar}\int_{b_{n+1}}^{x} |p|\, dx\right) \left[\frac{C_n}{2} \exp\left(-\frac{1}{\hbar}\int_{b_n}^{a_{n+1}} |p|\, dx\right) \cos\left(\frac{1}{\hbar}\int_{a_{n+1}}^{b_{n+1}} p\, dx\right)\right.\\ &\left. + D_n \exp\left(\frac{1}{\hbar}\int_{b_n}^{a_{n+1}} |p|\, dx\right) \sin\left(\frac{1}{\hbar}\int_{a_{n+1}}^{b_{n+1}} p\, dx\right)\right]\\ + \frac{1}{\sqrt{|p|}} &\exp\left(\frac{1}{\hbar}\int_{b_{n+1}}^{x} |p|\, dx\right) \left[-C_n \exp\left(-\frac{1}{\hbar}\int_{b_n}^{a_{n+1}} |p|\, dx\right) \sin\left(\frac{1}{\hbar}\int_{a_{n+1}}^{b_{n+1}} p\, dx\right)\right.\\ &\left. + 2D_n \exp\left(\frac{1}{\hbar}\int_{b_n}^{a_{n+1}} |p|\, dx\right) \cos\left(\frac{1}{\hbar}\int_{a_{n+1}}^{b_{n+1}} p\, dx\right)\right].\end{aligned}$$

We introduce the following notation:

$$\frac{1}{\hbar}\int_{b_1}^{a_2}|p|\,dx = \frac{1}{\hbar}\int_{b_2}^{a_3}|p|\,dx = \ldots = \frac{1}{\hbar}\int_{b_{N-1}}^{a_N}|p|\,dx = \tau,$$

$$\frac{1}{\hbar}\int_{a_1}^{b_1}p\,dx = \frac{1}{\hbar}\int_{a_2}^{b_2}p\,dx = \ldots = \frac{1}{\hbar}\int_{a_N}^{b_N}p\,dx = \sigma.$$

The earlier expression for ψ in the region of the $(n+1)$th barrier is then transformed to the form

$$\psi = \frac{1}{\sqrt{|p|}}\exp\left(-\frac{1}{\hbar}\int_{b_{n+1}}^{x}|p|\,dx\right)\left[\frac{C_n}{2}\exp(-\tau)\cos\sigma + D_n\exp(\tau)\sin\sigma\right]$$
$$+\frac{1}{\sqrt{|p|}}\exp\left(\frac{1}{\hbar}\int_{b_{n+1}}^{x}|p|\,dx\right)[-C_n\exp(-\tau)\sin\sigma + 2D_n\exp(\tau)\cos\sigma]$$
$$= \frac{C_{n+1}}{\sqrt{|p|}}\exp\left(-\frac{1}{\hbar}\int_{b_{n+1}}^{x}|p|\,dx\right) + \frac{D_{n+1}}{\sqrt{|p|}}\exp\left(\frac{1}{\hbar}\int_{b_{n+1}}^{x}|p|\,dx\right),$$

where

$$C_{n+1} = \tfrac{1}{2}C_n\,e^{-\tau}\cos\sigma + D_n\,e^{\tau}\sin\sigma,$$
$$D_{n+1} = -C_n\,e^{-\tau}\sin\sigma + 2D_n\,e^{\tau}\cos\sigma.$$

The connection between the coefficients C_{n+1} and D_{n+1} and C_n and D_n can conveniently be written in matrix form:

$$\begin{pmatrix}C_{n+1}\\ D_{n+1}\end{pmatrix} = \begin{pmatrix}\frac{1}{2}e^{-\tau}\cos\sigma & e^{\tau}\sin\sigma\\ -e^{-\tau}\sin\sigma & 2\,e^{\tau}\cos\sigma\end{pmatrix}\begin{pmatrix}C_n\\ D_n\end{pmatrix} = A\begin{pmatrix}C_n\\ D_n\end{pmatrix}. \quad (1)$$

Applying relation (1) N times in succession we get a connection between C_N, D_N and C_0, D_0,

$$\begin{pmatrix}C_N\\ D_N\end{pmatrix} = \begin{pmatrix}\frac{1}{2}e^{-\tau}\cos\sigma & e^{\tau}\sin\sigma\\ -e^{-\tau}\sin\sigma & 2e^{\tau}\cos\sigma\end{pmatrix}^N\begin{pmatrix}C_0\\ D_0\end{pmatrix} = A^N\begin{pmatrix}C_0\\ D_0\end{pmatrix}.$$

The wave function of a stationary state must decrease for both $x < a_1$ and $x > b_N$ so that we must require that $C_0 = D_N = 0$. It is easily seen that this entails that the matrix element $(A^N)_{22}$ is equal to zero. The condition $(A^N)_{22} = 0$ determines the energy spectrum of our problem. To evaluate this matrix element we consider the matrix

$$S = e^{At} = 1 + tA + \frac{t^2}{2!}A^2 + \frac{t^3}{3!}A^3 + \ldots + \frac{t^N}{N!}A^N + \ldots.$$

It can easily be shown directly that the matrix S satisfies the equation

$$\frac{dS}{dt} = AS \quad (2)$$

with the initial condition

$$S(0) = 1.$$

We write equation (2) in more detail

$$\frac{d}{dt}\begin{pmatrix} S_{11} & S_{12} \\ S_{21} & S_{22} \end{pmatrix} = \begin{pmatrix} \alpha & \beta \\ \gamma & \delta \end{pmatrix}\begin{pmatrix} S_{11} & S_{12} \\ S_{21} & S_{22} \end{pmatrix}$$

or

$$\begin{cases} \dfrac{dS_{11}}{dt} = \alpha S_{11} + \beta S_{21}, \\ \dfrac{dS_{21}}{dt} = \gamma S_{11} + \delta S_{21}, \end{cases} \qquad \begin{cases} \dfrac{dS_{12}}{dt} = \alpha S_{12} + \beta S_{22}, \\ \dfrac{dS_{22}}{dt} = \gamma S_{12} + \delta S_{22}. \end{cases}$$

Since the condition which determines the energy spectrum can be written in the form

$$(A^N)_{22} = \left(\frac{d^N S_{22}}{dt^N}\right)_{t=0} = 0,$$

it is sufficient to consider the second pair of equations.

If we put $S_{12} = f e^{\lambda t}$, $S_{22} = g e^{\lambda t}$, we get

$$f\lambda = \alpha f + \beta g,$$
$$g\lambda = \gamma f + \delta g.$$

The value of λ is determined from the equation

$$\begin{vmatrix} \alpha - \lambda & \beta \\ \gamma & \delta - \lambda \end{vmatrix} = 0, \quad \lambda^2 - \lambda(2\,e^{\tau} + \tfrac{1}{2}\,e^{-\tau})\cos\sigma + 1 = 0,$$

which gives two roots λ_1 and λ_2. Since $\lambda_1\lambda_2 = 1$ we can write $\lambda_{1,2}$ in the form

$$\lambda_{1,2} = e^{\pm iu},$$

where

$$\cos u = (e^{\tau} + \tfrac{1}{4}\,e^{-\tau})\cos\sigma.$$

If $\lambda_1 \neq \lambda_2$, the solution which satisfies the initial conditions $S_{12}(0) = 0$, $S_{22}(0) = 1$ has the form:

$$S_{12} = \beta(e^{\lambda_1 t} - e^{\lambda_2 t})/(\lambda_1 - \lambda_2),$$

$$S_{22} = \frac{(\lambda_1 - \alpha)\,e^{\lambda_1 t} - (\lambda_2 - \alpha)\,e^{\lambda_2 t}}{\lambda_1 - \lambda_2}.$$

The condition which determines the energy spectrum of our problem can now be written as follows:

$$(A^N)_{22} = \left(\frac{d^N S_{22}}{dt^N}\right)_{t=0} = \frac{1}{\lambda_1 - \lambda_2}\{(\lambda_1 - \alpha)\,\lambda_1^N - (\lambda_2 - \alpha)\,\lambda_2^N\} = 0.$$

We can substitute in this expression the values:

$$\lambda_{1,2} = e^{\pm iu}$$

and neglect $e^{-\tau}$ in the equation which determines $\cos u$ (this is equivalent to assuming that the penetrability is small):

$$\cos u \approx e^{\tau} \cos \sigma.$$

Under this assumption the condition determining the energy levels becomes very simply:

$$\frac{\sin (N+1) u}{\sin u} = 0.$$

This equation has the following roots:

$$u = \frac{n\pi}{N+1},$$

with $u = 0, \pi, 2\pi$ forbidden.

Thus, $\cos u$ has N different values,

$$\cos u = \cos \frac{n\pi}{N+1} \approx e^{\tau} \cos \sigma \quad (n = 1, 2, \ldots, N).$$

More explicitly,

$$\cos\left(\frac{1}{\hbar}\int_{a_1}^{b_1} p\,dx\right) = \exp\left(-\frac{1}{\hbar}\int_{b_1}^{a_2} |p|\,dx\right) \cos \frac{\pi n}{N+1} \quad (n = 1, 2, \ldots, N).$$

Since

$$\exp\left(-\frac{1}{\hbar}\int_{b_1}^{a_2} |p|\,dx\right)$$

is a small quantity, this last equation can be rewritten in the form:

$$\frac{1}{\hbar}\int_{a_1}^{b_1} p\,dx = \pi(m+\tfrac{1}{2}) + \exp\left(-\frac{1}{\hbar}\int_{b_1}^{a_2} |p|\,dx\right) \cos \frac{\pi n}{N+1}$$

$$(m = 0, 1, 2, \ldots) \quad (n = 1, 2, \ldots, N). \qquad (3)$$

This is the condition which determines the energy levels in the field $V(x)$. It is very similar to the quantisation condition for the field of a separate well. By considering equation (3) we can conclude that the energy spectrum in the field $V(x)$ is, roughly speaking, the energy spectrum of a separate well, with all its levels split into N sublevels. Let us determine the value of the shift, ΔE_m,

$$\frac{1}{\hbar}\int_{a_1}^{b_1} \sqrt{[2\mu(E_m^{(0)} - V)]}\,dx + \frac{\mu}{\hbar}\int_{a_1}^{b_1} \frac{dx}{\sqrt{[2\mu(E_m^{(0)} - V)]}} \Delta E_m$$

$$= \pi(m+\tfrac{1}{2}) + \exp\left(-\frac{1}{\hbar}\int_{b_1}^{a_2} |p|\,dx\right) \cos \frac{\pi n}{N+1} \quad (n = 1, 2, \ldots, N);$$

if we now introduce the notation

$$\frac{\pi}{\omega} = \mu \int_{a_1}^{b_1} \frac{dx}{p} = \mu \int_{a_1}^{b_1} \frac{dx}{\sqrt{[2\mu(E_m^{(0)} - V)]}},$$

we find

$$\Delta E_m = \frac{\hbar\omega}{\pi} \exp\left(-\frac{1}{\hbar}\int_{b_1}^{a_2} |p|\, dx\right) \cos\frac{n\pi}{N+1} \quad (n = 1, 2, \ldots, N).$$

The distance between the highest and lowest sublevel is equal to:

$$\frac{2\hbar\omega}{\pi} \exp\left(-\frac{1}{\hbar}\int_{b_1}^{a_2} |p|\, dx\right) \cos\frac{\pi}{N+1}.$$

15*. In the region $x < -b$ we have under the circumstances of our problem only a wave which goes to $-\infty$, or

$$\psi = \frac{c}{\sqrt{p}} \exp\left(\frac{i}{\hbar}\int_x^{-b} p\, dx\right).$$

If we extend this solution into the region $x > b$, we get the following expression for the wave function:

$$\begin{aligned}\psi = \frac{c}{\sqrt{p}} \exp\left(\frac{i}{\hbar}\int_b^x p\, dx\right) &\left[\tfrac{1}{8} \exp\left(-\frac{2}{\hbar}\int_a^b |p|\, dx + i\frac{\pi}{2}\right) \cos\left(\frac{1}{\hbar}\int_{-a}^{+a} p\, dx\right)\right.\\ &\left. + 2\exp\left(\frac{2}{\hbar}\int_a^b |p|\, dx - i\frac{\pi}{2}\right) \cos\left(\frac{1}{\hbar}\int_{-a}^{+a} p\, dx\right)\right]\\ &+ \frac{c}{\sqrt{p}} \exp\left(-\frac{i}{\hbar}\int_b^x p\, dx\right)\\ \times \left[\tfrac{1}{8} \exp\left(-\frac{2}{\hbar}\int_a^b |p|\, dx\right) \cos\left(\frac{1}{\hbar}\int_{-a}^{+a} p\, dx\right)\right. &- i \sin\left(\frac{1}{\hbar}\int_{-a}^{+a} p\, dx\right)\\ &\left. + 2\exp\left(\frac{2}{\hbar}\int_a^b |p|\, dx\right) \cos\left(\frac{1}{\hbar}\int_{-a}^{+a} p\, dx\right)\right].\end{aligned}$$

The quasi-stationary level is determined by the condition that there is no wave coming from $+\infty$.

If we put the second term in the last expression equal to zero, we get

$$\cot\left(\frac{1}{\hbar}\int_{-a}^{+a} p\, dx\right) = i\left[\tfrac{1}{8} \exp\left(-\frac{2}{\hbar}\int_a^b |p|\, dx\right) + 2\exp\left(\frac{2}{\hbar}\int_a^b |p|\, dx\right)\right]^{-1}.$$

Considering

$$\exp\left(-\frac{2}{\hbar}\int_a^b |p|\, dx\right)$$

to be a small quantity, we find that

$$\frac{1}{\hbar}\int_{-a}^{+a} p\,dx = \pi(n+\tfrac{1}{2}) - \frac{i}{2}\exp\left(-\frac{2}{\hbar}\int_a^b |p|\,dx\right),$$

from which follows the condition to determine the quasi-stationary levels $E_n^{(0)}$ and their width Γ,

$$\frac{1}{\hbar}\int_{-a}^{+a}\sqrt{[2\mu(E_n^0 - V)]}\,dx = (n+\tfrac{1}{2})\pi \quad (n = 0, 1, 2, \ldots),$$

$$\frac{\hbar\omega}{2\pi}\exp\left(-\frac{2}{\hbar}\int_a^b |p|\,dx\right) = \Gamma,$$

where

$$\omega = \pi\left(\mu\int_a^b \frac{dx}{\sqrt{[2\mu(E_n^0 - V)]}}\right)^{-1}.$$

We find the following value for the transmission coefficient:

$$D(E) = \left[4\exp\left(\frac{4}{\hbar}\int_a^b |p|\,dx\right)\cos^2\left(\frac{1}{\hbar}\int_{-a}^{+a} p\,dx\right) + \sin^2\left(\frac{1}{\hbar}\int_{-a}^{+a} p\,dx\right)\right]^{-1}.$$

For values of E coinciding with one of the quasi-levels we have $D(E_n^0) = 1$. For $|\Delta E| < |E_n^0|$ we have

$$D(E_n^0 + \Delta E) = \frac{\Gamma^2}{\Gamma^2 + (\Delta E)^2}.$$

The behaviour of $D(E)$ near a quasi-level is depicted in fig. 30.

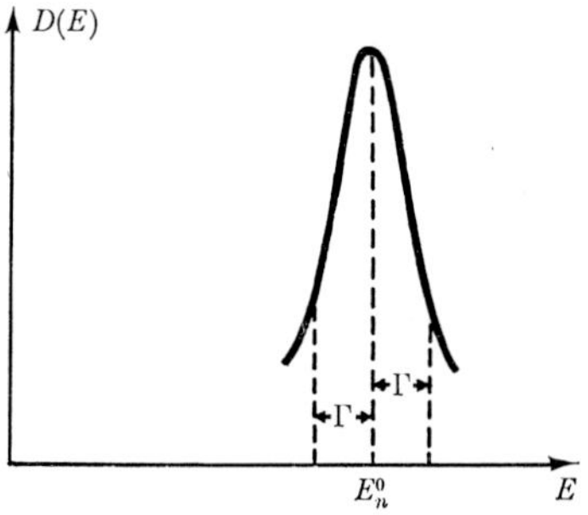

Fig. 30.

16. (i) We consider a one-dimensional barrier of arbitrary shape (see fig. 31). Let a particle of energy $E + V_0$ (V_0 is the barrier height) fall onto the barrier from the left. For $x \to +\infty$ we have only an outgoing wave

$$\psi = A\,e^{ik_2 x} \quad \left\{k_2 = \frac{1}{\hbar}\sqrt{[2\mu(E - V_0)]}\right\}, \tag{1}$$

and for $x \to -\infty$ a superposition of an incoming and a reflected wave

$$\psi = e^{ik_1 x} + B e^{-ik_1 x} \qquad \left[k_1 = \frac{1}{\hbar}\sqrt{(2\mu E)}\right]. \tag{2}$$

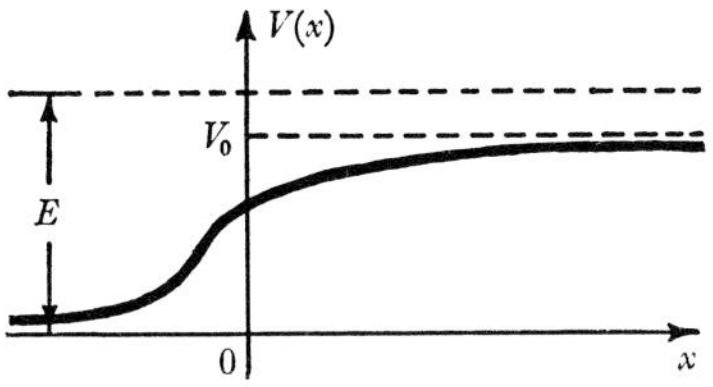

Fig. 31.

To prove the required relation we use the law of conservation of number of particles (equation of continuity):

$$\frac{\partial |\psi|^2}{\partial t} + \operatorname{div} \boldsymbol{j} = 0, \tag{3}$$

where

$$\boldsymbol{j} = -\frac{i\hbar}{2\mu}(\psi^* \nabla\psi - \psi\nabla\psi^*) \tag{4}$$

is the probability current density.

In the stationary problem under consideration we have $\partial|\psi|^2/\partial t = 0$, so that also div $\boldsymbol{j} = 0$. This equation is for our one-dimensional motion of the form

$$\frac{dj_x}{dx} = 0 \quad \text{or} \quad j \equiv j_x = \text{const.} \tag{5}$$

The law of conservation of number of particles means thus that the current density $j \equiv j_x$ is the same for all x.

We evaluate j for $x = +\infty$ and $x = -\infty$. Using equations (1), (2), and (4) we find easily:

$$j(+\infty) = \frac{\hbar k_2 |A|^2}{\mu}, \quad j(-\infty) = \frac{\hbar k_1}{\mu}(1 - |B|^2).$$

If we equate these two quantities, by using equation (5) we have:

$$\frac{k_2}{k_1}|A|^2 + |B|^2 = 1. \tag{6}$$

The first term of equation (6) is nothing but the ratio of the current density j in the outgoing and incoming waves, that is, by definition the transmission coefficient D. The second term is equal to the ratio of the values of j for the reflected and the incoming wave, that is, it is the reflection coefficient R. This concludes our proof.

17. The proof of the integral equation follows by using the fact that

$$\frac{d^2}{dx^2}|x| = 2\delta(x).$$

The reflection coefficient R is to lowest order in V given by the expression

$$R = \frac{\mu^2}{k^2\hbar^2}\left|\int V(y)e^{2iky}\,dy\right|^2.$$

3

Commutation relations; Heisenberg relations; spreading of wave packets; operators

1. $$\hat{A}\hat{B}\hat{C}-\hat{B}\hat{C}\hat{A} = \hat{A}\hat{B}\hat{C}-\hat{B}\hat{A}\hat{C}+\hat{B}\hat{A}\hat{C}-\hat{B}\hat{C}\hat{A},$$

or

$$[\hat{A},\hat{B}\hat{C}]_- = [\hat{A},\hat{B}]_-\hat{C}+\hat{B}[\hat{A},\hat{C}]_-.$$

We emphasise the similarity of the equation obtained to the formula for the differentiation of a product of two functions, which also retains the order in which the operators which form the original product [$\hat{B}$ and $\hat{C}$] occur in the different terms.

2. $$[\sum_i \hat{A}_i, \sum_k \hat{B}_k]_- \equiv (\sum_i \hat{A}_i)(\sum_k \hat{B}_k)-(\sum_k \hat{B}_k)(\sum_i \hat{A}_i)$$

$$= \sum_i\sum_k \hat{A}_i\hat{B}_k - \sum_i\sum_k \hat{B}_k\hat{A}_i = \sum_{i,k}[\hat{A}_i, \hat{B}_k]_-$$

as had to be proved.

3. By induction one can prove that, for $n \geqslant 1$,

$$[\hat{p}, \hat{x}^n]_- = -ni\hbar\hat{x}^{n-1}, \quad [\hat{x}, \hat{p}^n]_- = ni\hbar\hat{p}^{n-1}.$$

The given relations then follow by expanding $f(x)$ and $g(p)$ in Taylor series.

4. Let us consider the operator

$$\hat{a}(s) = \exp(s\hat{L})\,\hat{a}\exp(-s\hat{L}),$$

where s is an auxiliary parameter, and let us find the differential equation which is satisfied by $\hat{a}(s)$:

$$\frac{d\hat{a}(s)}{ds} = \hat{L}\exp(s\hat{L})\,\hat{a}\exp(-s\hat{L})-\exp(s\hat{L})\,\hat{a}\exp(-s\hat{L})\,\hat{L} = [\hat{L}, \hat{a}(s)]_-.$$

We shall differentiate this equation once more,

$$\frac{d^2\hat{a}(s)}{ds^2} = \left[\hat{L}, \frac{d\hat{a}(s)}{ds}\right] = [\hat{L}, [\hat{L}, \hat{a}(s)]_-]_-.$$

One sees easily that the derivative $d^n\hat{a}(s)/ds^n$ is equal to n successive commutators of the operator $\hat{L}$ with the operator $\hat{a}(s)$.

If we now write the operator

$$\exp(\hat{L})\,\hat{a}\exp(-\hat{L}) = \hat{a}(1)$$

as a Taylor series,

$$\hat{a}(1) = \hat{a}(0) + \frac{d\hat{a}(0)}{ds} + \frac{1}{2!}\frac{d^2\hat{a}(0)}{ds^2} + \dots,$$

and express the derivatives with respect to s at $s = 0$ in terms of successive commutators of the operator $\hat{L}$ with $\hat{a}(0) = \hat{a}$, we get the relation which had to be proved.

5. Let $\hat{\varphi}(\lambda) \equiv \exp(\lambda\hat{A})\exp(\lambda\hat{B})$. We then have

$$\frac{\partial\hat{\varphi}(\lambda)}{\partial\lambda} = \hat{A}\exp(\lambda\hat{A})\exp(\lambda\hat{B}) + \exp(\lambda\hat{A})\,\hat{B}\exp(\lambda\hat{B})$$
$$= (\hat{A}+\hat{B})\hat{\varphi}(\lambda) + [\exp(\lambda\hat{A}), \hat{B}]_-\exp(\lambda\hat{B}). \quad (1)$$

For $[\exp(\lambda\hat{A}), \hat{B}]_-$ we find

$$[\exp(\lambda\hat{A}), \hat{B}]_- = \sum_{n=0}^{\infty}\frac{\lambda^n}{n!}[\hat{A}^n, \hat{B}]_- = \sum_{n=0}^{\infty}\frac{\lambda^n}{(n-1)!}K\hat{A}^{n-1} = \lambda K\exp(\lambda\hat{A}), \quad (2)$$

where we have used the fact that $[\hat{A}, \hat{B}]_-$ is a c-number and thus commutes with $\hat{A}$.

Combining (1) and (2) we have

$$\frac{\partial\hat{\varphi}(\lambda)}{\partial\lambda} = (\hat{A}+\hat{B}+\lambda K)\hat{\varphi}(\lambda)$$

with the general solution

$$\hat{\varphi}(\lambda) = \exp[\lambda(\hat{A}+\hat{B}) + \tfrac{1}{2}\lambda^2 K]\,\hat{C},$$

where $\hat{C}$ is a constant, arbitrary operator. By letting $\lambda \to 0$, we find that $\hat{C}$ is the unit operator, which concludes the proof.

6. The proof follows, as both the left-hand and the right-hand sides of the equation satisfy the same differential equation

$$\frac{d\hat{\Omega}}{d\alpha} = -\hat{B}\hat{\Omega} + [\hat{A}, \hat{B}]_- e^{-\alpha\hat{B}},$$

and both sides vanish for $\alpha = 0$.

7. We consider first of all the case of a discrete set of wave functions ψ_i. The average values of the operators $\hat{A}$ and $\hat{B}$ in the state characterised by the function ψ ($\psi = \Sigma\, a_i\psi_i$) are equal to

$$\bar{\hat{A}} = \sum_{i,k} a_i^* A_{ik}\, a_k,$$

$$\bar{\hat{B}} = \sum_{i,k} a_i^* B_{ik}\, a_k.$$

We consider the non-negative quantity

$$J(\lambda) = \sum_i \left[\sum_k (A_{ik} + i\lambda B_{ik})\, a_k\right]^* \left[\sum_l (A_{il} + i\lambda B_{il})\, a_l\right] \geqslant 0,$$

where λ is a real parameter.

Collecting terms of the same power in λ and using the fact that $\hat{A}$ and $\hat{B}$ are Hermitean ($A_{ik} = A^*_{ki}$, $B_{ik} = B^*_{ki}$), we find

$$J(\lambda) = \sum_{i,k,l} [a^*_k A_{ki} A_{il}\, a_l + i\lambda a^*_k (A_{ki} B_{il} - B_{ki} A_{il})\, a_l$$
$$+ \lambda^2 a^*_k B_{ki} B_{il}\, a_l] = \overline{\hat{A}^2} - \lambda \overline{\hat{C}} + \lambda^2 \overline{\hat{B}^2}.$$

Here $\hat{C}$ is the Hermitean operator

$$\hat{C} = \frac{1}{i}(\hat{A}\hat{B} - \hat{B}\hat{A}).$$

The quadratic form $J(\lambda)$ is non-negative and thus $4\overline{\hat{A}^2}\,\overline{\hat{B}^2} \geqslant (\overline{\hat{C}})^2$. If we note that the operators $\Delta\hat{A} = \hat{A} - \overline{\hat{A}}$ and $\Delta\hat{B} = \hat{B} - \overline{\hat{B}}$ satisfy the same commutation relations as $\hat{A}$ and $\hat{B}$,

$$\Delta\hat{A}\,\Delta\hat{B} - \Delta\hat{B}\,\Delta\hat{A} = i\hat{C},$$

we get

$$\sqrt{[\overline{(\Delta\hat{A})^2}\,\overline{(\Delta\hat{B})^2}]} \geqslant \frac{|\overline{\hat{C}}|}{2}.$$

The proof of this relation for a continuum set can proceed along similar lines. The expression

$$J(\lambda) = \int [(\hat{A} + i\lambda\hat{B})\psi]^* [(\hat{A} + i\lambda\hat{B})\psi]\, d\tau$$

with real λ is non-negative and can be transformed as follows:

$$J(\lambda) = \int [(\hat{A}\psi)^* - i\lambda(\hat{B}\psi)^*] [\hat{A}\psi + i\lambda\hat{B}\psi]\, d\tau$$
$$= \int [\psi^* \hat{A}^2 \psi + i\lambda\psi^*(\hat{A}\hat{B} - \hat{B}\hat{A})\psi + \lambda^2 \psi^* \hat{B}^2 \psi]\, d\tau,$$

where we use the fact that, as the operator $\hat{A}$ is Hermitean,

$$\int (\hat{A}\psi)^* \varphi\, d\tau = \int \psi^* \hat{A}\varphi\, d\tau.$$

The rest of the proof proceeds as before.

8. $$\overline{(\Delta\hat{q})^2}\,\overline{(\Delta\hat{F})^2} \geqslant \frac{\hbar^2}{4} \left|\overline{\frac{\partial\hat{F}}{\partial\hat{p}}}\right|^2.$$

9. The energy of the oscillator in a stationary state is

$$E = \int \psi(x)\left(\frac{\hat{p}^2}{2\mu}+\frac{kx^2}{2}\right)\psi(x)\,dx = \frac{\overline{\hat{p}^2}}{2\mu}+\frac{k\overline{\hat{x}^2}}{2};$$

since

$$\overline{\hat{p}^2} = \overline{(\hat{p}-\bar{\hat{p}})^2}+(\bar{\hat{p}})^2 = \overline{(\Delta\hat{p})^2}+(\bar{\hat{p}})^2,$$
$$\overline{\hat{x}^2} = \overline{(\Delta\hat{x})^2}+(\bar{\hat{x}})^2,$$

and

$$\bar{\hat{p}} = 0, \quad \bar{\hat{x}} = 0.$$

we have

$$E = \frac{\overline{(\Delta\hat{p})^2}}{2\mu}+\frac{k\overline{(\Delta\hat{x})^2}}{2}.$$

From the Heisenberg relation $\overline{(\Delta\hat{p})^2}.\overline{(\Delta\hat{x})^2} \geqslant \hbar^2/4$ it follows that

$$E \geqslant \frac{\hbar^2}{8\mu\overline{(\Delta\hat{x})^2}}+\frac{k\overline{(\Delta\hat{x})^2}}{2}.$$

The expression on the right is minimum for

$$\overline{(\Delta\hat{x})^2} = \frac{\hbar}{2}\sqrt{\frac{1}{\mu k}},$$

so that

$$E_{\text{min}} \sim \frac{\hbar}{2}\sqrt{\left(\frac{k}{\mu}\right)} = \frac{\hbar\omega}{2},$$

where $\omega = \sqrt{(k/\mu)}$ is the oscillator frequency.

10. In the case under consideration we can neglect the screening of the field of the nucleus by the other electrons.

The energy of a K-electron is

$$E = \frac{p^2}{2\mu} - \frac{Ze^2}{r}.$$

Since $p \sim \hbar/r$, where r is of the dimensions of the region of the localisation, we have

$$E \sim \frac{\hbar^2}{2\mu r^2} - \frac{Ze^2}{r}. \tag{1}$$

This expression is minimum for $r = \hbar^2/Ze^2\mu = a/Z$ ($a = 0{\cdot}529.10^{-8}$ cm is the radius of the first Bohr orbit).

We have thus for the energy

$$E \sim -\frac{Z^2}{2}\frac{\mu e^4}{\hbar^2} = -Z^2.13{\cdot}5 \text{ eV}.$$

If we take into account the relativistic correction due to the change in mass, expression (1) will be of the form

$$E \geqslant (\mu_0^2 c^4 + c^2 p^2)^{\frac{1}{2}} - \frac{Ze^2}{r} - \mu_0 c^2 \geqslant \left(\mu_0^2 c^4 + \frac{c^2 \hbar^2}{r^2}\right)^{\frac{1}{2}} - \frac{Ze^2}{r} - \mu_0 c^2.$$

We find thus for the energy

$$E \geqslant \mu_0 c^2[(1 - \alpha^2 Z^2)^{\frac{1}{2}} - 1] \quad \text{where} \quad \alpha = e^2/\hbar c.$$

11. Let the regions of localisation of the first and second electron be of the dimensions r_1 and r_2. Using the Heisenberg relations we have then for the momenta of the electrons

$$p_1 \sim \frac{\hbar}{r_1}, \quad p_2 \sim \frac{\hbar}{r_2},$$

so that the kinetic energy is of the order of magnitude

$$\frac{\hbar^2}{2\mu}\left(\frac{1}{r_1^2} + \frac{1}{r_2^2}\right).$$

The potential energy of the interaction of the electrons with a nucleus of charge Z is given by

$$-Ze^2\left(\frac{1}{r_1} + \frac{1}{r_2}\right)$$

and the mutual interaction energy of the electrons is equal to $e^2/(r_1 + r_2)$. To find the energy of the ground state, we look for the minimum of the total energy,

$$E(r_1, r_2) = \frac{\hbar^2}{2\mu}\left(\frac{1}{r_1^2} + \frac{1}{r_2^2}\right) - Ze^2\left(\frac{1}{r_1} + \frac{1}{r_2}\right) + \frac{e^2}{r_1 + r_2}.$$

The minimum is realised for the values

$$r_1 = r_2 = \frac{\hbar^2}{\mu e^2} \frac{1}{Z - \frac{1}{4}}.$$

The ground state energy of an ion with two electrons and nuclear charge Z is thus equal to

$$E \sim -(Z - \tfrac{1}{4})^2 \frac{\mu e^4}{\hbar^2} = -2(Z - \tfrac{1}{4})^2 Ry,$$

$$Ry = \frac{1}{2} \frac{\mu e^4}{\hbar^2} = 13{\cdot}5 \text{ eV}.$$

A comparison with experimental data shows excellent agreement in view of the extreme simplicity of the calculation:

	H^-	He	Li^+	Be^{++}	B^{+++}	C^{++++}
E_{calc} Ry	−1·125	−6·125	−15·12	−28·12	−45·12	−66·12
E_{exp} Ry	−1·05	−5·807	−14·56	−27·31	−44·06	−64·8

12. It is not possible.

13. The error in the reasoning given lies in the identification (as to order of magnitude) of the precision Δx with which the particle is localised in space with the dimensions of the potential well a. In reality the precision of spatial localisation of the particle is determined by the "width" of the coordinate probability distribution $|\psi(x)|^2$; (similarly the indeterminacy of the momentum is given by the width of the momentum probability distribution $|G(p)|^2$). From a consideration of the normalisation integral $\int |\psi|^2\, dx$ it follows that the probability to find the particle inside or outside the well is respectively κa and $(1-\kappa a)$, where $\kappa a = \mu a^2 V_0/\hbar^2 \ll 1$.

Using a rough classical analogy one can say that the particle spends by far more of its time outside the well than inside it. This is a typical quantum effect since the region outside the well is classically inaccessible. We can use the deuteron as a three-dimensional example; its radius is appreciably larger than the range of the neutron–proton interaction, so that the particle spends a considerable part of its time outside the potential well.

The effective width of the coordinate distribution is of the order of magnitude (compare problem 7 of section 1)

$$(\Delta x)_{eff} \sim \frac{1}{\kappa} \gg a. \tag{1}$$

On the other hand, one can show that the effective width of the momentum distribution is equal to (compare problem 3 of section 1)

$$(\Delta p)_{eff} \sim \hbar\kappa. \tag{2}$$

Multiplying equations (1) and (2) we get, as should be the case, the Heisenberg relation,

$$(\Delta p)_{eff}(\Delta x)_{eff} \sim \hbar. \tag{3}$$

Let us now evaluate more precisely the coefficient of $\hbar$, in equation (3), by evaluating the product of the average square fluctuations in momentum and coordinate:

$$\Delta p \equiv \sqrt{[\overline{(p-\bar{p})^2}]} \quad \text{and} \quad \Delta x \equiv \sqrt{[\overline{(x-\bar{x})^2}]}. \tag{4}$$

We have, of course,

$$\overline{(p-\bar{p})^2} = \overline{p^2} - (\bar{p})^2, \tag{5'}$$

$$\overline{(x-\bar{x})^2} = \overline{x^2} - (\bar{x})^2. \tag{5''}$$

From the known wave function it follows that

$$\left.\begin{aligned} \bar{x} &= \int_{-\infty}^{\infty} x[\psi(x)]^2\, dx = 0, \\ \bar{p} &= \int_{-\infty}^{\infty} p[G(p)]^2\, dp = 0. \end{aligned}\right\} \tag{6}$$

If we assume $a^2 \ll \hbar^2/\mu V_0$ one can easily show that

$$\overline{x^2} = \int_{-\infty}^{\infty} x^2[\psi(x)]^2\, dx \approx \frac{1}{2\kappa^2}. \tag{7}$$

The quantity

$$\overline{p^2} = \int_{-\infty}^{\infty} p^2[G(p)]^2\, dp$$

can also be readily evaluated in the cordinate representation, that is, by using the equation

$$\overline{p^2} = \int_{-\infty}^{\infty} \psi(x)\,\hat{p}^2\,\psi(x)\, dx = \int_{-\infty}^{\infty} \psi(x)\left(\frac{\hbar}{i}\frac{\partial}{\partial x}\right)^2 \psi(x)\, dx = -\hbar^2 \int_{-\infty}^{\infty} \psi(x)\,\frac{d^2\psi}{dx^2}\, dx.$$

Integration by parts simplifies this equation to

$$\overline{p^2} = \hbar^2 \int_{-\infty}^{\infty} \left(\frac{d\psi}{dx}\right)^2 dx$$

and by substituting the wave function we find, to the same approximation,

$$\overline{p^2} \approx \hbar^2 \kappa^2. \tag{8}$$

We note that this value of $\overline{p^2}$ does not at all correspond to the kinetic energy in the classically accessible region (that is, inside the well),

$$V_0 - |E| \approx V_0,$$

but to the much smaller value $|E|$. This result could have been expected qualitatively since the particle spends most of its time in an inaccessible region (where its "kinetic energy" is negative), which leads to a large negative contribution to the average kinetic energy $(\overline{E-V}) = (1/2\mu)\overline{p^2}$.

From equations (4) and (8) it follows that

$$\Delta p = \hbar\kappa, \quad \Delta x = \frac{1}{\sqrt{(2)}\,\kappa},$$

and thus

$$\Delta p\,\Delta x = \frac{\hbar}{\sqrt{2}}.$$

It must be remembered that the least possible value of $\Delta p\,\Delta x$ is equal to $\hbar/2$ corresponding to a "Gaussian packet" $\psi \sim \exp(-bx^2)$.

14. The ratio of the momentum component perpendicular to the trajectory to the total momentum is in this case very small: $\Delta p/p \sim 6.10^{-7}$. Under such conditions the concept of a trajectory is completely reasonable and one should not expect any deviations from the laws of classical mechanics.

15. From the Heisenberg relation for the azimuthal angle φ,

$$\Delta\varphi . \Delta M \geqslant \hbar,$$

with M the angular momentum, we find that φ becomes completely undetermined ($\Delta\varphi \geqslant \pi$) if $\Delta M/M \leqslant 32\%$ (see also next problem).

16. It seems that we can violate the Heisenberg relation

$$\Delta M_z . \Delta\varphi \geqslant \hbar \tag{1}$$

as follows. We can choose an eigenstate with a well-defined value of $\hat{M}_z$ so that $\Delta M_z = 0$. As $\Delta\varphi \leqslant \pi$ from physical considerations, it looks as if (1) is violated.

There are two ways out. First of all, we can argue (see problem 15 of section 3) that $\Delta\varphi > \pi$ means that φ is completely undetermined and thus essential $\Delta\varphi \geqslant \pi$ corresponds to $\Delta\varphi = \infty$. This is a physical argument, but it does not completely remove the difficulty. This difficulty is removed when we bear in mind that (1) is derived by assuming that $\hat{M}_z$ and $\hat{\varphi}$ are Hermitean operators (see problem 1 of section 3). However, $\hat{M}_z$ is not a Hermitean operator, since the relation

$$\int_0^{2\pi} f(\varphi)\,\hat{M}_z g(\varphi)\,d\varphi = \int_0^{2\pi} g(\varphi)\,\hat{M}_z^* f(\varphi)\,d\varphi$$

only holds for $f(\varphi)$ and $g(\varphi)$ which are periodic functions of φ, but not for all functions $f(\varphi)$ and $g(\varphi)$. For a more detailed discussion of this problem we refer to short notes by Judge and by Judge and Lewis (*Physics Letters* **5**, 189 and 190, 1963).

17. The wave function $\psi(x, t)$ of a free particle depends on $\psi(x, 0)$ as follows:

$$\psi(x,t) = \frac{1}{(2\pi\hbar)^{\frac{1}{2}}}\int_{-\infty}^{+\infty} a(p)\exp\left[\frac{i}{\hbar}\left(px-\frac{p^2}{2\mu}t\right)\right]dp,$$

where

$$a(p) = \frac{1}{(2\pi\hbar)^{\frac{1}{2}}}\int_{-\infty}^{+\infty}\psi(x,0)\exp\left(-i\frac{p}{\hbar}x\right)dx$$

$$= \frac{1}{(2\pi\hbar)^{\frac{1}{2}}}\int_{-\infty}^{+\infty}\varphi(x)\exp\left(\frac{i(p_0-p)x}{\hbar}\right)dx.$$

The function $a(p)$ differs appreciably from zero only for those values of p which satisfy the condition

$$\frac{|p_0-p|}{\hbar}\delta \lesssim 1. \tag{A}$$

If x varies within the interval

$$-\delta < x < +\delta,$$

the oscillating factor $\exp[i(p_0-p)x/\hbar]$ changes little provided condition (A) is satisfied, and $\psi(x, t)$ can thus approximately be written in the form

$$\psi(x,t) \approx \frac{1}{(2\pi\hbar)^{\frac{1}{2}}}\int_{p_0-\hbar/\delta}^{p_0+\hbar/\delta} a(p)\exp\left[\frac{i}{\hbar}\left(px-\frac{p^2}{2\mu}t\right)\right]dp,$$

or

$$\psi(x,t) \approx \frac{\exp\left[\frac{i}{\hbar}\left(p_0x-\frac{p_0^2}{2\mu}t\right)\right]}{(2\pi\hbar)^{\frac{1}{2}}}\int_{-\hbar/\delta}^{+\hbar/\delta} a(p+p_0)$$
$$\times\exp\left\{+\frac{i}{\hbar}\left[p\left(x-\frac{p_0}{\mu}t\right)-\frac{p^2}{2\mu}t\right]\right\}dp.$$

From the last equation it follows that the wave function $\psi(x, t)$ will be appreciably different from zero only when the oscillating factor

$$\exp\left\{\frac{i}{\hbar}\left[p\left(x-\frac{p_0}{\mu}t\right)-\frac{p^2}{2\mu}t\right]\right\}$$

varies little when p varies within the interval $-\hbar/\delta < p < +\hbar/\delta$. The dimensions of the wave packet at time t are thus of the order of magnitude:

$$\delta_t \sim \delta + \frac{\hbar t}{2\mu\delta}.$$

18. To solve this problem, it is necessary to determine the wave function $\psi(x, t)$, which satisfies the Schrödinger equation

$$i\hbar\frac{\partial\psi}{\partial t} = \hat{H}\psi \tag{1}$$

and which at $t = 0$ has the given value $\psi(x, 0)$. If $\hat{H}$ does not contain the time explicitly, equation (1) has the solutions

$$\psi_n(x, t) = \psi_n(x) \exp -i\frac{E_n}{\hbar}t, \tag{2}$$

where $\psi_n(x)$ is the time-independent eigenfunction of the operator $\hat{H}$

$$\hat{H}\psi_n(x) = E_n \psi_n(x).$$

If we find the coefficients of the expansion of $\psi(x, 0)$ in terms of the set of functions $\psi_n(x)$,

$$\psi(x, 0) = \sum_n a_n \psi_n(x), \quad a_n = \int \psi_n^*(x)\psi(x, 0)\, dx,$$

the function $\Sigma_n a_n \psi_n(x) \exp[-i(E_n/\hbar)t]$ will satisfy equation (1) and is at $t = 0$ equal to $\psi(x, 0)$.

We have thus

$$\psi(x, t) = \sum_n a_n \psi_n(x) \exp -i\frac{E_n}{\hbar}t$$

or

$$\psi(x, t) = \int G_t(\xi, x)\psi(\xi, 0)\, d\xi, \tag{3}$$

where

$$G_t(\xi, x) = \sum_n \psi_n^*(\xi)\psi_n(x) \exp\left(-i\frac{E_n}{\hbar}t\right).$$

To solve our problem, it is thus necessary to evaluate the so-called Green function $G_t(\xi, x)$ and to use equation (3).

(a) In the case of a free particle, the eigenfunctions are

$$\psi_p(\boldsymbol{r}) = \frac{1}{(2\pi\hbar)^{\frac{3}{2}}} \exp i\frac{(\boldsymbol{p}\,.\,\boldsymbol{r})}{\hbar}, \quad E_p = \frac{p^2}{2\mu}$$

and the corresponding Green function

$$G_t(\boldsymbol{\rho}, \boldsymbol{r}) = \iiint \frac{d^3\boldsymbol{p}}{(2\pi\hbar)^3} \exp\left\{\frac{i}{\hbar}\left[(\boldsymbol{p}\,.\,\boldsymbol{r} - \boldsymbol{\rho}) - \frac{p^2 t}{2\mu}\right]\right\}$$

$$= \left(\frac{\mu}{2\pi i\hbar t}\right)^{\frac{3}{2}} \exp\left[\frac{i\mu}{2\hbar t}(r - \rho)^2\right].$$

Since the initial function was

$$\psi(\boldsymbol{r}, 0) = \frac{1}{(\pi\delta^2)^{\frac{3}{4}}} \exp\left[\frac{i(\boldsymbol{p}_0\,.\,\boldsymbol{r})}{\hbar} - \frac{r^2}{2\delta^2}\right],$$

we have:

$$\psi(\boldsymbol{r}, t) = \iiint \left(\frac{\mu}{2\pi i\hbar t}\right)^{\frac{3}{2}} \frac{1}{(\pi\delta^2)^{\frac{3}{4}}} \exp\left[-\frac{\rho^2}{2\delta^2} + \frac{i(\boldsymbol{p}_0\cdot\boldsymbol{\rho})}{\hbar} + \frac{i\mu}{2\hbar t}(\boldsymbol{r} - \boldsymbol{\rho})^2\right] d^3\boldsymbol{\rho},$$

whence we find:

$$\psi(\boldsymbol{r}, t) = \frac{1}{(\pi\delta^2)^{\frac{3}{4}}\left(1+\dfrac{\hbar^2 t^2}{\mu^2\delta^4}\right)^{\frac{3}{4}}}$$

$$\times \exp\left[-\frac{\left(\boldsymbol{r}-\dfrac{\boldsymbol{p}_0 t}{\mu}\right)^2}{2\delta^2\left(1+\dfrac{\hbar^2 t^2}{\mu^2\delta^4}\right)}\left(1-\frac{i\hbar t}{\mu\delta^2}\right)-\frac{ip_0^2 t}{2\mu\hbar}+\frac{i(\boldsymbol{p}_0\cdot\boldsymbol{r})}{\hbar}\right].$$

For the probability density we find:

$$\psi^*(\boldsymbol{r}, t)\,\psi(\boldsymbol{r}, t) = \frac{1}{\left[\pi\delta^2\left(1+\dfrac{\hbar^2 t^2}{\mu^2\delta^4}\right)\right]^{\frac{3}{2}}}\exp\left[-\frac{\left(\boldsymbol{r}-\dfrac{\boldsymbol{p}_0 t}{\mu}\right)^2}{\delta^2\left(1+\dfrac{\hbar^2 t^2}{\mu^2\delta^4}\right)}\right].$$

From this expression it is clear that the centre of gravity of the wave packet moves with a velocity $\boldsymbol{p}_0/\mu$. The dimensions of the packet δ_t which originally were of the order of δ increase with time according to the formula

$$\delta_t = \delta\sqrt{\left(1+\frac{\hbar^2 t^2}{\mu^2\delta^4}\right)},$$

but the distribution in r remains Gaussian as before. Let us estimate the time t during which the dimensions of the wave packet change in magnitude by an amount of the order of magnitude of its original dimensions,

$$\tau \sim \frac{\delta^2\mu}{\hbar}.$$

For $t \gg \tau$ the linear dimensions of the wave packet increase linearly with time

$$\delta_t \sim \frac{\hbar}{\mu\delta}t.$$

Let us consider some concrete examples.

For an electron initially localised within a region $\delta \sim 10^{-8}$ cm, τ is of the order of 10^{-16} sec. For a "classical" particle of $\mu = 1$ g, $\delta = 10^{-5}$ cm, $\tau = 10^{17}$ sec $\sim$ 3000 million years.

(b) The wave function for the one-dimensional motion of a particle in a uniform field $V = -Fx$ is of the form (see problem 19 of section 1),

$$\psi_E(x) = A\int_{-\infty}^{+\infty}\exp\left[i\left(\frac{u^3}{3}-uq\right)\right]du, \quad q = \left(x+\frac{E}{F}\right)\alpha,$$

where

$$\alpha = \left(\frac{2\mu F}{\hbar^2}\right)^{\frac{1}{3}}, \quad A = \frac{\alpha}{2\pi\sqrt{F}}.$$

Let us evaluate the Green function:

$$G_t(\xi, x) = \int_{-\infty}^{+\infty} dE \exp\left(-i\frac{Et}{\hbar}\right)\psi_E^*(\xi)\psi_E(x)$$

$$= A^2\int_{-\infty}^{+\infty} dE \exp\left(-i\frac{Et}{\hbar}\right)\int_{-\infty}^{+\infty}\int du\, dv \exp\left[-i\left(\frac{v^3}{3}-v\eta\right)+i\left(\frac{u^3}{3}-uq\right)\right],$$

where

$$\eta = \left(\xi+\frac{E}{F}\right)\alpha.$$

Let us first of all integrate over E:

$$G_t(\xi, x) = A^2\int_{-\infty}^{+\infty}\int du\, dv \exp\left(-i\frac{v^3}{3}+i\frac{u^3}{3}+iv\xi\alpha-iux\alpha\right)$$

$$\times\int_{-\infty}^{+\infty} dE \exp\left[-i\frac{E\alpha}{F}\left(u+\frac{Ft}{\alpha\hbar}-v\right)\right]$$

$$= A^2\int_{-\infty}^{+\infty}\int du\, dv \exp\left(-i\frac{v^3}{3}+i\frac{u^3}{3}+iv\xi\alpha-iux\alpha\right).\frac{2\pi F}{\alpha}\,\delta\left(u+\frac{Ft}{\alpha\hbar}-v\right).$$

If we use the properties of the δ-function we can integrate over v; we can then put the expression in the exponent in a form convenient for the integration over u,

$$G_t(\xi, x) = \frac{2\pi F}{\alpha}A^2\int_{-\infty}^{+\infty} du \exp\left\{-i\frac{Ft}{\alpha\hbar}\left[u+\frac{Ft}{2\alpha\hbar}+\frac{\alpha^2\hbar}{2Ft}(x-\xi)\right]^2-\frac{i}{12}\left(\frac{Ft}{\alpha\hbar}\right)^3\right.$$

$$\left.+\frac{i}{2}\frac{Ft}{\hbar}(x+\xi)+\frac{i\alpha^3\hbar}{4Ft}(x-\xi)^2\right\}.$$

We finally get for the Green function:

$$G_t(\xi, x) = \left(\frac{\mu}{2\pi i\hbar t}\right)^{\frac{1}{2}}\exp\left[-\frac{i}{12}\left(\frac{Ft}{\hbar\alpha}\right)^3+\frac{iFt}{2\hbar}(x+\xi)+\frac{i\mu}{2\hbar t}(x-\xi)^2\right].$$

As $F\to 0$ this expression goes over into the Green function for free one-dimensional motion. Using expression (3) we can determine the change with time of a wave function given at $t = 0$

$$\psi(x, 0) = \frac{1}{(\pi\delta^2)^{\frac{1}{4}}}\exp\left(-\frac{x^2}{2\delta^2}+\frac{ip_0 x}{\hbar}\right).$$

The result of this evaluation is

$$\psi(x,t) = \frac{1}{\left[\pi\delta^2\left(1+\dfrac{\hbar^2 t^2}{\mu^2\delta^4}\right)\right]^{\frac{1}{4}}} \exp\left[-\frac{\left(x-\dfrac{p_0 t}{\mu}-\dfrac{Ft^2}{2\mu}\right)^2}{2\delta^2\left(1+\dfrac{\hbar^2 t^2}{\mu^2\delta^4}\right)}\left(1-\frac{i\hbar t}{\mu\delta^2}\right) + \frac{i}{\hbar}(p_0+Ft)\,x - \frac{i}{\hbar}\int_0^t \frac{(p_0+Ft)^2}{2\mu}\,dt\right].$$

In the general case of a three-dimensional motion in a uniform field with an initial wave function

$$\psi(\boldsymbol{r},0) = \frac{1}{(\pi\delta^2)^{\frac{3}{4}}}\exp\left[-\frac{\boldsymbol{r}^2}{2\delta^2}+\frac{i(\boldsymbol{p}_0\cdot\boldsymbol{r})}{\hbar}\right]$$

we get

$$\psi(\boldsymbol{r},t) = \frac{1}{\left[\pi\delta^2\left(1+\dfrac{\hbar^2 t^2}{\mu^2\delta^4}\right)\right]^{\frac{3}{4}}} \exp\left[-\frac{\left(\boldsymbol{r}-\dfrac{\boldsymbol{p}_0 t}{\mu}-\dfrac{\boldsymbol{F}t^2}{2\mu}\right)^2}{2\delta^2\left(1+\dfrac{\hbar^2 t^2}{\mu^2\delta^4}\right)}\left(1-\frac{i\hbar t}{\mu\delta^2}\right) + \frac{i}{\hbar}(\boldsymbol{p}_0+\boldsymbol{F}t\,.\,\boldsymbol{r}) - \frac{i}{\hbar}\int_0^t \frac{(\boldsymbol{p}_0+\boldsymbol{F}t)^2}{2\mu}\,dt\right].$$

From this expression it follows that the probability density distribution remains Gaussian and that the centre of mass of the wave packet moves according to the law of classical mechanics with a uniform acceleration. The change with time in the size of the wave packet is the same as in the absence of a field (see above).

(c) The eigenfunctions of the Schrödinger equation

$$-\frac{\hbar^2}{2\mu}\psi_n'' + \frac{\mu\omega^2}{2}x^2\psi_n = E_n\psi_n$$

are of the following form:

$$\psi_n(x) = c_n \exp\left(-\frac{\alpha^2 x^2}{2}\right) H_n(\alpha x),$$

where

$$\alpha = \left(\frac{\mu\omega}{\hbar}\right)^{\frac{1}{2}}, \quad c_n^2 = \frac{1}{2^n\, n!}\frac{\alpha}{\sqrt{\pi}}, \quad E_n = \hbar\omega(n+\tfrac{1}{2}).$$

The required wave function $\psi(x,t)$ satisfies according to equation (2) the relation

$$\psi(x,t) = \sum_n a_n \psi_n(x) \exp\left(-i\frac{E_n}{\hbar}t\right), \tag{4}$$

where

$$a_n = c_n c \int_{-\infty}^{+\infty} H_n(\alpha x) \exp\left[-\alpha^2 \frac{(x-x_0)^2}{2} + \frac{ip_0 x}{\hbar} - \frac{\alpha^2 x^2}{2}\right] dx.$$

In order to evaluate the a_n we use an expression for the generating function for the Hermite polynomials,

$$\exp(-\lambda^2 + 2\lambda\eta) = \sum_{n=0}^{\infty} \frac{\lambda^n}{n!} H_n(\eta). \tag{5}$$

One sees easily that $a_n/c_n c$ is the coefficient of $\lambda^n/n!$ in the expansion in powers of λ of the expression

$$\int_{-\infty}^{+\infty} \exp\left(-\lambda^2 + 2\lambda\alpha x - \frac{\alpha^2(x-x_0)^2}{2} + \frac{ip_0 x}{\hbar} - \frac{\alpha^2 x^2}{2}\right) dx.$$

From this it follows that

$$a_n = c_n c \sqrt{(\pi)} \left(\alpha x_0 + \frac{ip_0}{\alpha\hbar}\right)^n \exp\left[-\frac{\alpha^2 x_0^2}{2} + \tfrac{1}{4}\left(\alpha x_0 + \frac{ip_0}{\alpha\hbar}\right)^2\right].$$

After substituting this expression into equation (4) it turns out that one can perform the summation over n, using again equation (5). If we use the notation

$$x_0 + \frac{ip_0}{\hbar\alpha^2} = Q\exp(-i\delta),$$

we get:

$$\psi(x,t) = c\exp\left\{-\frac{\alpha^2}{2}[x - Q\cos(\omega t+\delta)]^2 - ixQ\alpha^2 \sin(\omega t + \delta) - \frac{i\omega t}{2} + \frac{\alpha^2 Q^2}{4} i[\sin 2(\omega t+\delta) - \sin 2\delta]\right\}.$$

In this case the wave packet does not spread during the motion. The centre of gravity moves as before according to the laws of classical mechanics, performing a harmonic oscillation with amplitude Q and frequency ω. From the expression we have obtained for ψ it follows also that the average value of the momentum at time t is equal to

$$P(t) = \hbar Q\alpha^2 \sin(\omega t + \delta),$$

that is, to the classical momentum of the particle in the oscillator.

The expression

$$\exp\left\{-\frac{i\omega t}{2}+\frac{\alpha^2 Q^2}{4}i[\sin 2(\omega t+\delta)-\sin 2\delta]\right\}$$

can be written as follows:

$$\exp\left[-\frac{i}{\hbar}\int_0^t \frac{P^2(t)}{2\mu}\,dt\right].$$

19. The Schrödinger equation for an oscillator with a perturbing field is of the form

$$i\hbar\frac{\partial\psi}{\partial t}=-\frac{\hbar^2}{2\mu}\frac{\partial^2\psi}{\partial x^2}+\frac{\mu\omega^2 x^2}{2}\psi-f(t)\,x\psi.$$

Introducing a new coordinate $x_1 = x-\xi(t)$, we have

$$i\hbar\frac{\partial\psi}{\partial t}=i\hbar\dot{\xi}\frac{\partial\psi}{\partial x_1}-\frac{\hbar^2}{2\mu}\frac{\partial^2\psi}{\partial x_1^2}+\tfrac{1}{2}\mu\omega^2(x_1+\xi)^2\psi-f(t)\,(x_1+\xi)\,\psi.$$

If we put

$$\psi=\exp\left(\frac{i\mu\dot{\xi}x_1}{\hbar}\right)\varphi(x_1, t),$$

we get for φ the equation

$$i\hbar\frac{\partial\varphi}{\partial t}=-\frac{\hbar^2}{2\mu}\frac{\partial^2\varphi}{\partial x_1^2}+\tfrac{1}{2}\mu\omega^2 x_1^2\varphi+(\mu\ddot{\xi}+\mu\omega^2\xi-f\,)\,x_1\varphi-L\varphi,$$

where L is the Lagrangian, $L=\frac{1}{2}\mu\dot{\xi}^2-\frac{1}{2}\mu\omega^2\xi^2+f(t)\,\xi$. In the last expression the term $(\mu\ddot{\xi}+\mu\omega^2\xi-f\,)\,x_1\varphi$ is equal to zero, if we require ξ as a function of time to satisfy the classical equation of motion of an oscillator acted upon by a perturbing force:

$$\mu\ddot{\xi}+\mu\omega^2\xi=f(t).$$

If we introduce still another function χ by $\varphi=\chi\exp\left(\frac{i}{\hbar}\int_0^t L\,dt\right)$, we get for the function χ an equation which is the same as the equation of motion of a free oscillator

$$i\hbar\frac{\partial\chi}{\partial t}=-\frac{\hbar^2}{2\mu}\frac{\partial^2\chi}{\partial x_1^2}+\frac{\mu\omega^2 x_1^2}{2}\chi.$$

The wave functions of the oscillator acted upon by a perturbing force can thus be put in the form

$$\psi(x, t)=\chi[x-\xi(t), t]\exp\left[\frac{i\mu}{\hbar}\dot{\xi}(x-\xi)+\frac{i}{\hbar}\int_0^t L\,dt\right].$$

20*. We shall write the solution of the Schrödinger equation

$$i\hbar \frac{\partial \psi}{\partial t} = -\frac{\hbar^2}{2\mu}\frac{\partial^2 \psi}{\partial x^2} + \frac{\mu\omega^2(t)}{2} x^2 \psi \tag{1}$$

in the form

$$\psi(x, t) = \int G(x, t;\, x', \tau)\, \psi(x', \tau)\, dx'.$$

One can easily show that the Green function $G(x, t;\, x', \tau)$ must satisfy equation (1) and the initial condition

$$\lim_{t \to \tau+0} G(x, t;\, x', \tau) = \delta(x - x'). \tag{2}$$

We shall try to satisfy these two conditions by putting

$$G(x, t;\, x', \tau) \sim \exp\left\{\frac{i}{2\hbar}[a(t)\, x^2 + 2b(t)\, x + c(t)]\right\}. \tag{3}$$

Substituting expression (3) into equation (1) we get the following equations to determine a, b, and c:

$$\left.\begin{aligned} \frac{1}{\mu}\frac{da}{dt} &= -\frac{a^2}{\mu^2} - \omega^2(t), \\ \frac{db}{dt} &= -\frac{a}{\mu} b, \\ \frac{dc}{dt} &= i\hbar \frac{a}{\mu} - \frac{b^2}{\mu}. \end{aligned}\right\} \tag{4}$$

The solution of the set of equations (4) is of the form

$$a = \mu \frac{\dot{Z}}{Z}, \quad b = \frac{\text{const}}{Z}, \quad c = i\hbar \ln Z - \frac{1}{\mu}\int^t b^2\, dt, \tag{5}$$

where Z is a solution of the equation

$$\ddot{Z} = -\omega^2(t)\, Z.$$

We try to satisfy the initial conditions (2) by suitably choosing the integration constants. To do this we use one of the possible expressions for the δ-function,

$$\delta(x - x') = \lim_{t \to \tau} \sqrt{\left[\frac{\mu}{2\pi i\hbar(t - \tau)}\right]} \exp\left[\frac{i\mu}{2\hbar(t - \tau)}(x - x')^2\right] \tag{6}$$

(see problem 18a of section 3).

In order that expression (3) for $t \to \tau$ goes over into expression (6) it is necessary and sufficient that

$$Z = 0,\ \dot{Z} = 1 \quad \text{for} \quad t = \tau,$$

$$b = -\frac{\mu x'}{Z},$$

$$c = i\hbar \ln Z + \mu x'^2 \frac{Y}{Z},$$

where Y is a solution of the equation $\ddot{Y} = -\omega^2(t)\, Y$ with the boundary conditions $Y = 1$, $\dot{Y} = 0$ for $t = \tau$.

We note that since $Z\dot{Y} - Y\dot{Z} = -1$, we have

$$\frac{Y}{Z} = -\frac{dt}{Z^2}.$$

We thus find for the Green function of our problem the following expression:

$$G(x, t;\, x', \tau) = \sqrt{\left(\frac{\mu}{2\pi i\hbar Z}\right)} \exp\left[\frac{i\mu}{2\hbar Z}(\dot{Z}x^2 - 2xx' + Yx'^2)\right].$$

For the case where ω = constant, we have:

$$Z = \frac{1}{\omega}\sin\omega(t-\tau), \quad Y = \cos\omega(t-\tau),$$

and the Green function in that case is equal to:

$$G(x, t;\, x', \tau) = \sqrt{\left[\frac{\mu\omega}{2\pi\hbar i \sin\omega(t-\tau)}\right]} \exp\left\{\frac{i\mu\omega}{2\hbar\sin\omega(t-\tau)} \times \left[\cos\omega(t-\tau)\, x^2 - 2xx' + \cos\omega(t-\tau)\, x'^2\right]\right\}.$$

21*.

$$|\psi(x, t)|^2 = \frac{\alpha}{\sqrt{(2\pi)}} \frac{1}{\sqrt{\left(\cos^2\omega t + \frac{\alpha^4\hbar^2}{\mu^2\omega^2}\sin^2\omega t\right)}} \exp\left\{-\frac{\alpha^2[x - Q\cos(\omega t + \delta)]^2}{\cos^2\omega t + \frac{\alpha^4\hbar^2}{\mu^2\omega^2}\sin^2\omega t}\right\}$$

where

$$x_0 + \frac{ip_0}{\mu\omega} = Q\exp(-i\delta).$$

In the case where $\alpha = \sqrt{(\mu\omega/\hbar)}$, we get the result of problem 18c of section 3.

22*.

$$G(x, t; x', \tau) = \sqrt{\left(\frac{\mu}{2\pi i\hbar Z}\right)} \exp\left\{\frac{i\mu}{2\hbar Z}[\dot{Z}(x-\xi)^2 - 2(x-\xi)x' + Yx'^2] + \frac{i\mu}{\hbar}\dot{\xi}(t)(x-\xi) + \frac{i}{\hbar}\int_\tau^t L\,dt\right\}.$$

In this expression ξ satisfies the equation $\mu\ddot{\xi} = -\mu\omega^2\xi + f(t)$ and the initial condition $\xi(\tau) = 0, \dot{\xi}(\tau) = 0$, and L is the Lagrangian

$$L = \tfrac{1}{2}\mu\dot{\xi}^2 - \tfrac{1}{2}\mu\omega^2\xi^2 + f\xi.$$

23*. The probability of a transition from the state n to the state m is given by the relation (we use the system of units where $\hbar = 1, \mu = 1, \omega = 1$)

$$P_{mn}(t, 0) = |G_{mn}(t, 0)|^2, \qquad (1)$$

where

$$G_{mn}(t, 0) = \iint \psi_m^*(x)\, G(x, t; x', 0)\, \psi_n(x')\, dx\, dx'.$$

Using the generating function

$$\exp[-\tfrac{1}{2}(2z^2 + x^2 - 4zx)] = \sum_{n=0}^{\infty} \sqrt{\left[\frac{\sqrt{(\pi)}\,2^n}{n!}\right]} z^n \psi_n(x)$$

we can construct a function $G(u, v)$

$$G(u, v) = \iint \exp[-\tfrac{1}{2}(2v^2 + x^2 - 4vx) - \tfrac{1}{2}(2u^2 + x'^2 - 4ux')] \times G(x, t; x', 0)\, dx\, dx' = \sum_{m=0}^{\infty}\sum_{n=0}^{\infty} \sqrt{\left(\frac{\pi 2^{n+m}}{n!\,m!}\right)} v^m u^n G_{mn}(t, 0). \quad (2)$$

From expression (2) it follows that the quantities $G_{ij}(t, 0)$, the absolute squares of which determine the transition probabilities, are apart from a factor $\sqrt{(\pi 2^{i+j}/i!\,j!)}$ the coefficients in the series expansion of $G(u, v)$ in powers of u and v.

Let us evaluate $G(u, v)$; to do this we substitute into equation (2) the expression for the Green function G (see problem 22 of section 3).

After this substitution we get

$$G(u, v) = \frac{1}{\sqrt{(2\pi i Z)}} \exp\left(i\int_0^t L\,dt - i\xi\dot{\xi} + \frac{i\dot{Z}}{2Z}\xi^2 - u^2 - v^2\right) \times \iint \exp\left\{-\frac{1}{2}\left[\left(1 - \frac{i\dot{Z}}{Z}\right)x^2 + \frac{2i}{Z}xx' + \left(1 - \frac{iY}{Z}\right)x'^2 - 2\left(2v - \frac{i\dot{Z}}{Z}\xi + i\dot{\xi}\right)x - 2\left(2u + \frac{i\xi}{Z}\right)x'\right]\right\} dx\, dx'.$$

When ω = constant, ξ, Z, Y are of the form

$$\xi(t) = \int_0^t \sin(t-t')f(t')\,dt', \quad Z = \sin t, \quad Y = \cos t.$$

To evaluate the integral we use the formula

$$\int_{-\infty}^{+\infty}\int dx\,dy \exp\left[-\tfrac{1}{2}(ax^2+2bxy+cy^2-2px-2qy)\right]$$

$$= \frac{2\pi}{\sqrt{(ac-b^2)}}\exp\left[\frac{aq^2-2bpq+cp^2}{2(ac-b^2)}\right].$$

After some simple calculations we find the following expression for the function $G(u, v)$:

$$G(u,v) = \sqrt{(\pi)}\exp[iF(t)]\exp(-w/2)\exp\left(-\frac{AB}{w}uv+Au+Bv\right),$$

where

$$A = i\int_0^t e^{-it'}f(t')\,dt', \quad B = e^{-it}A, \quad 2w = |A|\,.\,|B| = \xi^2+\xi^2,$$

and $F(t)$ is some real function of the time.

To expand $G(u, v)$ in a power series, we use the relation

$$\exp\left(\alpha+\beta-\frac{\alpha\beta}{w}\right) = \sum_{m,n=0}^{\infty} c(m,n|w)\frac{\alpha^m}{m!}\frac{\beta^n}{n!},$$

where

$$c(m,n|w) = \sum_{l=0}^{\min(m,n)} \frac{m!\,n!}{l!\,(m-l)!\,(n-l)!}(-w)^{-l}.$$

Expanding, we now get

$$G(u,v) = \sqrt{(\pi)}\,e^{iF(t)}\,e^{-\frac{1}{2}w}\sum_{m,n=0}^{\infty} c(m,n|w)\frac{(Bv)^m}{m!}\frac{(Au)^n}{n!}. \tag{3}$$

From equations (2) and (3) it follows that

$$G_{mn}(t,0) = \frac{e^{-\frac{1}{2}w}\,.\,A^n\,B^m}{\sqrt{(2^{n+m}\,.\,m!\,n!)}}c(m,n|w)\,e^{iF(t)},$$

and the required transition probability is equal to

$$P_{mn}(t,0) = \frac{e^{-w}\,.\,w^{m+n}}{m!\,n!}\{c(m,n|w)\}^2.$$

For the particular case $n = 0$ the transition probability is given by

$$P_{m0}(t,0) = \frac{e^{-w}\,.\,w^m}{m!}, \quad \text{since} \quad c(m,0|w) = 1.$$

After we have evaluated the transition probability we can determine the average value of the energy and the square of the energy of an oscillator at time t.

These average values are given by the equations

$$\overline{E} = \sum_{m=0}^{\infty} P_{mn}(t,0).(m+\tfrac{1}{2}),$$

$$\overline{E^2} = \sum_{m=0}^{\infty} P_{mn}(t,0).(m+\tfrac{1}{2})^2.$$

To evaluate the two sums which are similar in character we consider the expression

$$\left(1-\frac{\alpha}{w}\right)^m e^{\alpha} = \Phi(m,\alpha\,|\,w).$$

One shows easily that

$$\Phi(m,\alpha\,|\,w) = \sum_{n=0}^{\infty} \frac{\alpha^n}{n!} c(m,n\,|\,w).$$

From the equation

$$\sum_{m=0}^{\infty} \frac{e^{-w}.w^m}{m!} \Phi(m,\alpha\,|\,w)\,\Phi(m,\beta\,|\,w) = e^{\alpha\beta/w}$$

it follows that

$$\sum_{m=0}^{\infty} \frac{e^{-w}.w^m}{m!} c(m,n\,|\,w).c(m,n'\,|\,w) = \delta_{nn'}\,n!\,w^{-n},$$

so that the physically obvious relation

$$\sum_{m=0}^{\infty} P_{mn} = 1$$

follows directly. Consider now the equation

$$\sum_{m=0}^{\infty} m\frac{e^{-w}.w^m}{m!} \Phi(m,\alpha\,|\,w)\,\Phi(m,\beta\,|\,w) = \left\{w-\alpha-\beta+\frac{\alpha\beta}{w}\right\} e^{\alpha\beta/w},$$

and differentiate its left-hand and right-hand side n times with respect to α and m times with respect to β and after that put $\alpha = \beta = 0$. We thus get

$$\sum_{m=0}^{\infty} mP_{mn} = n+w.$$

In this way we see that the average value of the energy of the oscillator at time t is equal to $\overline{E} = E_n + w$. In this expression w is the work done by the force $f(t)$ over a period t,

$$w = \int_0^t f(t)\,\dot{\xi}dt = \int_0^t (\ddot{\xi}+\xi)\,\dot{\xi}dt = \tfrac{1}{2}(\dot{\xi}^2+\xi^2)_{t=t} - \tfrac{1}{2}(\dot{\xi}^2+\xi^2)_{t=0}.$$

In a similar manner we find

$$\overline{E^2} = 2wE_n.$$

24*. The Hamiltonian of the system is of the form

$$\hat{H} = \frac{\hat{p}^2}{2\mu} + \frac{\mu\omega^2}{2}[x - x_0(t)]^2.$$

Our problem is to find the wave function $\psi(x, t)$, satisfying the equation

$$\hat{H}\psi = i\hbar\frac{\partial\psi}{\partial t} \tag{1}$$

and the boundary condition

$$\psi(x, 0) = \psi_0(x).$$

Let $\hat{U}(t, t_0)$ be the operator transforming $\psi(t_0)$ into $\psi(t)$:

$$\psi(t) = \hat{U}\psi(t_0).$$

Substituting this equation into the Schrödinger equation (1) and taking into account the fact that $\psi(t_0)$ can be any function, we get

$$\hat{H}\hat{U} = i\hbar\frac{\partial\hat{U}}{\partial t}.$$

Taking the Hermitean conjugate of this equation, we have

$$\hat{U}^+\hat{H} = -i\hbar\frac{\partial\hat{U}^+}{\partial t}.$$

We now introduce the operators $\hat{x}$ and $\hat{p}$ in the Heisenberg representation,

$$\hat{x}(t, t_0) = \hat{U}^+(t, t_0)\hat{x}\hat{U}(t, t_0),$$

$$\hat{p}(t, t_0) = \hat{U}^+(t, t_0)\hat{p}\hat{U}(t, t_0).$$

We easily get the following equations for $\hat{x}(t, t_0)$ and $\hat{p}(t, t_0)$ (the dot indicates differentiating with respect to t):

$$\dot{\hat{x}}(t, t_0) = \frac{i}{\hbar}\hat{U}^+[\hat{H}, \hat{x}]_-\hat{U},$$

$$\dot{\hat{p}}(t, t_0) = \frac{i}{\hbar}\hat{U}^+[\hat{H}, \hat{p}]_-\hat{U}.$$

Evaluating the commutators we find a set of equations for $\hat{x}(t, t_0)$ and $\hat{p}(t, t_0)$:

$$\dot{\hat{x}} = \frac{1}{\mu}\hat{p}, \quad \dot{\hat{p}} = -\mu\omega^2[\hat{x} - x_0(t)] \quad \text{or} \quad \ddot{\hat{x}} + \omega^2\hat{x} = \omega^2 x_0. \tag{2}$$

These equations have the "classical" form.

Since they are linear we can solve them as if they were c-number equations,

$$\hat{x} = \hat{c}_1 \sin \omega(t-t_0) + \hat{c}_2 \cos \omega(t-t_0) + x_1,$$

$$\hat{p} = \mu\omega\hat{c}_1 \cos \omega(t-t_0) - \mu\omega\hat{c}_2 \sin \omega(t-t_0) + p_1,$$

where x_1 and p_1 are the solutions of the inhomogeneous equations (2) which are zero at $t = t_0$:

$$\left.\begin{aligned} x_1(t, t_0) &= \omega \int_{t_0}^{t} x_0(t') \sin \omega(t-t')\, dt', \\ p_1(t, t_0) &= \mu\omega^2 \int_{t_0}^{t} x_0(t') \cos \omega(t-t')\, dt'. \end{aligned}\right\} \tag{3}$$

From the definition of the operators in the Heisenberg representation, it follows that

$$\hat{x}(t_0, t_0) = \hat{x}, \quad \hat{p}(t_0, t_0) = \hat{p}.$$

Satisfying the initial conditions we get

$$\left.\begin{aligned} \hat{x}(t, t_0) &= \hat{x} \cos \omega(t-t_0) + \frac{\hat{p}}{\mu\omega} \sin \omega(t-t_0) + x_1(t, t_0), \\ \hat{p}(t, t_0) &= \hat{p} \cos \omega(t-t_0) - \mu\omega\hat{x} \sin \omega(t-t_0) + p_1(t, t_0). \end{aligned}\right\} \tag{4}$$

From these equations we easily get the solution of our problem.

We write down the equation satisfied by the initial function,

$$\hat{H}_0 \psi_0 = E_0 \psi_0. \tag{5}$$

Here

$$\hat{H}_0 = \frac{\hat{p}^2}{2\mu} + \frac{\mu\omega^2 \hat{x}^2}{2}.$$

To determine $\hat{U}(t, t_0)$ we can write

$$\psi(t) = \hat{U}(t, 0) \psi_0 \quad \text{or} \quad \psi_0 = \hat{U}(0, t) \psi(t),$$

where $\psi(t)$ is the function we want to find.

Substituting this expression for ψ_0 into equation (5) we get

$$\hat{U}^+(0, t) \hat{H}_0 \hat{U}(0, t) \psi(t) = E_0 \psi(t). \tag{6}$$

The Hamiltonian in this equation is the Hamiltonian $\hat{H}_0$ which is constructed from our operators $\hat{x}(0, t)$, $\hat{p}(0, t)$,

$$\hat{U}^+(0, t) \hat{H}_0 \hat{U}(0, t) = \frac{\hat{p}^2(0, t)}{2\mu} + \frac{\mu\omega^2 \hat{x}^2(0, t)}{2}.$$

Using the relations following from equations (4) and (3),

$$\hat{x} = \hat{x}(0, t) \cos \omega t + \frac{\hat{p}(0, t)}{\mu\omega} \sin \omega t + x_1(t, 0),$$

$$\hat{p} = \hat{p}(0, t) \cos \omega t - \mu\omega\hat{x}(0, t) \sin \omega t + p_1(t, 0),$$

or, by a simple substitution, we can verify that

$$\frac{\hat{p}^2(0, t)}{2\mu} + \frac{\mu\omega^2 \hat{x}^2(0, t)}{2} = \frac{[\hat{p} - p_1(t, 0)]^2}{2\mu} + \frac{\mu\omega^2[\hat{x} - x_1(t, 0)]^2}{2}.$$

Equation (6) for $\psi(t)$ is thus of the form

$$\frac{[\hat{p} - p_1(t, 0)]^2}{2\mu} \psi + \frac{\mu\omega^2[\hat{x} - x_1(t, 0)]^2}{2} \psi = E_0 \psi$$

with the solution

$$\psi(x, t) = \exp\{ip_1(t, 0)\,[x - x_1(t, 0)]/\hbar\}\, \psi_0[x - x_1(t, 0)]. \tag{7}$$

(Since in our equation t is a parameter, ψ can contain a phase factor which depends in an arbitrary manner on t. This will, however, not affect the required probability.)

We could have chosen a different way of determining $\psi(t)$, for instance, by finding the Green function of the Schrödinger equation (compare problem 20 of section 3).

We chose the method used above since the initial state was a state with a well-defined energy value.

The function (7) which we obtained is the previous, zeroth order, distribution of the classical position of the particle [see equations (2) and (3)] in a system of coordinates connected to the moving particles.

To find the probability w_n that the nth state is excited, we must expand $\psi(t)$ in terms of the eigenfunctions of the Hamiltonian at time t $(t > T)$. These normalised eigenfunctions are of the form

$$\psi_n(x) = \left(\frac{\mu\omega}{\pi\hbar}\right)^{\frac{1}{4}} \frac{1}{\sqrt{(2^n\, n!)}} \exp[-\mu\omega(x - x_0)^2/2\hbar]\, H_n\left[\sqrt{\left(\frac{\mu\omega}{\hbar}\right)}\,(x - x_0)\right], \tag{8}$$

where

$$H_n(\xi) = (-1)^n \exp(\xi^2) \frac{d^n \exp(-\xi^2)}{d\xi^n} \tag{9}$$

are the Hermite polynomials. The coefficients of the required expansion are equal to

$$c_n(t) = \int_{-\infty}^{\infty} \psi(x,t)\,\psi_n^*(x)\,dx.$$

Substituting $\psi(x,t)$ from expression (7) and $\psi_n^* = \psi_n$ from equations (8) and (9) and integrating over $\xi = (x-x_0)\sqrt{(\mu\omega/\hbar)}$, we find finally

$$c_n = \frac{1}{\sqrt{(2^n\,n!\,\pi)}} \exp\left[i\frac{p_1(t,0)\,(x_0-x_1)}{\hbar}\right]$$

$$\times \int_{-\infty}^{\infty} \exp\left[i\frac{p_1(t,0)}{\sqrt{(\mu\hbar\omega)}}\,\xi\right] \exp\left\{-\frac{1}{2}\left[\xi+\frac{x_0-x_1}{\sqrt{(\hbar/\mu\omega)}}\right]\right\}^2 \exp\left(-\tfrac{1}{2}\xi^2\right) H_n(\xi)\,d\xi$$

$$= \frac{(-1)^n}{\sqrt{(2^n\,n!\,\pi)}} \exp\left[-\frac{\mu\omega(x_1-x_0)^2}{2\hbar} + i\frac{p_1(x_0-x_1)}{\hbar}\right]$$

$$\times \int_{-\infty}^{\infty} d\xi \exp\left\{\xi\left[i\frac{p_1}{\sqrt{(\mu\omega\hbar)}} + \sqrt{\left(\frac{\mu\omega}{\hbar}\right)}(x_1-x_0)\right]\right\} \frac{d^n \exp(-\xi^2)}{d\xi^n}$$

$$= \frac{1}{\sqrt{(2^n\,n!\,\pi)}} \exp\left[-\frac{\mu\omega(x_1-x_0)^2}{2\hbar} + i\frac{p_1(x_0-x_1)}{\hbar}\right]$$

$$\times \left[i\frac{p_1}{\sqrt{(\mu\omega\hbar)}} + \sqrt{\left(\frac{\mu\omega}{\hbar}\right)}(x_1-x_0)\right]^n$$

$$\times \int_{-\infty}^{\infty} \exp\left\{-\xi^2 + \xi\left[i\frac{p_1}{\sqrt{(\mu\omega\hbar)}} + \sqrt{\left(\frac{\mu\omega}{\hbar}\right)}(x_1-x_0)\right]\right\} d\xi$$

$$= \frac{1}{\sqrt{(2^n\,n!)}} \left[i\frac{p_1}{\sqrt{(\mu\omega\hbar)}} + \sqrt{\left(\frac{\mu\omega}{\hbar}\right)}(x_1-x_0)\right]^n$$

$$\times \exp\left\{-\frac{\mu\omega(x_1-x_0)^2}{2\hbar} + i\frac{p_1(x_0-x_1)}{\hbar}\right.$$

$$\left. + \frac{1}{4}\left[\sqrt{\left(\frac{\mu\omega}{\hbar}\right)}(x_1-x_0) + i\frac{p_1}{\sqrt{(\mu\omega\hbar)}}\right]^2\right\}$$

$$= \frac{1}{\sqrt{(2^n\,n!)}} \left[\sqrt{\left(\frac{\mu\omega}{\hbar}\right)}(x_1-x_0) + i\frac{p_1}{\sqrt{(\mu\omega\hbar)}}\right]^n$$

$$\times \exp\left[-\frac{\dfrac{p_1^2}{2\mu} + \dfrac{\mu\omega^2(x_1-x_0)^2}{2}}{2\hbar\omega} - i\frac{p_1(x_1-x_0)}{2\hbar}\right]$$

and thus

$$w_n = |c_n|^2 = \frac{1}{n!}\left[\frac{\dfrac{p_1^2}{2\mu}+\dfrac{\mu\omega^2(x_1-x_0)^2}{2}}{\hbar\omega}\right]^n \exp\left[-\frac{\dfrac{p_1^2}{2\mu}+\dfrac{\mu\omega^2(x_1-x_0)^2}{2}}{\hbar\omega}\right].$$

We note now that $[p_1^2/2\mu+\frac{1}{2}\mu\omega^2(x_1-x_0)^2]$ is the energy transferred to the classical particle which was initially at rest, during the process considered. If we denote this quantity by ϵ, we get

$$w_n = \frac{(\epsilon/\hbar\omega)^n}{n!}\exp(-\epsilon/\hbar\omega),$$

that is, a Poisson distribution with an average value of n equal to

$$\bar{n} = \frac{\epsilon}{\hbar\omega}.$$

We now consider limiting cases.

If the point of suspension is changed fast, x_1 and p_1 are approximately equal to

$$x_1 = x_0(1-\cos\omega t), \quad p_1 = \mu\omega x_0 \sin\omega t.$$

If we substitute these values into the expression for c_n we get

$$c_n = \frac{(-1)^n}{\sqrt{(2^n n!)}}\left[x_0\sqrt{\left(\frac{\mu\omega}{\hbar}\right)}\right]^n \exp\left(-in\omega t-\frac{\mu\omega^2 x_0^2}{2\hbar\omega}+i\frac{\mu\omega x_0^2}{\hbar}\sin\omega t\cos\omega t\right).$$

This result differs only by a phase factor $\exp[(i\mu\omega x_0^2/\hbar)\sin\omega t\cos\omega t]$ from the result obtained by expanding the initial function in terms of the eigenfunctions of the Hamiltonian for $t>0$. (This expression can be obtained by putting $x_1 = 0$, $p_1 = 0$, in the general expression for the c_n.)

If the change is slow we have in zeroth approximation $x_1 \approx x_0$, $p_1 \approx 0$, and thus

$$\psi(x,t) \approx \psi_0[x-x_0(t)].$$

This method can be applied to solving the oscillator problem when other parameters vary with time.

25*. Classically the equation of motion of a particle described by the given Hamiltonian is

$$\mu\ddot{x}+\mu\gamma\dot{x} = F(t),$$

so that we are dealing with a particle under the influence of a time-dependent force $F(t)$ which is subject to friction.

We now proceed as in problem 19 of section 3 and introduce a new variable

$$x_1 = x-\xi(t),$$

and a new wavefunction

$$\psi = \chi \exp\frac{i}{\hbar}\left\{\mu\xi x_1 e^{\gamma t} + \int^t L(\tau)d\tau\right\},$$

where $L(\tau)$ is the classical Lagrangian,

$$L = e^{\gamma t}[\tfrac{1}{2}\mu\dot{\xi}^2 + F(t)\xi].$$

We can then find for χ the equation

$$-\frac{\hbar^2}{2\mu}e^{-\gamma t}\frac{\partial^2\chi}{\partial x_1^2} = i\hbar\frac{\partial\chi}{\partial t}, \tag{A}$$

which is the Schrödinger equation for a free particle subject to friction.

Separating the variables we find the following solution of equation (A):

$$\chi_k = A\exp\left[\pm ikx_1 + \frac{i\hbar k^2}{2\mu\gamma}e^{-\gamma t}\right].$$

In the case when $F(t)$, we have $x_1 \equiv x$ and the χ_k form a complete orthonormal set with the normalization constant $A = L^{-1/2}$, where L is the one-dimensional volume. Expanding ψ_0 in terms of the χ_k, we find finally

$$\psi(x, t) = B(t)\exp\left\{-\frac{[x-(\hbar k_0/\mu)(1-e^{-\gamma t})/\gamma]^2}{\sigma^2 + \hbar^2(1-e^{-\gamma t})^2/\mu^2\gamma^2\sigma^2}\right\},$$

where $B(t)$ is a (time-dependent) normalizing constant. We see that the centre of the wavepacket moves like a classical point particle with velocity $\hbar k_0/\mu$ in a viscous medium.

As $t \to \infty$, we get

$$\psi(x, t\to\infty) = B_\infty \exp\left[-\frac{\{x-(\hbar k_0/\gamma\mu)\}^2}{\sigma^2 + \hbar^2/\mu^2\gamma^2\sigma^2}\right].$$

We refer to a paper by L.H.Buch and H.H.Denman (*Am.J.Phys.*, **42**, 304, 1974) for a general discussion of the problem treated here.

28. We shall denote the operators under consideration by $\hat{A}$ and $\hat{B}$ and the Hamiltonian of the system by $\hat{H}$. We have then

$$\frac{d}{dt}(\hat{A}\hat{B}) = \frac{\partial}{\partial t}(\hat{A}\hat{B}) + \frac{i}{\hbar}[\hat{H}, \hat{A}\hat{B}]_-.$$

After an obvious transformation, using the results of problem 1 of section 3, combining the corresponding terms in $d\hat{A}/dt \equiv \dot{\hat{A}}$ and $d\hat{B}/dt \equiv \dot{\hat{B}}$ we find the required rule for differentiation

$$\frac{d}{dt}(\hat{A}\hat{B}) = \dot{\hat{A}}\hat{B} + \hat{A}\dot{\hat{B}}.$$

29. In the Heisenberg representation we have

$$\hat{x}(t) = \exp(i\hat{H}t/\hbar)\,\hat{x}\exp(-i\hat{H}t/\hbar). \tag{1}$$

In our case $\hat{H} = \hat{p}^2/2\mu$, so that

$$\hat{x}(t) = \exp[i(\hat{p}^2 t/2\mu\hbar)]\,\hat{x}\exp[-i(\hat{p}^2 t/2\mu\hbar)].$$

We shall work in the momentum representation. In that case

$$\hat{x} = i\hbar\,(\partial/\partial p)$$

and

$$\hat{x}\exp(-i\hat{p}^2 t/2\mu\hbar) = i\hbar\frac{\partial}{\partial p}\exp(-i\hat{p}^2 t/2\mu\hbar) = \exp(-i\hat{p}^2 t/2\mu\hbar)\,i\hbar\frac{\partial}{\partial p} + \frac{pt}{\mu}\exp(-i\hat{p}^2 t/2\mu\hbar),$$

or

$$\hat{x}\exp(-i\hat{p}^2 t/2\mu\hbar) = \exp(-i\hat{p}^2 t/2\mu\hbar)\left(\hat{x} + t\,\frac{\hat{p}}{\mu}\right).$$

Substituting this in equation (1) we get

$$\hat{x}(t) = \hat{x} + t\,\frac{\hat{p}}{\mu},$$

which is the same relation between the operators as exists between the classical quantities.

30. The equations of motion for the operators are of the form

$$\left.\begin{aligned}\frac{d\hat{x}}{dt} &= \frac{i}{\hbar}[\hat{H}, \hat{x}(t)]_-,\\ \frac{d\hat{p}}{dt} &= \frac{i}{\hbar}[\hat{H}, \hat{p}(t)]_-.\end{aligned}\right\} \tag{1}$$

The derivatives on the left are in the Heisenberg representation taken as derivatives with respect to the time coordinate which explicitly enters in the Heisenberg operators. Since the operators $\exp(\pm i\hat{H}t/\hbar)$ commute with the Hamiltonian we can write equation (1) in the form

$$\left.\begin{aligned}\frac{d\hat{x}}{dt} &= \frac{i}{\hbar}\exp(i\hat{H}t/\hbar)\,[\hat{H}, \hat{x}]_-\exp(-i\hat{H}t/\hbar),\\ \frac{d\hat{p}}{dt} &= \frac{i}{\hbar}\exp(i\hat{H}t/\hbar)\,[\hat{H}, \hat{p}]_-\exp(-i\hat{H}t/\hbar).\end{aligned}\right\} \tag{2}$$

After having evaluated the commutators we get

$$\left.\begin{aligned}\frac{d\hat{x}}{dt} &= \frac{\hat{p}(t)}{\mu},\\ \frac{d\hat{p}}{dt} &= -\mu\omega^2\hat{x}(t).\end{aligned}\right\} \tag{3}$$

Because these equations are linear they can be solved as any other set of equations for ordinary, c-number quantities. We get then

$$\left.\begin{aligned}\hat{x}(t) &= \hat{c}_1\cos\omega t+\hat{c}_2\sin\omega t,\\ \hat{p}(t) &= \hat{c}_2\,\mu\omega\cos\omega t-\hat{c}_1\,\mu\omega\sin\omega t.\end{aligned}\right\} \tag{4}$$

The constants of integration $\hat{c}_1$ and $\hat{c}_2$ are only constants as far as the time is concerned. They can be determined from the initial conditions. Indeed, from the definition of the Heisenberg operator it follows that

$$\hat{L}(t)|_{t=0} = \exp(i\hat{H}t/\hbar)\,\hat{L}\exp(-i\hat{H}t/\hbar)|_{t=0} = \hat{L}.$$

Hence we have

$$\hat{x}(t)|_{t=0} = \hat{x} = \hat{c}_1,$$
$$\hat{p}(t)|_{t=0} = \hat{p} = \hat{c}_2\,\mu\omega.$$

We have thus finally

$$\left.\begin{aligned}\hat{x}(t) &= \hat{x}\cos\omega t+\frac{\hat{p}}{\mu\omega}\sin\omega t,\\ \hat{p}(t) &= \hat{p}\cos\omega t-\mu\omega\hat{x}\sin\omega t.\end{aligned}\right\} \tag{5}$$

These equations are valid in any Heisenberg representation. In particular, for the coordinate representation we get

$$\left.\begin{aligned}\hat{x}_x(t) &= x\cos\omega t+\frac{\hbar}{i\mu\omega}\sin\omega t\frac{\partial}{\partial x},\\ \hat{p}_x(t) &= \frac{\hbar}{i}\cos\omega t\frac{\partial}{\partial x}-\mu\omega x\sin\omega t.\end{aligned}\right\} \tag{6}$$

31. $i\hbar\dfrac{\partial\psi_{\text{int}}}{\partial t} = (\hat{H}_1)_{\text{int}}\psi_{\text{int}}, \quad i\hbar\dot{\hat{L}}_{\text{int}} = [\hat{L}_{\text{int}}, \hat{H}_0]_-.$

33. The operator corresponding to the time derivative of a quantity A which does not explicitly depend on the time is of the form

$$\dot{\hat{A}} = \frac{i}{\hbar}[\hat{H},\hat{A}]_- \equiv \frac{i}{\hbar}(\hat{H}\hat{A}-\hat{A}\hat{H}). \tag{1}$$

The average value of $\dot{A}$ in a stationary state of the discrete spectrum with a wave function ψ_n is equal to the corresponding diagonal element of the matrix of the operator (1),

$$\bar{\dot{A}} = \int \psi_n^* \hat{\dot{A}} \psi_n \, d\tau \equiv \dot{A}_{nn} = \frac{i}{\hbar}(HA - AH)_{nn}. \tag{2}$$

By virtue of the fact that the Hamiltonian is diagonal in the energy representation and that the matrix elements of A are finite in the case of the discrete spectrum, the right-hand side of equation (2) will tend to zero identically, which proves the required relation $\bar{\dot{A}} = 0$. (The classical analogue of this property of $\dot{A}$ is the fact that the time average of any bounded quantity A vanishes.)

A concrete example of this equation can be found in problem 16 of section 5.

34. The virial theorem in classical mechanics states that if the potential energy of a mechanical system is a homogeneous function of its (Cartesian) coordinates, and if the motion of the system is bounded to a finite region of space, the time averages of the kinetic energy T and the potential energy V are connected as follows:

$$n\bar{V} = 2\bar{T}. \tag{1}$$

The bars indicate here time averages and n is the degree of homogeneity of the function $V(x_1, x_2, \ldots, x_i, \ldots, x_{3N})$, where N is the number of point particles in the system, so that we have from Euler's theorem on homogeneous functions

$$nV = \sum_{i=1}^{3N} x_i \frac{\partial V}{\partial x_i}. \tag{2}$$

Theorem (1) remains valid also in quantum mechanics if we understand by averaging the quantum mechanical averages (over the wave functions); we shall denote them here by $\langle \rangle$.

It is convenient for a proof to use matrix methods. (We note that a proof using the Schrödinger equation in the coordinate representation is appreciably more cumbersome, the same applies to the proof of the sum rule in problem 83 of section 7.) We have

$$2\hat{T} - n\hat{V} = \sum_{i=1}^{3N} \left(\frac{\hat{p}_i^2}{\mu_i} - \hat{x}_i \frac{\partial \hat{V}}{\partial x_i} \right). \tag{3}$$

Using the operator relations

$$\hat{\dot{x}}_i = \frac{\hat{p}_i}{\mu_i}, \quad \hat{\dot{p}}_i = -\frac{\partial \hat{V}}{\partial x_i}$$

we write equation (3) in the following form:

$$2\hat{T} - n\hat{V} = \sum_{i=1}^{3N} (\dot{\hat{x}}_i \hat{p}_i + \hat{x}_i \dot{\hat{p}}_i). \tag{3'}$$

This operator relation is, of course, valid in any representation.

We now go over to the energy representation, and use as a base the orthonormal set of eigenfunctions ψ_m of the Hamiltonian $\hat{H} = T + \hat{V}$. Since the motion is supposed to be bounded these functions refer to a discrete energy spectrum so that we are dealing with matrices all of whose matrix elements are finite—this is essential for our proof.

Let us write down the diagonal matrix element (m, m) of the operator (3′) or, in other words, the average value of the operator (3′) with respect to the wave function of the stationary state ψ_m.

From the rules for multiplication and addition of matrices we get

$$\langle 2T - nV \rangle_{\psi_m} \equiv (2T - nV)_{mm} = \sum_{i=1}^{3N} \sum_{l} [(\dot{x}_i)_{ml} (p_i)_{lm} + (x_i)_{ml} (\dot{p}_i)_{lm}]. \tag{4}$$

We have also the well-known matrix relations

$$(\dot{x}_i)_{ml} = \frac{i}{\hbar} (E_m - E_l) (x_i)_{ml}, \quad (\dot{p}_i)_{lm} = \frac{i}{\hbar} (E_l - E_m) (p_i)_{lm},$$

where E_m and E_l are energy levels of the system.

The expression under the double sum sign in equation (4) is thus equal to zero and we have found the required quantum mechanical generalisation of relation (1),

$$n\langle V \rangle = 2\langle T \rangle.$$

In particular, if $n = 2$ (harmonic oscillations) we have $\langle V \rangle = \langle T \rangle$ and if $n = -1$ (Coulomb interaction) $\langle V \rangle = -2\langle T \rangle$.

35. The quantity p_0 is the average value of the momentum of the particle.

36. For the proof we use the following operator relation which is valid when the Hamiltonian does not explicitly depend on the time and when there is no magnetic field:

$$\hat{\boldsymbol{p}} = \frac{i\mu}{\hbar} (\hat{H}\boldsymbol{r} - \boldsymbol{r}\hat{H}).$$

The average value of $\hat{\boldsymbol{p}}$ in the state ψ of the discrete spectrum is given by

$$\bar{\boldsymbol{p}} = \frac{i\mu}{\hbar} \int \psi^* (\hat{H}\boldsymbol{r} - \boldsymbol{r}\hat{H}) \psi \, d\tau.$$

Since $\hat{H}$ is Hermitean, we have

$$\bar{\hat{p}} = \frac{i\mu}{\hbar}\int(\hat{H}^*\psi^*.\mathbf{r}\psi - \psi^*\mathbf{r}\hat{H}\psi)\,d\tau.$$

Since for a stationary state

$$\hat{H}\psi = E\psi, \quad \hat{H}^*\psi^* = E\psi^*,$$

we find finally

$$\bar{\hat{p}} = 0.$$

37. (a) $\hat{x}(t) = x - \frac{i\hbar}{\mu}t\frac{\partial}{\partial x}$,

(b) $\hat{x}(t) = x\cos\omega t - \frac{i\hbar}{\mu\omega}\sin\omega t\frac{\partial}{\partial x}$.

38. $\overline{(\Delta x)_t^2} = \overline{(\Delta x)_0^2} + \frac{t}{\mu}\overline{[(\Delta p)(\Delta x)+(\Delta x)(\Delta p)]_0} + \frac{t^2}{\mu^2}\overline{(\Delta p)_0^2}$.

Note. From the relation we have derived one can easily determine the moment τ when the quantity $\overline{(\Delta x)^2}$ has a minimum value. The function $\overline{(\Delta x)_t^2}$ is symmetrical with respect to the point τ. In the case where the wave function at $t = 0$ has the form $\psi(x) = \varphi(x)\exp(ip_0 x/\hbar)$ [$\varphi(x)$ is a real function] we have $\tau = 0$ (see problem 18a of section 3).

39. (a) $\overline{(\Delta x)_t^2} = \overline{(\Delta x)_{t=0}^2} + \frac{\hbar^2 t^2}{\mu^2}\int_{-\infty}^{+\infty}\left(\frac{\partial\varphi}{\partial x}\right)^2 dx$,

(b) $\overline{(\Delta x)_t^2} = \overline{(\Delta x)_{t=0}^2}\cdot\cos^2\omega t + \frac{\hbar^2}{\mu^2\omega^2}\int_{-\infty}^{+\infty}\left(\frac{\partial\varphi}{\partial x}\right)^2\sin^2\omega t\,dx$.

40. The equations of motion for $\hat{a}(t)$ and $\hat{a}^+(t)$ are (see problem 27 of section 3)

$$i\hbar\dot{\hat{a}} = [\hat{a},\hat{H}]_-, \quad i\hbar\dot{\hat{a}}^+ = [\hat{a}^+,\hat{H}]_-,$$

and using the expression

$$\hat{H} = \hbar\omega(\hat{a}^+\hat{a} + \tfrac{1}{2})$$

for the Hamiltonian and the commutation relation

$$[\hat{a},\hat{a}^+]_- = 1,$$

both of which follow from the equation of problem 8 of section 1, we get

$$i\hbar\dot{\hat{a}} = \hbar\omega\hat{a}, \quad i\hbar\dot{\hat{a}}^+ = -\hbar\omega\hat{a}^+,$$

from which follows

$$\hat{a}(t) = \hat{a}\,e^{-i\omega t}, \quad \hat{a}^+(t) = \hat{a}^+ e^{i\omega t}.$$

41. The Hamiltonian of the system,

$$\hat{H} = \frac{\hat{p}^2}{2\mu} + \tfrac{1}{2}\mu\omega^2 \hat{x}^2 - f(t)\,\hat{x},$$

can be expressed in terms of $\hat{a}$ and $\hat{a}^+$ (see problem 8 of section 1) as follows:

$$\hat{H} = \hat{H}_0 + \hat{H}_1,$$

where

$$\hat{H}_0 = \hbar\omega(\hat{a}^+\hat{a} + \tfrac{1}{2}), \quad \hat{H}_1 = -f(t)\sqrt{(\hbar/2\mu\omega)}\,(\hat{a} + \hat{a}^+).$$

We look for a solution of the Schrödinger equation in the form

$$\psi(t) = \exp(-i\hat{H}_0 t/\hbar)\,\varphi(t). \tag{1}$$

In that case φ will satisfy the equation

$$i\hbar\dot{\varphi} = \hat{H}_1(t)\,\varphi, \tag{2}$$

where (see preceding problem)

$$\begin{aligned}\hat{H}_1(t) &= \exp(i\hat{H}_0 t/\hbar)\,\hat{H}_1 \exp(-i\hat{H}_0 t/\hbar)\\ &= -f(t)\sqrt{(\hbar/2\mu\omega)}\,[\hat{a}\exp(-i\omega t) + \hat{a}^+\exp(i\omega t)].\end{aligned} \tag{3}$$

From (2) and (3) it follows that we can write

$$\varphi(t) = \exp\left\{\frac{-i}{\hbar}\left[\int_{-\infty}^{t} f(t')\sqrt{\left(\frac{\hbar}{2\mu\omega}\right)}\,\hat{a}\exp(-i\omega t')\,dt' + \int_{-\infty}^{t} f(t')\sqrt{\left(\frac{\hbar}{2\mu\omega}\right)}\,\hat{a}^+\exp(-i\omega t')\,dt'\right]\right\}\varphi(-\infty).$$

At $t = +\infty$ we have

$$\varphi(\infty) = \exp[iC(\hat{a} + \hat{a}^+)]\,\varphi(-\infty), \tag{4}$$

where

$$C = \sqrt{\left(\frac{1}{2\mu\hbar\omega}\right)}\int_{-\infty}^{+\infty} f(t')\,e^{i\omega t'}\,dt',$$

which is real because $f(t)$ is an even function of t.

If we now use the result of problem 5 of section 3 and the commutation relation $[\hat{a}, \hat{a}^+]_- = 1$, we get from (4)

$$\varphi(\infty) = \exp(-\tfrac{1}{2}C^2)\exp(iC\hat{a}^+)\exp(iC\hat{a})\,\varphi(-\infty). \tag{5}$$

If the ψ_n are the oscillator eigenfunctions, we have (see problem 8 of section 1)

$$(\hat{a}^+)^n\,\psi_0 = \sqrt{(n!)}\,\psi_n, \quad \hat{a}\psi_0 = 0. \tag{6}$$

At $t = -\infty$ the oscillator is in its ground state, and it follows from (1) that we can thus write $\varphi(-\infty) = \psi_0$, and using (5) and (6) we find

$$\varphi(\infty) = \exp(-\tfrac{1}{2}C^2)\sum_n \frac{(iC)^n}{\sqrt{(n!)}}\psi_n,$$

and the probability W_{n0} that at $t = +\infty$ the nth excited state is occupied is thus given by a Poisson distribution

$$W_{n0} = e^{-\nu}\frac{\nu^n}{n!},$$

with

$$\nu = \frac{1}{2\hbar\mu\omega}\left|\int_{-\infty}^{+\infty} f(t)\, e^{i\omega t}\, dt\right|^2.$$

(a) For $f(t) = f_0 \exp(-t^2/\tau^2)$ we find

$$\nu = \frac{\pi f_0^2 \tau^2}{2\hbar\mu\omega}\exp(-\tfrac{1}{2}\omega^2\tau^2).$$

(b) For $f(t) = \dfrac{f_0}{(t/\tau)^2+1}$ we find

$$\nu = \frac{\pi^2 f_0^2 \tau^2}{2\hbar\mu\omega}e^{-2\omega\tau}.$$

42. We shall denote the first of the required operators by $\hat{M}_{(p)}$ and have thus:

in the $\boldsymbol{r}$-representation $\quad \left(\dfrac{\hat{1}}{r} = \dfrac{1}{r}\right)$: $\quad \dfrac{\hat{1}}{r}\psi(\boldsymbol{r}) = \varphi(\boldsymbol{r}),$ (1)

in the $\boldsymbol{p}$-representation $\quad \left(\dfrac{\hat{1}}{r} = \hat{M}_{(p)}\right)$: $\quad \hat{M}_{(p)}g(\boldsymbol{p}) = f(\boldsymbol{p}),$ (2)

where $g(\boldsymbol{p})$ and $f(\boldsymbol{p})$ are the Fourier components of the functions $\psi(\boldsymbol{r})$ and $\varphi(\boldsymbol{r})$ (the wave functions in the $\boldsymbol{p}$-representation) which are connected to them as follows:

$$g(\boldsymbol{p}) = \frac{1}{(2\pi\hbar)^{\frac{3}{2}}}\int \psi(\boldsymbol{r}) \exp\left[-\frac{i(\boldsymbol{p}.\boldsymbol{r})}{\hbar}\right] d^3\boldsymbol{r}, \tag{3}$$

$$f(\boldsymbol{p}) = \frac{1}{(2\pi\hbar)^{\frac{3}{2}}}\int \varphi(\boldsymbol{r}) \exp\left[-\frac{i(\boldsymbol{p}.\boldsymbol{r})}{\hbar}\right] d^3\boldsymbol{r}. \tag{4}$$

Apart from here, it is also necessary for other problems to know the Fourier transform of $1/r$. We shall find it in the simplest possible way. We write:

$$\frac{1}{r} = \int a(\boldsymbol{k}) \exp[i(\boldsymbol{k}.\boldsymbol{r})]\, d^3\boldsymbol{k}. \tag{5}$$

Operate on both sides of this equation with the Laplacian ∇^2. We have clearly the identity

$$\nabla^2 \exp[i(\boldsymbol{k}.\boldsymbol{r})] = -k^2 \exp[i(\boldsymbol{k}.\boldsymbol{r})], \tag{6}$$

and also

$$\nabla^2 \frac{1}{r} = -4\pi\delta(\boldsymbol{r}), \tag{7}$$

where $\delta(\boldsymbol{r})$ is the δ-function. Equation (7) is easily proved by writing ∇^2 as div grad, integrating both sides over a small volume and using Gauss's theorem. Applying the Laplacian operator on to equation (5) leads then to

$$-4\pi\delta(\boldsymbol{r}) = -\int k^2 a(\boldsymbol{k}) \exp[i(\boldsymbol{k}.\boldsymbol{r})]\, d^3\boldsymbol{k}.$$

From the Fourier theorem and the definition of the δ-function it then follows that

$$k^2 a(\boldsymbol{k}) = \frac{1}{(2\pi)^3}\int 4\pi\delta(\boldsymbol{r}) \exp[-i(\boldsymbol{k}.\boldsymbol{r})]\, d^3\boldsymbol{r} = \frac{1}{2\pi^2},$$

so that

$$a(\boldsymbol{k}) = a(k) = \frac{1}{2\pi^2 k^2};$$

and finally

$$\frac{1}{r} = \frac{1}{2\pi^2}\int \frac{\exp[i(\boldsymbol{k}.\boldsymbol{r})]\, d^3\boldsymbol{k}}{k^2}. \tag{8}$$

To determine the form of $\hat{M}_{(p)}$ we substitute expression (1) into equation (4), use equation (8), substituting $\boldsymbol{k} = \boldsymbol{p}'/\hbar$ in it. We then get

$$f(\boldsymbol{p}) = \frac{1}{(2\pi\hbar)^{\frac{3}{2}}}\int \frac{1}{r}\psi(\boldsymbol{r}) \exp\left[-\frac{i(\boldsymbol{p}.\boldsymbol{r})}{\hbar}\right] d^3\boldsymbol{r}$$

$$= \frac{1}{(2\pi\hbar)^{\frac{3}{2}}}\cdot\frac{1}{2\pi^2\hbar}\iint \frac{\psi(\boldsymbol{r})}{(p')^2} \exp\left[-\frac{i}{\hbar}(\boldsymbol{p}-\boldsymbol{p}'.\boldsymbol{r})\right] d^3\boldsymbol{p}'\, d^3\boldsymbol{r}.$$

Integrating over $\boldsymbol{r}$ and using equation (3) we have

$$f(\boldsymbol{p}) = \frac{1}{2\pi^2\hbar}\int \frac{g(\boldsymbol{p}-\boldsymbol{p}')\, d^3\boldsymbol{p}'}{(p')^2}.$$

Choosing $\boldsymbol{p}-\boldsymbol{p}'$ as a new variable of integration and expressing $f(\boldsymbol{p})$ through equation (2) we get finally:

$$\hat{M}_{(p)}\, g(\boldsymbol{p}) = \frac{1}{2\pi^2\hbar}\int \frac{g(\boldsymbol{p}')\, d^3\boldsymbol{p}'}{(\boldsymbol{p}-\boldsymbol{p}')^2}$$

(it is clear from the derivation that the integration is over the whole of $\boldsymbol{p}'$-space).

We see thus that $\hat{\frac{1}{r}} \equiv \hat{M}_{(\boldsymbol{p})}$ is an integral operator with the kernel

$$G(\boldsymbol{p}, \boldsymbol{p}') = \frac{1}{2\pi^2 \hbar (\boldsymbol{p} - \boldsymbol{p}')^2}.$$

In a completely similar way we find for the $\hat{\frac{1}{p}}$ operator in the $\boldsymbol{r}$-representation, which we denote by $\hat{L}_{(\boldsymbol{r})}$, the equation

$$\hat{L}_{(\boldsymbol{r})} \psi(\boldsymbol{r}) = \frac{1}{2\pi^2 \hbar} \int \frac{\psi(\boldsymbol{r}')\, d^3 \boldsymbol{r}'}{(\boldsymbol{r} - \boldsymbol{r}')^2},$$

that is, $\left(\hat{\frac{1}{p}}\right) \equiv \hat{L}_{(\boldsymbol{r})}$ is an integral operator with kernel

$$G(\boldsymbol{r}, \boldsymbol{r}') = \frac{1}{2\pi^2 \hbar (\boldsymbol{r} - \boldsymbol{r}')^2}.$$

43*. Let $\psi(x)$ be a function, acted upon by $\hat{\frac{1}{p}}$ (we shall here and henceforth drop the index x of p). Let $\varphi(x)$ be the result of this operation.

$$\varphi = \hat{\frac{1}{p}} \psi. \tag{1}$$

Our problem is now to find a way to determine $\varphi(x)$ for given $\psi(x)$. To solve this problem we go over to the momentum representation. Let $f(p)$ and $g(p)$ be the functions corresponding to ψ and φ in the momentum representation, so that equation (1) can be written in the form

$$g(p) = \frac{1}{p} f(p). \tag{1'}$$

Equation (1′) shows that the function $g(p)$ has in general a pole at $p = 0$, so that it does not satisfy the general requirement asked of the wave functions in quantum mechanics. In order that this requirement is not violated, it is necessary that $f(p)$ tends to zero for $p = 0$:

$$f(0) = 0. \tag{2}$$

Since $f(p)$ is the pth Fourier component of the function $\psi(x)$, we can write this condition in the form

$$\int_{-\infty}^{\infty} \psi(x)\, dx = 0. \tag{2'}$$

We shall assume in the following that condition (2) is satisfied, because only then can the operator $\widehat{\frac{1}{p}}$ be determined uniquely.

To solve the problem it is necessary to transform equation (1′) to the coordinate representation. We multiply thereto equation (1′) by $\exp(ipx/\hbar)/\sqrt{(2\pi\hbar)}$ and integrate over all values of p:

$$\varphi(x) = \frac{1}{\sqrt{(2\pi\hbar)}} \int_{-\infty}^{\infty} \frac{f(p)}{p} \exp(ipx/\hbar)\, dp. \tag{3}$$

It is now convenient to consider the integration in equation (3) as an integration along the real axis in the complex p-plane. Since the function under the integral sign does not have a pole at $p = 0$ by virtue of condition (2), we can displace the contour of integration in the region of this point downwards (to the region $\operatorname{Im} p < 0$):

$$\varphi(x) = \frac{1}{\sqrt{(2\pi\hbar)}} \int_{(\smile)} \frac{f(p)}{p} \exp(ipx/\hbar)\, dp. \tag{3′}$$

We now express in this equation $f(p)$ in terms of $\psi(x)$:

$$f(p) = \frac{1}{\sqrt{(2\pi\hbar)}} \int_{-\infty}^{\infty} \psi(x') \exp(-ipx/\hbar)\, dx'. \tag{4}$$

Changing the order of integration, we have

$$\varphi(x) = \frac{1}{2\pi\hbar} \int_{-\infty}^{\infty} \psi(x')\, dx' \int_{(\smile)} \frac{\exp[ip(x-x')/\hbar]}{p}\, dp. \tag{5}$$

Evaluating first of all the integral over the momentum in equation (5) we have

$$\int_{(\smile)} \frac{\exp[ip(x-x')/\hbar]}{p}\, dp. \tag{6}$$

The value of this integral will depend on the sign of $x - x'$. For $x > x'$ we close the contour by a semicircle of infinite radius in the upper half-plane. The integral over this half-plane is equal to zero, so that the integral (6) is equal to $2\pi i$ times the residue at the only pole $p = 0$, which lies inside the contour. The residue is equal to unity so that

$$\int_{(\smile)} \frac{\exp[ip(x-x')/\hbar]}{p}\, dp = 2\pi i \quad (x > x'). \tag{7}$$

For $x < x'$ we close the contour with an infinite semicircle in the lower half-plane. As there are no poles within the contour we get clearly

$$\int_{(\smile)} \frac{\exp[ip(x-x')/\hbar]}{p}\, dp = 0 \quad (x < x'). \tag{8}$$

Substituting expressions (7) and (8) into equation (5) we find

$$\varphi(x) = \frac{i}{\hbar}\int_{-\infty}^{x} \psi(x')\, dx', \tag{9}$$

which is the solution of our problem.

Similarly $\frac{\hat{1}}{p}$ is an integral operator of the form

$$\frac{\hat{1}}{p} = \frac{i}{\hbar}\int_{-\infty}^{x} \dots dx'. \tag{10}$$

This result is natural since we looked for the inverse of the momentum operator $\hat{p} = (\hbar/i)(\partial/\partial x)$. Indeed, letting the operator $\hat{p}$ act upon equation (1), we have

$$\hat{p}\varphi(x) = \hat{p}\frac{\hat{1}}{p}\psi(x) = \psi(x). \tag{11}$$

Verifying this result by using equation (9) we find

$$\hat{p}\varphi(x) = \frac{\hbar}{i}\frac{\partial}{\partial x}\left[\frac{i}{\hbar}\int_{-\infty}^{x} \psi(x')\, dx'\right] = \psi(x). \tag{11'}$$

It is necessary to note the following regarding equations (9) and (10). The contour of integration in equation (3) can be changed also in a different manner, for instance, displacing it into the upper half-plane. In the corresponding evaluation we are then instead of equation (9) led to the equation

$$\varphi(x) = -\frac{i}{\hbar}\int_{x}^{\infty} \psi(x')\, dx'. \tag{9'}$$

Expression (9′) for $\varphi(x)$ differs from expression (9) by

$$\frac{i}{\hbar}\int_{-\infty}^{\infty} \psi(x')\, dx', \tag{12}$$

which vanishes for functions satisfying condition (2′). The definition (9) of the action of the operator $\frac{\hat{1}}{p}$ is thus the same as (9′) and is generally unique.

When condition (2′) is not satisfied, the definitions (9) and (9′) are different which shows the non-uniqueness of the action of the operator $\frac{\hat{1}}{p}$.

Similarly we find easily for the operator $\frac{\hat{1}}{x}$ in the p-representation the following expression

$$\frac{\hat{1}}{x} = -\frac{i}{\hbar}\int_{-\infty}^{p} dp'. \tag{13}$$

This operator is thus also an integral operator; the operator equation (13) is equivalent to the relation

$$\frac{\hat{1}}{x} f(p) = -\frac{i}{\hbar}\int_{-\infty}^{p} f(p')\,dp'.$$

44. We choose our origin at the left-hand edge of the well; the wave functions of the stationary states of the particle inside the well are then of the form

$$\psi_n(x) = \sqrt{\left(\frac{2}{a}\right)} \sin\frac{\pi n x}{a} \quad (n = 1, 2, 3, \ldots), \tag{1}$$

where a is the width of the well; outside the well $\psi \equiv 0$.

The required matrix elements of the coordinate x are equal to

$$x_{mn} = \int_{-\infty}^{\infty} \psi_m^*(x)\, x\psi_n(x)\,dx = \int_0^a x\psi_m\psi_n\,dx. \tag{2}$$

Substituting expression (1) into equation (2) we find after a simple calculation

$$x_{mn} = \frac{2}{a}\cdot\frac{1}{2}\int_0^a x\left[\cos\frac{\pi(m-n)x}{a} - \cos\frac{\pi(m+n)x}{a}\right]dx$$

$$= \frac{a}{\pi^2}\left\{\frac{1-\cos[\pi(m+n)]}{(m+n)^2} - \frac{1-\cos[\pi(m-n)]}{(m-n)^2}\right\} = \frac{4a}{\pi^2}\,\frac{[(-1)^{m-n}-1]\,mn}{(m^2-n^2)^2}. \tag{3}$$

From equation (3) it is clear that the only non-diagonal $(m \neq n)$ matrix elements of the coordinate which are different from zero are those with an odd difference $(m-n)$, that is, the elements corresponding to transitions with a change of parity. [We remind ourselves that all wave functions (1) have a well-defined parity (with respect to the centre of the well), namely $n = 1, 3, 5, \ldots$ correspond to even and $n = 2, 4, 6, \ldots$ to odd parity.] The diagonal elements $(m = n)$ are different from zero. One can find their value, for instance, by letting $(m-n) \to 0$ in expression (3), and using the fact that $1 - \cos x \approx \frac{1}{2}x^2$. The result is

$$x_{mm} = \tfrac{1}{2}a.$$

(This result is immediately obvious, since x_{mm} is the average value of x in the mth stationary state which is equal to $\frac{1}{2}a$.)

The selection rule obtained can be visualised by the following simple considerations. The matrix of the coordinate x is the sum of the matrices of the quantities $x - \frac{1}{2}a$ and $\frac{1}{2}a$. The elements of the first of these matrices are different from zero only if the product $\psi_m \psi_n$ is of odd parity, that is, for odd $m-n$ since $x - \frac{1}{2}a$ is an odd function with respect to the centre of the well. The elements of the other matrix are clearly equal to $\frac{1}{2}a\, \delta_{mn}$, that is, it is a diagonal matrix. The selection rules found follow thus immediately.

To find the matrix of the momentum $(p_x)_{mn} \equiv p_{mn}$ we use the well-known relation,

$$p_{mn} = \mu(\dot{x})_{mn} = \mu \frac{i}{\hbar}(E_m - E_n)\, x_{mn}, \tag{4}$$

where μ is the particle mass, and $E_n = (\pi^2 \hbar^2 / 2\mu a^2)\, n^2$ are the energy levels of the particle in the well.

Using the result of equation (3) for x_{mn}, we get

$$p_{mn} = \frac{2i\hbar}{a} \cdot \frac{[(-1)^{m-n} - 1]\, mn}{(m^2 - n^2)}. \tag{5}$$

One can easily verify that now the diagonal elements are equal to zero; this should, of course, be the case since these elements are the average value of the momentum in the corresponding stationary states which should be equal to zero, since the motion is bounded. The selection rule for the non-diagonal elements of p_{mn} is clearly the same as for x_{mn}.

We note finally that it follows from equation (5) that $p^*_{mn} = p_{nm}$, that is, that the momentum operator is, as it should be, Hermitean. The same is true for x_{mn}.

45. In the formula

$$\overline{L^2} = \frac{\int \psi^* \hat{L}^2 \psi \, d\tau}{\int \psi^* \psi \, d\tau}$$

the denominator is, of course, positive so that we must prove that the enumerator is positive.

Since $\hat{L}$ is Hermitean, we get

$$\int \psi^* \hat{L}\hat{L}\psi \, d\tau = \int (\hat{L}\psi)(\hat{L}^* \psi^*)\, d\tau = \int |\hat{L}\psi|^2 \, d\tau > 0,$$

as had to be proved.

46. (1) It is not linear. Indeed, denoting the operator of complex conjugation by $\hat{K}$, we have

$$\hat{K}(c_1\psi_1+c_2\psi_2) = c_1^*\hat{K}\psi_1+c_2^*\hat{K}\psi_2 \neq c_1\hat{K}\psi_1+c_2\hat{K}\psi_2.$$

(2) It is not Hermitean.

Any Hermitean operator satisfies the condition

$$\int\psi^*\hat{L}\varphi\,d\tau = \int\varphi\hat{L}^*\psi^*\,d\tau. \tag{1}$$

In the present case we have

$$\int\psi^*\hat{K}\varphi\,d\tau = \int\psi^*\varphi^*\,d\tau,$$

$$\int\varphi\hat{K}^*\psi^*\,d\tau = \int\varphi\psi\,d\tau,$$

so that condition (1) is not satisfied.

(3) The operator K is its own complex conjugate, that is, it is real. Indeed,

$$\hat{K}\psi = \psi^*,$$

so that

$$\hat{K}^*\psi^* = \psi.$$

Denoting ψ^* by φ we have

$$\hat{K}^*\varphi = \varphi^* = \hat{K}\varphi,$$

or

$$\hat{K}^* = \hat{K}.$$

(The operator of complex conjugation changes i to $-i$, and its complex conjugate changes $-i$ to i, which is the same.)

47. $\hat{T} = e^{ia\hat{p}/h}$. The proof follows by using a Taylor expansion for $f(x)$.

48. Let us write the required expansion as a power series,

$$\sum_{n=0}^{\infty}\lambda^n\hat{L}_n,$$

where $\hat{L}_n$ is an operator to be determined.

We have thus

$$(\hat{A}-\lambda\hat{B})^{-1} = \sum_{n=0}^{\infty}\lambda^n\hat{L}_n.$$

Multiplying by $\hat{A}-\lambda\hat{B}$, for instance, from the left we get

$$1 = \sum_{n=0}^{\infty}\lambda^n(\hat{A}-\lambda\hat{B})\hat{L}_n = A\hat{L}_0+\sum_{n=1}^{\infty}\lambda^n(\hat{A}\hat{L}_n-\hat{B}\hat{L}_{n-1}).$$

Comparing the coefficients of the same power of λ on the left and on the right we get:

$$\hat{A}\hat{L}_0 = 1, \quad \hat{A}\hat{L}_n - \hat{B}\hat{L}_{n-1} = 0,$$

whence

$$\hat{L}_0 = \hat{A}^{-1}, \quad \hat{L}_n = \hat{A}^{-1}\hat{B}\hat{L}_{n-1}.$$

We have thus

$$(\hat{A} - \lambda\hat{B})^{-1} = \hat{A}^{-1} + \lambda\hat{A}^{-1}\hat{B}\hat{A}^{-1} + \lambda^2\hat{A}^{-1}\hat{B}\hat{A}^{-1}\hat{B}\hat{A}^{-1} + \ldots .$$

In the case where A and B are numbers, this expression goes over into the usual one

$$\frac{1}{A - \lambda B} = \frac{1}{A} + \lambda\frac{B}{A^2} + \lambda^2\frac{B^2}{A^3} \ldots .$$

49. It is, generally speaking, impossible to give a definite answer to the question on the basis of commutability or non-commutability of the operators $\hat{A}$ and $\hat{B}$ only.

Indeed, on the one hand, commutability is, strictly speaking, not necessary in order that the corresponding quantities have simultaneously well-defined values. For the motion in a spherically symmetric field, for instance, we have simultaneously well-defined values (equal to zero) of all three components of the angular momentum for all states where the square of the total angular momentum is equal to zero, although none of these three operators commutes with the others.

The non-commutability of the operators $\hat{A}$ and $\hat{B}$ (first case) does thus not exclude the simultaneous realisation of some of their eigenvalues.

On the other hand, from the fact alone that $\hat{A}$ and $\hat{B}$ commute (second case) it is, generally speaking, impossible to conclude that the quantity B has a well-defined value in the state ψ_A.

The fact is that commutability of $\hat{A}$ and $\hat{B}$ is sufficient only for the existence of a complete set of states with simultaneously well-defined values of A and B (we shall denote the wave functions of such states by ψ_{AB}), but not to guarantee that any given wave function ψ_A belongs to the set of functions ψ_{AB}. This is connected with the phenomenon of degeneracy, where one value of A corresponds to a number of values of B so that the wave function ψ_A is in the general case a linear superposition of functions ψ_{AB} of the form

$$\psi_A = \sum_{B'} c_{B'}\psi_{AB'},$$

which reduces to ψ_{AB} only for a very definite choice of the coefficients $(c_{B'} = \delta_{B'B})$.

We shall consider two simple cases:

(1) For the motion of a system of particles in a spherically symmetric field in a state with a well-defined value of the square of the total angular momentum M^2 the value of its z-component M_z can either have a well-defined value, or not.

Indeed, all non-zero values of M^2 are degenerate with respect to different values of M_z (this is connected with the existence of the operators $\hat{M}_x$ and $\hat{M}_y$ which commute with $\hat{M}^2$, but not with $\hat{M}_z$); ψ_M is thus a superposition of the form

$$\psi_{M^2} = \sum_{M_z} c_{M_z} \psi_{M^2, M_z}.$$

In particular, one can find such states ψ_{M^2} where M_z has definitely not a well-defined value, notwithstanding the fact that the operator $\hat{M}_z$ commutes with $\hat{M}^2$; such states are those with well-defined values of M_x or M_y (except for the state where $M^2 = 0$, when also $M_x = M_y = M_z = 0$).

(2) Each energy level E, except the ground state level, of the hydrogen atom corresponds to several values of the square of the angular momentum M^2 so that the states with a well-defined energy are, generally speaking, not characterised by a well-defined value of M^2, but are a superposition of states with different values of M^2, and only for a definite choice of the coefficients of this superposition do they reduce to the functions ψ_{E,M^2}. From the fact that E is well defined it does not therefore follow that M^2 is well defined in spite of the commutability of $\hat{M}^2$ with the energy operator $\hat{H}$.

The commutability of $\hat{A}$ and $\hat{B}$ therefore does not in general yield an answer to our original question. To solve it we must let the operator $\hat{B}$ act upon the function ψ_A. If the equation

$$\hat{B}\psi_A \doteq \lambda\psi_A$$

is satisfied (λ is a number), ψ_A is a wave function of $\hat{B}$ and the quantity B has in the state ψ_A a well-defined value (λ); in the opposite case B has in this state no well-defined value.

Only in the particular case of no degeneracy (for instance, for a one-dimensional motion bounded at least at one side) it follows automatically from the commutability of $\hat{A}$ and $\hat{B}$ that any eigenfunction of $\hat{A}$ is at the same time an eigenfunction of $\hat{B}$, and vice versa, so that for any state either both the quantities A and B or neither of them have well-defined values.

50. Let us denote the totality of the coordinates (or, more generally, the totality of the independent variables of the chosen representation) of the system by one letter x, the Hamiltonian of the system by $\hat{H}(x)$

and its wave function by $\psi(x)$. Let the transformation under consideration change the set of coordinates x to the set x':

$$x \to x'. \tag{1}$$

Let $\hat{O}$ be the operator of the transformation (1). The invariance of the Hamiltonian with respect to this transformation means that

$$\hat{H}(x') = \hat{H}(x). \tag{2}$$

Let $\hat{O}$ act upon $\hat{H}\psi$.

From the definition of $\hat{O}$ and equation (2) it follows that

$$\hat{O}\hat{H}(x)\,\psi(x) = \hat{H}(x')\,\psi(x') = \hat{H}(x)\,\psi(x') = \hat{H}(x)\,\hat{O}\psi(x),$$

which is equivalent to the operator equation

$$\hat{O}\hat{H} = \hat{H}\hat{O}. \tag{3}$$

We introduce the operator $\hat{O}^{-1}$ which is the inverse of $\hat{O}$,

$$\hat{O}\hat{O}^{-1} = \hat{1}, \tag{4}$$

where $\hat{1}$ is the unit operator, that is, the operator of the identical transformation. Multiplying equation (3) from the right by $\hat{O}^{-1}$ we get

$$\hat{O}\hat{H}\hat{O}^{-1} = \hat{H}. \tag{3'}$$

This relation, which is identical with the commutation relation (3) expresses the invariance of $\hat{H}$ under the transformation $\hat{O}$ in operator form.

51. We are concerned here with finding the combined integrals of motion (excluding parity) of a system of interacting point particles in given external fields. In classical mechanics this problem is solved by finding the coordinate (or time) transformations under which the Lagrangian of the system under consideration is invariant.

In quantum mechanics the solution of this problem is formally even somewhat simpler: the integrals of motion are those mechanical quantities, the operators of which commute with the Hamiltonian of the system (and also which do not depend explicitly on the time). The commutability of those operators among themselves depends in general not on the form of the external field. The pairs $\hat{M}_x$ and $\hat{M}_y$, $\hat{M}_x$ and $\hat{p}_y$, $\hat{p}_x$ and $\hat{I}$ ($\hat{I}$ is the inversion operator), and so on, do not commute, while $\hat{M}_x$ and $\hat{I}$, $\hat{M}_x$ and $\hat{p}_x$, and so on, do always commute.

We shall restrict ourselves to the case where the external field can be derived from a potential (excluding thus magnetic fields, and so on) and we shall number the particles by an index $i = 1, 2, \ldots, N$. We can then write the Hamiltonian of the system in the form ($i, k = 1, 2, \ldots, N$)

$$\hat{H} = \sum_{i=1}^{N}\left(-\frac{\hbar^2}{2\mu_i}\nabla_i^2\right) + U_{\text{int}}(\ldots, |\boldsymbol{r}_i - \boldsymbol{r}_k|, \ldots) + U_{\text{ext}}(\boldsymbol{r}_1, \ldots, \boldsymbol{r}_N) \equiv \hat{H}_0 + U_{\text{ext}},$$

where U_{int} is the potential energy of the interaction between the particles in the system.

All operators of interest to us, namely:

$$\hat{\boldsymbol{P}}(\hat{P}_x, \hat{P}_y, \hat{P}_z), \quad \hat{\boldsymbol{M}}(\hat{M}_x, \hat{M}_y, \hat{M}_z), \quad \hat{M}^2, \quad I,$$

commute clearly with $\hat{H}_0$ (which corresponds to a bound system of point particles; the non-commutability of the operators $\hat{M}_x, \hat{M}_y, \ldots$ among themselves leads clearly to the degeneracy of the energy eigenvalues and the total angular momentum eigenvalues with respect to these quantities, $M_x, \ldots$) and we thus have to evaluate only the commutability with the operator of the potential energy of the external field,

$$U_{\text{ext}} = \sum_{i=1}^{N} U_{i\,\text{ext}}(\boldsymbol{r}_i).$$

We shall write down $A = \text{const}$ for every integral of motion A.

(1) $$U_{\text{ext}} = 0, \quad \partial\hat{H}/\partial t = 0.$$

The integrals of motion of the total system are: the energy E, the three components of the angular momentum M_x, M_y, and M_z, the total angular momentum M^2, the three components of the linear momentum P_x, P_y, and P_z (that is, $\boldsymbol{P}$), and the parity I. The energy levels are degenerate so that each state of the system is characterised by a complete set of a smaller number of (commuting) mechanical quantities, for instance, by

$$P_x, P_y, P_z$$

or

$$E, M^2, M_z \quad \text{and so on.}$$

A similar situation also holds when external fields are present.

(2) Take the z-axis along the axis of the cylinder.

$$U_{\text{ext}} = \sum_{i=1}^{N} U_i(\rho_i),$$

so that the operators

$$\hat{P}_z = \frac{\hbar}{i}\sum_{i=1}^{N}\frac{\partial}{\partial z_i}, \quad \hat{M}_z = \frac{\hbar}{i}\sum_{i=1}^{N}\frac{\partial}{\partial \varphi_i} \quad \text{and} \quad \hat{I}$$

commute with U_{ext} (and thus with $\hat{H}$).

Since $\partial\hat{H}/\partial t = 0$, we get finally

$$E = \text{const}, \quad P_z = \text{const}, \quad M_z = \text{const}, \quad I = \text{const}.$$

(3) Take the xy-plane in the plane.

$$U_{\text{ext}} = \sum_{i=1}^{N} U_i(|z_i|), \quad \frac{\partial\hat{H}}{\partial t} = 0.$$

The operators

$$\hat{P}_x = \frac{\hbar}{i} \sum_{i=1}^{N} \frac{\partial}{\partial x_i}, \quad \hat{P}_y = \frac{\hbar}{i} \sum_{i=1}^{N} \frac{\partial}{\partial y_i},$$

$$\hat{M}_z = \frac{\hbar}{i} \sum_{i=1}^{N} \frac{\partial}{\partial \varphi_i} = \frac{\hbar}{i} \sum_{i=1}^{N} \left(x_i \frac{\partial}{\partial y_i} - y_i \frac{\partial}{\partial x_i} \right),$$

and also $\hat{I}$ commute with U_{ext} (that is, also with $\hat{H}$). We have thus $E = \text{const}$, $M_z = \text{const}$, $P_x = \text{const}$, $P_y = \text{const}$, $I = \text{const}$.

(4) Take the origin at the centre of the sphere.

$$U_{\text{ext}} = \sum_{i=1}^{N} U_i(r_i), \quad \frac{\partial \hat{H}}{\partial t} = 0.$$

The operators $\hat{M}_x, \hat{M}_y, \hat{M}_z, \hat{M}^2$ when expressed in polar coordinates act only upon the variables ϑ_i, φ_i and commute thus with U_{ext} and $\hat{H}$.

Hence: $E = \text{const}$, $M_x = \text{const}$, $M_y = \text{const}$, $M_z = \text{const}$, $M^2 = \text{const}$, $I = \text{const}$.

(5) Let the half-plane be the xy-plane on one side of the y-axis.

$$U_{\text{ext}} = \sum_{i=1}^{N} U_i(x_i, |z_i|), \quad \hat{P}_y = \frac{\hbar}{i} \sum_{i=1}^{N} \frac{\partial}{\partial y_i}.$$

Thus we have

$$E = \text{const}, \quad P_y = \text{const}.$$

(6) Let the points be along the z-axis.

$$U_{\text{ext}} = \sum_{i=1}^{N} U_i(\rho_i, z_i), \quad \hat{M}_z = \frac{\hbar}{i} \sum_{i=1}^{N} \frac{\partial}{\partial \varphi_i}.$$

We have thus: $E = \text{const}$, $M_z = \text{const}$.

In the particular case where both points carry the same charge (both in sign and in magnitude), $U_i = U_i(\rho_i, |z_i|)$ and parity I is thus also conserved.

(7) Let the field be along the z-axis.

Denote the field intensity by $\mathscr{E}(t)$ and the charges of the particles in the system by e_i, so that we have

$$U_{\text{ext}}(t) = -\mathscr{E}(t) \sum_{i=1}^{N} e_i z_i$$

(the same expression holds for the gravitational field, if we replace $\mathscr{E}$ by g and the charges e_i by the masses m_i).

$$U_{\text{ext}} \text{ commutes with } \quad \hat{M}_z = \frac{\hbar}{i} \sum_{i=1}^{N} \left(x_i \frac{\partial}{\partial y_i} - y_i \frac{\partial}{\partial x_i} \right),$$

$$\hat{P}_x = \frac{\hbar}{i} \sum_{i=1}^{N} \frac{\partial}{\partial x_i}, \quad \hat{P}_y = \frac{\hbar}{i} \sum_{i=1}^{N} \frac{\partial}{\partial y_i}.$$

Hence

$$M_z = \text{const}, \quad P_x = \text{const}, \quad P_y = \text{const}.$$

(8) Take the z-axis along the axis of the conductor.

$$U_{\text{ext}}(t) = f(t) \sum_{i=1}^{N} U_i(\rho_i), \quad \hat{M}_z = \frac{\hbar}{i} \sum_{i=1}^{N} \frac{\partial}{\partial \varphi_i}, \quad \hat{P}_z = \frac{\hbar}{i} \sum_{i=1}^{N} \frac{\partial}{\partial z_i}.$$

In this way we get

$$M_z = \text{const}, \quad P_z = \text{const}, \quad \text{and also} \quad I = \text{const}.$$

(9) Take the origin at the centre of the ellipsoid, so that we get

$$U_{\text{ext}} = \sum_{i=1}^{N} U_i(|x_i|, |y_i|, |z_i|),$$

since the ellipsoid possessed three, mutually perpendicular, planes of symmetry.

The Hamiltonian $\hat{H}$ commutes thus with the reflection operator $\hat{I}$ (and does not contain the time explicitly).

Hence,

$$E = \text{const}, \quad I = \text{const}.$$

(10) Take the z-axis along the axis of the screw, let its pitch be a and denote by φ the angle of rotation around the axis.

The function $U_{\text{ext}} = \sum_{i=1}^{N} U_i(\rho_i, \varphi_i, z_i)$ is invariant under the transformation

$$\varphi_i \to \varphi_i + \delta\varphi \quad (t = 1, 2, \ldots, N), \quad z_i \to z_i + \frac{\delta\varphi}{2\pi} a$$

because $\delta\varphi = 2\pi$ should correspond to $\delta z = a$, with ρ_i fixed,

$$\delta U_{\text{ext}} = \sum_{i=1}^{N} \frac{\partial U_{\text{ext}}}{\partial \varphi_i} \delta\varphi + \sum_{i=1}^{N} \frac{\partial U_{\text{ext}}}{\partial z_i} \delta z$$

$$= \delta\varphi \left(\sum_{i=1}^{N} \frac{\partial}{\partial \varphi_i} + \frac{a}{2\pi} \sum_{i=1}^{N} \frac{\partial}{\partial z_i} \right) U_{\text{ext}} = 0.$$

In other words, the operator $\hat{U}_{\text{ext}}$ (and thus the Hamiltonian $\hat{H}$) commutes with the operator

$$\sum_i \frac{\partial}{\partial \varphi_i} + \frac{a}{2\pi} \sum_i \frac{\partial}{\partial z_i} = \frac{i}{\hbar} \left(\hat{M}_z + \frac{a}{2\pi} \hat{P}_z \right).$$

We have thus

$$M_z + \frac{a}{2\pi} P_z = \text{const},$$

and also $E = \text{const}$, since $\partial \hat{H} / \partial t = 0$.

(11) Take the lateral edges of the prism in the z-direction.

$$U_{\text{ext}} = \sum_{i=1}^{N} U_i(x_i, y_i), \quad \frac{\partial \hat{H}}{\partial t} = 0, \quad \hat{P}_z = \frac{\hbar}{i} \sum_{i=1}^{N} \frac{\partial}{\partial z_i},$$

so that

$$E = \text{const}, \quad P_z = \text{const}.$$

In the particular case where the cross-section of the prism is a regular polygon of order $2n$ (with n an integer) the prism possesses two mutually perpendicular axes of symmetry so that

$$U_{\text{ext}} = \sum_i U_i(|x_i|, |y_i|)$$

(the z-axis is taken through the centre of the polygon) and because $\hat{I}$ and $\hat{H}$ commute, parity I is conserved.

(12) Take the z-axis along the axis of the cone.

If the cone is a single one,

$$U_{\text{ext}} = \sum_{i=1}^{N} U_i(\rho_i, z_i),$$

so that $E = \text{const}$ and $M_z = \text{const}$.

If the cone is double we have (the origin is taken at the vertex)

$$U_{\text{ext}} = \sum_i U_i(\rho_i, |z_i|),$$

and therefore also $I = \text{const}$.

(13) Let the z-axis be along the axis of the torus and the xy-plane in its equatorial plane.

We have in cylindrical coordinates

$$U_{\text{ext}} = \sum_{i=1}^{N} U_i(\rho_i, |z_i|),$$

and thus

$$E = \text{const}, \quad M_z = \text{const}, \quad I = \text{const}.$$

52. Let us consider two inertial systems of reference, $K(x, t)$ and $K'(x', t')$ (we shall only consider one-dimensional motion for the sake of simplicity; it is clear how one can generalise the result to a three-dimensional case), which move with a relative velocity u (see fig. 32).

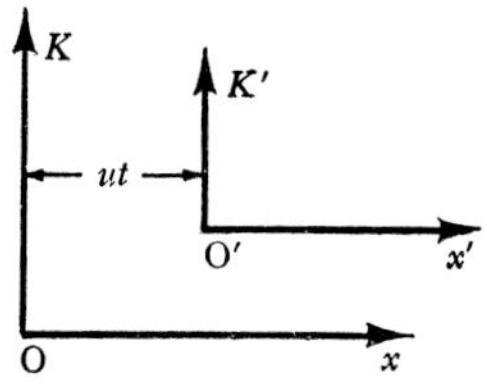

Fig. 32.

Let the potential energy of a particle in the field of a force acting upon it be $V'(x', t')$ in the system K'. From the Galilean transformation

$$\left.\begin{aligned} x &= x' + ut' \\ t &= t' \end{aligned}\right\} \tag{1}$$

we find the potential energy in the system K

$$V'(x-ut, t) \equiv V(x, t). \tag{2}$$

The Schrödinger equation for a particle of mass μ in the system K' is of the form

$$-\frac{\hbar^2}{2\mu}\frac{\partial^2 \psi'}{\partial x'^2} + V'\psi' = i\hbar\frac{\partial \psi'}{\partial t'}. \tag{3}$$

We must show that in the system K the Schrödinger equation

$$-\frac{\hbar^2}{2\mu}\frac{\partial^2 \psi}{\partial x^2} + V\psi = i\hbar\frac{\partial \psi}{\partial t} \tag{4}$$

is also valid, with V defined by equation (2) while the function ψ is because of its physical meaning completely analogous to ψ'; the quantity $|\psi(x, t)|^2 \equiv w(x, t)$ is the density of the probability of finding the particle at time t at the point x; this follows from the initially defined connection between ψ and ψ'.

Indeed, the fact that the particle is found at a given time at a given point in space does not depend on the choice of the system of reference. We can thus put the two corresponding probabilities equal to one another:

$$w'(x', t') = w(x, t), \tag{5}$$

or, using equation (1),

$$w'(x-ut, t) = w(x, t). \tag{5'}$$

It follows thus that the wave functions can differ from one another only by a phase factor of absolute magnitude unity,

$$\psi(x, t) = \exp(iS)\,\psi'(x', t') = \exp[iS(x, t)]\,\psi'(x-ut, t). \tag{6}$$

Moreover, a particle velocity v' in K' corresponds to a velocity $v = v' + u$ in K, so that we get for the momenta, $p = mv$ and $p' = mv'$,

$$p = p' + \mu u. \tag{7}$$

From this connection between p and p' it follows that we must have a relation analogous to equation (5),

$$w'(p', t') = w(p, t), \tag{8}$$

or, using equations (7) and (1),

$$w'(p-\mu u, t) = w(p, t). \tag{8'}$$

We now introduce an equation for the function $\psi(x, t)$ determined by equation (6).

Substituting into equation (3)

$$\psi' = \exp[-iS(x, t)]\,\psi(x, t), \tag{9}$$

using equation (2), and introducing by means of equation (1) the independent variables (x, t) we get

$$-\frac{\hbar^2}{2\mu}\frac{\partial^2\psi}{\partial x^2}+i\hbar\left(\frac{\hbar}{\mu}\frac{\partial S}{\partial x}-u\right)\frac{\partial\psi}{\partial x}+\left[V(x, t)+i\frac{\hbar^2}{2\mu}\frac{\partial^2 S}{\partial x^2}\right.$$
$$\left.+\frac{\hbar^2}{2\mu}\left(\frac{\partial S}{\partial x}\right)^2-\hbar u\frac{\partial S}{\partial x}-\hbar\frac{\partial S}{\partial t}\right]\psi = i\hbar\frac{\partial\psi}{\partial t}. \tag{10}$$

Up to now the function $S(x, t)$ {or rather $\exp[iS(x, t)]$} played the role of an arbitrary phase factor. We choose this function now in such a way that equation (10) goes over into the Schrödinger equation (4). For this it is necessary, as can be seen by comparing equations (10) and (4) that S satisfies the equations

$$\frac{\hbar}{\mu}\frac{\partial S}{\partial x}-u = 0, \tag{11}$$

$$i\frac{\hbar^2}{2\mu}\frac{\partial^2 S}{\partial x^2}+\frac{\hbar^2}{2\mu}\left(\frac{\partial S}{\partial x}\right)^2-\hbar u\frac{\partial S}{\partial x}-\hbar\frac{\partial S}{\partial t} = 0. \tag{11'}$$

These equations can easily be integrated. From equation (11) it follows that

$$S = \frac{\mu u}{\hbar}x+\varphi(t),$$

where $\varphi(t)$ is an arbitrary function of t.

We determine $\varphi(t)$ by substituting this expression into equation (11′), and we get

$$S = \frac{\mu u}{\hbar}x-\frac{\mu u^2}{2\hbar}t, \tag{12}$$

where we have omitted an arbitrary additive constant in S, since it would only lead to an inessential phase factor in ψ [see equation (6)].

From equations (6) and (12) it follows finally that the wave function in K is equal to

$$\psi(x, t) = \exp[i(\mu ux-\tfrac{1}{2}\mu u^2 t)/\hbar]\,\psi'(x-ut, t). \tag{13}$$

As should have been expected this wave function is the product of the wave function in K' and the function $\exp(i\mu ux/\hbar - i\mu u^2 t/2\hbar)$ which describes the free motion of a particle fixed in the system K' with respect to the system K.

We note that the wave function we have found satisfies condition (8′) automatically. Indeed, the momentum probability density at time t is equal to

$$w(p, t) = |c(p, t)|^2,$$

with

$$c(p, t) = \frac{1}{\sqrt{(2\pi\hbar)}} \int_{-\infty}^{\infty} \psi(x, t) \exp(-ipx/\hbar)\, dx. \tag{14}$$

Substituting here $\psi(x, t)$ from equation (13) we find easily

$$c(p, t) = \exp[i(\tfrac{1}{2}\mu u^2 - pu)t/\hbar] \cdot \frac{1}{\sqrt{(2\pi\hbar)}}$$
$$\times \int_{-\infty}^{\infty} \psi'(x', t) \exp[-i(p - \mu u)\, x'/\hbar]\, dx'.$$

The integral on the right-hand side is clearly [see equation (14)] $c'(p - \mu u, t)$. We have thus

$$c(p, t) = \exp[i(\tfrac{1}{2}\mu u^2 - pu)\, t/\hbar]\, c'(p - \mu u, t).$$

By taking the absolute square of both sides of this equation we get the required relation (8′).

We have thus shown the invariance of the non-relativistic Schrödinger equation under a Galilean transformation.

53. From the Schrödinger equation

$$i\hbar \frac{\partial \psi}{\partial t} = \hat{H}\psi,$$

it follows that any state changes with time according to

$$\psi(\xi, t) = \exp -(i\hat{H}t/\hbar)\, \psi(\xi, 0), \tag{1}$$

where the set of independent variables on which the wave function of the system depends is denoted by ξ.

The operator $\hat{A}$ in the Heisenberg representation is of the form

$$\hat{A}(t) = \exp(i\hat{H}t/\hbar)\, \hat{A} \exp(-i\hat{H}t/\hbar), \tag{2}$$

where $\hat{A}$ is the operator corresponding to the physical quantity A in the Schrödinger representation. For our problem we have

$$\hat{A}\psi(\xi, 0) = A\psi(\xi, 0). \tag{3}$$

Substituting here $\psi(\xi, 0) = \exp(i\hat{H}t/\hbar)\,\psi(\xi, t)$ and operating on it from the left with $\exp(-i\hat{H}t/\hbar)$, we get

$$\exp(-i\hat{H}t/\hbar)\,\hat{A}\exp(i\hat{H}t/\hbar)\,\psi(\xi, t) = A\psi(\xi, t). \tag{4}$$

Comparing the operator on the left-hand side of this equation with expression (2) we find

$$\hat{A}(-t)\,\psi(\xi, t) = A\psi(\xi, t).$$

Our proof is valid for the case where the Hamiltonian $\hat{H}$ of the system does not depend explicitly on the time. To generalise it for the case of a time-dependent Hamiltonian it is necessary to replace in this proof $\exp(-i\hat{H}t/\hbar)$ by the operator $\hat{U}(t, t_0)$ which transforms the wave function at time t_0 into the wave function at time t.

54. The states of the system are described by wave functions $\psi(t)$ satisfying the wave equation

$$i\hbar\frac{\partial\psi}{\partial t} = \hat{H}(t)\,\psi. \tag{1}$$

We have also

$$\hat{H}(t)\,\psi_n(t) = E_n(t)\,\psi_n(t). \tag{2}$$

Since the motion is bounded the spectrum of the "levels" E_n is discrete (the concept of an energy "level" has here a purely formal meaning, since the energy is not conserved for a time-dependent Hamiltonian), and since we are dealing with a one-dimensional case, there is no degeneracy. This last fact means in turn that the functions $\psi_n(t)$ are real (apart from an inessential phase factor).

In line with the problem we are considering we shall expand the wave function $\psi(t)$ in a series of the instantaneous eigenfunctions of the Hamiltonian $\psi_n(t)$. Assuming that we know the wave function at $t = 0$, we can write its expansion for $t \geqslant 0$ in the form

$$\psi = \sum_n c_n(t)\,\psi_n(t)\exp\left[-\frac{i}{\hbar}\int_0^t E_n(t')\,dt'\right]. \tag{3}$$

The coefficients $c_n(t)$ of this expansion also represent the wave function in the required representation. To write the wave equation in that representation means to find a set of equations which connect the coefficients c_n and their time derivatives $\dot{c}_n$.

Substituting expression (3) into equation (1) we get

$$i\hbar\sum_n\left(\dot{c}_n\psi_n + c_n\dot{\psi}_n - \frac{i}{\hbar}c_n\psi_n E_n\right)\exp\left[-\frac{i}{\hbar}\int_0^t E_n(t')\,dt'\right]$$
$$= \hat{H}\sum_n c_n\psi_n\exp\left[-\frac{i}{\hbar}\int_0^t E_n(t')\,dt'\right].$$

By virtue of equation (2) the right-hand side of this equation cancels against the last term on the left-hand side. Multiplying both sides from the left by $\psi_k^* = \psi_k$ and integrating over the coordinates of the system q we find, using the orthonormality of the ψ_n,

$$\dot{c}_k = -\sum_n c_n \exp\left[\frac{i}{\hbar}\int_0^t (E_k - E_n)\,dt'\right]\int \psi_k \dot{\psi}_n\,dq. \tag{4}$$

We shall transform the last integral in that equation. First of all we have for $n = k$,

$$\int \psi_k \dot{\psi}_k\,dq = \frac{1}{2}\frac{d}{dt}\int \psi_k^2\,dq = 0,$$

since

$$\int \psi_k^2\,dq = 1.$$

Further, by differentiating equation (2) with respect to the time, multiplying it from the left by $\psi_k^* = \psi_k$ and integrating over q, we have for $n \neq k$:

$$\int \psi_k \frac{\partial \hat{H}}{\partial t}\psi_n\,dq + \int \psi_k \hat{H}\dot{\psi}_n\,dq = E_n \int \psi_k \dot{\psi}_n\,dq. \tag{5}$$

Because the Hamiltonian is Hermitean and real we have, using equation (2),

$$\int \psi_k \hat{H}\dot{\psi}_n\,dq = \int (\hat{H}\psi_k)_n \dot{\psi} dq = E_k \int \psi_k \dot{\psi}_n\,dq.$$

Substituting this result into equation (5) we get

$$\int \psi_k \dot{\psi}_n\,dq = -\frac{\int \psi_k (\partial \hat{H}/\partial t)\psi_n\,dq}{E_k - E_n} \quad (n \neq k). \tag{6}$$

Finally, substituting expression (6) into equation (4) (we remember that the analogous term with $n = k$ is equal to zero) and also introducing the notation

$$E_k - E_n = \hbar\omega_{kn}, \quad \int \psi_k \left(\frac{\partial \hat{H}}{\partial t}\right)\psi_n\,dq = \left(\frac{\partial H}{\partial t}\right)_{kn}, \tag{7}$$

we find the required set of equations

$$\dot{c}_k = \sum_n{}' \frac{c_n}{\hbar\omega_{kn}}\left(\frac{\partial H}{\partial t}\right)_{kn} \exp\left(i\int_0^t \omega_{kn}\,dt'\right), \tag{8}$$

where the prime on the summation sign indicates that there is no term with $n = k$ in the sum. The quantity $(\partial H/\partial t)_{kn}$ is according to equation

(7) the matrix element of the time derivative of the Hamiltonian corresponding to the transition $k \to n$.

It is clear from our derivation that the equation (8) is completely equivalent to the original wave equation (1).

Let us compare the equations we have obtained with the original equations of the usual time-dependent perturbation theory. Then we have $\hat{H}(t) = \hat{H}_0 + \hat{H}_1(t)$, where $\hat{H}_0$ is the time-independent part of the Hamiltonian with the eigenvalues $E_n^{(0)}$ and the corresponding eigenfunctions $\psi_n^{(0)}(q)$. The latter can be chosen as the base for our representation [expansion (3) is clearly the natural generalisation of this expansion],

$$\psi = \sum_n a_n(t)\,\psi_n^{(0)} \exp\left(-\frac{i}{\hbar} E_n^{(0)}\, t\right). \tag{9}$$

Substituting expression (9) into equation (1), and so on, we get for the coefficients a_n the following set of equations:

$$\dot{a}_k = \frac{1}{i\hbar} \sum_n a_n\, (H_1)_{kn} \exp\,(i\omega_{kn}^{(0)}\, t), \tag{10}$$

where

$$(H_1)_{kn} = \int \psi_k^{(0)*}\, \hat{H}_1(t)\, \psi_n^{(0)}\, dq, \quad \hbar\omega_{kn}^{(0)} = E_k^{(0)} - E_n^{(0)}. \tag{11}$$

In contradistinction to the analogous equation (8) the summation includes now also a term with $n = k$, and the frequencies of the transitions $\omega_{kn}^{(0)}$ do not depend upon the time.

As with equation (8), the set of equations (10) is also completely equivalent to the original wave equation (1). It is clear that the exact solution of either of these sets is in general just as difficult as the solution of the original equation (1). Equations (8) and (10) are, however, very convenient for approximate solutions if there are small factors on their right-hand sides which make it possible to obtain the unknown c_n or a_n from their unperturbed (or initial) values, that is, from the given quantities, up to second-order quantities.

It is obvious that equation (10) is an adequate set of equations when the matrix elements $(H_1)_{kn}$ are sufficiently small, that is, when the time-dependent term $H_1(t)$ of the Hamiltonian is small ("ordinary" perturbation theory), while the "adequacy" of equation (8) lies in the smallness of the matrix elements $(\partial H/\partial t)_{kn}$, that is, in a sufficiently slow change of the Hamiltonian $\hat{H}(t)$. The application of the method of successive approximations in the latter case is sometimes called the adiabatic perturbation theory (see next problem) which was developed in 1926–8 by Born and Fock.

55. We start from the exact equation (8) of the preceding problem. In general, if the Hamiltonian of the system does not depend on the time ($\partial\hat{H}/\partial t = 0$) we get from equations (7) and (8) c_k = const for all k. If the derivative $\partial H/\partial t$ is sufficiently small, though different from zero, we can put the c_n in the right-hand side of equation (8) approximately equal to a constant, and for our problem put $c_n \approx \delta_{nm}$. We have thus for all $k \neq m$

$$\dot{c}_k \approx \frac{1}{\hbar\omega_{km}}\left(\frac{\partial H}{\partial t}\right)_{km} \exp\left(i\int_0^t \omega_{km}\,dt'\right) \tag{1}$$

(for $k = m$ we have, of course, to the same approximation, $\dot{c}_m = 0$). Integrating equation (1) for the given boundary conditions we get

$$c_k(t) = \frac{1}{\hbar}\int_0^t \frac{1}{\omega_{km}}\left(\frac{\partial H}{\partial t}\right)_{km} \exp i\int_0^{t'} \omega_{km}\,dt''\,dt' \quad (k \neq m). \tag{2}$$

From the derivation it follows that this adiabatic formula is valid provided the probability amplitudes for the states which are different from the initial state are small,

$$|c_k(t)| \ll 1 \quad \text{for} \quad k \neq m. \tag{3}$$

To estimate the order of magnitude of $|c_k(t)|$ it is sufficient to assume ω_{km} and $(\partial H/\partial t)_{km}$ to be approximately constant in equation (2); this equation then simplifies and we get

$$|c_k| \sim \frac{(\partial H/\partial t)(1/\omega_{km})}{E_k - E_m} \quad (k \neq m).$$

On the right-hand side we have the ratio of the change in the Hamiltonian over a time of the order of magnitude of the Bohr period $1/\omega_{km}$ for the $m \to k$ transition to the difference in the energy of the states m and k. From equation (3) it follows that if that ratio is small, one may apply the adiabatic perturbation theory.

Thus, if over a period of the order of the Bohr period, the Hamiltonian varies for some transition by an amount which is small compared to the "energy of the transition", the probability for that transition will be small; this is sometimes called adiabatic invariance from the correspondence with its classical counterpart.

56. The Hamiltonian of the oscillator is for $t \geqslant 0$ of the form

$$\hat{H}(t) = \frac{1}{2\mu}\hat{p}^2 + \tfrac{1}{2}\mu\omega^2[x - a(t)]^2, \tag{1}$$

where $a(t)$ is the coordinate of the point of suspension which for our problem is equal to $v_0 t$, with v_0 = const, the velocity of the motion of the point of suspension. The "instantaneous" eigenfunctions of the Hamiltonian (1) are of the form

$$\psi_n = \left(\frac{\mu\omega}{\pi\hbar}\right)^{\frac{1}{4}} \frac{1}{\sqrt{(2^n n!)}} \exp\left\{-\frac{\mu\omega}{2\hbar}[x-a(t)]^2\right\} H_n\left\{\sqrt{\left(\frac{\mu\omega}{\hbar}\right)}[x-a(t)]\right\},$$

and the matrix elements of the operator

$$\frac{\partial \hat{H}}{\partial t} = -\mu\omega^2[x-a(t)]\,\dot{a} = -\mu\omega^2 v_0[x-a(t)],$$

evaluated with these functions, are clearly different from zero only for the transition $n = 0 \rightarrow n = 1$ (we remember that the initial state was the ground state) and then they are equal to

$$\left(\frac{\partial H}{\partial t}\right)_{10} = -\mu\omega^2 v_0 \sqrt{\left(\frac{\hbar}{2\mu\omega}\right)}. \tag{2}$$

It is clear from equation (1) that the spectrum of the oscillator levels for a moving point of suspension will in general not change, that is, all ω_{km} are constant.

Owing to this and also to the fact that the quantity (2) is constant we can completely evaluate expression (2) of the preceding problem. For the probability amplitude of the first excited state we get thus, putting $\omega_{10} = \omega$,

$$c_1(t) \approx -\frac{1}{i\hbar\omega^2}\mu\omega^2 v_0 \sqrt{\left(\frac{\hbar}{2\mu\omega}\right)}[\exp(i\omega t)-1] = iv_0\sqrt{\left(\frac{\mu}{2\hbar\omega}\right)}[\exp(i\omega t)-1].$$

The probability to find the oscillator at time t in the first excited state is thus equal to

$$w_1(t) = |c_1(t)|^2 = \frac{\mu v_0^2}{\hbar\omega}(1-\cos\omega t);$$

that is, it oscillates in time. The probability that excitation has taken place during the process under consideration (that is for $t \geqslant T$) is equal to

$$w_1(T) = \frac{\mu v_0^2}{\hbar\omega}(1-\cos wT).$$

In order that the adiabatic approximation, which we have used, can be applied, the inequality $w_1(t) \ll 1$ must be satisfied for all t, or

$$v_0 \ll \sqrt{\left(\frac{\hbar\omega}{\mu}\right)}. \tag{3}$$

Expressed differently, the adiabatic approximation can be applied provided the velocity of motion of the point of suspension is small compared to the characteristic velocity of the oscillator in its ground state.

57. (i) The wave function $\psi(x, t)$ satisfies the equation

$$i\hbar\frac{\partial\psi}{\partial t} = -\frac{\hbar^2}{2\mu}\frac{\partial^2\psi}{\partial x^2} \quad (0 \leqslant x \leqslant a) \tag{1}$$

and the boundary conditions

$$\psi(0, t) = \psi(a, t) = 0,$$

while

$$a = a(t) = a_0 f(t) \quad (f(t) \text{ is a given function}).$$

Introducing a new variable $y = x/f(t)$, equation (1) is transformed to

$$i\hbar[f(t)]^2\frac{\partial\psi}{\partial t} + \frac{\hbar^2}{2\mu}\frac{\partial^2\psi}{\partial y^2} = \frac{i\hbar y}{2}\frac{d}{dt}(f^2)\frac{\partial\psi}{\partial y}, \tag{2}$$

and $\psi(y, t)$ satisfies the usual boundary conditions for non-moving walls,

$$\psi(0, t) = \psi(a_0, t) = 0. \tag{3}$$

Once again introducing a new variable $\tau = \int_c^t dt/[f(t)]^2$, with c a constant, and putting $f(t) \equiv \varphi(\tau)$, we get

$$i\hbar\frac{\partial\psi}{\partial\tau} = -\frac{\hbar^2}{2\mu}\frac{\partial^2\psi}{\partial y^2} + i\hbar\frac{d}{d\tau}[\ln\varphi(\tau)]\,y\frac{\partial\psi}{\partial y}. \tag{4}$$

We have thus reduced the problem to the solution of equation (4) with a Hamiltonian which depends explicitly on the "time" τ,

$$\hat{H}(y, \tau) = -\frac{\hbar^2}{2\mu}\frac{\partial^2}{\partial y^2} + i\hbar\frac{d}{d\tau}[\ln\varphi(\tau)]\,y\frac{\partial}{\partial y} \tag{4'}$$

with the normal boundary conditions (3).

It is clear that it is impossible to solve equation (4) for any arbitrary function $f(t) \equiv \varphi(\tau)$. An approximate solution can be obtained from perturbation theory, if $f(t)$ depends only weakly on the time. The given problem of slowly changing boundary conditions is thus reduced to a small time-dependent perturbation of the Hamiltonian.

We note that we must generalise the perturbation theory equations slightly because of the non-Hermitean character of the Hamiltonian (4′).

This question of the non-Hermitean character also arises because the integral of the function $\psi(y, t)$ or $\psi(y, \tau)$ depends explicitly on the time. Indeed, at any time t we have

$$\int_0^{a(t)} |\psi(x, t)|^2 \, dx = 1, \quad dx = f(t)\, dy,$$

so that

$$\int_0^{a_0} |\psi(y, t)|^2 \, dy = \frac{1}{f(t)}. \tag{5}$$

(ii) From equations (2) and (4) it follows directly for what choice of $f(t)$, y and t (or y and τ) can be separated. This particular case arises when

$$f(t) = \sqrt{\left(1+\frac{t}{t_0}\right)}, \tag{6}$$

where t_0 is an arbitrary constant. (This choice of $f(t)$ satisfies, in particular, the initial condition $f(0) = 1$ or $a(0) = a_0$.) Indeed, substituting expression (6) into equation (2) and changing the zero of the time scale by taking $t+t_0 \equiv t'$, we get

$$i\hbar t' \frac{\partial \psi}{\partial t'} + \frac{\hbar^2 t_0}{2\mu} \frac{\partial^2 \psi}{\partial y^2} = \frac{i\hbar}{2} y \frac{\partial \psi}{\partial y}. \tag{7}$$

In this equation the variables can clearly be separated and we have

$$\psi_n(y, t') = Y_n(y)\, T_n(t'), \tag{8}$$

where n is the number of the eigenfunction.

We shall not solve here the equations for $Y_n(y)$ and $T_n(t')$. We note only that the Y_n turn out to be superpositions of Hermite functions satisfying the boundary conditions $Y_n(0) = Y_n(a_0) = 0$, and that the time dependence of the absolute magnitude of the T_n follows easily from equations (5), (6), and (8).

$$|T_n(t')| = \text{const}\,(t')^{-\frac{1}{4}} = \text{const}\left(1+\frac{t}{t_0}\right)^{-\frac{1}{4}}. \tag{9}$$

Because of the orthogonality of the $Y_n(y)$ the time dependence (9) is also valid for the average absolute square $[\overline{|\psi|^2}]^{\frac{1}{2}}$ of the total wave function $\psi(x, t)$, which is a superposition of partial solutions (8),

$$\psi(x, t) = \sum_{n=1}^{\infty} C_n Y_n \left[\frac{x}{f(t)}\right] T_n(t+t_0).$$

It is clear that the corresponding decrease of the probability density follows simply from the increase of the well-width using the equation

$$\int_0^{a(t)} |\psi(x, t)|^2 \, dx = \text{const}.$$

58*. Let $x_1 = 0$ and $x_2 = a(t)$ be the coordinates of the walls of the well. The "instantaneous" wave functions and energy levels of the stationary states are of the form

$$\psi_n = \sqrt{\left[\frac{2}{a(t)}\right]} \sin\frac{\pi n x}{a(t)}, \quad E_n = \frac{\pi^2 \hbar^2 n^2}{2\mu[a(t)]^2} \quad (n = 1, 2, 3, \ldots). \tag{1}$$

For computational reasons it is convenient to consider first of all a well of very large, but finite depth V_0 and only afterwards go to the limit $V_0 \to \infty$ (compare problem 6 of section 1). If we take only the dominant terms of the expansion in $1/\sqrt{V_0}$ of the solution of the problem of a symmetric well of finite depth, we find easily

$$\psi_n(x) \approx \sqrt{\left(\frac{2}{a}\right)} \sin\left[\frac{n\pi}{a} x - \left(2\frac{x}{a} - 1\right)\frac{n\pi}{a}\frac{\hbar}{\sqrt{(2\mu V_0)}}\right] \tag{2}$$

[for the levels we can directly use equation (1)]. Let us evaluate the matrix elements of $\partial\hat{H}/\partial t$ corresponding to the transition $n \to m$. Clearly we have $\partial\hat{H}/\partial t = \partial V_{\rm rh}/\partial t$ where $V_{\rm rh} = V(x-a)$ is the potential of the right-hand wall. Furthermore, we have

$$\frac{\partial V_{\rm rh}}{\partial x} = V_0\, \delta[x - a(t)],$$

$$\frac{\partial V_{\rm rh}}{\partial t} = \frac{\partial V_{\rm rh}}{\partial a}\frac{da}{dt} = -\frac{\partial V_{\rm rh}}{\partial x}\dot{a} = -V_0\,\delta[x-a(t)]\,\dot{a}, \tag{3}$$

so that the matrix elements of $\partial H/\partial t$ taken between the "instantaneous" functions (2) are equal to ($\epsilon \to 0$)

$$\left(\frac{\partial H}{\partial t}\right)_{mn} \equiv \int_0^{a+\epsilon} \psi_m(x)\frac{\partial \hat{H}}{\partial t}\psi_n(x)\,dx = -V_0\,\dot{a}\int_0^{a+\epsilon}\psi_m(x)\,\psi_n(x)\,\delta(x-a)\,dx$$

$$= -V_0\, a\psi_m(a)\,\psi_n(a). \tag{4}$$

Putting $x = a$ in equation (2) we get

$$\psi_n(a) = \sqrt{\left(\frac{2}{a}\right)}(-1)^{n+1}\frac{n\pi}{a}\frac{\hbar}{\sqrt{(2\mu V_0)}},$$

which gives us after substitution into equation (4)

$$\left(\frac{\partial H}{\partial t}\right)_{mn} = (-1)^{n+m+1}\, nm\frac{\pi^2\hbar^2\,\dot{a}(t)}{\mu[a(t)]^3}. \tag{5}$$

We thus get for the coefficients c_m of the instantaneous functions in the first approximation of the adiabatic perturbation theory (see problem 55 of section 3)

$$\dot{c}_m = 2(-1)^{n+m+1}\frac{nm}{m^2-n^2}\frac{\dot{a}}{a}\exp\left[i\frac{\pi^2\hbar}{2\mu}(m^2-n^2)\int_0^t\frac{dt'}{a^2}\right] \quad (m\neq n). \tag{6}$$

Integrating we get for $c_m(t)$, if we use the initial conditions $c_m(0) = \delta_{mn}$, the expression

$$c_m(t) = 2(-1)^{n+m+1}\frac{nm}{m^2-n^2}\int_0^t dt'\frac{\dot{a}}{a}\exp\left[i\frac{\pi^2\hbar}{2\mu}(m^2-n^2)\int_0^{t'}\frac{dt''}{a^2}\right] \quad (m\neq n). \tag{7}$$

Let us now consider the problem under what conditions expression (7) leads to a transition probability from the initial state n to the mth state which is well defined. We note first of all that there are two possibilities: (a) the motion of the wall stops at $t = T$; (b) the motion continues for all values of t.

In the first case, equation (7) is only valid for $t \leqslant T$. For large values of t, after the wall has come to rest, the instantaneous functions are the exact ones and the coefficients c_m will no longer depend on the time. Since $\psi(x, t)$ must be continuous at $t = T$ we have

$$c_m(t > T) = c_m(T), \tag{8}$$

where the $c_m(T)$ are the values of the c_m obtained from equation (7) for $t = T$. Since the probability of finding the particle in the nth state is equal to $|c_m(t > T)|^2$ and does not change with time, there is in this case always a well-defined transition probability. This probability is equal to

$$w_{n\to m} = |c_m(T)|^2 = \frac{4n^2m^2}{(m^2-n^2)^2}\left|\int_0^T dt'\frac{\dot{a}}{a}\exp\left[i\frac{\pi^2\hbar}{2\mu}(m^2-n^2)\int_0^{t'}\frac{dt''}{a^2}\right]\right|^2. \tag{9}$$

Let us now consider the other case where the motion of the wall continues right up to $t\to\infty$. We shall restrict our considerations to motions where the width of the well tends to a finite limit, $a(t)\to a_\infty$ as $t\to\infty$.

The spectrum of the system at any time t remains then discrete and the relative position of the levels does not change (no so-called crossing-over of terms). The transition probability is determined by equation (9) in which T will tend to infinity. To solve the problem of the existence of this probability we assume that the wall approaches its limiting position as follows:

$$a_\infty - a(t) \sim \frac{1}{t^\gamma} \quad (\gamma > 0). \tag{10}$$

In this case we can everywhere replace a by a_∞ for sufficiently large values of t. We are thus led to an integral of the form

$$\int^{\infty} \frac{dt}{t^{\gamma+1}} \exp(i\omega_\infty t), \tag{11}$$

where $\omega_\infty = \pi^2\hbar(m^2-n^2)/2\mu a_\infty^2$. We only gave the upper limit, to stress that the expression under the integral sign is valid for $t\to\infty$. This integral will in our case ($\gamma > 0$) always tend to an upper limit which ensures the existence of the transition probability.

Let us discuss our earlier result (9) and find the conditions under which we can apply the adiabatic approximation which we have used. If T is much less than the Bohr period of the transition $n\to m$, that is, if

$$T \ll \frac{\mu a^2}{\hbar|m^2-n^2|},$$

the exponent in expression (9) is clearly ≈ 1 so that

$$w_{n\to m} \approx \frac{4n^2m^2}{(m^2-n^2)^2}\left(\int_0^T \frac{\dot a}{a}\,dt'\right)^2 \sim \frac{4n^2m^2}{(m^2-n^2)^2}\left(\frac{\Delta a}{a}\right)^2, \tag{9'}$$

where $\Delta a \equiv a(T)-a(0)$.

The condition of applicability of the adiabatic approximation is $w_{n\to m} \ll 1$ that is, in our case,

$$\frac{|\Delta a|}{a} \ll \frac{|m^2-n^2|}{mn}.$$

Of more interest is the case of protracted action,

$$T \gg \frac{\mu a^2}{\hbar|m^2-n^2|}.$$

In that case the exponent under the integral sign of equation (9) oscillates fast, so that the integral is of the order of

$$\frac{|\dot a|}{a}\,\frac{\mu a^2}{\hbar|m^2-n^2|}$$

and

$$w_{n\to m} \sim \frac{n^2m^2}{(m^2-n^2)^4}\left(\frac{\mu a\dot a}{\hbar}\right)^2. \tag{9''}$$

From equations (9) or (9″) we can see that the transition probability decreases steeply with increasing $|m-n|$. The condition of applicability of the adiabatic approximation, $w_{n\to m} \ll 1$, is now of the form

$$|\dot a| \ll \frac{(m^2-n^2)^2}{nm}\,\frac{\hbar}{\mu a}.$$

For $n \sim m \sim 1$ this last inequality means that the velocity with which the wall moves must be small compared to the characteristic velocity of a particle in the lowest level of the well $[v \sim \sqrt{(E_1/\mu)} \sim \hbar/\mu a]$ (compare the preceding problem).

We emphasise that the quantum number (n) of the initial state is an "adiabatic invariant" when the wall moves, while the energy of the particle in that level (E_n) may change very strongly [see equation (1)].

We note also that equation (9) is symmetric against a permutation of the initial and final states n and m, that is, it gives also the probability $w_{n\to m}$ for the transition $m \to n$. This happens because of the reversibility of quantum mechanics, that is, the symmetry of its equations with respect to a change in sign of the time.

59. 12·3 Å, 0·287 Å, and 0·0186 Å.

60. 150 eV and 0·082 eV.

61. As soon as $h/\mu v \sim d$.

62. $E = h^2/2\mu d^2 \approx 0{\cdot}2 \, . \, 10^{-2}$ eV to $1{\cdot}3 \, . \, 10^{-2}$ eV.

63. It is necessary that the dimensions of the regions or "obstacle" limiting the motion of the particles be appreciably larger than their de Broglie wavelengths (compare the situation in wave optics). We find thus that these dimensions must be $\gg 4 \, . \, 10^{-8}$ cm for the electron and $\gg 3 \, . \, 10^{-12}$ cm for the proton.

64. $\lambda = hc/E = 2{\cdot}07 \, . \, 10^{-14}$ cm. To study nuclear structure which has features of dimensions of the order of 10^{-12} cm, one needs particles with wavelengths much smaller than 10^{-12} cm.

4

Angular momentum; spin

1. (i) $[\hat{J}_+, \hat{J}_-]_- = 2\hbar\hat{J}_z, \quad [\hat{J}_\pm, \hat{J}_z]_- = \mp\hbar\hat{J}_\pm.$
(iii) $\hat{J}_z\{\hat{J}_\pm\,|a,b\rangle\} = (b \pm \hbar)\{\hat{J}_\pm\,|a,b\rangle\}, \quad \hat{J}^2\{\hat{J}_\pm\,|a,b\rangle\} = a\{\hat{J}_\pm\,|a,b\rangle\}.$ (1)
(iv) $\langle j', m'\,|\hat{J}_z\,|j,m\rangle = m\hbar\delta_{jj'}\,\delta_{mm'}.$ (2)
From (1) it follows that $\hat{J}_+|j,m\rangle = c\,|j,m+1\rangle$, and hence that $\langle j,m\,|\hat{J}_-\hat{J}_+|j,m\rangle = |c|^2$. As $\hat{J}_-\hat{J}_+ = \hat{J}^2 - \hat{J}_z(\hat{J}_z+1)$, we find that $|c|^2/\hbar^2 = j(j+1) - m(m+1)$, and hence, choosing the phase of c to be zero,

$$\langle j,m+1\,|\hat{J}_+|j,m\rangle = \hbar\{j(j+1)-m(m+1)\}^{1/2}, \tag{3}$$

while all other matrix elements vanish. Similarly we find

$$\langle j,m-1\,|\hat{J}_-|j,m\rangle = \hbar\{j(j+1)-m(m-1)\}^{1/2}. \tag{4}$$

From the definition of $\hat{J}_\pm$ we then find for the non-vanishing matrix elements of $\hat{J}_x$ and $\hat{J}_y$

$$\langle j,m+1\,|\hat{J}_x\,|j,m\rangle = \tfrac{1}{2}\hbar\sqrt{\{j(j+1)-m(m+1)\}} = \langle j,m\,|\hat{J}_x\,|j,m+1\rangle$$
$$= i\langle j\,.m+1\,|\hat{J}_y\,|j,m\rangle = -i\langle j,m\,|\hat{J}_y\,|j,m+1\rangle,$$

or in matrix form

$$\hat{J}_x/\hbar \equiv \begin{array}{c|cccccc} m \backslash m' & j & j-1 & j-2 & & -j+1 & -j \\ \hline j & 0 & \sqrt{[2j\,.\,1]} & 0 & \cdots & 0 & 0 \\ \cdot & \sqrt{[2j\,.\,1]} & 0 & \sqrt{[(2j-1)\,.\,2]} & \cdots & 0 & 0 \\ \cdot & 0 & \sqrt{[(2j-1)\,.\,2]} & 0 & \cdots & 0 & 0 \\ \cdot & \cdots & \cdots & \cdots & \cdots & \cdots & \cdots \\ \cdot & 0 & 0 & 0 & \cdots & 0 & \sqrt{[1\,.\,2j]} \\ -j & 0 & 0 & 0 & \cdots & \sqrt{[1\,.\,2j]} & 0 \end{array},$$

$$\hat{J}_y/\hbar \equiv \begin{array}{c|cccccc} m \backslash m' & j & j-1 & j-2 & & -j+1 & -j \\ \hline j & 0 & -i\sqrt{[2j\,.\,1]} & 0 & \cdots & 0 & 0 \\ \cdot & i\sqrt{[2j\,.\,1]} & 0 & -i\sqrt{[(2j-1)\,.\,2]} & \cdots & 0 & 0 \\ \cdot & 0 & i\sqrt{[(2j-1)\,.\,2]} & 0 & \cdots & 0 & 0 \\ \cdot & \cdots & \cdots & \cdots & \cdots & \cdots & \cdots \\ \cdot & 0 & 0 & 0 & \cdots & 0 & -i\sqrt{[1\,.\,2j]} \\ -j & 0 & 0 & 0 & \cdots & i\sqrt{[1\,.\,2j]} & 0 \end{array}.$$

$$\hat{J}_z/\hbar \equiv \begin{array}{c} {}^{m}\,{}^{m'} \\ j \\ \cdot \\ \cdot \\ \cdot \\ \cdot \\ -j \end{array} \begin{array}{c} \begin{array}{cccccc} j & j-1 & j-2 & & -j+1 & -j \end{array} \\ \begin{pmatrix} j & 0 & 0 & \dots & 0 & 0 \\ 0 & j-1 & 0 & \dots & 0 & 0 \\ 0 & 0 & j-2 & \dots & 0 & 0 \\ \dots & \dots & \dots & \dots & \dots & \dots \\ 0 & 0 & 0 & \dots & -j+1 & 0 \\ 0 & 0 & 0 & \dots & 0 & -j \end{pmatrix} \end{array}.$$

2. The required operator (which we shall denote by $\hat{R}_{\varphi_0}$) must transform any arbitrary function of the coordinates of the system $\psi(\boldsymbol{r}_1, \boldsymbol{r}_2, \dots, \boldsymbol{r}_i, \dots, \boldsymbol{r}_N)$ into the same function ψ but of coordinates which are rotated over the given angle.

For a rotation over an infinitesimal angle $d\varphi$ around the direction $\boldsymbol{n}$ (where $\boldsymbol{n}$ is a unit vector) the radius vector of the ith particle $\boldsymbol{r}_i$ will be increased by $d\boldsymbol{r}_i$ where

$$d\boldsymbol{r}_i = d\varphi\,[\boldsymbol{n} \wedge \boldsymbol{r}_i].$$

Under a rotation over a finite angle $\int d\varphi = \varphi_0$ around the same axis $\boldsymbol{n}$ we find for the finite change in $\boldsymbol{r}_i$

$$\int d\boldsymbol{r}_i = \int d\varphi\,[\boldsymbol{n} \wedge \boldsymbol{r}_i] = \left[\boldsymbol{n} \wedge \int \boldsymbol{r}_i\, d\varphi\right] \equiv \delta \boldsymbol{r}_i. \tag{1}$$

Let us consider the most general operator $\hat{R}$ which is the operator of arbitrary finite displacements of the particles in the system. (Strictly speaking these displacements must satisfy the condition $\operatorname{div} \delta\boldsymbol{r} = 0$.)

To determine $\hat{R}$ we have

$$\hat{R}\psi(\boldsymbol{r}_1, \dots, \boldsymbol{r}_i, \dots, \boldsymbol{r}_N) = \psi(\boldsymbol{r}_1 + \delta\boldsymbol{r}_1, \dots, \boldsymbol{r}_i + \delta\boldsymbol{r}_i, \dots, \boldsymbol{r}_N + \delta\boldsymbol{r}_N). \tag{2}$$

We shall expand the right-hand side of equation (2) in a Taylor series; this expansion can be written in the form

$$\begin{aligned} &\psi(\boldsymbol{r}_1 + \delta\boldsymbol{r}_1, \dots, \boldsymbol{r}_i + \delta\boldsymbol{r}_i, \dots, \boldsymbol{r}_N + \delta\boldsymbol{r}_N) \\ &\quad = \psi(\boldsymbol{r}_1, \dots, \boldsymbol{r}_i, \dots, \boldsymbol{r}_N) + \sum_{i=1}^{N} \left(\delta\boldsymbol{r}_i \cdot \frac{\partial\psi}{\partial\boldsymbol{r}_i}\right) + \frac{1}{2!}\left[\sum_{i=1}^{N}\left(\delta\boldsymbol{r}_i \cdot \frac{\partial}{\partial\boldsymbol{r}_i}\right)\right]^2 \psi + \dots \\ &\quad \equiv \left\{1 + \sum_{i=1}^{N}\left(\delta\boldsymbol{r}_i \cdot \frac{\partial}{\partial\boldsymbol{r}_i}\right) + \frac{1}{2!}\left[\sum_{i=1}^{N}\left(\delta\boldsymbol{r}_i \cdot \frac{\partial}{\partial\boldsymbol{r}_i}\right)\right]^2 + \dots\right\} \psi(\boldsymbol{r}_1, \dots, \boldsymbol{r}_i, \dots, \boldsymbol{r}_N). \end{aligned} \tag{3}$$

Comparing equations (3) and (2) we can convince ourselves that the expression within the square brackets is the operator $\hat{R}$.

Since that expression is the power-series expansion of the exponential function we can write $\hat{R}$ in the following symbolic form [we introduce the operators of the momenta of the particles $\hat{\boldsymbol{p}}_i = (\hbar/i)(\partial/\partial \boldsymbol{r}_i)]$:

$$\hat{R} = \exp\left[\sum_{i=1}^{N}\left(\delta\boldsymbol{r}_i\cdot\frac{\partial}{\partial\boldsymbol{r}_i}\right)\right] = \exp\left[\frac{i}{\hbar}\sum_{i=1}^{N}(\delta\boldsymbol{r}_i\cdot\hat{\boldsymbol{p}}_i)\right]. \tag{4}$$

In the particular case of a parallel translation of all particles over a distance $\boldsymbol{a}$ we have

$$\delta\boldsymbol{r}_1 = \delta\boldsymbol{r}_2 = \ldots = \delta\boldsymbol{r}_i = \ldots = \delta\boldsymbol{r}_N = \boldsymbol{a},$$

and equation (4) will be of the form

$$\hat{R}_a \equiv \hat{T}_a = \exp\left[\frac{i}{\hbar}(\boldsymbol{a}.\hat{\boldsymbol{P}})\right],$$

where $\hat{\boldsymbol{P}} = \sum_{i=1}^{N}\hat{\boldsymbol{p}}_i$ is the operator of the total momentum of the system of particles.

We can now go over to the required case of a rotation over a finite angle.

If we take the $\boldsymbol{n}$-axis along the z-axis of a system of cylindrical coordinates (ρ, φ, z) and substitute $\delta\varphi_i = \varphi_0$, $\delta\rho_i = \delta z_i = 0$, we have

$$\left(\delta\boldsymbol{r}_i\cdot\frac{\partial}{\partial\boldsymbol{r}_i}\right) = \delta\varphi_i\frac{\partial}{\partial\varphi_i} + \delta\rho_i\frac{\partial}{\partial\rho_i} + \delta z_i\frac{\partial}{\partial z_i} = \varphi_0\frac{\partial}{\partial\varphi_i}.$$

Substituting this into equation (4) and bearing in mind that $(\hbar/i)(\partial/\partial\varphi_i)$ is the operator of the z-component of the angular momentum of the ith particle, we have

$$\hat{R}_{\varphi_0} = \exp\left(\varphi_0\sum_i\frac{\partial}{\partial\varphi_i}\right) = \exp\left(\frac{i}{\hbar}\varphi_0\hat{M}_z\right), \tag{5}$$

where $\hat{M}_z = \sum_{i=1}^{N}\frac{\hbar}{i}\frac{\partial}{\partial\varphi_i}$ is the operator of the component of the angular momentum of the system of particles along the axis of rotation.

Finally we have in vector form $[M_z \equiv M_n = (\boldsymbol{n}.\boldsymbol{M})]$:

$$\hat{R}_{\varphi_0,n} = \exp\left(\frac{i}{\hbar}\varphi_0(\boldsymbol{n}.\hat{\boldsymbol{M}})\right).$$

3. It is well known that the wave function transforms as follows:

$$\psi'(r) = \{1 + i(\boldsymbol{d\alpha}.\hat{\boldsymbol{l}})\}\psi(r), \tag{1}$$

under an infinitesimal rotation of the system of coordinates. Here $\boldsymbol{d\alpha}$ is a vector directed along the axis of rotation and with magnitude equal to the angle of rotation and $\hat{\boldsymbol{l}}$ is the operator of the orbital angular momentum.

Let us consider first of all a rotation over an angle $d\alpha$ around the z-axis. For such a rotation we have

$$\psi'(r, \theta, \varphi) = \psi(r, \theta, \varphi + d\alpha) = \psi(r, \theta, \varphi) + \frac{\partial \psi}{\partial \varphi} d\alpha. \tag{2}$$

Comparing equations (2) and (1) we find

$$\hat{l}_z = -i \frac{\partial}{\partial \varphi}.$$

To find the form of the operator $\hat{l}_x$ in spherical coordinates we perform a rotation around the x-axis.

We have then

$$\psi'(r, \theta, \varphi) = \psi(r, \theta + d\theta, \varphi + d\varphi) = \left\{1 + \left(\frac{d\theta}{d\alpha}\frac{\partial}{\partial \theta} + \frac{d\varphi}{d\alpha}\frac{\partial}{\partial \varphi}\right) d\alpha\right\} \psi(r, \theta, \varphi),$$

from which it follows that

$$\hat{l}_x = -i \left(\frac{d\theta}{d\alpha}\frac{\partial}{\partial \theta} + \frac{d\varphi}{d\alpha}\frac{\partial}{\partial \varphi}\right).$$

Let us evaluate $d\theta/d\alpha$ and $d\varphi/da$. One sees easily that

$$z' - z = -y\, d\alpha,$$

$$y' - y = z\, d\alpha,$$

and since $z' = r \cos\theta$, $y' = r \sin\theta \sin\varphi$ we have

$$\frac{d\theta}{d\alpha} = -\sin\varphi, \quad \frac{d\varphi}{d\alpha} = -\cot\theta . \cos\varphi,$$

and thus

$$\hat{l}_x = i \left(\sin\varphi \frac{\partial}{\partial \theta} + \cot\theta . \cos\varphi \frac{\partial}{\partial \varphi}\right).$$

Similarly we find

$$\hat{l}_y = -i \left(\cos\varphi \frac{\partial}{\partial \theta} - \cot\theta . \sin\varphi \frac{\partial}{\partial \varphi}\right).$$

4. If the operator $\hat{\Omega}$ changes a function φ into a new function ψ,

$$\psi = \hat{\Omega}\varphi,$$

we must in the rotated frame have the relation

$$\psi' = \hat{\Omega}'\varphi',$$

and since [see equation (2) of problem 2 of section 4]

$$\hat{R}\psi = \psi', \quad \hat{R}\varphi = \varphi',$$

the required relation follows.

5. It can be shown (cf.problem 47 of section 3 and problem 2 of section 4) that momentum and angular momentum of a system are simply connected with the operators of infinitesimal translations and infinitesimal rotations of the system (more exactly, they are proportional to the difference of those operators and the unit operator).†

Any translation commutes with any other translation so that the operators of the different components of the momentum will also commute. Two rotations around two non-parallel axes will, however, not commute with one another which corresponds to the non-commutability of the operators of the different components of the angular momentum.

10. $\hat{l}_{z'} = \hat{l}_x \cos(xz') + \hat{l}_y \cos(yz') + \hat{l}_z \cos(zz')$.

12. If we use the $|j, m\rangle$ representation (see problem 1 of section 4) for the spin functions and if the rotation is characterized by the Euler angles ϑ, ψ, and φ, we can get the new spin function $\begin{pmatrix} \psi_1' \\ \psi_0' \\ \psi_{-1}' \end{pmatrix}$ from the old spin function $\begin{pmatrix} \psi_1 \\ \psi_0 \\ \psi_{-1} \end{pmatrix}$ by three successive rotations:

$$\begin{pmatrix} \psi_1' \\ \psi_0' \\ \psi_{-1}' \end{pmatrix} = \hat{R}_3 \hat{R}_2 \hat{R}_1 \begin{pmatrix} \psi_1 \\ \psi_0 \\ \psi_{-1} \end{pmatrix},$$

where $\hat{R}_1$ corresponds to a rotation around the z-axis over an angle φ, $\hat{R}_2$ to a rotation around the x-axis over an angle ϑ, and $\hat{R}_3$ to a rotation around the z-axis over an angle ψ.

In the representation used $\hat{R}_1$ and $\hat{R}_3$ are given by the relations [compare equation (5) of the solution to problem 2 of section 4 and the matrix representation of the angular momentum operators in problem 1 of section 4]

$$\hat{R}_1 = \begin{pmatrix} e^{i\varphi} & 0 & 0 \\ 0 & 1 & 0 \\ 0 & 0 & e^{-i\varphi} \end{pmatrix}, \quad \hat{R}_3 = \begin{pmatrix} e^{i\psi} & 0 & 0 \\ 0 & 1 & 0 \\ 0 & 0 & e^{-i\psi} \end{pmatrix}.$$

† See, for instance, H. A. Kramers, *Quantum Mechanics* (Amsterdam: North Holland Publishing Company, 1957) or A. S. Davydov, *Quantum Mechanics* (Oxford: Pergamon Press, 1965).

To find $\hat{R}_2$, we use the fact that $\hat{R}_2 = \exp(i\vartheta\hat{M}_x/\hbar)$. Using the matrix representation of $\hat{M}_x$ given in problem 1 of section 4, we find that

$$\hat{R}_2 = \begin{pmatrix} \cos^2\frac{1}{2}\vartheta & \frac{i}{\sqrt{2}}\sin\vartheta & -\sin^2\frac{1}{2}\vartheta \\ \frac{i}{\sqrt{2}}\sin\vartheta & \cos\vartheta & \frac{i}{\sqrt{2}}\sin\vartheta \\ -\sin^2\frac{1}{2}\vartheta & \frac{i}{\sqrt{2}}\sin\vartheta & \cos^2\frac{1}{2}\vartheta \end{pmatrix}.$$

We thus finally get

$$\begin{pmatrix} \psi_1' \\ \psi_0' \\ \psi_{-1}' \end{pmatrix} = \begin{pmatrix} e^{i(\psi+\varphi)}\cos^2\frac{1}{2}\vartheta & \frac{i}{\sqrt{2}}e^{i\psi}\sin\vartheta & -e^{i(\psi-\varphi)}\sin^2\frac{1}{2}\vartheta \\ \frac{i}{\sqrt{2}}e^{i\varphi}\sin\vartheta & \cos\vartheta & \frac{i}{\sqrt{2}}e^{-i\varphi}\sin\vartheta \\ -e^{-i(\psi-\varphi)}\sin^2\frac{1}{2}\vartheta & \frac{i}{\sqrt{2}}e^{-i\psi}\sin\vartheta & e^{-i(\psi+\varphi)}\cos^2\frac{1}{2}\vartheta \end{pmatrix} \begin{pmatrix} \psi_1 \\ \psi_0 \\ \psi_{-1} \end{pmatrix}.$$

13. Using the result of the preceding problem we find for the case $M = 1$,

$$w(+1) = \cos^4\frac{\theta}{2}, \quad w(0) = \tfrac{1}{2}\sin^2\theta, \quad w(-1) = \sin^4\frac{\theta}{2};$$

for $M = 0$,

$$w(+1) = \tfrac{1}{2}\sin^2\theta, \quad w(0) = \cos^2\theta, \quad w(-1) = \tfrac{1}{2}\sin^2\theta;$$

and finally for $M = -1$,

$$w(+1) = \sin^4\frac{\theta}{2}, \quad w(0) = \tfrac{1}{2}\sin^2\theta, \quad w(-1) = \cos^4\frac{\theta}{2}.$$

14.
$$\hat{\sigma}_x = \begin{pmatrix} 0 & 1 \\ 1 & 0 \end{pmatrix}, \quad \hat{\sigma}_y = \begin{pmatrix} 0 & -i \\ i & 0 \end{pmatrix}, \quad \hat{\sigma}_z = \begin{pmatrix} 1 & 0 \\ 0 & -1 \end{pmatrix}.$$

Comparing this with the representation in problem 1 of section 4 we see that we can interpret $\frac{1}{2}\hbar\hat{\sigma}$ as the operator of an angular momentum with $j = \frac{1}{2}$; that is, the operator of a spin-$\frac{1}{2}$ particle.

16. Using the result of problem 2 of section 4 we find for the transformation operator $\hat{T}$ (compare problem 12 of section 4)

$$\hat{T} = e^{\frac{1}{2}i\psi\hat{\sigma}_z}\, e^{\frac{1}{2}i\vartheta\hat{\sigma}_x}\, e^{\frac{1}{2}i\varphi\hat{\sigma}_z} = \begin{pmatrix} e^{\frac{1}{2}i(\varphi+\psi)}\cos\frac{1}{2}\vartheta & ie^{\frac{1}{2}i(\psi-\varphi)}\sin\frac{1}{2}\vartheta \\ ie^{\frac{1}{2}i(\varphi-\psi)}\sin\frac{1}{2}\vartheta & e^{-\frac{1}{2}i(\varphi+\psi)}\cos\frac{1}{2}\vartheta \end{pmatrix}.$$

17. The Euler angles ϑ, φ, and ψ are connected with α, β, γ, and Φ through the relations

$$\cos\tfrac{1}{2}\Phi = \cos\tfrac{1}{2}\vartheta\cos\tfrac{1}{2}(\varphi+\psi), \quad \alpha\sin\tfrac{1}{2}\Phi = \sin\tfrac{1}{2}\vartheta\cos\tfrac{1}{2}(\varphi-\psi),$$
$$\beta\sin\tfrac{1}{2}\Phi = \sin\tfrac{1}{2}\vartheta\sin\tfrac{1}{2}(\varphi-\psi), \quad \gamma\sin\tfrac{1}{2}\Phi = \cos\tfrac{1}{2}\Phi\sin\tfrac{1}{2}(\varphi+\psi).$$

Hence, using the results of the previous problem we find

$$\hat{T} = \cos\tfrac{1}{2}\Phi + i(\alpha\hat{\sigma}_x + \beta\hat{\sigma}_y + \gamma\hat{\sigma}_z)\sin\tfrac{1}{2}\Phi.$$

19. $w(+\tfrac{1}{2}) = \cos^2\dfrac{\theta}{2}, \quad w(-\tfrac{1}{2}) = \sin^2\dfrac{\theta}{2}.$

The average value of the spin component is equal to $\tfrac{1}{2}\cos\theta$.

20. We use the matrix of the transformation of the components of the spin function under a rotation of the coordinate axes. This matrix was given in the solution to problem 16 of section 4. Using this matrix we find the spin function in the new system of coordinates

$$\psi_1' = \exp[\tfrac{1}{2}i(\varphi+\psi)+i\alpha]\cos\frac{\theta}{2}.\cos\delta + i\exp[\tfrac{1}{2}i(\psi-\varphi)+i\beta]\sin\frac{\theta}{2}.\sin\delta,$$

$$\psi_2' = i\exp[\tfrac{1}{2}i(\varphi-\psi)+i\alpha]\sin\frac{\theta}{2}.\cos\delta + \exp[-\tfrac{1}{2}i(\varphi+\psi)+i\beta]\cos\frac{\theta}{2}.\sin\delta.$$

We find the probability that the spin is directed along the z'-axis,

$$w_1 = \psi_1'^{*}\psi_1' = \cos^2\frac{\theta}{2}.\cos^2\delta + \sin^2\frac{\theta}{2}.\sin^2\delta + \tfrac{1}{2}\sin\theta.\sin 2\delta.\sin(\varphi+\alpha-\beta).$$

From this formula it follows that the probability for the value of a spin component along the orbital direction depends only on the difference $\alpha-\beta$ and not on α and β separately.

21. The spin direction is determined by the angles

$$\theta = 2\delta, \quad \Phi = \tfrac{1}{2}\pi + \beta - \alpha.$$

22. It is possible. In the case of a mixed ensemble for every direction of the inhomogeneous magnetic field one will always get a splitting into two beams. In the case of a pure ensemble by a suitable alignment of the instrument one can obtain the disappearance of one of the beams.

23. The operator of the square of the total spin is given by the equation

$$\begin{aligned}\hat{S}^2 &= \hat{S}_1^2 + \hat{S}_2^2 + 2(\hat{S}_1.\hat{S}_2) = \tfrac{3}{2}\hbar^2\hat{1}_1\hat{1}_2 + \tfrac{1}{2}\hbar^2[\hat{\sigma}_{1x}\hat{\sigma}_{2x} + \hat{\sigma}_{1y}\hat{\sigma}_{2y} + \hat{\sigma}_{1z}\hat{\sigma}_{2z}]\\ &= \tfrac{3}{2}\hbar^2\begin{pmatrix}1&0\\0&1\end{pmatrix}_1\begin{pmatrix}1&0\\0&1\end{pmatrix}_2 + \tfrac{1}{2}\hbar^2\left[\begin{pmatrix}0&1\\1&0\end{pmatrix}_1\begin{pmatrix}0&1\\1&0\end{pmatrix}_2 + \begin{pmatrix}0&-i\\i&0\end{pmatrix}_1\begin{pmatrix}0&-i\\i&0\end{pmatrix}_2\right.\\ &\quad\left. + \begin{pmatrix}1&0\\0&-1\end{pmatrix}_1\begin{pmatrix}1&0\\0&-1\end{pmatrix}_2\right],\end{aligned}$$

where the indices 1 and 2 distinguish the two particles and the operators with index 1 (2) operate only on the spinor of the first (second) particle, while $\hat{1}$ is the unit operator $\begin{pmatrix} 1 & 0 \\ 0 & 1 \end{pmatrix}$.

We now introduce the unit spinors α and β:

$$\alpha \equiv \begin{pmatrix} 1 \\ 0 \end{pmatrix}, \quad \beta \equiv \begin{pmatrix} 1 \\ 0 \end{pmatrix}.$$

The general spinor of a system of two spin-$\frac{1}{2}$ particles can then be written in the form

$$\Psi = a\alpha_1\alpha_2 + b\alpha_1\beta_2 + c\beta_1\alpha_2 + d\beta_1\beta_2,$$

and one finds that Ψ is an eigenfunction of $\hat{S}_z = \hat{S}_{1z} + \hat{S}_{2z} = \frac{1}{2}\hbar(\hat{\sigma}_{1z} + \hat{\sigma}_{2z})$, provided $b = c$ or $b = -c$. Solving the eigenvalue problem $\hat{S}^2\Psi = \lambda\Psi$ we finally find the following joint eigenfunctions:

$$\begin{aligned}
\Psi_1 &= \alpha_1\alpha_2, & S^2 &= 2\hbar^2, & S_z &= \hbar; \\
\Psi_2 &= \alpha_1\beta_2 + \beta_1\alpha_2, & S^2 &= 2\hbar^2, & S_z &= 0; \\
\Psi_3 &= \beta_1\beta_2, & S^2 &= 2\hbar^2, & S_z &= -\hbar; \\
\Psi_4 &= \alpha_1\beta_2 - \beta_1\alpha_2, & S^2 &= 0, & S_z &= 0.
\end{aligned}$$

27. The possible wave functions for the two-particle system are

$$\psi_{1,1} = \alpha_1\alpha_2, \quad \psi_{1,-1} = \beta_1\beta_2, \quad \psi_{1,0} = \frac{1}{\sqrt{2}}(\alpha_1\beta_2 + \beta_1\alpha_2);$$

$$\psi_{0,0} = \frac{1}{\sqrt{2}}(\alpha_1\beta_2 - \beta_1\alpha_2).$$

The first three wave functions correspond to the triplet state and the last one to the singlet state. The dipole–dipole interaction energies corresponding to these wave functions are

$$\langle V\rangle_{1,1} = \langle V\rangle_{1,-1} = -2A/d^3; \quad \langle V\rangle_{1,0} = 4A/d^3; \quad \langle V\rangle_{0,0} = 0.$$

The wave function at $t = 0$ is $\sqrt{(2)}\,[\psi_{1,0} + \psi_{0,0}]$ so that at time t we have

$$\psi = \sqrt{(2)}\,[\psi_{1,0}\exp(-4iAt/d^3\hbar) + \psi_{0,0}],$$

and we see that after a time $\hbar d^3/4\pi A$ the wave function will be $\psi = -\sqrt{(2)}\,[\psi_{1,0} - \psi_{0,0}]$, corresponding to a spin-flop.

28. Proceeding as in problem 23 of section 4 we find four quartet state spinors:

$$\begin{aligned}
\Psi_i &= \alpha_1\alpha_2\alpha_3, & S^2 &= \tfrac{15}{4}\hbar^2, & S_z &= \tfrac{3}{2}\hbar; \\
\Psi_2 &= \alpha_1\alpha_2\beta_3 + \alpha_1\beta_2\alpha_3 + \beta_1\alpha_2\alpha_3, & S^2 &= \tfrac{15}{4}\hbar^2, & S_z &= \tfrac{1}{2}\hbar; \\
\Psi_3 &= \alpha_1\beta_2\beta_3 + \beta_1\alpha_2\beta_3 + \beta_1\beta_2\alpha_3, & S^2 &= \tfrac{15}{4}\hbar^2, & S_z &= -\tfrac{1}{2}\hbar; \\
\Psi_4 &= \beta_1\beta_2\beta_3, & S^2 &= \tfrac{15}{4}\hbar^2, & S_z &= -\tfrac{3}{2}\hbar;
\end{aligned}$$

and two, degenerate, doublet state spinors (λ and μ are arbitrary constants)

$$\Psi_5 = \alpha_1\alpha_2\beta_3 + \lambda\alpha_1\beta_2\alpha_3 - (1+\lambda)\beta_1\alpha_2\alpha_3, \qquad S^2 = \tfrac{3}{4}\hbar^2, \qquad S_z = \tfrac{1}{2}\hbar;$$
$$\Psi_6 = \alpha_1\beta_2\beta_3 + \mu\beta_1\alpha_2\beta_3 - (1+\mu)\beta_1\beta_2\alpha_3, \qquad S^2 = \tfrac{3}{4}\hbar^2, \qquad S_z = -\tfrac{1}{2}\hbar.$$

31. By writing out explicitly the corresponding 4×4 matrices one finds that

$$(\hat{\boldsymbol{\sigma}}_k \cdot \hat{\boldsymbol{\sigma}}_l) = 2\hat{P}_{kl} - \hat{I}_{kl}, \tag{1}$$

where $\hat{I}_{kl}$ is the 4×4 unit matrix.

Using (1) and the results obtained in the preceding problem, the proof follows in a straightforward way.

32. Between the components of a spinor function and the components of a symmetric spinor there exists the following relation:

$$\psi^{11} = \psi_1, \quad \psi^{12} = \psi^{21} = \frac{1}{\sqrt{2}}\psi_0, \quad \psi^{22} = \psi_{-1}, \tag{1}$$

where ψ_1, ψ_0, and ψ_{-1} are the components of the spin-1 spinor (cf.problem 12 of section 4). A spinor of the second rank transforms as the product of two spinors of the first rank, that is

$$\psi'^{11} = \alpha^2\psi^{11} + 2\alpha\beta\psi^{12} + \beta^2\psi^{22},$$
$$\psi'^{12} = \alpha\gamma\psi^{11} + (\alpha\delta + \beta\gamma)\psi^{12} + \beta\delta\psi^{22},$$
$$\psi'^{22} = \gamma^2\psi^{11} + 2\gamma\delta\psi^{12} + \delta^2\psi^{22}.$$

If we now replace the components of the spinor by the components of a spin function we get by applying equation (1)

$$\psi_1' = \alpha^2\psi_1 + \sqrt{(2)}\,\alpha\beta\psi_0 + \beta^2\psi_{-1},$$
$$\psi_0' = \sqrt{(2)}\,\alpha\gamma\psi_1 + (\alpha\delta + \beta\gamma)\psi_0 + \sqrt{(2)}\,\beta\delta\psi_{-1},$$
$$\psi_{-1}' = \gamma^2\psi_1 + \sqrt{(2)}\,\gamma\delta\psi_0 + \delta^2\psi_{-1}.$$

If we then substitute in this relation the values of the coefficients

$$\alpha = \exp\left[\tfrac{1}{2}i(\varphi+\psi)\right]\cos\frac{\theta}{2}, \qquad \beta = i\sin\frac{\theta}{2}.\exp\left[-\tfrac{1}{2}i(\varphi-\psi)\right],$$

$$\gamma = i\sin\frac{\theta}{2}.\exp\left[\tfrac{1}{2}i(\varphi-\psi)\right], \qquad \delta = \exp\left[-\tfrac{1}{2}i(\varphi+\psi)\right]\cos\frac{\theta}{2},$$

we find the result of problem 12 of section 4.

33. To find the required probability we use a formal method which consists in considering instead of a particle with angular momentum j a

system consisting of $2j$ particles of spin $\frac{1}{2}$. Since in our problem the angular momentum component of the particle is equal to j, all particles in the equivalent system of $2j$ particles must have a z-component of the spin equal to $+\frac{1}{2}$. The probability for a spin component $+\frac{1}{2}$ (or $-\frac{1}{2}$) along the z-axis is for each of the particles equal to $\cos^2(\theta/2)$ [or $\sin^2(\theta/2)$] (problem 19 of section 4). In order that the value of the z-component of the total angular momentum of these particles is equal to m, it is necessary that $j+m$ particles have a z-component $+\frac{1}{2}$ and the remaining $j-m$ particles a z-component $-\frac{1}{2}$. The required probability $w(m)$ is obtained by multiplying

$$\left(\cos^2\frac{\theta}{2}\right)^{j+m} \times \left(\sin^2\frac{\theta}{2}\right)^{j-m}$$

by the number of ways of distributing $2j$ particles into two such groups, that is, by $(2j)!/(j+m)!\,(j-m)!$. In this way we get

$$w(m) = \frac{(2j)!}{(j+m)!\,(j-m)!}\left(\cos^2\frac{\theta}{2}\right)^{j+m}\left(\sin^2\frac{\theta}{2}\right)^{j-m}.$$

One verifies easily that $\sum_{-j}^{+j} w(m) = 1$.

34. We shall consider the case $s = 1$. The wave function is then a 3-component column matrix.

(i) Yes.

(ii) No, not all states correspond to the spin oriented in a definite direction. The reason is that if we wish to find the *two* angles characterising the direction of the spin—if it exists—we have *three* equations to be satisfied, and that will in general be impossible.

35*. The states of a system with angular momentum J can be described by a symmetric spinor of rank $2J$. To solve the present problem we must establish a connection between the components

$$\psi^{\overbrace{11\ldots1}^{J+M}\ \overbrace{22\ldots2}^{J-M}}, \quad \psi'^{\overbrace{11\ldots1}^{J+M'}\ \overbrace{22\ldots2}^{J-M'}}.$$

From Fig. 33 one sees easily that

$$\psi'^{\overbrace{11\ldots1}^{J+M'}\ \overbrace{22\ldots2}^{J-M'}} = \frac{(J+M')!\,(J-M')!}{(2J)!}$$

$$\times \sum_{\nu=0}(2J)!\,\frac{\gamma^{\nu}\,\beta^{M'-M+\nu}\,\alpha^{J+M-\nu}\,\delta^{J-M'-\nu}}{\nu!\,(M'-M+\nu)!\,(J+M-\nu)!\,(J-M'-\nu)!}\,\psi^{\overbrace{11\ldots1}^{J+M}\ \overbrace{22\ldots2}^{J-M}},$$

where α, β, γ, and δ are the Cayley–Klein parameters.

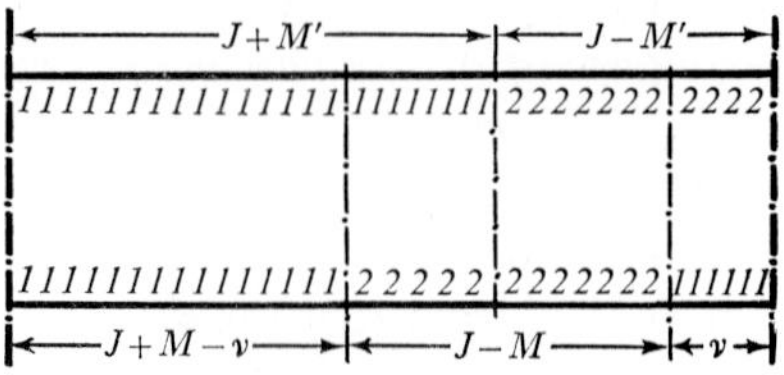

Fig. 33.

Since

$$\psi_{\overbrace{11\ldots1}^{J+M}\ \overbrace{22\ldots2}^{J-M}} = \sqrt{\left[\frac{(J+M)!\,(J-M)!}{(2J)!}\right]}\,\psi(M),$$

$$\psi'_{\overbrace{11\ldots1}^{J+M'}\ \overbrace{22\ldots2}^{J-M'}} = \sqrt{\left[\frac{(J+M')!\,(J-M')!}{(2J)!}\right]}\,\psi'(M'),$$

and $\psi(M) = 1$ we have

$$\psi'(M') = \sqrt{[(J+M')!\,(J-M')!\,(J+M)!\,(J-M)!]}$$
$$\times \sum_{\nu=0} \frac{\gamma^{\nu}\,\beta^{M'-M+\nu}\,\alpha^{J+M-\nu}\,\delta^{J-M'-\nu}}{\nu!\,(M'-M+\nu)!\,(J+M-\nu)!\,(J-M'-\nu)!}.$$

It follows then that:

$$P(M,M') = (J+M')!\,(J-M')!\,(J+M)!\,(J-M)!\left(\cos\frac{\vartheta}{2}\right)^{4J}$$
$$\times\left\{\sum_{\nu=0}\frac{(-1)^{\nu}\left(\tan\frac{\vartheta}{2}\right)^{2\nu-M+M'}}{\nu!\,(M'-M+\nu)!\,(J+M-\nu)!\,(J-M'-\nu)!}\right\}^2.$$

In the summation we must put all terms which lead to a negative number in the factorials equal to zero, that is, in other words, we must sum over ν in such a way that ν satisfies the inequalities

$$\nu \geqslant M - M',$$
$$\nu \leqslant J + M,$$
$$\nu \leqslant J - M'.$$

36*. We find first of all the eigenfunctions of the operator $\hat{j}_z$. To do this we write the operator $\hat{j}_z$ in matrix form

$$\hat{j}_z = \begin{pmatrix} \hat{l}_z + \frac{1}{2} & 0 \\ 0 & \hat{l}_z - \frac{1}{2} \end{pmatrix};$$

since $\hat{l}_z = -i(\partial/\partial\varphi)$, the equation determining the eigenfunctions and eigenvalues of $\hat{j}_z$ is of the form

$$\begin{pmatrix} -i\dfrac{\partial}{\partial\varphi}+\frac{1}{2} & 0 \\ 0 & -i\dfrac{\partial}{\partial\varphi}-\frac{1}{2} \end{pmatrix}\begin{pmatrix}\psi_1\\ \psi_2\end{pmatrix} = m\begin{pmatrix}\psi_1\\ \psi_2\end{pmatrix}$$

or

$$-i\frac{\partial\psi_1}{\partial\varphi}+\tfrac{1}{2}\psi_1 = m\psi_1,$$

$$-i\frac{\partial\psi_2}{\partial\varphi}-\tfrac{1}{2}\psi_2 = m\psi_2.$$

It follows that

$$\psi_1 = f_1(r,\vartheta)\exp[i(m-\tfrac{1}{2})\varphi], \quad \psi_2 = f_2(r,\vartheta)\exp[i(m+\tfrac{1}{2})\varphi],$$

where f_1 and f_2 are arbitrary functions of r and ϑ and m is a half-odd-integer.

From all possible functions of the form

$$\begin{pmatrix} f_1(r,\vartheta)\exp[i(m-\frac{1}{2})\varphi] \\ f_2(r,\vartheta)\exp[i(m+\frac{1}{2})\varphi] \end{pmatrix}$$

we must select those which are at the same time eigenfunctions of the operator $\hat{l}^2$. Such eigenfunctions are of the form

$$\begin{pmatrix}\psi_1\\ \psi_2\end{pmatrix} = \begin{pmatrix} R_1(r)\,Y_{l,m-\frac{1}{2}}(\vartheta,\varphi) \\ R_2(r)\,Y_{l,m+\frac{1}{2}}(\vartheta,\varphi) \end{pmatrix}.$$

The last stage in this construction will be to make this function an eigenfunction of the operator of the square of the total angular momentum by suitably choosing R_1 and R_2. To do this we write the equation $\hat{j}^2\psi = j(j+1)\psi$ in matrix form

$$\begin{pmatrix} \hat{l}^2+\frac{3}{4}+\hat{l}_z & \hat{l}_x-i\hat{l}_y \\ \hat{l}_x+i\hat{l}_y & \hat{l}^2+\frac{3}{4}-\hat{l}_z \end{pmatrix}\begin{pmatrix} R_1(r)\,Y_{l,m-\frac{1}{2}}(\vartheta,\varphi) \\ R_2(r)\,Y_{l,m+\frac{1}{2}}(\vartheta,\varphi) \end{pmatrix} = j(j+1)\begin{pmatrix} R_1(r)\,Y_{l,m-\frac{1}{2}}(\theta,\varphi) \\ R_2(r)\,Y_{l,m+\frac{1}{2}}(\theta,\varphi) \end{pmatrix},$$

and take into account the properties of the spherical harmonics

$$(\hat{l}_x+i\hat{l}_y)\,Y_{lm} = \sqrt{[(l+m+1)(l-m)]}\,Y_{l,m+1},$$

$$(\hat{l}_x-i\hat{l}_y)\,Y_{lm} = \sqrt{[(l-m+1)(l+m)]}\,Y_{l,m-1}.$$

It follows then from the matrix relation which we have written down that R_1 and R_2 must satisfy two homogeneous equations

$$[l(l+1)-j(j+1)+m+\tfrac{1}{4}]\,R_1+\sqrt{[(l+\tfrac{1}{2})^2-m^2]}\,R_2=0,$$
$$\sqrt{[(l+\tfrac{1}{2})^2-m^2]}\,R_1+[l(l+1)-j(j+1)-m+\tfrac{1}{4}]\,R_2=0.$$

In order that these equations can be solved, it is necessary that j is equal to either $l+\frac{1}{2}$ or $l-\frac{1}{2}$.

If we put $j=l+\frac{1}{2}$ we get

$$R_1=\sqrt{(l+\tfrac{1}{2}+m)}\,R(r),\quad R_2=\sqrt{(l-m+\tfrac{1}{2})}\,R(r)$$

and thus

$$\psi(l,j=l+\tfrac{1}{2},m)=R(r)\begin{pmatrix}\sqrt{\left(\dfrac{l+m+\frac{1}{2}}{2l+1}\right)}\,Y_{l,m-\frac{1}{2}}\\ \sqrt{\left(\dfrac{l-m+\frac{1}{2}}{2l+1}\right)}\,Y_{l,m+\frac{1}{2}}\end{pmatrix}\quad(l=0,1,2,\ldots);$$

similarly we get for $j=l-\frac{1}{2}$:

$$\psi(l,j=l-\tfrac{1}{2},m)=R(r)\begin{pmatrix}-\sqrt{\left(\dfrac{l-m+\frac{1}{2}}{2l+1}\right)}\,Y_{l,m-\frac{1}{2}}\\ \sqrt{\left(\dfrac{l+m+\frac{1}{2}}{2l+1}\right)}\,Y_{l,m+\frac{1}{2}}\end{pmatrix}\quad(l=1,2,3,\ldots).$$

The factor $1/\sqrt{(2l+1)}$ follows from the normalisation.

37.

	probability	
	$j=l+\frac{1}{2}$	$j=l-\frac{1}{2}$
orbital ang. mom. $m-\frac{1}{2}$ spin $\frac{1}{2}$	$\frac{l+m+\frac{1}{2}}{2l+1}$	$\frac{l-m+\frac{1}{2}}{2l+1}$
orbital ang. mom. $m+\frac{1}{2}$ spin $-\frac{1}{2}$	$\frac{l-m+\frac{1}{2}}{2l+1}$	$\frac{l+m+\frac{1}{2}}{2l+1}$

$$\bar{l}_z(j=l+\tfrac{1}{2})=\frac{2ml}{2l+1},\qquad \bar{s}_z(j=l+\tfrac{1}{2})=\frac{m}{2l+1},$$

$$\bar{l}_z(j=l-\tfrac{1}{2})=\frac{2m(l+1)}{2l+1},\quad \bar{s}_z(j=l-\tfrac{1}{2})=-\frac{m}{2l+1}.$$

38. The eigenfunctions of the operator of the spin component in the direction Θ, Φ can be found from the relation

$$(\sigma_x \sin\Theta\cos\Phi + \sigma_y \sin\Theta\sin\Phi + \sigma_z\cos\Theta)\begin{pmatrix}\alpha\\ \beta\end{pmatrix} = \begin{pmatrix}\alpha\\ \beta\end{pmatrix},$$

from which it follows that

$$\alpha\sin\Theta\exp(i\Phi) - \beta\cos\Theta = \beta. \qquad (1)$$

From the last equation we find the ratio α/β:

$$\frac{\alpha}{\beta} = \cot\frac{\Theta}{2}\exp(-i\Phi). \qquad (2)$$

On the other hand, from the explicit form of the functions

$$\psi(l, j = l+\tfrac{1}{2}, m) \quad \text{and} \quad \psi(l, j = l-\tfrac{1}{2}, m)$$

we find

$$\frac{\alpha}{\beta} = c_j\frac{Y_{l,m-\frac{1}{2}}(\vartheta,\varphi)}{Y_{l,m+\frac{1}{2}}(\vartheta,\varphi)} = c_j\frac{P_l^{m-\frac{1}{2}}(\cos\vartheta)}{P_l^{m+\frac{1}{2}}(\cos\vartheta)}\exp(-i\varphi), \qquad (3)$$

where

$$c_j = \sqrt{\left(\frac{l+m+\frac{1}{2}}{l-m+\frac{1}{2}}\right)} \quad \text{for} \quad j = l+\tfrac{1}{2},$$

$$c_j = -\sqrt{\left(\frac{l-m+\frac{1}{2}}{l+m+\frac{1}{2}}\right)} \quad \text{for} \quad j = l-\tfrac{1}{2}.$$

If we compare equations (2) and (3) we find that $\Phi = \varphi$, that is, the spin direction in the given point of space lies in the plane through the z-axis and the given point. The angle Θ is determined from the condition

$$\cot\frac{\Theta}{2} = c_j\frac{P_l^{m-\frac{1}{2}}(\cos\vartheta)}{P_l^{m+\frac{1}{2}}(\cos\vartheta)}.$$

39. The wave function of the system $\Psi(J, M)$ has the form of a sum of products of functions of the separate particles,

$$\Psi(J, M) = c_1\psi_1^{(1)}\psi_{M-1}^{(2)} + c_0\psi_0^{(1)}\psi_M^{(2)} + c_{-1}\psi_{-1}^{(1)}\psi_{M+1}^{(2)},$$

where the lower index on the wave functions indicates the value of the angular momentum component.

The coefficients c_i must be determined from the condition

$$\hat{J}^2\Psi(J, M) = J(J+1)\Psi(J, M). \qquad (1)$$

The wave functions of the first particle can be written conveniently in the form

$$\psi_1^{(1)} = \begin{pmatrix}1\\0\\0\end{pmatrix}, \quad \psi_0^{(1)} = \begin{pmatrix}0\\1\\0\end{pmatrix}, \quad \psi_{-1}^{(1)} = \begin{pmatrix}0\\0\\1\end{pmatrix}.$$

If we use that notation for the wave functions, the operator $\hat{\boldsymbol{l}}$ will be a 3×3 matrix

$$\hat{l}_{1x} = \frac{1}{\sqrt{2}}\begin{pmatrix}0 & 1 & 0\\ 1 & 0 & 1\\ 0 & 1 & 0\end{pmatrix}, \quad \hat{l}_{1y} = \frac{1}{\sqrt{2}}\begin{pmatrix}0 & -i & 0\\ i & 0 & -i\\ 0 & i & 0\end{pmatrix}, \quad \hat{l}_{1z} = \begin{pmatrix}1 & 0 & 0\\ 0 & 0 & 0\\ 0 & 0 & -1\end{pmatrix},$$

and we find for the operator $\hat{\boldsymbol{J}}^2$:

$$\hat{\boldsymbol{J}}^2 = \hat{\boldsymbol{l}}_1^2 + \hat{\boldsymbol{l}}_2^2 + 2(\hat{\boldsymbol{l}}_1 . \hat{\boldsymbol{l}}_2)$$

$$= \begin{pmatrix} l(l+1)+2+2\hat{l}_{2z} & \sqrt{(2)}\,\hat{l}_{2-} & 0\\ \sqrt{(2)}\,\hat{l}_{2+} & l(l+1)+2 & \sqrt{(2)}\,\hat{l}_{2-}\\ 0 & \sqrt{(2)}\,\hat{l}_{2+} & l(l+1)+2-2\hat{l}_{2z}\end{pmatrix},$$

where

$$\hat{l}_+ = \hat{l}_x + i\hat{l}_y, \quad \hat{l}_- = \hat{l}_x - i\hat{l}_y.$$

If we use the properties of the operators $\hat{l}_+$ and $\hat{l}_-$,

$$\hat{l}_+\psi_m = \sqrt{[(l+m+1)(l-m)]}\,\psi_{m+1},$$

$$\hat{l}_-\psi_m = \sqrt{[(l+m)(l-m+1)]}\,\psi_{m-1},$$

we find that condition (1) leads to two equations,

$$[J(J+1)-l(l+1)-2M]\,c_1 = \sqrt{(2)}\sqrt{[(l+M)(l-M+1)]}\,c_0,$$

$$[J(J+1)-l(l+1)+2M]\,c_{-1} = \sqrt{(2)}\sqrt{[(l+M+1)(l-M)]}\,c_0$$

(the third equation is satisfied identically).

If we solve these equations we find for $c(J, M)$:

$$\begin{pmatrix} c_1(l+1,M) & c_0(l+1,M) & c_{-1}(l+1,M) \\ c_1(l,M) & c_0(l,M) & c_{-1}(l,M) \\ c_1(l-1,M) & c_0(l-1,M) & c_{-1}(l-1,M) \end{pmatrix}$$

$$= \begin{pmatrix} \sqrt{\left[\dfrac{(l+M)(l+M+1)}{2(2l+1)(l+1)}\right]} & \sqrt{\left[\dfrac{(l+M+1)(l-M+1)}{(2l+1)(l+1)}\right]} & \sqrt{\left[\dfrac{(l-M)(l-M+1)}{2(2l+1)(l+1)}\right]} \\ -\sqrt{\left[\dfrac{(l+M)(l-M+1)}{2l(l+1)}\right]} & \dfrac{M}{\sqrt{[l(l+1)]}} & \sqrt{\left[\dfrac{(l+M+1)(l-M)}{2l(l+1)}\right]} \\ \sqrt{\left[\dfrac{(l-M)(l-M+1)}{2l(2l+1)}\right]} & -\sqrt{\left[\dfrac{(l+M)(l-M)}{l(2l+1)}\right]} & \sqrt{\left[\dfrac{(l+M)(l+M+1)}{2l(2l+1)}\right]} \end{pmatrix}$$

Since this matrix is orthogonal, its inverse is the same as its transposed and therefore each of the functions $\psi_1^{(1)}\psi_{M-1}^{(2)}$, $\psi_0^{(1)}\psi_M^{(2)}$, $\psi_{-1}^{(1)}\psi_{M+1}^{(2)}$ can be expressed as a linear combination of $\Psi(l+1, M)$, $\Psi(l, M)$, $\Psi(l-1, M)$ with coefficients which are in the columns of this matrix.†

40. The states with well-determined values of J ($\hat{J}^2$ is an integral of motion) can be constructed from the states with $L = J - \frac{1}{2}$, $L = J + \frac{1}{2}$. An inversion ($x \to -x$, $y \to -y$, and $z \to -z$) will not change the Hamiltonian operator of the total system (parity is an integral of motion). The parity is different in the states with $L = J - \frac{1}{2}$ and $L = J + \frac{1}{2}$. It follows from this that in a state with given value of J the orbital angular momentum L corresponding to the relative motion of the particles must have a completely well-determined value.

44. $\overline{\hat{\mu}_x} = \overline{\hat{\mu}_y} = 0$,

$$\overline{\hat{\mu}_z} = M_J \left[\tfrac{1}{2}(g_1 + g_2) + \tfrac{1}{2}(g_1 - g_2)\frac{J_1(J_1+1) - J_2(J_2+1)}{J(J+1)}\right].$$

45. $-0{\cdot}26$.

46. $-1{\cdot}91$.

47. (a) 0·88; (b) 0·5; (c) 0·69; (d) 0·31.

48. The weight of the D wave is equal to 0·05.

49. The states $^2P_{1/2}$, $^3P_0$, and $^3P_1$ can occur, the others are nonsense.

50. The spin of the α-particle and the spin of the nucleus B are equal to zero so that the orbital angular momentum, L, corresponding to the relative motion of the α-particle and the product nucleus is equal to unity. It follows that this system is in a state of odd parity (the α-particle has even parity) if the original nucleus had even parity.

† We refer to a note by R. S. Dorfman (*Am.J.Phys.*, **40**, 356, 1972) for a discussion of possible sign-ambiguities when using the method given here.

52. It is impossible.

54. Using the laws of conservation of energy, momentum, and angular momentum and the fact that a photon has a spin 1 directed along its direction of motion, we find that the answer to (i) is no and to (ii) yes.

55. Since for any of the particles there are possible $(2I+1)$ spin orientations, the total number of independent spin functions for the system of the two particles A and B is equal to $(2I+1)^2$.

These functions, which are not symmetrised with respect to the spins, have the form $\chi_i^{(A)}\chi_k^{(B)}$, where $i, k = -I, -I+1, \ldots, I-1, I$. If we symmetrise them we get the following result: for $i = k$ there are $(2I+1)$ functions of the form $\chi_i^{(A)}\chi_i^{(B)}$ (symmetrical); for $i \neq k$ there are $(2I+1)^2-(2I+1) = 2I(2I+1)$ functions of which one half, that is, $I(2I+1)$ functions are of the form $\chi_i^{(A)}\chi_k^{(B)}+\chi_k^{(A)}\chi_i^{(B)}$ (symmetrical) and $I(2I+1)$ functions of the antisymmetrical form $\chi_i^{(A)}\chi_k^{(B)}-\chi_k^{(A)}\chi_i^{(B)}$. In this way we get altogether $(I+1)(2I+1)$ functions which are symmetric with respect to permutation of the spins of the particles A and B and $I(2I+1)$ antisymmetric functions, which leads to the required ratio $(1+I)/I$.

56. -1.

57. All $\langle Q_{ik}\rangle$ vanish except $\langle Q_{zz}\rangle$. $\langle Q_{zz}\rangle = \langle r^2\rangle \dfrac{2l(l+1)-6m^2}{(2l-1)(2l+3)}$.

58*. The averages of the off-diagonal elements vanish and $\langle Q_{xx}\rangle = \langle Q_{yy}\rangle = -\frac{1}{2}\langle Q_{zz}\rangle$ with

$$\langle Q_{zz}\rangle = \langle r^2\rangle \tfrac{1}{2}\left[1-\frac{3m_j^2}{j(j+1)}\right].$$

60. The quadrupole moment is defined by the equation

$$Q_0 = \int (3\cos^2\theta - 1)\, r^2 |\psi_{j,m_j=j}|^2\, d^3\boldsymbol{r}. \tag{1}$$

If $j = l+\frac{1}{2}$, we have

$$\psi_{l+\frac{1}{2},m_j=j} = R(r)\, Y_{l,l}(\theta,\varphi)\begin{pmatrix}1\\0\end{pmatrix},$$

and if $j = l-\frac{1}{2}$

$$\psi_{l-\frac{1}{2},m_j=j} = \frac{R(r)}{\sqrt{(2l+1)}}\begin{pmatrix}-Y_{l,l-1}(\theta,\varphi)\\ \sqrt{(2l)}\, Y_{l,l}(\theta,\varphi)\end{pmatrix}.$$

We thus get from (1)

$$Q_0 = -\langle r^2\rangle_{\text{Av}} \frac{2j-1}{2j+2}.$$

5

Central field of force

1. $\hat{p}_1+\hat{p}_2 \equiv \hat{P} = -i\hbar\nabla_R$; $\hat{l}_1+\hat{l}_2 \equiv \hat{L} = [R \wedge \hat{P}]+[r \wedge \hat{p}]$,

where

$$\hat{p} = -i\hbar\nabla_r.$$

2. Putting $R_{nl} = \chi_{nl}/r$ the equation for R_{nl} goes over into

$$-\frac{\hbar^2}{2\mu}\chi''_{nl}-\left[E_{nl}-V(r)-\frac{\hbar^2 l(l+1)}{2\mu r^2}\right]\chi_{nl} = 0. \tag{1}$$

This equation is formally the same as the Schrödinger equation for a one-dimensional motion in the region $0 \leqslant r < \infty$ with an effective potential

$$V_{\text{eff}}(r) = V(r)+\frac{\hbar^2 l(l+1)}{2\mu r^2}.$$

Since $\chi_{nl} = rR_{nl}$ vanishes for $r = 0$, we can assume that $V = +\infty$ for $r < 0$ for this one-dimensional problem.

3. We shall write the Hamiltonian operator in the following form:

$$\hat{H} = \hat{H}_0+\frac{\hbar^2}{2\mu}\frac{l(l+1)}{r^2}, \quad \text{where} \quad \hat{H}_0 = -\frac{\hbar^2}{2\mu r^2}\frac{\partial}{\partial r}\left(r^2\frac{\partial}{\partial r}\right)+V(r).$$

The smallest values of the energy and the corresponding eigenfunctions are related by the equations

$$E_l^{\text{min}} = \int \psi_l^*\left\{\hat{H}_0+\frac{\hbar^2}{2\mu}\frac{l(l+1)}{r^2}\right\}\psi_l\,d\tau,$$

$$E_{l+1}^{\text{min}} = \int \psi_{l+1}^*\left[\hat{H}_0+\frac{\hbar^2}{2\mu}\frac{(l+1)(l+2)}{r^2}\right]\psi_{l+1}\,d\tau.$$

The last expression can be written in the form

$$E_{l+1}^{\text{min}} = \int \psi_{l+1}^*\left\{\hat{H}_0+\frac{\hbar^2 l(l+1)}{2\mu r^2}\right\}\psi_{l+1}\,d\tau+\int \frac{\hbar^2}{\mu}\frac{(l+1)}{r^2}\psi_{l+1}^*\psi_{l+1}\,d\tau.$$

Let us compare the first term in this expression with E_l^{min}. Since ψ_l corresponds to the minimum eigenvalue of the operator

$$\hat{H}_0+\frac{\hbar^2}{2\mu}\frac{l(l+1)}{r^2}$$

we have

$$\int \psi_{l+1}^{*}\left\{\hat{H}_0+\frac{\hbar^2}{2\mu}\frac{l(l+1)}{r^2}\right\}\psi_{l+1}\,d\tau > \int \psi_{l}^{*}\left\{\hat{H}_0+\frac{\hbar^2}{2\mu}\frac{l(l+1)}{r^2}\right\}\psi_{l}\,d\tau.$$

As far as the integral $\int \frac{\hbar^2}{\mu}\frac{(l+1)}{r^2}\psi_{l+1}^{*}\psi_{l+1}\,d\tau$ is concerned it is always larger than zero. Hence $E_l^{\min} < E_{l+1}^{\min}$, that is, we have proved the statement given above.

4. The potential energy is $V(r) = \frac{1}{2}\mu\omega^2 r^2$.

The radial part R of the wave function satisfies the equation

$$R'' + \frac{2}{r}R' + \left[\frac{2\mu E}{\hbar^2} - \frac{\mu^2\omega^2 r^2}{\hbar^2} - \frac{l(l+1)}{r^2}\right]R = 0.$$

If we substitute $\chi = Rr$ and use the notation

$$k = \frac{\sqrt{(2\mu E)}}{\hbar}, \quad \frac{\mu\omega}{\hbar} = \lambda,$$

we have

$$\chi'' + \left\{k^2 - \lambda^2 r^2 - \frac{l(l+1)}{r^2}\right\}\chi = 0. \tag{1}$$

If we consider the asymptotic behaviour of χ as $r\to 0$ and as $r\to\infty$, we can put the solution for χ in the form

$$\chi = r^{l+1}\exp\left(-\tfrac{1}{2}\lambda r^2\right)u(r). \tag{2}$$

If we substitute expression (2) into equation (1) we get an equation to determine the function $u(r)$:

$$u'' + 2\left(\frac{l+1}{r} - \lambda r\right)u' - \left[2\lambda\left(l+\tfrac{3}{2}\right) - k^2\right]u = 0. \tag{3}$$

By introducing a new independent variable $\xi = \lambda r^2$ equation (3) goes over into the following differential equation:

$$\xi\frac{d^2u}{d\xi^2} + \left[\left(l+\tfrac{3}{2}\right)-\xi\right]\frac{du}{d\xi} + \left[\tfrac{1}{2}\left(l+\tfrac{3}{2}\right) - \tfrac{1}{2}s\right]u = 0,$$

where

$$s = \frac{k^2}{2\lambda} = \frac{E}{\hbar\omega}.$$

The solution of this equation is the confluent hypergeometric function

$$u = F\left[\tfrac{1}{2}\left(l+\tfrac{3}{2}-s\right), l+\tfrac{3}{2};\ \xi\right].$$

Requiring that R vanishes as $r \to \infty$ we get

$$\tfrac{1}{2}(l+\tfrac{3}{2}-s) = -n_r \quad (n_r = 0, 1, 2, \ldots)$$

and hence we find that the energy levels are equal to $E_{n_r l} = \hbar\omega(l+2n_r+\tfrac{3}{2})$, and the wave functions given by

$$\psi_{n_r lm} = r^l \exp(-\tfrac{1}{2}\lambda r^2)\, F\{-n_r, l+\tfrac{3}{2}, \lambda r^2\}\, Y_{lm}(\vartheta, \varphi).$$

5. The wave functions are

$$\Phi_{n_1 n_2 n_3}(x, y, z) = \varphi_{n_1}(x)\,\varphi_{n_2}(y)\,\varphi_{n_3}(z),$$

where

$$\varphi_n(x) = \frac{1}{\sqrt{(2^n \lambda^{n-\frac{1}{2}} n!)}} \frac{1}{\sqrt[4]{\pi}} \left(\lambda x - \frac{\partial}{\partial x}\right)^n \exp(-\tfrac{1}{2}\lambda x^2).$$

The corresponding energy levels are (see problem 5 of section 1)

$$E_{n_1 n_2 n_3} = \hbar\omega(n_1+n_2+n_3+\tfrac{3}{2}).$$

The connection between $\psi_{n_r lm}$ and $\Phi_{n_1 n_2 n_3}$ for $n_r = 0$, $l = 1$ is of the form:

$$\psi_{011} = \frac{1}{\sqrt{2}}(\Phi_{100}+i\Phi_{010}),$$

$$\psi_{010} = \Phi_{001},$$

$$\psi_{01,-1} = \frac{1}{\sqrt{2}}(\Phi_{100}-i\Phi_{010}).$$

6. $$Z_n = (n+1)(n+2),$$

where $$n = 2n_r + l.$$

7. For $^4_2$He we have

$$\rho(\boldsymbol{r}) = \frac{4}{[r_0\sqrt{(2\pi)}]^3} \exp\left[-\frac{1}{2}\left(\frac{r}{r_0}\right)^2\right],$$

where

$$r_0 = \sqrt{\left(\frac{\hbar}{2\mu\omega}\right)}; \quad R = r_0.$$

For $^{16}_{8}$O we have

$$\rho(\boldsymbol{r}) = \frac{4}{[r_0\sqrt{(2\pi)}]^3} \left(1+\frac{r^2}{r_0^2}\right) \exp\left[-\frac{1}{2}\left(\frac{r}{r_0}\right)^2\right],$$

$$R = 1{\cdot}94 r_0.$$

8. The equation for the radial function is of the form

$$-\frac{\hbar^2}{2\mu}\frac{1}{r^2}\frac{d}{dr}\left(r^2\frac{dR}{dr}\right) + \frac{\hbar^2}{2\mu}\frac{l(l+1)}{r^2}R + \{V(r)-E\}R = 0,$$

where μ is the reduced mass, $\mu = M_p M_n/(M_p+M_n) \approx M/2$, since $M_p \approx M_n = M$.

If we put $l = 0$ and $R = \chi(r)/r$, we find

$$\frac{d^2\chi}{dr^2}+\frac{2\mu}{\hbar^2}[E+A\exp(-r/a)]\chi = 0.$$

Introducing a new variable

$$\xi = \exp(-r/2a),$$

we find

$$\frac{d^2\chi}{d\xi^2}+\frac{1}{\xi}\frac{d\chi}{d\xi}+\left(c^2-\frac{k^2}{\xi^2}\right)\chi = 0,$$

where

$$c^2 = \frac{8\mu}{\hbar^2}Aa^2;$$

$$k^2 = -\frac{8\mu}{\hbar^2}Ea^2 > 0.$$

The general solution of this equation is

$$\chi = B_1 J_k(c\xi)+B_2 J_{-k}(c\xi),$$

where J_k is a Bessel function. As $r \to \infty$ ($\xi = 0$) the wave function of a stationary state must tend to zero so that $B_2 = 0$ and hence

$$R = \frac{B_1}{r}J_k[c\exp(-r/2a)].$$

In order that R is finite for $r = 0$ we must have

$$J_k(c) = 0.$$

This equation gives us the connection between a and A. To obtain values for a and A referring to the ground state, it is necessary that c is the first root of the Bessel function (the radial wave function must not have any nodes):

$a.10^{13}$, cm	k	c	A, MeV
1	0·45	3·1	100
2	0·91	3·7	36
4·4	2·02	5·1	14

9. The average value of the energy E in the state described by the wave function $\psi(\mathbf{r})$ is given by the following expression:

$$E = \frac{\hbar^2}{2\mu}\int(\nabla\psi)^2\,d\tau+\int V\psi^2\,d\tau.$$

According to the variational principle the quantity E takes on the value of the ground state energy if ψ is the exact ground state function. If we take for ψ a function depending on one or more parameters $\alpha, \beta, \ldots$, the energy E will be a function of those parameters, $E(\alpha, \beta, \ldots)$, and the best approximation to the energy and the wave function of the ground state will be obtained for the values $\alpha = \alpha_0$, $\beta = \beta_0, \ldots$, which satisfy the conditions

$$\left(\frac{\partial E(\alpha, \beta, \ldots)}{\partial \alpha}\right)_{\substack{\alpha=\alpha_0\\ \beta=\beta_0\\ \cdots}} = 0; \quad \left(\frac{\partial E(\alpha, \beta, \ldots)}{\partial \beta}\right)_{\substack{\alpha=\alpha_0\\ \beta=\beta_0\\ \cdots}} = 0, \ldots .$$

The quantity $E(\alpha_0, \beta_0, \ldots)$ is always larger than the ground state energy and is the nearer to it the larger and more expedient the class of trial functions taken.

In our case $\psi = [1/\sqrt{(4\pi)}]\, R(r)$, where $R(r) = c\exp(-\alpha r/2a)$. From the normalisation condition it follows that $c^2 = \alpha^3/2a^3$ so that

$$E(\alpha) = c^2 \frac{\hbar^2}{2\mu} \int_0^\infty \left(\frac{\alpha}{2a}\right)^2 \exp(-\alpha r/a)\, r^2\, dr - c^2 A \int_0^\infty \exp\left(-\frac{\alpha r}{a} - \frac{r}{a}\right) r^2\, dr$$

$$= \frac{\hbar^2}{2\mu}\left(\frac{\alpha}{2a}\right)^2 - A\left(\frac{\alpha}{\alpha+1}\right)^3.$$

We find the minimum of $E(\alpha)$

$$\frac{dE(\alpha)}{d\alpha} = \frac{\hbar^2 \alpha}{4\mu a^2} - \frac{3A\alpha^2}{(\alpha+1)^4} = 0.$$

Hence,

$$\frac{(\alpha_0+1)^4}{\alpha_0} = \frac{12A\mu a^2}{\hbar^2} = 22{\cdot}3; \quad \alpha_0 = 1{\cdot}34.$$

The value of the energy for this value of the parameter is

$$E = -2{\cdot}14 \text{ MeV}.$$

The exact solution of this problem gives for the given values of A and a the value $E = -2{\cdot}2$ MeV (see preceding problem).

10. The equation for the radial part of the wave function for $r < a$ is of the form

$$\frac{1}{r^2}\frac{d}{dr}\left(r^2 \frac{dR}{dr}\right) - \frac{l(l+1)}{r^2} R + k^2 R = 0, \tag{1}$$

where

$$k^2 = \frac{2\mu E}{\hbar^2},$$

while for $r = a$, $R = 0$.

We introduce instead of R a new function $\chi(r)$ through the formula

$$\frac{\chi(r)}{\sqrt{r}} = R(r).$$

Substituting this into equation (1) we get for $\chi(r)$ the equation

$$\chi'' + \frac{1}{r}\chi' + \left\{k^2 - \frac{(l+\frac{1}{2})^2}{r^2}\right\}\chi = 0,$$

the solution of which are Bessel functions of half odd integral order,

$$\chi(r) = J_{l+\frac{1}{2}}(kr),$$

$$R(r) = \frac{c}{\sqrt{r}} J_{l+\frac{1}{2}}(kr).$$

The value of the energy $E = \hbar^2 k^2/2\mu$ of the stationary states is obtained by requiring that the Bessel functions tend to zero at $r = a$

$$J_{l+\frac{1}{2}}(ka) = 0,$$

while c follows from the normalisation condition.

The simplest energy levels to determine are those for particles with angular momentum $l = 0$. In that case

$$J_{\frac{1}{2}}(kr) = \sqrt{\left(\frac{2}{\pi kr}\right)} \sin kr$$

and the energy is given by

$$E_{n0} = \frac{\hbar^2}{2\mu}\frac{n^2\pi^2}{a^2}.$$

11. The problem reduces to the solution of the one-dimensional problem with potential

$$V(r) = \begin{cases} -V_0 & 0 < r < a, \\ 0 & r > a, \\ \infty & r < 0. \end{cases}$$

If we put in problem 7 of section 1 $V_1 = \infty$, $V_2 = V_0$, we get the equation which determines the energy levels of the discrete spectrum

$$ka = n\pi - \arc\sin\frac{\hbar k}{\sqrt{(2\mu V_0)}}, \quad k = \frac{\sqrt{(2\mu E)}}{\hbar}.$$

These energy levels can easily be obtained using graphical methods (see fig. 34).

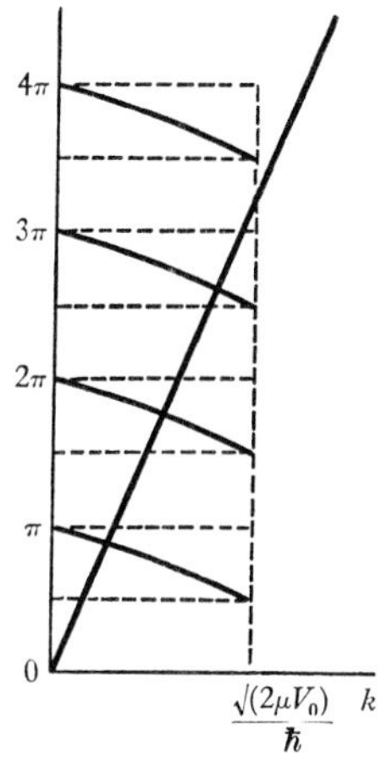

Fig. 34.

The depth of the well for which the first discrete level occurs is equal to

$$V_0 = \frac{\pi^2 \hbar^2}{8\mu a^2}.$$

12. The wavefunction is $\psi = f(r)\cos\vartheta$ with

$$f(r) = A\left[\frac{\sin\alpha r}{\alpha^2 r^2} - \frac{\cos\alpha r}{\alpha r}\right], \quad r < a;$$

$$= Be^{-\beta r}\left[\frac{1}{\beta r} + \frac{1}{\beta^2 r^2}\right], \quad r > a,$$

where $\alpha^2 = (2\mu/\hbar^2)(E + V_0)$, $\beta^2 = -2\mu E/\hbar^2$, and A and B are constants which are related through the requirement that $f(r)$ be continuous at $r = a$.

13. When the well edge is rounded off all levels are changed upwards, that is, $\Delta E > 0$. States with large values of l will suffer large shifts of the levels, since the particles in a state with a large value of the angular moment spent a relatively large part of their time near the edge of the well.

14*. The radial wave function satisfies the equation

$$\chi'' + \frac{2\mu}{\hbar^2}(E - V)\chi = 0.$$

In the region I, where $V = 0$, the solution which tends to zero for $r = 0$ is

$$\chi = A \sin kr, \quad k^2 = \frac{2\mu E}{\hbar^2}.$$

In the region II ($V = V_0$) the general solution is of the form

$$\chi = B_+ \exp[\kappa(r - r_1)] + B_- \exp[-\kappa(r - r_1)], \quad \kappa^2 = \frac{2\mu(V_0 - E)}{\hbar^2}.$$

The coefficients B_+ and B_- are determined from the condition of continuity of χ and χ' at the boundary of the regions I and II,

$$A \sin kr_1 = B_+ + B_-,$$
$$Ak \cos kr_1 = \kappa(B_+ - B_-).$$

Hence

$$\left.\begin{aligned} B_+ &= \frac{A}{2}\left(\sin kr_1 + \frac{k}{\kappa}\cos kr_1\right), \\ B_- &= \frac{A}{2}\left(\sin kr_1 - \frac{k}{\kappa}\cos kr_1\right). \end{aligned}\right\} \quad (1)$$

The solution in region III, where again $V = 0$, is

$$\chi = C_+ \exp[ik(r - r_2)] + C_- \exp[-ik(r - r_2)].$$

The condition of continuity on the boundary between regions II and III leads to

$$B_+ \exp[\kappa(r_2 - r_1)] + B_- \exp[-\kappa(r_2 - r_1)] = C_+ + C_-,$$
$$\kappa\{B_+ \exp[\kappa(r_2 - r_1)] - B_- \exp[-\kappa(r_2 - r_1)]\} = ik(C_+ - C_-).$$

We find thus

$$C_+ = \tfrac{1}{2}B_+\left(1 + \frac{\kappa}{ik}\right)\exp[\kappa(r_2 - r_1)] + \tfrac{1}{2}B_-\left(1 - \frac{\kappa}{ik}\right)\exp[-\kappa(r_2 - r_1)],$$

$$C_- = \tfrac{1}{2}B_+\left(1 - \frac{\kappa}{ik}\right)\exp[\kappa(r_2 - r_1)] + \tfrac{1}{2}B_-\left(1 + \frac{\kappa}{ik}\right)\exp[-\kappa(r_2 - r_1)].$$

Using equation (1) we can express C_+ and C_- in terms of A,

$$C_+ = \tfrac{1}{4}A\sin kr_1\left(1+\frac{\kappa}{ik}\right)\exp\left[\kappa(r_2-r_1)\right]\left\{1+\frac{1-\frac{\kappa}{ik}}{1+\frac{\kappa}{ik}}\exp\left[-2\kappa(r_2-r_1)\right]\right.$$

$$\left.+\frac{k}{\kappa}\cot kr_1\left[1-\frac{1-\frac{\kappa}{ik}}{1+\frac{\kappa}{ik}}\exp\left[-2\kappa(r_2-r_1)\right]\right]\right\}. \qquad (2)$$

Expressions (1) and (2) determine the form of the stationary wave functions of the particle. The behaviour of the wave function depends essentially on the particle energy. Let us consider the dependence of C_+ and C_- on the energy. We will assume that $\kappa(r_2-r_1) \gg 1$. We can then neglect all terms which contain a factor $\exp\left[-2\kappa(r_2-r_1)\right]$, and we have

$$C_+ \approx \frac{A}{4}\sin kr_1\left(1+\frac{\kappa}{ik}\right)\exp\left[\kappa(r_2-r_1)\right]\left\{1+\frac{k}{\kappa}\cot kr_1\right\}, \quad C_- = C_+^*.$$

If the quantity within the braces is not too small the coefficients C_+ and C_- are appreciably larger than A, that is the wave function is appreciably different from zero only in region III (fig. 35a). For some values of the energy when the expression within braces in equation (2) is small C_+ and C_- can be anomalously small. Such energies will be in the neighbourhood of the values E_n which satisfy the transcendental

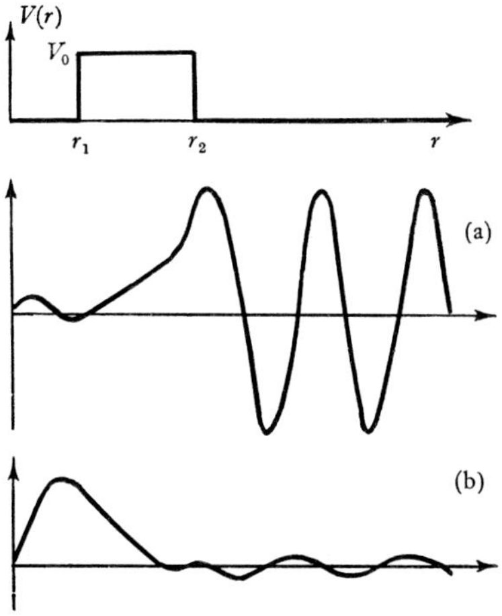

Fig. 35.

equation

$$1+\sqrt{\left(\frac{E_n}{V_0-E_n}\right)}\cot\sqrt{\left(\frac{2\mu E_n}{\hbar^2}\right)}r_1=0,$$

and which are the so-called quasi-stationary levels.

One can easily show that the values E_n are the true discrete energy levels of the problem with a potential depicted in fig. 36 ($r_2 \to \infty$).

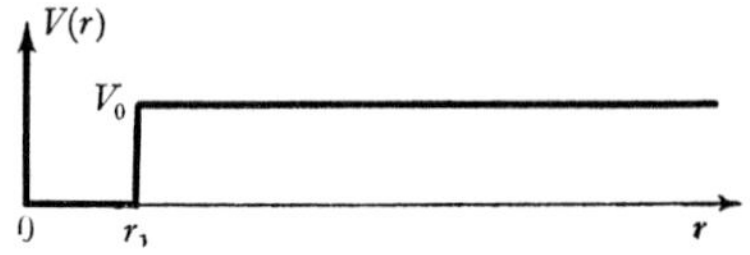

Fig. 36.

The values of the energy which are lying in a narrow band near the quasi-level correspond thus to wave functions which are vanishingly small in region III (fig. 35b).

The probability to find particles with a strictly defined energy within region I is equal to zero. Indeed, the wave function of a particle with a well-defined energy belongs to the continuous spectrum and the integral over region III of $|\psi(r, E)|^2$ diverges; at the same time the integral over region I is finite. This statement remains valid for states near the quasi-level. To find, therefore, the probability of emergence of the particle from region I it is necessary to consider states which are a super-position of a number of stationary states with nearly the same energy, that is a "wave packet" which is localised in region I, and to investigate its "spreading" with time. We shall take for the wave function at $t = 0$ a function χ_0 which is practically equal to zero in region III and which in the regions I and II is the same as the wave function of the quasi-stationary state.

We expand $\chi_0(r)$ in terms of stationary wave functions,

$$\chi_0(r) = \int_0^\infty \varphi(E)\,\chi_E(r)\,dE. \tag{3}$$

The functions $\chi_E(r)$ will be supposed to be normalised in the energy scale. The state of the particle at time t will be

$$\chi_0(r, t) = \int \varphi(E)\,\chi_E(r)\exp(-iEt/\hbar)\,dE.$$

We consider the probability that the particle during a period t will have gone into the initial state $\chi_0(r)$,

$$W(t) = \left|\int_0^\infty \chi_0(r)\,\chi_0(r, t)\,dr\right|^2 = \left|\int_0^\infty |\varphi(E)|^2 \exp(-iEt/\hbar)\,dE\right|^2. \tag{4}$$

The problem is thus reduced to finding the distribution in energy of the initial state, $|\varphi(E)|^2$.

From equation (3) it follows that

$$\varphi(E) = \int_0^\infty \chi_0(r)\,\chi_E(r)\,dr. \tag{5}$$

In agreement with what we have said, one can take for $\chi_0(r)$ an eigenfunction of the auxiliary problem with the potential depicted in fig. 36 for $r < r_1$,

$$\chi_0(r) = a \sin k_0 r \quad \left(k_0^2 = \frac{2\mu E_0}{\hbar^2}\right)$$

and for $r > r_1$

$$\chi_0(r) = -\frac{k_0}{\kappa_0} a \exp\left[-2\kappa(r-r_1)\right], \quad \kappa^2 = \kappa_0^2 - k_0^2.$$

The value of k_0 is determined by the condition

$$\sin k_0 r_1 = -\frac{k_0}{\kappa_0}, \quad \cos k_0 r_1 = \frac{\kappa}{\kappa_0},$$

and the normalisation constant is given by $a = \sqrt{[2\kappa/(1+\kappa r_1)]}$.

The functions $\chi_E(r)$ are in regions I, II, III determined apart from a general multiplying constant A. We must now choose A in such a way that the $\chi_E(r)$ are normalised in the energy scale. The asymptotic form of $\chi_E(r)$ is determined by the values of the coefficients C_+ and C_-. The normalisation

$$\int_0^\infty \chi_E(r)\,\chi_{E'}^*(r)\,dr = \delta(E-E')$$

leads to

$$|C_+| = |C_-| = \frac{1}{\hbar}\sqrt{\left(\frac{\mu}{2\pi k}\right)}.$$

We can thus determine the dependence of A on energy using equation (2). We noted earlier that the ratio

$$\frac{|C_\pm|}{A(E)} = \frac{(1/\hbar)\sqrt{(\mu/2\pi k)}}{A(E)}$$

was large for practically all values of the energy and small only when E is near one of the quasi-levels. The function $\varphi(E)$ has thus a steep maximum near $E_0 = \hbar^2 k_0^2/2\mu$. In the region of the other quasi-levels because of the nearly total orthogonality of the function $\chi_0(r)$ to the eigenfunctions $\chi_E(r)$, which refer to the other quasi-levels, the integral in equation (5) will be nearly equal to zero, even though $A(E)$ will again increase. Thus, in expression (4) for the probability $W(t)$ only the region of values E which are near to E_0 will give an appreciable contribution.

After these preliminary considerations we can attack again the evaluation of $\varphi(E)$. First of all we find the dependence of A on E. From equations (2) for C_+ and C_- it follows that

$$C_+ = C_-^* = \tfrac{1}{2}B_+\left(1+\frac{\kappa}{ik}\right)\exp[\kappa(r_2-r_1)]+\tfrac{1}{2}B_-\left(1-\frac{\kappa}{ik}\right)\exp[-\kappa(r_2-r_1)].$$

Further, B_+ and B_- can be expressed in terms of A:

$$B_+ = \frac{A}{2}\left(\sin kr_1+\frac{k}{\kappa}\cos kr_1\right),$$

$$B_- = \frac{A}{2}\left(\sin kr_1-\frac{k}{\kappa}\cos kr_1\right).$$

We can put $k-k_0 = \Delta k$ near the quasi-level and assume that the following inequalities are satisfied:

$$|\Delta k| \ll k_0 \quad \text{and} \quad |\Delta k| \ll \kappa. \tag{6}$$

The dominant terms in B_+ and B_- will then be

$$B_+ = \frac{A}{2}\frac{\kappa_0}{\kappa^2}(1+\kappa r_1)\,\Delta k,$$

$$B_- = -A\frac{k_0}{\kappa_0}.$$

Assuming that $\exp[-\kappa(r_2-r_1)] \ll 1$, we have

$$|C_\pm| = \frac{A}{4}\frac{\kappa_0^2}{k\kappa^2}(1+\kappa r_1)\exp[\kappa(r_2-r_1)] \times \sqrt{\left[(\Delta k)^2+\left(\frac{4\kappa^3}{\kappa_0^4}\cdot\frac{k^2}{1+\kappa r_1}\exp[-2\kappa(r_2-r_1)]\right)^2\right]}$$

and since $|C_+| = (1/\hbar)\sqrt{(\mu/2\pi k)}$, we get

$$A(E) = \frac{\dfrac{1}{\hbar}\sqrt{\left(\dfrac{\mu}{2\pi k}\right)}\dfrac{4k\kappa^2}{\kappa_0^2(1+\kappa r_1)}\exp[-\kappa(r_2-r_1)]}{\sqrt{\left[(\Delta k)^2+\left(\dfrac{4\kappa^3}{\kappa_0^4}\cdot\dfrac{k^2}{1+\kappa r_1}\exp[-2\kappa(r_2-r_1)]\right)^2\right]}}.$$

We can now easily evaluate the integral (5) which determines $\varphi(E)$ if we assume that as before the inequalities (6) are satisfied. The function $\chi_E(r)$ differs in the regions I and II little from $[A(E)/a]\chi_0$, and in region III is in general not of any importance for finding $\varphi(E)$, since $\chi_0(r)$ decreases exponentially for $r > r_1$.

We have thus

$$\varphi(E) = \frac{A(E)}{a} \int_0^\infty \chi_0^2(r)\,dr = \frac{A(E)}{a}.$$

A simple transformation leads to

$$\varphi^2(E) = \frac{\hbar}{2\pi\tau}\,\frac{1}{(E-E_0)^2+\hbar^2/4\tau^2},$$

where

$$E-E_0 = \frac{\hbar^2}{\mu} k_0\,\Delta k$$

and

$$\tau = \frac{\mu}{8\hbar}\,\frac{\kappa_0^4}{\kappa^3 k^3} \exp\,[2\kappa(r_2-r_1)]\,(1+\kappa r_1).$$

If we perform the integration in equation (4) we find the decay law

$$W(t) = \exp\,(-t/\tau).$$

The probability to find the particle still in the initial state inside the barrier $W(t)$ decreases by a factor e during a period

$$\tau = \frac{1}{16}\,\frac{\hbar}{V_0}\left(\frac{V_0^2}{E(V_0-E)}\right)^{\frac{3}{2}} \exp\,[2\kappa(r_2-r_1)]\,(1+\kappa r_1).$$

15*. We write the Schrödinger equation for the stationary state of a muon in the field of a nucleus of charge $+Ze$ (in this case the motion of the centre of mass of the system is of no interest and can be eliminated in the usual way) in the following form:

$$\nabla^2\psi(\boldsymbol{r}) + \frac{2\mu'}{\hbar^2}\,[E-V(r)]\,\psi(\boldsymbol{r}) = 0. \tag{1}$$

(We note that the use of the non-relativistic wave equation is justified since, as we shall see in a moment, the kinetic energy of the meson is of the order of magnitude 10 MeV, which is small compared with its rest mass, $\mu c^2 \sim 100$ MeV.)

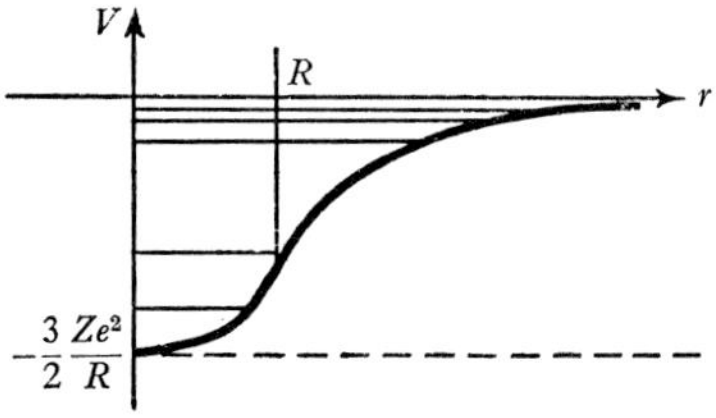

Fig. 37.

In equation (1) $\boldsymbol{r}$ is the radius vector of the meson with respect to the centre of the nucleus, $\mu' = \mu M/(\mu+M)$ is the reduced mass of the system of the meson and the nucleus, E is the energy of the relative motion (since even for light nuclei $M_{\text{nucl}}/\mu \gtrsim 10$, μ' is practically the same as the meson mass μ; we shall, therefore, everywhere in the following omit the prime on the μ).

The potential energy $V(r)$, which we shall normalise by putting it equal to zero for $r = \infty$, will be of the form

$$V(r) = \begin{cases} -\dfrac{Ze^2}{R}\left(\dfrac{3}{2} - \dfrac{1}{2}\dfrac{r^2}{R^2}\right) & \text{for} \quad r \leqslant R, \\ -\dfrac{Ze^2}{r} & \text{for} \quad r \geqslant R. \end{cases} \tag{2}$$

[Expression (2) follows at once, if we take into account the relation

$$V(r) = -e\varphi(r)$$

from the solution of the elementary electrostatic problem of finding the potential $\varphi(r)$ inside and outside a uniformly charged sphere.]

The potential curve $V(r)$ and also the energy levels of the meson are given in fig. 37.

We can see from equation (2) that the muon moves outside the nucleus in an attractive Coulomb field corresponding to a charge $+Ze$ and inside the nucleus in the potential of an isotropic harmonic oscillator of frequency

$$\omega = \sqrt{\left(\frac{Ze^2}{\mu R^3}\right)}. \tag{3}$$

This follows immediately by comparing the oscillator potential $\frac{1}{2}\mu\omega^2 r^2$ with the expression (2) for the potential, $Ze^2 r^2/2R^3$; the term $-\frac{3}{2}Ze^2/R$ in this potential is clearly an inessential constant term.

We note that for all real nuclei this characteristic frequency is approximately the same. Indeed it is well known that

$$R \approx r_0 A^{\frac{1}{3}}, \tag{4}$$

where $r_0 \approx 1{\cdot}2 . 10^{-13}$ cm and A is the mass number of the nucleus, so that

$$\frac{Z}{R^3} \sim \frac{Z}{A};$$

but the last ratio is approximately constant for all nuclei (and practically equal to $\frac{1}{2}$). In agreement with this we shall assume that the ratio Z/R^3 is constant when we consider later the limiting cases of $Z \to 0$ and $Z \to \infty$. The solution of equation (1) for arbitrary values of R is rather cumbersome; we shall not obtain it, but restrict ourselves to consider two limiting cases (and at the same time we shall restrict ourselves to the bound states of the meson which are of the greatest interest and from which clearly the discrete spectrum of negative values of the energy E follows),

(a) $R \to 0$ and (b) $R \to \infty$ (while $Z/R^3 = \text{const!}$).

The first case (a) corresponds to a "hydrogen-like" system of the nucleus + meson, so that the energy levels of the meson are given by the Balmer formula

$$E_n^{\text{Coul}} = -\frac{Z^2 \mu e^4}{2\hbar^2 n^2} \quad (n = 1, 2, 3, \ldots), \tag{5}$$

and the wave functions have "a hydrogen-like" form (with the electron mass replaced by the meson mass).

In the second case (b) the system is that of an isotropic harmonic oscillator whose energy levels are given by the equations,

$$E_n^{\text{osc}} = \hbar\omega(n + \tfrac{3}{2}) \quad (n = 0, 1, 2, 3, \ldots), \tag{6}$$

where ω is determined by equation (3).

Let us now write down the normalised ground state wave function of the oscillator

$$\psi_0 = \left(\frac{\mu\omega}{\pi\hbar}\right)^{\frac{3}{4}} \exp\left(-\frac{\mu\omega r^2}{2\hbar}\right),$$

and also the average value of the radius r in this state,

$$\bar{r} = \int_0^\infty r|\psi_0|^2 4\pi r^2\, dr = \frac{2}{\sqrt{\pi}}\sqrt{\left(\frac{\hbar}{\mu\omega}\right)}. \tag{7}$$

One can easily verify that for real nuclei (for which R is finite) the region of applicability of the case (a) is determined by the inequality $R \ll a$, where

$$a = \frac{\hbar^2}{Z\mu e^2} \tag{8}$$

is the radius of the first Bohr orbit of a meson of mass μ in the field of a nucleus of charge Ze, while the region of applicability of case (b) is determined by the inequality $R \gg a$.

The first of these statements is obvious; it means that the dimensions of the "oscillator" region are small compared with the distances which are of importance in the motion in the attractive Coulomb field. The second statement can be obtained as follows: "the oscillatory behaviour" means that the limiting radius R is large compared to the dimensions of the region of motion, $\sim\sqrt{(\hbar/\mu\omega)}$; this characteristic quantity which is the so-called zero point amplitude of the vibrations of the oscillator is, as can be seen from its meaning, the distance r over which the potential energy $\frac{1}{2}\mu\omega^2 r^2$ is of the same order of magnitude as the zero point energy of the oscillator, $\frac{3}{2}\hbar\omega$:

$$r = \sqrt{\left(\frac{3\hbar}{\mu\omega}\right)} \sim \sqrt{\left(\frac{\hbar}{\mu\omega}\right)}.$$

The quantity $\bar{r}$ is also, of course, of the same order of magnitude. From the inequality $R \gg \sqrt{(\hbar/\mu\omega)}$ it follows also by using equations (3) and (8) that $R \gg a$.

We examine how far each of the limiting cases (a) and (b) are applicable to real nuclei for which we shall evaluate the ratio R/a.

From equations (4) and (8) we have

$$\frac{R}{a} = \frac{r_0 A^{\frac{1}{3}} Z \mu e^2}{\hbar^2} \approx 2^{\frac{1}{3}} \frac{\mu}{\mu_{el}} \cdot \frac{r_0}{a_0} Z^{\frac{4}{3}}, \tag{9}$$

where we have introduced for the sake of simplicity the electron mass μ_{el} and where

$$a_0 = \frac{\hbar^2}{\mu_{el} e^2} \approx 0{\cdot}53 . 10^{-8} \text{ cm}$$

and where we have also put $A \approx 2Z$ which is sufficiently accurate for our estimate.

Substituting $r_0/a_0 \approx 2{\cdot}3 . 10^{-5}$ and $\mu/\mu_{el} \approx 210$, we find that

$$\frac{R}{a} = 1 \quad \text{for} \quad Z \approx 45.$$

For light nuclei, that is $Z \ll 45$, we have thus the "hydrogen-like" case of the motion of the muon. The other limiting case $Z \gg 45$ (oscillatory behaviour) begins to be of importance and then only to a rough approximation for very heavy nuclei with $Z \approx 80$–100.

If we put $A = 2Z$ and evaluate r by using equations (7) and (3) we find $\bar{r} \approx 7 . 10^{-13}$ cm: this quantity is not small but approximately the same as the radius of a heavy nucleus from which it is clear that we have still not a limiting case but an intermediate one.

Let us note that the solution obtained of the problem gives for nuclei of $Z \approx 80$–90 a characteristic distance between the lower levels of about 6 MeV (this result is independent of any approximation) while the oscillatory $\hbar\omega$ is approximately equal to 10 MeV, as can be seen from equation (3). That part of the normalisation integral $\int |\psi_0|^2 \, dV = 1$ (for the ground state) which refers to the region $r \leqslant R$ is approximately equal to 0·55 so that even for heavy nuclei the muon is roughly speaking about 55 per cent of the time inside the nucleus.

Finally we see from the level scheme that our assumption about the non-relativistic character of the meson motion was justified.

16. The probability for transition from a state characterised by the quantum numbers n_1, l_1, m_1 (energy, angular momentum, and z-component of angular momentum) to a state n_2, l_2, m_2 is proportional to the absolute square of the corresponding matrix element of the ∇ operator,

$$\nabla_{n_1 l_1 m_1, n_2 l_2 m_2} = \int \psi^*_{n_1 l_1 m_1} \nabla \psi_{n_2 l_2 m_2} \, d\tau, \tag{1}$$

where $d\tau = dx\,dy\,dz$. The matrix elements (1) are those of the gradient operator (or, apart from a multiplying factor, the momentum operator) in the E, M^2, M_z representation.

According to the operator equation

$$\nabla = \frac{i}{\hbar}\hat{\boldsymbol{p}} = \frac{i}{\hbar}\mu\dot{\hat{\boldsymbol{r}}},$$

the matrix element (1) is proportional to the matrix element of the velocity operator $\dot{\hat{\boldsymbol{r}}}$ but it follows from a well-known matrix equation (see problems 42 and 44 of section 3) that

$$(\dot{\boldsymbol{r}})_{n_1 n_2} = i\frac{E_{n_1} - E_{n_2}}{\hbar}(\boldsymbol{r})_{n_1 n_2},$$

where E_{n_1} and E_{n_2} are the energy levels, so that the matrix of the gradient reduces to the matrix of the radius vector $\boldsymbol{r}$,

$$\int \psi^*_{n_1 l_1 m_1} \nabla \psi_{n_2 l_2 m_2} \, d\tau = -\frac{(E_{n_1} - E_{n_2})\,\mu}{\hbar^2} \int \psi^*_{n_1 l_1 m_1} \boldsymbol{r} \psi_{n_2 l_2 m_2} \, d\tau. \tag{2}$$

It is thus immediately clear that in the matrix of the gradient (or momentum) only those elements can be different from zero which are non-diagonal in n, that is, in the energy, and the average value of the gradient or of the momentum of a particle in a stationary state of the discrete spectrum (which is the diagonal matrix element in the energy representation) will be equal to zero. (This fact is clearly connected with the fact that the motion in the system is finite: if $p_{av} \neq 0$ the particle would go to infinity.) In the case of the continuous spectrum, when the motion is infinite, that is, proceeds to non-bounded regions of space, the integral in expression (2) diverges and our conclusions do no longer hold. For a plane wave $\psi \sim \exp[i(\boldsymbol{p}.\boldsymbol{r})/\hbar]$ the average value of the momentum is given by

$$\boldsymbol{p}_{av} = \boldsymbol{p} \neq 0.$$

If $E_{n_1} \neq E_{n_2}$ the gradient matrix (∇) obeys the same selection rules in l and m as the matrix $\boldsymbol{r}$, namely: the only matrix elements which are different from zero are those for which $l_1 - l_2 = \pm 1$, while $m_1 - m_2 = 0$ (selection rule for z and $\partial/\partial z$) or $m_1 - m_2 = \pm 1$ (selection rule for $x, \partial/\partial x$, y and $\partial/\partial y$).

17*. The wavefunction in the state considered can be written in the form

$$\psi_{nlm} = \frac{\chi_{nl}(r)}{r} Y_{lm}(\vartheta, \varphi),$$

where $\chi_{nl}(r)$ satisfies equation (1) of problem 2 of section 5 with $V(r) = -e^2/r$. Putting $\rho = r/a_B$ where $a_B = \hbar^2/\mu e^2$ is the Bohr radius and using the fact that $E_{nl} = -e^2/2a_B n^2$, we can write the equation for χ_{nl} in the form

$$\chi''_{nl} + \left[\frac{2}{\rho} - \frac{1}{n^2} - \frac{l(l+1)}{\rho^2}\right]\chi_{nl} = 0, \tag{1}$$

where the prime now indicates differentiation with respect to ρ.

If we multiply equation (1) by $\rho^{p+1}\chi'_{nl} - \frac{1}{2}(p+1)\rho^p\chi_{nl}$ and integrate over ρ, we get after a few partial integrations the recursion formula

$$\frac{p+1}{n^2}\langle\rho^p\rangle - (2p+1)\langle\rho^{p-1}\rangle + \frac{p}{4}[(2l+1)^2 - p^2]\langle\rho^{p-2}\rangle = 0. \tag{2}$$

$p = 0$ gives $\langle\rho^{-1}\rangle = n^{-2}$.
$p = 1$ gives $\langle\rho\rangle = \frac{1}{2}[3n^2 - l(l+1)]$.
$p = 2$ gives $\langle\rho^2\rangle = \frac{1}{2}[5n^2 + 1 - 3l(l+1)]n^2$.

To find the average of ρ^p for $p < -1$, we must first find the average of ρ^{-2}. This can be done by considering χ_{nl} to be a function of the parameter l in equation (1). From the theory of the hydrogen atom it follows that for a bound state $n - l$ must be an integer so that in equation (1) we must consider $n - l$ to be fixed so that n is a linear function of l. We now can take the derivative of equation (1) with respect to l, after dividing by χ_{nl}:

$$\frac{\partial}{\partial l}\frac{\chi''_{nl}}{\chi_{nl}} = \left[-\frac{\partial}{\partial l}\quad -\frac{1}{n^2} + \frac{2}{\rho} - \frac{l(l+1)}{\rho^2}\right] = -\frac{2}{n^3} + \frac{2l+1}{\rho^2}. \tag{3}$$

Multiplying equation (3) by χ_{nl}^2 and integrating over ρ from 0 to ∞, we find

$$\langle\rho^{-2}\rangle = n^{-3}(l+\tfrac{1}{2})^{-1}.$$

From the recurrence relation (2) we then find

$$\langle\rho^{-3}\rangle = [n^3 l(l+\tfrac{1}{2})(l+1)]^{-1}, \quad \langle\rho^{-4}\rangle = \frac{3n^2 - l(l+1)}{n^5 l(l+\frac{1}{2})(l+1)(2l^2+2l-\frac{3}{2})}.$$

18. In the equation for $\chi = Rr$

$$\chi'' + \left\{\frac{2\mu}{\hbar^2}[E - V(r)] - \frac{l(l+1)}{r^2}\right\}\chi = 0$$

we make the substitution

$$\chi = A \exp[i(S/\hbar)],$$

where A and S are real functions.

If we put the real and imaginary part of the equation separately equal to zero we have

$$2A'S' + S''A = 0, \tag{1}$$

$$S'^2 - \frac{\hbar^2 A''}{A} = 2\mu[E - V(r)] - \frac{\hbar^2 l(l+1)}{r^2}. \tag{2}$$

From the first equation we get

$$A = \frac{\text{const}}{\sqrt{S'}}.$$

The second equation we can solve approximately, assuming $\hbar^2$ to be a small quantity; it is necessary, however, to bear in mind that when we make the transition to classical mechanics ($\hbar \to 0$) one must assume that $\hbar l$ is finite since $\hbar l$ is the angular momentum in classical mechanics. For small values of r, when the dominating term on the right-hand side of equation (2) is $\hbar^2 l(l+1)/r^2$ we have $S' \approx i\hbar\sqrt{[l(l+1)]}/r$, $A \sim \sqrt{r}$, so that we find approximately $\hbar^2 A''/A \approx -\hbar^2/4r^2$. We thus obtain a better approximation for S if we take this term into account by substituting this approximate equality into equation (2) (such a correction does not exist for large values of r). We find thus

$$S = \int\sqrt{\left\{2\mu[E - V(r)] - \frac{\hbar^2(l+\frac{1}{2})^2}{r^2}\right\}}\,dr, \tag{3}$$

$$A = \frac{\text{const}}{\sqrt[4]{\left\{2\mu[E - V(r)] - \frac{\hbar^2(l+\frac{1}{2})^2}{r^2}\right\}}}.$$

19. (i) The solution of equation (1) of problem 1 of section 5 for $V(r) = 0$ is

$$\chi_l(r) = j_l(kr),$$

where we have put $E = \hbar^2 k^2/2\mu$ and where j_l is a spherical Bessel function. For large values of r we find

$$\chi_l(r) \to \sin(kr - \tfrac{1}{2}l\pi).$$

(ii) Using equation (3) of the preceding problem we find

$$\chi_l = \frac{\text{const}}{\left[1 - \dfrac{(l+\frac{1}{2})^2}{k^2r^2}\right]^{1/4}} \sin\left\{k\int_{r_t}^{r} dr\sqrt{\left[1 - \frac{(l+\frac{1}{2})^2}{k^2r^2}\right]} + \frac{\pi}{4}\right\}, \tag{1}$$

where r_t is the turning point, $r_t = (l+\frac{1}{2})/k$, and the phase $\pi/4$ comes from the usual argument (compare problem 21 of section 1).

For large values of r we get from expression (1)

$$\chi_l(r) \to \sin(kr - \tfrac{1}{2}\pi kr_t + \tfrac{1}{4}\pi) = \sin(kr - \tfrac{1}{2}l\pi),$$

which is the same as the exact expression. We note that, if we had not made the correction for $\hbar^2 A''/A$ in equation (2) of the preceding problem, we would not have got this exact agreement.

20*. The quantization condition is

$$\int_{r_1}^{r_2} \sqrt{\left\{\frac{2\mu}{\hbar^2}\left[E + \frac{e^2}{r} - \frac{\hbar^2}{2\mu}\frac{(l+\frac{1}{2})^2}{r^2}\right]\right\}} dr = (n_r + \tfrac{1}{2})\pi,$$

or, putting $E = -\hbar^2\kappa^2/2\mu a_B^2$, $a_B = \hbar^2/\mu e^2$, $\rho = r/a_B$,

$$\kappa\int_{\rho_1}^{\rho_2} \frac{d\rho}{\rho}\sqrt{\{(\rho - \rho_1)(\rho_2 - \rho)\}} = (n_r + \tfrac{1}{2})\pi, \tag{1}$$

where the turning points ρ_1 and ρ_2 are given by the equation

$$\rho_{1,2} = \kappa^{-2}[1 \mp \sqrt{\{1 - \kappa^2(l+\tfrac{1}{2})^2\}}]. \tag{2}$$

The integral in equation (1) can be evaluated either by straightforward means or by taking a contour integral in the complex r-plane (see D. ter Haar, *Elements of Hamiltonian Mechanics,* Pergamon, Oxford, 1971, p.138), and the result is

$$\tfrac{1}{2}\kappa\rho_1\left[\sqrt{\frac{\rho_2}{\rho_1}} - 1\right]^2 = (n_r + \tfrac{1}{2})\pi,$$

which, by using equations (2), can be rewritten in the form

$$\kappa = \frac{1}{n_r + l + 1},$$

whence we find

$$E = -\frac{\mu e^4}{2\hbar^2 n^2}, \quad n = n_r + l + 1 .$$

21. Because of degeneracy in a central field the number of states is not the same as the number of discrete levels. We therefore first determine the number of levels for a given angular momentum M; this is a one-dimensional problem and we can replace $V(x)$ in equation (1) of problem 22 of section 1 by

$$V_{\text{eff}}(r) = V(r) + \frac{M^2}{2\mu r^2} .$$

For a given M the total number of states is thus

$$\frac{\sqrt{(2\mu)}}{h}\int\left[-V(r) - \frac{M^2}{2\mu r^2}\right]^{1/2} dr ,$$

and the total number of levels, N, is found by integrating this expression over $M/\hbar$ (this integration replaces in the semi-classical case the summation over l):

$$N = \int\left(d\frac{M}{\hbar}\right)\frac{\sqrt{(2\mu)}}{h}\int\left[-V(r) - \frac{M^2}{2\mu r^2}\right]^{1/2} dr$$
$$= -\frac{\mu}{4\hbar^2}\int r V(r) dr ,$$

where the integration is over those values of $V(r)$ for which $V(r) < 0$.

Motion of particles in a magnetic field

1. An electron will describe a spiral in a uniform magnetic field if we consider this motion in classical mechanics; the axis of this spiral will be in the direction of the magnetic field. The motion in a plane perpendicular to the magnetic field will be periodic with frequency equal to twice the Larmor frequency ω_L ($\omega_L = e\mathscr{H}/2\mu c$). Let us consider the motion of a wave packet in quantum mechanics. The Schrödinger equation for a particle in a magnetic field can be written as follows:

$$i\hbar\frac{\partial\Psi}{\partial t} = -\frac{\hbar^2}{2\mu}\nabla^2\Psi + \frac{\hbar}{i}\omega_L\left(x\frac{\partial\Psi}{\partial y} - y\frac{\partial\Psi}{\partial x}\right) + \tfrac{1}{2}\mu\omega_L^2(x^2+y^2)\Psi.$$

To find a solution to this problem it is convenient to go over to a rotating system of reference

$$\left.\begin{aligned} x &= x'\cos\omega_L t' - y'\sin\omega_L t', \\ y &= x'\sin\omega_L t' + y'\cos\omega_L t', \\ z &= z', \\ t &= t', \end{aligned}\right\} \qquad (1)$$

$$\Psi(x, y, z, t) = \Psi'(x', y', z', t').$$

We have then

$$\frac{\partial}{\partial t'} = \frac{\partial}{\partial t} + \omega_L\left(x\frac{\partial}{\partial y} - y\frac{\partial}{\partial x}\right), \qquad \nabla'^2 = \nabla^2.$$

The Schrödinger equation in the new variables has the following form:

$$i\hbar\frac{\partial\Psi'}{\partial t'} = -\frac{\hbar^2}{2\mu}\nabla'^2\Psi' + \frac{\mu\omega_L^2}{2}(x'^2+y'^2)\Psi'.$$

The solution of this equation can be obtained by separating the variables x', y', z'. The equation for the function $\varphi(z')$ describes free motion along the z-axis. The solution of the equation determining the function $\psi(x, y, t)$ is of the form

$$\psi(x, y, t) = \Sigma A_{nm}\chi_n\left[x'\sqrt{\left(\frac{\mu\omega_L}{\hbar}\right)}\right]\chi_m\left[y'\sqrt{\left(\frac{\mu\omega_L}{\hbar}\right)}\right]\exp[-i\omega_L t(n+m+1)].$$

Here x' and y' are functions of the coordinates x, y and the time determined by equation (1), the χ_n are the wave functions of the harmonic oscillator and the A_{nm} are coefficients which must be chosen in such a way that the initial conditions are satisfied. This expression for $\psi(x, y, t)$ changes sign only if t increases over a period of the classical motion $T = \pi/\omega_L$. Indeed, x' and y' change sign and if we take into account the property of an eigenfunction of the oscillator,

$$\chi_n(-\xi) = (-1)^n \chi_n(\xi),$$

we get

$$\psi(x, y, t+\pi/\omega_L) = \Sigma A_{nm}(-1)^n \chi_n\left[x'\sqrt{\left(\frac{\mu\omega_L}{\hbar}\right)}\right](-1)^m \chi_m\left[y'\sqrt{\left(\frac{\mu\omega_L}{\hbar}\right)}\right]$$

$$\times \exp[-i\omega_L t(n+m+1)] \exp[-i\pi(n+m+1)]$$

$$= -\psi(x, y, t).$$

In a plane perpendicular to the magnetic field the wave packet will thus change its form periodically with a period equal to the period of the classical motion of a particle in the magnetic field. Along the magnetic field the packet will spread exactly in the same way as for free motion. The wave function $\psi(x, y, t)$ can be found explicitly if we take for the initial wave function

$$\psi(x, y, 0) = \exp\left\{-\frac{\alpha^2}{2}[(x-x_0)^2+(y-y_0)^2]+\frac{ip_{0x}}{\hbar}x+\frac{ip_{0y}}{\hbar}y\right\}.$$

If we put all $A_{nm} = 0$, except $A_{00} = 1$, we find that such a wave packet does not spread in the xy-plane, and that its centre of gravity describes a classical trajectory.

2. To find the operator $\hat{v}$ it is necessary to take the commutator of the vector $\boldsymbol{r}$ with the Hamiltonian

$$\hat{v} = \frac{i}{\hbar}(\hat{H}\boldsymbol{r} - \boldsymbol{r}\hat{H});$$

since

$$\hat{H} = \frac{1}{2\mu}\left(\hat{\boldsymbol{p}} - \frac{e}{c}\boldsymbol{A}\right)^2 + V(\boldsymbol{r}),$$

we find

$$\mu\hat{v} = \left(\hat{\boldsymbol{p}} - \frac{e}{c}\boldsymbol{A}\right).$$

We can now find the commutation rules for these operators

$$\hat{v}_x\hat{v}_y - \hat{v}_y\hat{v}_x = \frac{e}{\mu^2 c}[-(\hat{p}_x A_y - A_y\hat{p}_x) + (\hat{p}_y A_x - A_x\hat{p}_y)]$$

$$= \frac{ie\hbar}{\mu^2 c}\left(\frac{\partial A_y}{\partial x} - \frac{\partial A_x}{\partial y}\right) = \frac{ie\hbar}{\mu^2 c}\mathscr{H}_z.$$

By cyclic permutation we obtain the other two relations.

3. Let us take the z-axis along the direction of the magnetic field, the intensity of which we shall denote by $\mathscr{H}$.

The velocity components of the particle satisfy the following commutation rules (see preceding problem):

$$\hat{v}_x\hat{v}_y-\hat{v}_y\hat{v}_x=\frac{ie\hbar}{\mu^2c}\mathscr{H},\quad \hat{v}_y\hat{v}_z-\hat{v}_z\hat{v}_y=0,\quad \hat{v}_z\hat{v}_x-\hat{v}_x\hat{v}_z=0.$$

The energy operator is equal to

$$\hat{H}=\frac{\mu\hat{v}_x^2}{2}+\frac{\mu\hat{v}_y^2}{2}+\frac{\mu\hat{v}_z^2}{2}.$$

We shall put $\hat{H}$ in the form of a sum of two commuting operators,

$$\hat{H}_1=\frac{\mu\hat{v}_x^2}{2}+\frac{\mu\hat{v}_y^2}{2},\quad \hat{H}_2=\frac{\mu\hat{v}_z^2}{2}.$$

The eigenvalues of $\hat{H}$ are equal to the sum of the eigenvalues of $\hat{H}_1$ and $\hat{H}_2$. Let us first find the eigenvalues of $\hat{H}_1$. We introduce a new notation $\hat{v}_x=\alpha\hat{Q}$, $\hat{v}_y=\alpha\hat{P}$, where $\alpha=\sqrt{(e\hbar\mathscr{H}/\mu^2c)}$. In the variables $\hat{P}$, $\hat{Q}$ the commutation rule is of the form $\hat{P}\hat{Q}-\hat{Q}\hat{P}=-i$, and the operator $\hat{H}_1$ is given by $\hat{H}_1=\hbar(e\mathscr{H}/2\mu c)(\hat{P}^2+\hat{Q}^2)$. From problem 8 of section 1 we find the eigenvalues of $\hat{H}_1$

$$E_{1n}=\hbar\frac{e\mathscr{H}}{\mu c}(n+\tfrac{1}{2})\quad (n=0,1,2,\ldots).$$

The eigenvalues of $\hat{H}_2$ form a continuous spectrum. The energy for the motion in a magnetic field is thus given by

$$E_{nv_z}=\hbar\frac{e\mathscr{H}}{\mu c}(n+\tfrac{1}{2})+\frac{\mu v_z^2}{2}.$$

4. We shall take the z-axis along the direction of the magnetic field and the x-axis along the electrical field. We shall take the vector potential of the magnetic field in the form $A_y=\mathscr{H}x$, $A_x=A_z=0$. The Hamiltonian operator can in that case be written in the following form:

$$\hat{H}=\frac{\hat{p}_x^2}{2\mu}+\frac{\left(\hat{p}_y-\frac{e}{c}\mathscr{H}x\right)^2}{2\mu}+\frac{\hat{p}_z^2}{2\mu}-e\mathscr{E}x.$$

If we introduce the following notation,

$$\frac{e\mathscr{H}}{c}x-\hat{p}_y-\frac{\mu c\mathscr{E}}{\mathscr{H}}=\hat{\pi},$$

we get for $\hat{H}$ the expression

$$\hat{H}=\frac{\hat{p}_x^2}{2\mu}+\frac{\hat{\pi}^2}{2\mu}-\frac{\hat{p}_y c\mathscr{E}}{\mathscr{H}}-\frac{\mu c^2\mathscr{E}^2}{2\mathscr{H}^2}+\frac{\hat{p}_z^2}{2\mu}.$$

The commutation relation between $\hat{p}_x$ and $\hat{\pi}$ is

$$\hat{p}_x\hat{\pi}-\hat{\pi}\hat{p}_x = -i\frac{\hbar e\mathscr{H}}{c}.$$

We find thus that the eigenvalues of the operator $\hat{H}_1 = \hat{p}_x^2/2\mu+\hat{\pi}^2/2\mu$ are the same as the energy levels of an oscillator vibrating with twice the Larmor frequency,

$$E_{1n} = \hbar\frac{e\mathscr{H}}{\mu c}(n+\tfrac{1}{2}).$$

Since the operators $\hat{p}_y$ and $\hat{p}_z$, which occur in the last terms of the Hamiltonian operator commute with $\hat{H}_1$, the operator

$$\hat{H}_2 = \frac{\hat{p}_z^2}{2\mu}-\frac{\hat{p}_y c\mathscr{E}}{\mathscr{H}}-\frac{\mu c^2\mathscr{E}^2}{2\mathscr{H}^2}$$

can be put on diagonal form at the same time as $\hat{H}_1$.

The energy spectrum of the particle is thus

$$E_{np_yp_z} = \hbar\frac{e\mathscr{H}}{\mu c}(n+\tfrac{1}{2})+\frac{\hat{p}_z^2}{2\mu}-\frac{p_y c\mathscr{E}}{\mathscr{H}}-\frac{\mu c^2\mathscr{E}^2}{2\mathscr{H}^2}.$$

Comparing this result with the result of the preceding problem shows that the electrical field lifts the degeneracy which occurs for the case of a magnetic field only: the energy levels in an electrical field depend on three quantum numbers.

5. $\psi_{np_yp}(x, y, z)$

$$= \exp\left[\frac{i}{\hbar}(p_y y+p_z z)\right]\exp\left[-\frac{e\mathscr{H}}{2\hbar c}\left(x-\frac{cp_y}{e\mathscr{H}}-\frac{\mu c^2\mathscr{E}}{e\mathscr{H}^2}\right)^2\right]$$
$$\times H_n\left[\sqrt{\left(\frac{e\mathscr{H}}{c\hbar}\right)}\left(x-\frac{cp_y}{e\mathscr{H}}-\frac{\mu c\mathscr{E}}{e\mathscr{H}^2}\right)\right].$$

6. $E_{nmk} = \hbar\sqrt{(\omega_L^2+\omega_0^2)}\,(2n+|m|+1)+m\hbar\omega_L+\hbar\omega_0(k+\tfrac{1}{2}),$

where

$$\omega_L = \frac{e\mathscr{H}}{2\mu c},\quad n = 0, 1, 2, \ldots,\ m = 0, \pm 1, \pm 2, \ldots,\ k = 0, 1, 2, \ldots.$$

7. $$\hat{x}(t) = \left(\frac{i\hbar}{\mu\omega}\frac{\partial}{\partial y}+\frac{x}{2}\right)\cos\omega t+\left(-\frac{i\hbar}{\mu\omega}\frac{\partial}{\partial x}+\frac{y}{2}\right)\sin\omega t+\left(-\frac{i\hbar}{\mu\omega}\frac{\partial}{\partial y}+\frac{x}{2}\right),$$

$$\hat{y}(t) = \left(-\frac{i\hbar}{\mu\omega}\frac{\partial}{\partial y}-\frac{x}{2}\right)\sin\omega t+\left(-\frac{i\hbar}{\mu\omega}\frac{\partial}{\partial x}+\frac{y}{2}\right)\cos\omega t+\left(\frac{i\hbar}{\mu\omega}\frac{\partial}{\partial x}+\frac{y}{2}\right),$$

where $\omega = eH/\mu c$ (twice the Larmor precession frequency).

8*. The Schrödinger equation in cylindrical coordinates ρ, φ, z is of the form

$$-\frac{\hbar^2}{2\mu}\left(\frac{\partial^2\psi}{\partial z^2}+\frac{\partial^2\psi}{\partial\rho^2}+\frac{1}{\rho}\frac{\partial\psi}{\partial\rho}+\frac{1}{\rho^2}\frac{\partial^2\psi}{\partial\varphi^2}\right)-\frac{ie\hbar}{2\mu c}\mathscr{H}\frac{\partial\psi}{\partial\varphi}+\frac{e^2\mathscr{H}^2}{8\mu c^2}\rho^2\psi = E\psi.$$

We write its solution in the form

$$\psi(\rho, \varphi, z) = \frac{1}{\sqrt{(2\pi)}}R(\rho)\exp(ik_z z)\exp(im\varphi).$$

We introduce the quantities γ and β

$$\gamma = \frac{e\mathscr{H}}{2c\hbar}, \quad \beta = \frac{2\mu E}{\hbar^2} - k_z^2.$$

In the equation which determines the radial function $R(\rho)$,

$$R'' + \frac{1}{\rho}R' + \left(\beta - \gamma^2\rho^2 - 2\gamma m - \frac{m^2}{\rho^2}\right)R = 0,$$

we introduce a new independent variable $\xi = \gamma\rho^2$, so that

$$\xi R'' + R' + \left(-\frac{\xi}{4} + \lambda - \frac{m^2}{4\xi}\right)R = 0, \tag{1}$$

where

$$\lambda = \frac{\beta}{4\gamma} - \frac{m}{2}.$$

The required function behaves as $\exp(-\xi/2)$, as $\xi \to \infty$ and is for small ξ proportional to $\xi^{|m|/2}$. We write the solution of the differential equation (1) in the form

$$R = \exp(-\xi/2)\,\xi^{|m|/2}\,w(\xi).$$

The function $w(\xi)$ follows from the equation

$$\xi w'' + (1 + |m| - \xi)w' + \left(\lambda - \frac{|m|+1}{2}\right)w = 0,$$

the solution of which is a confluent hypergeometric function,

$$w = F\left[-\left(\lambda - \frac{|m|+1}{2}\right), |m|+1, \xi\right].$$

In order that the wave function remains finite it is necessary that the quantity $\lambda - \frac{1}{2}(|m|+1)$ is equal to a non-negative integer n. The energy levels are thus determined by the expression

$$E = \hbar\frac{e\mathscr{H}}{\mu c}\left(n + \frac{|m|}{2} + \frac{m}{2} + \tfrac{1}{2}\right) + \frac{\hbar^2 k_z^2}{2\mu}.$$

9. In cylindrical coordinates we have

$$J_\rho = 0,$$

$$J_\varphi = \left(\frac{e\hbar m}{\mu\rho} - \frac{e^2\mathscr{H}}{2\mu c}\rho\right)|\psi_{nmk_z}|^2,$$

$$J_z = \frac{e\hbar k_z}{\mu}|\psi_{nmk_z}|^2.$$

10. The equation for the radial wave function R goes over to the form

$$u'' + \left[\frac{2\mu}{\hbar^2}E - k_z^2 - \frac{1}{\rho^2}\left(m + \frac{e\mathscr{H}}{2\hbar c}\rho^2\right)^2\right]u = 0,$$

where $u = \sqrt{(\rho)}\,R$ and where $m^2 - \frac{1}{4}$ is changed to m^2. (This change is analogous to the change $l(l+1) \to (l+\frac{1}{2})^2$ and can be based on the method discussed in the solution of problem 2 of section 5.) The expression

$$V_{\text{eff}}(\rho) = \frac{\hbar^2}{2\mu\rho^2}\left(m + \frac{e\mathscr{H}}{2\hbar c}\rho^2\right)^2$$

can be considered to be the effective potential energy for a one-dimensional motion.

From the quantisation condition

$$\int_{\rho_1}^{\rho_2}\sqrt{\left[\frac{2\mu}{\hbar^2}E - k_z^2 - \frac{1}{\rho^2}\left(m + \frac{e\mathscr{H}}{2\hbar c}\rho^2\right)^2\right]}\,d\rho = \pi(n+\tfrac{1}{2})$$

we get the energy spectrum

$$E - \frac{\hbar^2 k_z^2}{2\mu} = \frac{e\hbar\mathscr{H}}{\mu c}\left(n + \frac{|m|}{2} + \frac{m}{2} + \tfrac{1}{2}\right).$$

The energy calculated from the minimum of $V_{\text{eff}}(\rho)$ is equal to

$$E' = \frac{e\hbar\mathscr{H}}{\mu c}(n+\tfrac{1}{2})$$

and is the energy of the radial motion, while the energy

$$E'' = \frac{e\hbar\mathscr{H}}{2\mu c}(m + |m|)$$

corresponds to the energy of the rotational motion. The transition to the classical circular orbit is realised provided $E' \ll E''$ or $n \ll (m+|m|)/2$. This condition is clearly only satisfied for positive values of m and can thus be written in the form $n \ll m$.

11. If we put the expression under the square root sign equal to zero we get for $m>0$,

$$\rho_{1,2}=\sqrt{\left(\frac{2\hbar c}{e\mathscr{H}}\right)}\,[\sqrt{(n+m+\tfrac{1}{2})}\pm\sqrt{(n+\tfrac{1}{2})}].$$

12.
$$\Delta\rho\sim\sqrt{\left(\frac{\hbar c}{e\mathscr{H}}\right)}.$$

14. The Pauli equation is of the form

$$i\hbar\frac{\partial}{\partial t}\begin{pmatrix}\psi_1\\ \psi_2\end{pmatrix}=\hat{H}_0\begin{pmatrix}\psi_1\\ \psi_2\end{pmatrix}-\mu_0(\hat{\boldsymbol{\sigma}}.\mathscr{H})\begin{pmatrix}\psi_1\\ \psi_2\end{pmatrix},$$

where $\hat{H}_0=(1/2\mu)[\boldsymbol{p}-(e/c)A]^2+eV$, and where μ_0 is the magnetic moment of the particle. We shall write the wave function in the form

$$\begin{pmatrix}\psi_1\\ \psi_2\end{pmatrix}=\varphi(x,y,z,t)\begin{pmatrix}s_1(t)\\ s_2(t)\end{pmatrix}.$$

The function φ is a solution of the equation

$$i\hbar\frac{\partial\varphi}{\partial t}=\hat{H}_0\varphi.$$

For the spin function $\begin{pmatrix}s_1\\ s_2\end{pmatrix}$ we then get the equation

$$i\hbar\frac{\partial}{\partial t}\begin{pmatrix}s_1\\ s_2\end{pmatrix}=-\mu_0(\hat{\boldsymbol{\sigma}}.\mathscr{H})\begin{pmatrix}s_1\\ s_2\end{pmatrix}.$$

15. Since $\mathscr{H}_x=\mathscr{H}_y=0$, $\mathscr{H}_z=\mathscr{H}(t)$, we have

$$i\hbar\frac{\partial s_1}{\partial t}=-\mu_0\mathscr{H}(t)\,s_1,$$

$$i\hbar\frac{\partial s_2}{\partial t}=\mu_0\mathscr{H}(t)\,s_2.$$

The solutions of these equations are of the form

$$s_1=c_1\exp\left[\frac{i\mu_0}{\hbar}\int_0^t\mathscr{H}(t)\,dt\right],$$

$$s_2=c_2\exp\left[-\frac{i\mu_0}{\hbar}\int_0^t\mathscr{H}(t)\,dt\right].$$

The constants c_1 and c_2 can be found from the initial conditions,

$$c_1=\exp(-i\alpha)\cos\delta,\quad c_2=\exp(i\alpha)\sin\delta.$$

It follows from the form of the functions s_1 and s_2 that the probability of one or the other orientation of the spin along the z-axis does not change with time. The average value of the x-component of the spin follows from the equation

$$\bar{\bar{s}}_x = \frac{\hbar}{2}\sin 2\delta . \cos\left[\frac{2\mu_0}{\hbar}\int_0^t \mathscr{H}(t)\,dt - 2\alpha\right]$$

and similarly

$$\bar{\bar{s}}_y = -\frac{\hbar}{2}\sin 2\delta . \sin\left[\frac{2\mu_0}{\hbar}\int_0^t \mathscr{H}(t)\,dt - 2\alpha\right].$$

The direction along which the spin has a component equal to $+\frac{1}{2}$ is characterised by the polar coordinates

$$\Theta = 2\delta, \quad \Phi = 2\left[\alpha - \frac{\mu_0}{\hbar}\int_0^t \mathscr{H}(t)\,dt\right].$$

This direction will thus describe a conical surface in time. For a constant field strength the line "along which the spin is directed" will rotate uniformly round the direction of the magnetic field with a frequency $2\mu_0\mathscr{H}/\hbar$.

16. The states for arbitrary polarisation of the incoming beam can always be considered to be a superposition of two states, one of which, $\begin{pmatrix}1\\0\end{pmatrix}$, has the spin along the $+z$-axis while the other, $\begin{pmatrix}0\\1\end{pmatrix}$, has the spin in the opposite direction. Let us first of all consider the case when the neutron spins in the incoming beam are directed along the z-axis. The incoming, reflected, and diffracted waves will be of the form

$$A\begin{pmatrix}1\\0\end{pmatrix}\exp[i(\boldsymbol{k}.\boldsymbol{r})], \quad B\begin{pmatrix}1\\0\end{pmatrix}\exp[i(\boldsymbol{k}_1.\boldsymbol{r})], \quad C\begin{pmatrix}1\\0\end{pmatrix}\exp[i(\boldsymbol{k}_2.\boldsymbol{r})].$$

The quantities $\boldsymbol{k}, \boldsymbol{k}_1, \boldsymbol{k}_2$ are connected with the total energy E and the magnetic moment μ_0 of the neutron as follows:

$$\boldsymbol{k} = \frac{\boldsymbol{p}}{\hbar}, \quad \frac{\hbar^2 k^2}{2\mu} = E, \quad \frac{\hbar^2 k_1^2}{2\mu} = E, \quad \frac{\hbar^2 k_2^2}{2\mu} = E + \mu_0\mathscr{H}.$$

From the condition of continuity at the scattering plane ($x = 0$) of both the wave function and its derivative with respect to x it follows that:

$$k_y = k_{1y} = k_{2y}, \quad k_z = k_{1z} = k_{2z},$$
$$A + B = C,$$
$$k_x A + k_{1x} B = k_{2x} C.$$

From these relations it follows that $k_{1x} = -k_x$, that is, the angle of incidence φ is equal to the angle of reflection φ_1. For the sake of simplicity we shall put $k_y = 0$. Solving the equations we have

$$\frac{B}{A} = \frac{k_x - k_{2x}}{k_x + k_{2x}}, \quad \frac{C}{A} = \frac{2k_x}{k_x + k_{2x}},$$

$$k_{2x} = k_x \sqrt{\left(1 + \frac{2\mu}{\hbar^2 k_x^2} \mu_0 \mathscr{H}\right)}.$$

The coefficient of reflection R is thus equal to

$$R = \left(\frac{B}{A}\right)^2 = \left(\frac{k_x - k_{2x}}{k_x + k_{2x}}\right)^2.$$

If the neutron spins are directed along the $-z$-axis we get

$$k_{2x} = k_x \sqrt{\left(1 - \frac{2\mu}{\hbar^2 k_x^2} \mu_0 \mathscr{H}\right)}$$

and the other results remain the same. Since μ_0 is negative for a neutron, the angle of diffraction $\varphi_{2\uparrow} > \varphi > \varphi_{2\downarrow}$ (see fig. 38).

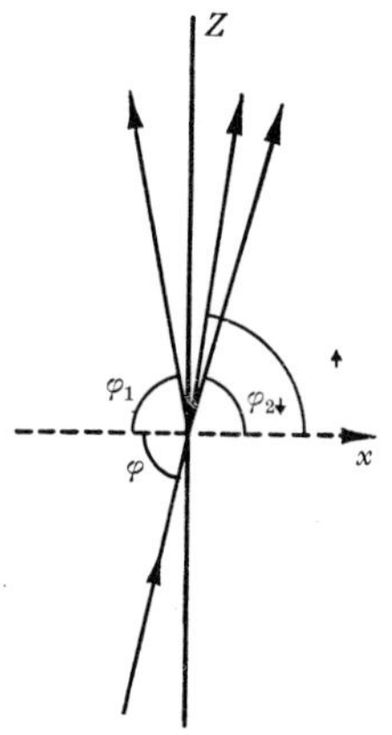

Fig. 38.

In the case of an arbitrary orientation of the neutron spin the wave function in the region $x > 0$ will be of the form

$$C_\uparrow \begin{pmatrix} 1 \\ 0 \end{pmatrix} \exp[-i(\boldsymbol{k}_\uparrow . \boldsymbol{r})] + C_\downarrow \begin{pmatrix} 0 \\ 1 \end{pmatrix} \exp[i(\boldsymbol{k}_\downarrow . \boldsymbol{r})],$$

where $C_\uparrow$ and $C_\downarrow$ are the expansion coefficients of the initial spin state in terms of the states $\begin{pmatrix} 1 \\ 0 \end{pmatrix}$ and $\begin{pmatrix} 0 \\ 1 \end{pmatrix}$. A simple estimate shows that even for $\mathscr{H} \sim 10^4$ Oe an appreciable reflection can only take place for very slow (thermal) neutrons ($\lambda \sim 1$ Å) and for angles of incidence φ which differ from $\pi/2$ only by a fraction of a degree.

17. The Schrödinger equation for the spin function in the z-representation, $\begin{pmatrix} s_1 \\ s_2 \end{pmatrix}$, is of the form

$$i\hbar \frac{\partial}{\partial t}\begin{pmatrix} s_1 \\ s_2 \end{pmatrix} = -\mu \begin{pmatrix} \mathscr{H}_z & \mathscr{H}_x - i\mathscr{H}_y \\ \mathscr{H}_x + i\mathscr{H}_y & -\mathscr{H}_z \end{pmatrix}\begin{pmatrix} s_1 \\ s_2 \end{pmatrix},$$

where μ is the magnetic moment of the particle.

We shall introduce the notation

$$\frac{\mu}{\hbar}\mathscr{H}\cos\vartheta = a, \quad \frac{\mu}{\hbar}\mathscr{H}\sin\vartheta = b.$$

In this notation the equations which determine the components s_1 and s_2 have the following form:

$$\frac{ds_1}{dt} = ias_1 + ib\exp(-i\omega t)\, s_2,$$

$$\frac{ds_2}{dt} = ib\exp(i\omega t)\, s_1 - ias_2.$$

The solution of this set of equations is

$$s_1 = A\exp(ip_1 t) + B\exp(ip_2 t),$$

$$s_2 = \exp(i\omega t)\left[\frac{-a+p_1}{b}A\exp(ip_1 t) + \frac{-a+p_2}{b}B\exp(ip_2 t)\right],$$

where

$$p_1 = \sqrt{\left(\frac{\omega^2}{4} + a^2 + b^2 + \omega a\right)} - \frac{\omega}{2},$$

$$p_2 = -\sqrt{\left(\frac{\omega^2}{4} + a^2 + b^2 + \omega a\right)} - \frac{\omega}{2}.$$

The quantities A and B are determined from the initial conditions and the normalisation condition $|s_1|^2 + |s_2|^2 = 1$. After some simple calculations we get for the transition probability the following value:

$$P(\tfrac{1}{2}, -\tfrac{1}{2}) = \frac{\sin^2\vartheta}{1 + q^2 - 2q\cos\vartheta}\sin^2\left[\frac{t}{2}\omega(1 - 2q\cos\vartheta + q^2)^{\frac{1}{2}}\right],$$

where q is the ratio of the Larmor precession frequency to the frequency ω of the rotating magnetic field,

$$q = -\frac{2\mu\mathscr{H}}{\hbar\omega} = \frac{\omega_L}{\omega}.$$

The quantity q is positive if the magnetic field rotates in the direction of the precession and negative if the rotation is in the opposite direction.

If the angle ϑ is small, that is, $\sqrt{(\mathscr{H}_x^2+\mathscr{H}_y^2)}/\mathscr{H}_z \ll 1$, the transition probability is approximately equal to

$$P(\tfrac{1}{2}, -\tfrac{1}{2}) = \frac{\vartheta^2}{(1-q)^2+q\vartheta^2}\sin^2\left\{\frac{t}{2}\omega[(1-q)^2+q\vartheta^2]^{\frac{1}{2}}\right\}.$$

From this formula it follows that at resonance $\omega = \omega_L$, or for $q = +1$ the probability for a re-orientation of the magnetic moment with respect to the magnetic field, which is equal to $P(\frac{1}{2}, -\frac{1}{2}) \approx \sin^2 \frac{1}{2}t\omega\vartheta$, can become practically equal to unity for certain values of t.

If in the case under consideration we change the direction of rotation of the magnetic field, or change the sign of $\mathscr{H}_z$, we find for the transition probability the result

$$P(\tfrac{1}{2}, -\tfrac{1}{2}) = \frac{\vartheta^2}{4}\sin^2 \omega t,$$

which is considerably less than unity. On the basis of such a sharp qualitative difference one can determine the sign of the magnetic moment of the particle.

18. It is simplest to work in the representation in which $\hat{\sigma}_x$ is diagonal so that the initial spinor is $\begin{pmatrix}1\\0\end{pmatrix}$. The wave equation for the spinor is then

$$i\hbar\frac{\partial}{\partial t}\begin{pmatrix}a\\b\end{pmatrix} = -\mu_0\begin{pmatrix}0 & -iH\\ iH & 0\end{pmatrix}\begin{pmatrix}a\\b\end{pmatrix},$$

with boundary condition $a = 1$, $b = 0$ at $t = 0$. The solution of this equation is (μ_0 is the magnetic moment of the particle)

$$a = \cos(\mu_0 Ht/\hbar), \quad b = \sin(\mu_0 Ht/\hbar),$$

so that the required probability is $\cos^2(\mu_0 Ht/\hbar)$.

19. $\tau = \dfrac{\pi\hbar}{4\mu_0 H}(4k+1), \quad k = 0, 1, 2, \ldots.$

20. $\sin^2\left\{\dfrac{\mu_0 H}{\gamma\hbar}(1-e^{-\gamma t})\right\}$, where μ_0 is the magnetic moment of the particle.

21.† (a) $\hat{x}(t) = \dfrac{\hat{p}_x}{m}t + x, \quad \hat{y}(t) = -\dfrac{\mu k}{2m}t^2\sigma_y + \dfrac{\hat{p}_y}{m}t + y,$

$$\hat{z}(t) = \frac{\mu k}{2m}t^2\sigma_z + \frac{\hat{p}_z}{m}t + z.$$

† We use m here for the particle mass in order to avoid confusion with the magnetic moment μ.

(b) $$\overline{\hat{z}(t)} = (\bar{z})_0 + \frac{\mu k}{2m} t^2(\alpha\alpha^* - \beta\beta^*),$$

$$\overline{\hat{y}(t)} = (\bar{y})_0 - \frac{\mu k}{2m} t^2(i\beta^* \alpha - i\beta\alpha^*),$$

$$\overline{\hat{x}(t)} = (\bar{x})_0 + \frac{p_0}{m} t,$$

$$\overline{(\Delta\hat{z})_t^2} = \overline{(\Delta z)_0^2} + \frac{\mu^2 k^2}{4m^2} t^4[1-(\alpha\alpha^* - \beta\beta^*)^2] + \frac{\hbar^2 t^2}{m^2}\int\left(\frac{\partial\varphi}{\partial z}\right)^2 d\tau,$$

$$\overline{(\Delta\hat{y})_t^2} = \overline{(\Delta y)_0^2} + \frac{\mu^2 k^2}{4m^2} t^4[1-(i\beta^* \alpha - i\beta\alpha^*)^2] + \frac{\hbar^2 t^2}{m^2}\int\left(\frac{\partial\varphi}{\partial y}\right)^2 d\tau,$$

$$\overline{(\Delta\hat{x})_t^2} = \overline{(\Delta x)_0^2} + \frac{\hbar^2 t^2}{m^2}\int\left(\frac{\partial\varphi}{\partial x}\right)^2 d\tau.$$

Note. We can consider, for instance, a system of particles the z-components of the spins of which at $t = 0$ have definitely the value $+\frac{1}{2}$, that is $\alpha = 1$, $\beta = 0$. From the results we have obtained, one easily sees that when such particles are moving in an inhomogeneous magnetic field, they will form two spots on a screen. The z-components of these spots will be the same but their y-coordinates will have opposite signs.

22. (i) The equation of motion for the two-component wave function is for $t > 0$ (we have $\hat{\boldsymbol{J}} = \frac{1}{2}\hat{\boldsymbol{\sigma}}$)

$$\tfrac{1}{2} g\mu_B \hat{\sigma}_x \mathscr{H}\psi = i\hbar\dot{\psi},$$

or

$$-i\omega_0\begin{pmatrix}0 & 1\\ 1 & 0\end{pmatrix}\begin{pmatrix}a\\ b\end{pmatrix} = \begin{pmatrix}\dot{a}\\ \dot{b}\end{pmatrix}, \tag{1}$$

where
$$\omega_0 = g\mu_B \mathscr{H}/2\hbar.$$

From (1) we find
$$a = \cos\omega_0 t, \quad b = -i\sin\omega_0 t.$$

(ii) $\langle \hat{J}_x \rangle = 0, \quad \langle \hat{J}_y \rangle = -\frac{1}{2}\sin 2\omega_0 t, \quad \langle \hat{J}_z \rangle = \frac{1}{2}\cos 2\omega_0 t.$

(iii) $T \gtrsim \omega_0^{-1}$.

23. The singlet energy eigenvalue is $-3A$.
The triplet energy eigenvalues are $A + \frac{1}{2} g\mu_B \mathscr{H} M_z$ ($M_z = 0, \pm 1$).

24. (i) From the results of the preceding problem for $\mathscr{H} = 0$ we see that

$$4A/\hbar = 2.10^5 \text{ Mc/s}, \quad \text{or} \quad A = 5.10^{-17} \text{ erg}.$$

(ii) Single-quantum decay is impossible because linear momentum cannot be conserved.

Two-quantum decay means that the two photons must go off into opposite directions to conserve linear momentum. As the angular momentum of a photon is $\hbar$ along its direction of propagation, the total angular momentum of the two photons must be 0 or 2. Hence, the singlet state can decay by two-photon emission, but for the decay of the triplet state we need at least three photons.

(iii) When the magnetic field is zero, the spin eigenfunctions are (see problem 27 of section 4)

$$\psi_{1,1} = \alpha_1\alpha_2, \quad \psi_{1,0} = \frac{1}{\sqrt{2}}(\alpha_1\beta_2+\beta_1\alpha_2), \quad \psi_{1,-1} = \beta_1\beta_2,$$

$$\psi_{0,0} = \frac{1}{\sqrt{2}}(\alpha_1\beta_2-\beta_1\alpha_2),$$

where $\alpha = \begin{pmatrix}1\\0\end{pmatrix}$ and $\beta = \begin{pmatrix}0\\1\end{pmatrix}$, with spin energies

$$E_{0,0} = -3A, \quad E_{1,0} = E_{1,\pm 1} = A.$$

When the field is non-vanishing we must solve the eigenvalue problem

$$\begin{pmatrix} A-\lambda & 0 & 0 & 0 \\ 0 & A-\lambda & 0 & 2\mu_B\mathscr{H} \\ 0 & 0 & A-\lambda & 0 \\ 0 & 2\mu_B\mathscr{H} & 0 & -3A-\lambda \end{pmatrix} = 0$$

with eigenvalues $\lambda_{1,2} = \lambda_{1,\pm 1} = A$, $\lambda_{3,4} = -A \pm 2[A^2+\mu_B^2\mathscr{H}^2]^{\frac{1}{2}}$, and the eigenfunctions are $\psi_{1,\pm 1}$ and

$$\psi_3 = \gamma\psi_{1,0}+[-2+2\sqrt{(1+\gamma^2)}]\,\psi_{0,0}, \quad \psi_4 = [2-2\sqrt{(1+\gamma^2)}]\,\psi_{1,0}+\gamma\psi_{0,0}, \qquad (1)$$

with $\gamma = 2\mu_B\mathscr{H}/A$.

(iv) The presence of a magnetic field mixes the singlet and triplet states [see (iii)], thus reducing the triplet lifetime.

To find the lifetime τ for a given value of $\mathscr{H}$, we write

$$\frac{1}{\tau} = \frac{P_1}{\tau_1}+\frac{P_3}{\tau_3}, \qquad (2)$$

where P_1 and P_3 are the probabilities that the system, originally in the triplet state when $\mathscr{H} = 0$, is in the singlet or the triplet state. As the three states $\psi_{1,0}$, $\psi_{1,\pm 1}$ are equally probable when $\mathscr{H} = 0$, and as for $\mathscr{H} \neq 0$, the admixture of the singlet state to $\psi_{1,0}$ is [see (1)]

$$[-2+2\sqrt{(1+\gamma^2)}]^2/\gamma^2,$$

we find that

$$P_3 = 1 - \frac{4[\sqrt{(1+\gamma^2)}-1]^2}{3\gamma^2}, \quad P_1 = \frac{4[-1+\sqrt{(1+\gamma^2)}]^2}{3\gamma^2}.$$

As $\tau_1 = 10^{-10}$ sec, $\tau_3 = 10^{-7}$ sec, and $A = 5.10^{-17}$ erg, we find for the value of $\mathscr{H}$ which reduces τ to 10^{-8} sec from (2): $\mathscr{H} \sim 400$ oersted.

25*. The direction of the magnetic field will be characterised by polar angles ϑ, φ. These angles are functions of the time. The Hamiltonian operator for a neutral particle can be written in the form

$$\hat{H} = -\mu\mathscr{H}(\hat{J}_x \sin\vartheta\cos\varphi + \hat{J}_y \sin\vartheta\sin\varphi + \hat{J}_z\cos\vartheta),$$

where $\mathscr{H}$ is the absolute value of the magnetic field intensity. We shall denote by $\hat{J}_\xi$ the angular momentum operator in the direction of the magnetic field

$$\hat{J}_\xi = \hat{J}_x \sin\vartheta\cos\varphi + \hat{J}_y \sin\vartheta\sin\varphi + \hat{J}_z\cos\vartheta,$$

and we shall introduce a function $\psi_m(t)$, which is the eigenfunction of the operator $\hat{J}_\xi$, or

$$\hat{J}_\xi\psi_m(t) = m\psi_m(t).$$

We shall write the solution of the Schrödinger equation

$$i\hbar\frac{\partial\psi}{\partial t} = \hat{H}\psi$$

in the form

$$\psi = \Sigma\, a_m(t)\,\psi_m(t).$$

It is well known (see problem 19 of section 4) that

$$\psi_m(t) = \exp(-i\hat{J}_z\varphi)\exp(-i\hat{J}_y\vartheta)\,\psi_m^{(0)},$$

where $\psi_m^{(0)}$ satisfies the relation

$$\hat{J}_z\psi_m^{(0)} = m\psi_m^{(0)}.$$

We shall first of all evaluate $\dot{\psi}_m(t)$. If we use the relations

$$\exp(i\hat{J}_y\vartheta)\hat{J}_z\exp(-i\hat{J}_y\vartheta) = \hat{J}_z\cos\vartheta - \hat{J}_x\sin\vartheta,$$

$$\hat{J}_x\psi_m^{(0)} = \tfrac{1}{2}\sqrt{[(j+m)(j-m+1)]}\,\psi_{m-1}^{(0)} + \tfrac{1}{2}\sqrt{[(j+m+1)(j-m)]}\,\psi_{m+1}^{(0)},$$

$$\hat{J}_y\psi_m^{(0)} = \frac{i}{2}\sqrt{[(j+m)(j-m+1)]}\,\psi_{m-1}^{(0)} - \frac{i}{2}\sqrt{[(j+m+1)(j-m)]}\,\psi_{m+1}^{(0)},$$

we have

$$\begin{aligned}\dot{\psi}_m(t) &= (-i\dot{\varphi}m\cos\vartheta)\,\psi_m(t)\\ &\quad + \tfrac{1}{2}\sqrt{[(j+m+1)(j-m)]}\,(i\dot{\varphi}\sin\vartheta - \dot{\vartheta})\,\psi_{m+1}(t)\\ &\quad + \tfrac{1}{2}\sqrt{[(j+m)(j-m+1)]}\,(i\dot{\varphi}\sin\vartheta + \dot{\vartheta})\,\psi_{m-1}(t).\end{aligned}$$

If we substitute this value of $\dot{\psi}_m(t)$ into the expression

$$i\hbar \sum_m \{\psi_m(t)\,\dot{a}_m(t) + a_m(t)\,\dot{\psi}_m(t)\} = -\mu\mathscr{H} \sum_m m a_m(t)\,\psi_m(t),$$

we obtain a set of equations which determine the change in the coefficients a_m with time,

$$i\hbar \frac{da_m}{dt} + m\mu\mathscr{H} a_m = -m\hbar\dot{\varphi}\cos\vartheta\, a_m$$
$$+\tfrac{1}{2}\hbar(\dot{\varphi}\sin\vartheta + i\dot{\vartheta})\sqrt{[(j+m)(j-m+1)]}\, a_{m-1}$$
$$+\tfrac{1}{2}\hbar(\dot{\varphi}\sin\vartheta - i\dot{\vartheta})\sqrt{[(j-m)(j+m+1)]}\, a_{m+1}.$$

If

$$\dot{\varphi} \ll \frac{\mu\mathscr{H}}{\hbar}, \quad \dot{\vartheta} \ll \frac{\mu\mathscr{H}}{\hbar},$$

that is, if the angular velocity of the change of direction of the magnetic field is much smaller than the frequency of precession, we can neglect in this set of equations the right-hand side and find thus

$$a_m \sim \exp\left(i\frac{m\mu\mathscr{H}}{\hbar}t\right).$$

In this case the probability for different values of the angular momentum components does not change with time.

26. If we write the Hamiltonian in the form

$$H = \frac{1}{2\mu}\left(\hat{\boldsymbol{p}} - \frac{e}{c}\hat{\boldsymbol{A}} \,.\, \hat{\boldsymbol{p}} - \frac{e}{c}\hat{\boldsymbol{A}}\right),$$

we verify easily that the condition that it is the same as $\hat{H}'$ is that the same components of the vector operators $\hat{\boldsymbol{p}} \equiv (\hbar/i)\nabla$ and $\hat{\boldsymbol{A}}(x, y, z)$ commute, that is,

$$(\nabla \,.\, \boldsymbol{A}) \equiv \operatorname{div} \boldsymbol{A} = 0.$$

27. If we use for the vector potential the gauge where $A_\varphi = \frac{1}{2}$, $A_\rho = A_z = 0$ (ρ, φ, and z are cylindrical polars; compare problem 8 of section 6), we get the Hamiltonian

$$-\frac{\hbar^2}{2\mu}\left(\frac{\partial^2\psi}{\partial z^2} + \frac{\partial^2\psi}{\partial\rho^2} + \frac{1}{\rho}\frac{\partial\psi}{\partial\rho} + \frac{1}{\rho^2}\frac{\partial^2\psi}{\partial\varphi^2}\right) - \frac{ie\hbar}{2\mu c}\mathscr{H}\frac{\partial\psi}{\partial\varphi} + \frac{e^2\mathscr{H}^2}{8\mu c^2}\rho^2\psi = E\psi,$$

which for a particle which is constrained to move along the circle $\rho = a$ reduces to

$$-\frac{\hbar^2}{2\mu a^2}\frac{d^2\psi}{d\varphi^2} - \frac{ie\hbar}{2\mu c}\mathscr{H}\frac{d\psi}{d\varphi} + \frac{e^2\mathscr{H}^2}{8\mu c^2}a^2\psi = E\psi$$

with the solution

$$\psi = \exp(im\varphi), \quad m = 0, \pm 1, \pm 2, \ldots .$$

The corresponding energy eigenvalues are

$$E_m = \frac{1}{2\mu}\left[\frac{m\hbar}{a} + \frac{ea\mathscr{H}}{2c}\right]^2,$$

and we see that the ground state energy as function of $\mathscr{H}$ is given by the equation

$$E_{\text{gr st}} = \frac{1}{2\mu}\left(\frac{m\hbar}{a} + \frac{ea\mathscr{H}}{2c}\right)^2, \quad -\frac{ea^2\mathscr{H}}{2\hbar c} - \tfrac{1}{2} \leqslant m \leqslant -\frac{ea^2\mathscr{H}}{2\hbar c} + \tfrac{1}{2}.$$

This means that $E_{\text{gr st}}$ is a periodic function of $\mathscr{H}$ with periodicity $2\hbar c/ea^2$ and with discontinuities in slope at $\mathscr{H} = (2m+1)\hbar c/ea^2$.

7

Atoms

1. From the given inequality we get

$$\int |\nabla\psi|^2\, d\tau + Z\int (\nabla|\psi|^2 . \nabla r)\, d\tau + Z^2\int (\nabla \boldsymbol{r} . \nabla \boldsymbol{r})|\psi|^2\, d\tau \geqslant 0.$$

Integrating the second term by parts and using the fact that $(\nabla \boldsymbol{r} . \nabla \boldsymbol{r}) = 1$, and $\nabla^2 r = 2/r$, we have

$$\frac{1}{2}\int |\nabla\psi|^2\, d\tau - \int \frac{Z}{r}|\psi|^2\, d\tau \geqslant -\frac{Z^2}{2}\int |\psi|^2\, d\tau.$$

The left-hand side of the inequality is the average value of the Hamiltonian operator (in units $e = \hbar = \mu = 1$) $\hat{H} = -\frac{1}{2}\nabla^2 - Z/r$ for the state ψ. The lower limit of the energy, $-Z^2/2$, is reached for the state with wave function ψ_0, which satisfies the first-order differential equation

$$\nabla\psi_0 + Z\psi_0 \nabla r = 0,$$

from which it follows that

$$\psi_0 \sim \exp(-Zr).$$

4. We evaluate first of all the wave function in the momentum representation from the general formula

$$\varphi(\boldsymbol{p}) = \frac{1}{2(2\pi\hbar)^{\frac{3}{2}}}\iiint \exp[-i(\boldsymbol{p} . \boldsymbol{r})/\hbar]\, \psi(\boldsymbol{r})\, d\tau.$$

For the $1s$-state we find

$$\varphi_{1s}(p) = \frac{1}{\pi}\left(\frac{2a}{\hbar}\right)^{\frac{3}{2}} \frac{1}{[(p^2 a^2/\hbar^2)+1]^2}.$$

Similarly we find for the $2s$-state

$$\varphi_{2s}(p) = \frac{1}{2\pi}\left(\frac{2a}{\hbar}\right)^{\frac{3}{2}} \frac{\dfrac{p^2 a^2}{\hbar^2} - \dfrac{1}{4}}{\left(\dfrac{p^2 a^2}{\hbar^2} + \dfrac{1}{4}\right)^3}.$$

For the $2p$-state there are three eigenfunctions ($m_z = -1, 0, +1$)

$$\varphi_{2p}^{(0)}(p) = -i\frac{1}{\pi}\left(\frac{a}{\hbar}\right)^{\frac{3}{2}} \frac{p_z a}{\hbar[(p^2 a^2/\hbar^2)+\frac{1}{4}]^3},$$

$$\varphi_{2p}^{(\pm 1)}(p) = -\frac{i}{\sqrt{(2)}\,\pi}\left(\frac{a}{\hbar}\right)^{\frac{3}{2}} \frac{(p_x \pm i p_y)\,a}{\hbar[(p^2 a^2/\hbar^2)+\frac{1}{4}]^3}.$$

From these expressions we can find the normalised momentum distribution

$$w(\mathbf{p}) = |\varphi(\mathbf{p})|^2.$$

5*. The Hamiltonian function in relativistic mechanics is of the form

$$H = \sqrt{(\mu^2 c^4 + p^2 c^2)} - \mu c^2 + V(r) \approx \frac{p^2}{2\mu} + V(r) + H_1,$$

where

$$H_1 = -\frac{p^4}{8\mu^3 c^2}.$$

We shall now take $\mathbf{p}$ to be the operator $\mathbf{p} = -i\hbar\nabla$, and H_1 will be considered to be a small perturbation. We then get in the approximation used by us in the Schrödinger equation $[\hat{p}^2/2\mu + V(r)]\psi = E\psi$, and the required correction for the energy in a state with n, l, m is

$$\Delta E_1 = -\frac{1}{8\mu^3 c^2}\int \psi^* \hat{p}^4 \psi \, d\tau = -\int \frac{1}{2\mu c^2}\int \psi^* [E - V(r)]^2 \psi \, d\tau$$

$$= \frac{3E^2}{2\mu c^2} - \frac{(\mu e^4/\hbar^2)^2}{n^3 \mu c^2 (2l+1)} = \left[\frac{3}{8n^4} - \frac{1}{(2l+1)\,n^3}\right] \frac{\mu e^4}{\hbar^2}\left(\frac{e^2}{\hbar c}\right)^2.$$

7. The tritium nucleus goes over into the ^{3}He nucleus during β-decay. The influence of β-decay on the atomic electron is essentially contained in the fact that during a short time interval, $t\ (\ll \hbar^3/\mu e^4)$, the potential energy of the electron in the atom changes: it becomes equal to $V = -e^2/r$ instead of $V = -2e^2/r$. The time t can be estimated as the time of flight of the β-particle through the atom,

$$t \sim \frac{a_0}{v},$$

where $a_0 = \hbar^2/\mu e^2$, v is the velocity of the β-particle. Since the energy of the β-particle is of the order of several keV, we find $t \sim 0{\cdot}1\,\hbar^3/\mu e^4$. The wave function of the electron does not have time to change during t, since we get from the Schrödinger equation

$$\delta\psi \sim \frac{e^2}{r}\frac{t}{i\hbar}\psi \ll \psi.$$

Let us expand the wave function of the electron ψ in terms of the eigenfunctions of the electron in the field with $Z = 2$,

$$\psi = \sum_n c_n \psi_n + \int c_k \psi_k \, d^3 \boldsymbol{k}.$$

The expansion coefficients

$$c_n = \int \psi \psi_n \, d\tau,$$

$$c_k = \int \psi \psi_k \, d\tau$$

determine the probability for excitation

$$w_n = \sum |c_n|^2$$

and ionisation

$$w_{\text{ion}} = \int |c_k|^2 \, d^3 \boldsymbol{k}.$$

Since ψ is spherically symmetric, c_n and c_k will be different from zero only if the states n and $\boldsymbol{k}$ are s-states ($l = 0$).

Since

$$R_{n0}^{(Z)} = 2\left(\frac{Z}{n}\right)^{\frac{3}{2}} \exp(-Zr/n) \, F\left(-n+1, 2, \frac{2Zr}{n}\right),$$

we find

$$c_n = \int_0^\infty R_{n0}^{(Z)} R_{10}^{(Z')} r^2 \, dr = \frac{8}{(Z'+Z/n)^3} \left(\frac{ZZ'}{n}\right)^{\frac{3}{2}} F\left(-n+1, 3, 2, \frac{2Z}{nZ'+Z}\right).$$

If we put $Z = 2$, $Z' = 1$, we get for $n = 1$:

$$c_1 = \frac{16\sqrt{2}}{27},$$

that is, the probability for the ^{3}He-ion to be in the ground state is equal to $w_1 = |c_1|^2 = (8/9)^3 = 0{\cdot}70$. The total probability of excitation and ionisation will thus be equal to $1 - w_1 = 0{\cdot}30$. For

$$n = 2, \quad c_2 = -\tfrac{1}{2}, \quad w_2 = 0{\cdot}25.$$

Using the equation

$$F(\alpha, \beta, \gamma, x) = (1-x)^{\gamma-\alpha-\beta} F(\gamma-\alpha, \gamma-\beta, \gamma, x)$$

we find

$$w_n = |c_n|^2 = \frac{2^9 n^5 (n-2)^{2n-4}}{(n+2)^{2n+4}}.$$

We find for the value of the probability for excitation of the first few levels, using this formula,

$$w_1 = \left(\frac{8}{9}\right)^3, \quad w_2 = \tfrac{1}{4}, \quad w_3 = \frac{2^9\,3^5}{5^{10}} \approx 1{\cdot}3\%, \quad w_4 = \frac{2^{23}}{6^{12}} \approx 0{\cdot}39\%.$$

8. Let the electron be in a stationary state with well-defined value m of the z-component of the angular momentum. The wave function of such a state is equal to:

$$\psi_{nlm}(r, \vartheta, \varphi) = R_{nl}(r)\, P_l^{(m)}(\cos\vartheta) \exp(im\varphi).$$

The electrical current density in the state ψ_{nlm} is in a system of polar coordinates, r, ϑ, φ, of the following form:

$$J_r = J_\vartheta = 0, \quad J_\varphi = -\frac{|e|\hbar m}{\mu r \sin\vartheta} |\psi_{nlm}|^2.$$

It is obvious that the magnetic field vector will be along the z-axis. A circular current of strength dJ will produce at the point O (see fig. 39) a magnetic field, the strength of which is given by

$$d\mathscr{H}_z = \frac{dJ}{rc} 2\pi \sin^2\vartheta;$$

since $dJ = J_\varphi r\, d\vartheta\, dr$,

$$\mathscr{H}_z = -\frac{2\pi |e|\hbar m}{\mu c} \int_0^\infty \frac{R^2}{r}\, dr \int_0^\pi |P_l^{(m)}|^2 \sin\vartheta\, d\vartheta = -\frac{m|e|\hbar}{\mu c} \int_0^\infty \frac{R^2}{r}\, dr.$$

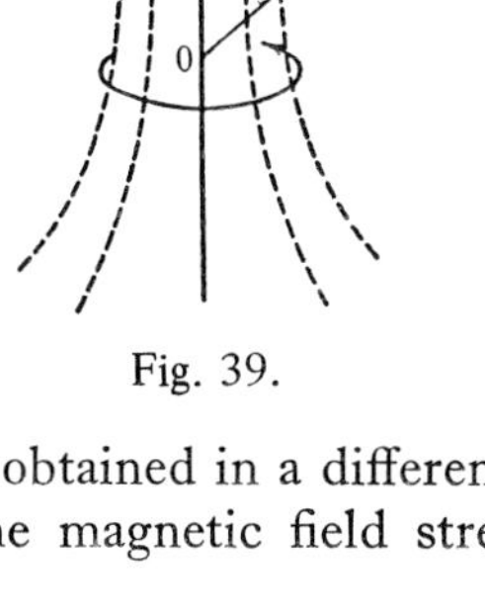

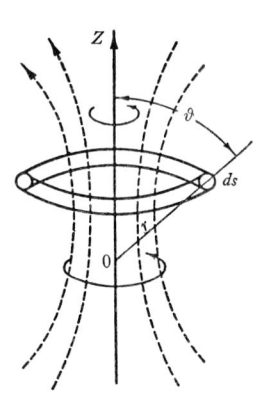

Fig. 39.

This result can also be obtained in a different way: it is well known in electrodynamics that the magnetic field strength produced by the motion of a charge is equal to (we neglect here retardation effects):

$$\mathscr{H} = \frac{e}{\mu c}\frac{1}{r^3}[\boldsymbol{r}\wedge\boldsymbol{p}] = \frac{e}{\mu c}\frac{1}{r^3}\boldsymbol{l},$$

where $\boldsymbol{r}$ is the radius vector from the point of observation and $\boldsymbol{l}$ the angular momentum.

In quantum theory, in order to obtain the average value of the magnetic field strength, we must evaluate an integral of the form

$$\overline{\mathscr{H}_z} = -\frac{|e|\hbar}{\mu c}\int \psi^* \frac{\hat{l}_z}{r^3}\psi\, d\tau;$$

since $\hat{l}_z\psi = m\psi$,

$$\overline{\mathscr{H}_z} = -m\frac{|e|\hbar}{\mu c}\overline{r^{-3}} = -m\frac{|e|\hbar}{\mu c}\left(\frac{\mu e^2}{\hbar^2}\right)^3 \frac{1}{n^3 l(l+\frac{1}{2})(l+1)}.$$

The average values of $\mathscr{H}_x$ and $\mathscr{H}_y$ will in our case be equal to zero since

$$\int \psi^*_{nlm}\hat{l}_y\psi_{nlm}\,d\tau = \int \psi^*_{nlm}\hat{l}_x\psi_{nlm}\,d\tau = 0.$$

For a $2p$-state ($m = 1$) we have

$$\overline{\mathscr{H}_z} = -\frac{1}{12}\frac{|e|\hbar}{2\mu c}\left(\frac{\mu e^2}{\hbar^2}\right)^3, \quad \text{or} \quad \overline{\mathscr{H}_z} \sim 10^4\,\text{Oe}.$$

9. The magnetic moment of the particle is equal to:

$$\mathfrak{M} = \frac{e}{2\mu c}\int \psi^* \hat{\boldsymbol{l}}\psi\, d\tau.$$

In the case of two particles we introduce new variables: the centre of mass coordinates (X, Y, Z) and the relative coordinates (x, y, z).

The average value of the magnetic moment expressed in the new variables will be

$$\overline{\mathfrak{M}_z} = \frac{e}{2c}\int \psi^* \left[\left(\frac{1}{m}+\frac{1}{M}\right)\left(X\frac{\partial}{\partial y} - Y\frac{\partial}{\partial x}\right) + \frac{1}{m+M}\left(x\frac{\partial}{\partial Y} - y\frac{\partial}{\partial X}\right) + \frac{m-M}{mM}\left(x\frac{\partial}{\partial y} - y\frac{\partial}{\partial x}\right)\right]\psi\, d\tau,$$

and similar expressions for $\overline{\mathfrak{M}_x}$ and $\overline{\mathfrak{M}_y}$.

In a stationary state, the average values of the coordinates x, y, z and of the momenta $-i\hbar(\partial/\partial x)$, $-i\hbar(\partial/\partial y)$, $-i\hbar(\partial/\partial z)$ are equal to zero. The above expression simplifies thus and becomes

$$\overline{\mathfrak{M}_z} = -\frac{e}{2mc}\left(1 - \frac{m}{M}\right)\int \psi^* \left[-i\hbar\left(x\frac{\partial}{\partial y} - y\frac{\partial}{\partial x}\right)\right]\psi\, d\tau.$$

In the present problem m is the mass of an electron and M the mass of the nucleus.

10. $\Delta E = 0{\cdot}00844 . 2{\cdot}79 \frac{Z^3}{n^3}$ cm^{-1}.

For the hydrogen atom ground state ($Z = n = 1$) we have

$$\Delta E = 0{\cdot}0235 \text{ cm}^{-1}.$$

11*. To determine the energy it is necessary to find the magnetic field strength produced by the electron. Owing to the orbital motion of the electron there will be at the position of the nucleus a magnetic field the strength of which is given by the Biot–Savart law,

$$\mathscr{H}_l = \frac{1}{c} \frac{[\boldsymbol{r} \wedge \boldsymbol{j}]}{r^3},$$

where $\boldsymbol{r}$ is the radius vector from the nucleus to the electron and $\boldsymbol{j} = -e\boldsymbol{v}$ ($-e$ is the electronic charge). We introduce the orbital angular momentum operator $\hat{\boldsymbol{l}}$. We then get for $\mathscr{H}_l$

$$\hat{\mathscr{H}}_l = -\frac{e\hbar}{\mu c} \frac{1}{r^3} \hat{\boldsymbol{l}}.$$

Since the electron possesses apart from an electric charge also a spin, the total magnetic field strength at the position of the nucleus will be equal to

$$\hat{\mathscr{H}} = \hat{\mathscr{H}}_l + \hat{\mathscr{H}}_s = \frac{e\hbar}{\mu c} \frac{1}{r^3} \left[\hat{\boldsymbol{s}} - \frac{3(\boldsymbol{r} . \hat{\boldsymbol{s}}) \boldsymbol{r}}{r^2} - \hat{\boldsymbol{l}} \right].$$

The operator of the hyperfine structure energy can thus be written in the following form:

$$\hat{w} = -\beta(\hat{\boldsymbol{i}} . \hat{\mathscr{H}}),$$

where $\hat{\boldsymbol{i}}$ is the operator of the nuclear spin and $\beta\hat{\boldsymbol{i}}$ its magnetic moment. We shall consider $\hat{w}$ to be a small perturbation. The unperturbed state is characterised by the quantum numbers n, j ($j = l + \frac{1}{2}$, $j = l - \frac{1}{2}$), l (we shall assume LS coupling).

To determine the hyperfine structure energy we must average the operator $\hat{w}$ over the state with quantum numbers

$$f, j, l (\hat{\boldsymbol{f}} = \hat{\boldsymbol{j}} + \hat{\boldsymbol{i}}).$$

If we use the equation

$$(\hat{\boldsymbol{A}})_{fm}^{fm'} = \frac{(\hat{\boldsymbol{f}} . \hat{\boldsymbol{A}})_f^f}{f(f+1)} (\hat{\boldsymbol{f}})_{fm}^{fm'}$$

we get

$$(\hat{\mathscr{H}})_{nljm_j}^{nljm_j'} = -\frac{e\hbar}{\mu c} \left(\frac{1}{r^3} \right)_{nljm}^{nljm_j} \frac{l(l+1)}{j(j+1)} (\hat{\boldsymbol{j}})_{jm}^{jm_j'}.$$

Using this relation we can easily show that the operator $\hat{w}$ can be written in the form:

$$\hat{w} = \frac{e\hbar}{\mu c}\beta\frac{1}{r^3}\frac{l(l+1)}{j(j+1)}(\hat{\boldsymbol{\imath}}.\hat{\boldsymbol{\jmath}}).$$

It follows thus that the hyperfine structure energy E is determined by the following expression:

$$E = \frac{e\hbar}{2\mu c}\beta\overline{\left(\frac{1}{r^3}\right)}\frac{l(l+1)}{j(j+1)}[f(f+1)-j(j+1)-i(i+1)].$$

We thus see that any term characterised by the numbers n, l, j, is split into $2i+1$ components (if $j>i$) because of the hyperfine structure. The different values of f give us the interval rule for the hyperfine structure multiplets.

We note that simply by counting the number of components of the hyperfine structure multiplet in the spectrum of a given isotope we can determine the nuclear spin. In fig. 40 we have given the hyperfine structure of the sodium D lines.

The fine structure, that is, the presence of a doublet (the line $D_1 = 5896$ Å, corresponding to a transition $3\,^2P_{\frac{1}{2}} \to 3\,^2S_{\frac{1}{2}}$ and $D_2 = 5890$ Å, corresponding to a transition $3\,^2P_{\frac{3}{2}} \to 3\,^2S_{\frac{1}{2}}$), is due to the spin-orbit interaction (see problem 23 of section 11).

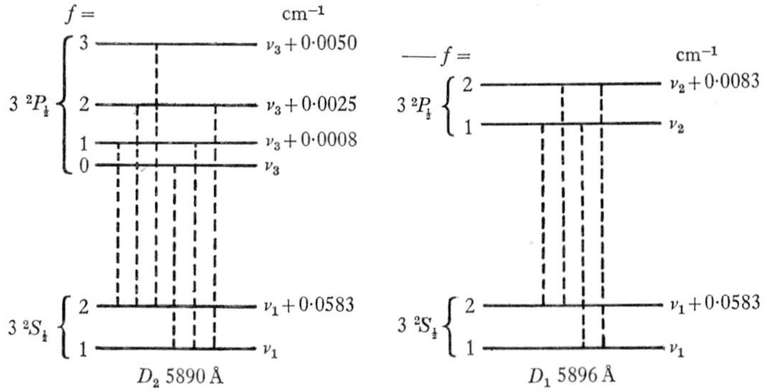

Fig. 40.

12. Let us consider the operator of the nuclear kinetic energy, $\hat{T}$. In the centre of mass system

$$\boldsymbol{P}+\Sigma \boldsymbol{p}_i = 0,$$

where $\boldsymbol{P}$ is the nuclear momentum and $\boldsymbol{p}_i$ are the electronic momenta,

$\hat{T}$ is of the form

$$\hat{T} = \frac{\hat{P}^2}{2M} = \frac{(\Sigma \hat{p}_i)^2}{2M} = \Sigma \frac{\hat{p}_i^2}{2M} + \sum_{i>k} \frac{(\hat{p}_i \cdot \hat{p}_k)}{M}.$$

Since the ratio of the electronic to the nuclear mass $m/M \ll 1$, we can write the required energy shift using perturbation theory as follows:

$$\Delta E = \int \psi^* \hat{T} \psi \, d\tau,$$

where ψ is the wave function of the electrons in the field of an infinitely heavy, fixed nucleus.

The first term in the expression for $\hat{T}$ differs only by a factor m/M from the kinetic energy of the electrons which is equal to the energy of the atom with the opposite sign by using the virial theorem. We get thus for ΔE a sum of two terms:

$$\Delta E = \Delta E_1 + \Delta E_2,$$

$$\Delta E_1 = -\frac{m}{M} E, \quad \Delta E_2 = \frac{1}{M} \int \psi^* \sum_{i>k} (\hat{p}_i \cdot \hat{p}_k) \psi \, d\tau.$$

We shall now consider in more detail the term ΔE_2. If we take for ψ a product of wave functions of the separate electrons, ΔE_2 will vanish, since the average value of the momentum of an electron in a bound state is always equal to zero. If, however, we symmetrise such a wave function in the appropriate manner, ΔE_2 will be different from zero. The symmetrised wave function of the helium atom can be written in the form

$$\psi(r_1, r_2) = \frac{1}{\sqrt{2}} [\psi_1(r_1) \psi_2(r_2) \pm \psi_1(r_2) \psi_2(r_1)],$$

where the upper sign refers to para-helium (total spin $S = 0$) and the lower sign to ortho-helium ($S = 1$).

If we put this expression into the equation for ΔE_2, we get

$$\Delta E_2 = \pm \frac{1}{M} \left| \int \psi_1^* \hat{p} \psi_2 \, d\tau \right|^2.$$

The matrix element of the momentum is different from zero only if $\Delta l = 0, \pm 1$, and ΔE_2 is therefore equal to zero for $1snd$, $1snf$, ..., states.

If we take for the $1s$-electron and for the np-electron hydrogenic wave functions with effective charges Z_1, Z_2

$$\psi_{1s}(r) = \sqrt{\left(\frac{Z_1^3}{\pi}\right)} \exp(-Z_1 r),$$

$$\psi_{np} = Y_{10} \frac{2Z_2^{\frac{5}{2}}}{3} \frac{\sqrt{(n^2-1)}}{n^2} r \exp(-Z_2 r/n) F\left(-n+2, 4; \frac{2Z_2 r}{n}\right),$$

we get

$$\Delta E_2 = \pm \frac{m}{M}\frac{64}{3}(Z_1 Z_2)^5 \frac{(Z_1 n - Z_2)^{2n-4}}{(Z_1 n + Z_2)^{2n+4}} n^3(n^2-1),$$

where the upper sign refers to paraterms $1snp\,{}^1P$ and the lower sign to orthoterms $1snp\,{}^3P$.

13. ${}^1S_0$, ${}^3S_1$, ${}^3P_{0,1,2}$, ${}^2D_{\frac{3}{2},\frac{5}{2}}$, ${}^4D_{\frac{1}{2},\frac{3}{2},\frac{5}{2},\frac{7}{2}}$.

14. (a) ${}^1S_0\,{}^3S_1$;
 (b) ${}^1P_1\,{}^3P_{012}$;
 (c) ${}^1D_2\,{}^3D_{123}$;
 (d) ${}^1S_0\,{}^3S_1\,{}^1P_1$, ${}^3P_{012}\,{}^1D_2\,{}^3D_{123}$.

15. (a) ${}^4S\,{}^2P\,{}^2D$;
 (b) ${}^1S\,{}^3P\,{}^1D\,{}^3F\,{}^1G$;
 (c) ${}^2S\,{}^2P\,{}^4P\,{}^2D$.

16. O ${}^3P_2$, Cl ${}^2P_{\frac{3}{2}}$, Fe ${}^5D_4$, Co ${}^4F_{\frac{3}{2}}$, As ${}^4S_{\frac{3}{2}}$, La ${}^2D_{\frac{3}{2}}$.

17. K, Zn, C, O are even; B, N, Cl are odd.

18. If all triples of quantum numbers are different then the number of states is equal to the number of ways in which we can take N from $N/2+M_s$, or

$$g(M_s) = C_N^{\frac{1}{2}N+M_s}.$$

If there are N' pairs of identical triples then we have

$$g(M_s) = C_{N-2N'}^{\frac{1}{2}(N-2N')+M_s}.$$

19. The number of states is equal to

$$\frac{N_l(N_l-1)\dots(N_l-x+1)}{x!}, \quad \text{where} \quad N_l = 2(2l+1).$$

21*. We shall denote the antisymmetric wave function

$$\Phi = \frac{1}{N!}\begin{vmatrix} \psi_{n^1 l^1 m_l^1 m_s^1}(\xi_1) & \psi_{n^1 l^1 m_l^1 m_s^1}(\xi_2) & \dots & \psi_{n^1 l^1 m_l^1 m_s^1}(\xi_N) \\ \psi_{n^2 l^2 m_l^2 m_s^2}(\xi_1) & \psi_{n^2 l^2 m_l^2 m_s^2}(\xi_2) & \dots & \psi_{n^2 l^2 m_l^2 m_s^2}(\xi_N) \\ \dots & \dots & \dots & \dots \\ \psi_{n^N l^N m_l^N m_s^N}(\xi_1) & \psi_{n^N l^N m_l^N m_s^N}(\xi_2) & \dots & \psi_{n^N l^N m_l^N m_s^N}(\xi_N) \end{vmatrix},$$

which is constructed from functions of the one-electron problem, simply by

$$\Phi(n^1\,l^1\,m^1\,m_s^1, n^2\,l^2\,m_l^2\,m_s^2, \dots, n^N\,l^N\,m_l^N\,m_s^N).$$

Let us consider the action of the symmetrical operator

$$(\hat{L}_x - i\hat{L}_y) = \sum_{i=1}^{N}(\hat{l}_x^i - i\hat{l}_y^i)$$

on the antisymmetric function

$$\Phi(n^1\, l^1\, m_l^1\, m_s^1, n^2\, l^2\, m_l^2\, m_s^2, \ldots, n^N\, l^N\, m_l^N\, m_s^N).$$

One easily sees that

$$\begin{aligned}(\hat{L}_x - i\hat{L}_y)\,&\Phi(n^1\, l^1\, m_l^1\, m_s^1, n^2\, l^2\, m_l^2\, m_s^2, \ldots)\\ &= \sqrt{[(l^1+m_l^1)(l^1-m_l^1+1)]}\,\Phi(n^1\, l^1\, m_l^1-1 m_s', n^2\, l^2\, m_l^2\, m_s^2, \ldots)\\ &\quad+\sqrt{[(l^2+m_l^2)(l^2-m_l^2+1)]}\,\Phi(n^1\, l^1\, m_l^1\, m_s^1, n^2\, l^2\, m_l^2-1 m_s^2, \ldots)\\ &\quad+\ldots+\sqrt{[(l^N+m_l^N)(l^N-m_l^N+1)]}\,\Phi(n^1\, l^1\, m_l^1\, m_s^1, \ldots, n^N\, l^N\, m_l^N-1 m_s^N).\end{aligned}$$

The action of the operator $\hat{S}_x - i\hat{S}_y$ gives a similar result with, however, m_s instead of m_l decreased by unity.

If the wave function is an eigenfunction of the four commuting operators $\hat{S}^2, \hat{L}^2, \hat{S}_z, \hat{L}_z$, then the action of the operators $(\hat{L}_x - i\hat{L}_y)$ and $(\hat{S}_x - i\hat{S}_y)$ leads to the following result:

$$\left.\begin{aligned}(\hat{L}_x - i\hat{L}_y)\,&\Phi(SLM_S M_L)\\ &= \sqrt{[(L+M_L)(L-M_L+1)]}\,\Phi(SLM_S M_L-1),\\ (\hat{S}_x - i\hat{S}_y)\,&\Phi(SLM_S M_L)\\ &= \sqrt{[(S+M_S)(S-M_S+1)]}\,\Phi(SLM_S-1 M_L).\end{aligned}\right\} \quad (1)$$

After these preliminaries we can immediately go over to the solution of our problem.

In the case under consideration we are dealing with equivalent electrons so that we can everywhere omit the quantum numbers *nl*. The value of the *z*-component of the spin angular momentum will be indicated by indices $(\pm)$ on m_l.

For the configuration p^3 we get the following classification of states according to the values of M_S and M_L (we limit ourselves to non-negative values) and we get the following term scheme:

M_S	M_L			
$\frac{3}{2}$	0	$\Phi(1^+, -1^+, 0^+)$	$[^4S]$	Φ_1
$\frac{1}{2}$	0	$\Phi(1^+, -1^+, 0^-)$	$[^4S$	Φ_2
		$\Phi(1^-, -1^+, 0^+)$	^2P	Φ_3
		$\Phi(1^+, -1^-, 0^+)$	$^2D]$	Φ_4
$\frac{1}{2}$	1	$\Phi(1^+, 1^-, -1^+)$	$[^2P$	Φ_5
		$\Phi(1^+, 0^+, 0^-)$	$^2D]$	Φ_6
$\frac{1}{2}$	2	$\Phi(1^+, 1^-, 0^+)$	$[^2D]$	Φ_7.

We now evaluate the action of the operators $\hat{L}_x - i\hat{L}_y$ and $\hat{S}_x - i\hat{S}_y$ on the states given a moment ago,

$$(\hat{L}_x - i\hat{L}_y)\Phi_5 = \sqrt{(2)}(\Phi_3 - \Phi_2),$$
$$(\hat{L}_x - i\hat{L}_y)\Phi_6 = \sqrt{(2)}(\Phi_2 - \Phi_4),$$
$$(\hat{L}_x - i\hat{L}_y)\Phi_7 = \sqrt{(2)}(\Phi_5 - \Phi_6),$$
$$(\hat{S}_x - i\hat{S}_y)\Phi_1 = \Phi_2 + \Phi_3 + \Phi_4,$$

where Φ_1 is the wave function of the state ${}^4S$ with $M_S = \frac{3}{2}, M_L = 0$.

The action of the operator $\hat{S}_x - i\hat{S}_y$ leads, according to equation (1), to the result

$$(\hat{S}_x - i\hat{S}_y)\Phi({}^4S, \tfrac{3}{2}, 0) = \sqrt{(3)}\,\Phi({}^4S, \tfrac{1}{2}, 0).$$

Since

$$(\hat{S}_x - i\hat{S}_y)\Phi_1 = \Phi_2 + \Phi_3 + \Phi_4,$$

we have

$$\Phi({}^4S, \tfrac{1}{2}, 0) = \frac{1}{\sqrt{3}}(\Phi_2 + \Phi_3 + \Phi_4).$$

Similarly we find for ${}^2D$ terms the following wave function:

$$\Phi({}^2D, \tfrac{1}{2}, 2) = \Phi_7,$$

$$\Phi({}^2D, \tfrac{1}{2}, 1) = \frac{1}{\sqrt{2}}(\Phi_5 - \Phi_6),$$

$$\Phi({}^2D, \tfrac{1}{2}, 0) = \frac{1}{\sqrt{6}}(\Phi_3 - 2\Phi_2 + \Phi_4).$$

The state ${}^2P, \frac{1}{2}, 1$ ($M_S = \frac{1}{2}, M_L = 1$) is a linear combination of the state Φ_5 and Φ_6, which is orthogonal to the state ${}^2D, \frac{1}{2}, 1$. From this we get

$$\Phi({}^2P, \tfrac{1}{2}, 1) = \frac{1}{\sqrt{2}}(\Phi_5 + \Phi_6),$$

and also

$$\Phi({}^2P, \tfrac{1}{2}, 0) = \frac{1}{\sqrt{2}}(\Phi_3 - \Phi_4).$$

In the same way we can obtain the wave functions corresponding to negative values of the z-components of the angular momentum.

22*. From the table which is given below it follows that to determine the eigenfunctions of the two ${}^2D$ terms we must first evaluate those of the $^2H, ^2G, ^4F, ^2F$ terms

$$
\begin{array}{cc}
M_S & M_L \\
\tfrac{1}{2} & 5 \quad \Phi(2^+, 2^-, \ 1^+) \ \Phi_1 \quad [^2H] \\
\tfrac{1}{2} & 4 \quad \begin{array}{l} \Phi(2^+, 2^-, \ 0^+) \ \Phi_2 \\ \Phi(2^+, 1^+, \ 1^-) \ \Phi_3 \end{array} \left[\begin{array}{c} ^2H \\ ^2G \end{array}\right] \\
\tfrac{1}{2} & 3 \quad \begin{array}{l} \Phi(2^+, 2^-, -1^+) \ \Phi_4 \\ \Phi(2^+, 1^+, \ 0^-) \ \Phi_5 \\ \Phi(2^+, 1^-, \ 0^+) \ \Phi_6 \\ \Phi(2^-, 1^+, \ 0^+) \ \Phi_7 \end{array} \left[\begin{array}{c} ^2H \\ ^2G \\ ^4F \\ ^2F \end{array}\right] \\
\tfrac{1}{2} & 2 \quad \begin{array}{l} \Phi(2^+, 2^-, -2^+) \ \Phi_8 \\ \Phi(2^+, 1^+, -1^-) \ \Phi_9 \\ \Phi(2^+, 1^-, -1^+) \ \Phi_{10} \\ \Phi(2^-, 1^+, -1^+) \ \Phi_{11} \\ \Phi(2^+, 0^+, \ 0^-) \ \Phi_{12} \\ \Phi(1^+, 1^-, \ 0^+) \ \Phi_{13} \end{array} \left[\begin{array}{c} ^2H \\ ^2G \\ ^4F \\ ^2F \\ ^2D \\ ^2D \end{array}\right] \\
\tfrac{3}{2} & 3 \quad \Phi(2^+, 1^+, \ 0^+) \ \Phi_{14} \quad [^4F] \\
\tfrac{3}{2} & 2 \quad \Phi(2^+, 1^+, -1^+) \ \Phi_{15} \quad [^4F].
\end{array}
$$

We first of all determine the result of the action of the operators $\hat{L}_x - i\hat{L}_y$ and $\hat{S}_x - i\hat{S}_y$ on these states:

$$(\hat{L}_x - i\hat{L}_y)\Phi_1 = -2\Phi_3 + \sqrt{(6)}\,\Phi_2,$$
$$(\hat{L}_x - i\hat{L}_y)\Phi_2 = -2\Phi_7 + 2\Phi_6 + \sqrt{(6)}\,\Phi_4,$$
$$(\hat{L}_x - i\hat{L}_y)\Phi_3 = -\sqrt{(6)}\,\Phi_6 + \sqrt{(6)}\,\Phi_5,$$
$$(\hat{S}_x - i\hat{S}_y)\Phi_{14} = \Phi_7 + \Phi_6 + \Phi_5,$$
$$(\hat{L}_x - i\hat{L}_y)\Phi_4 = -2\Phi_{11} + 2\Phi_{10} + 2\Phi_6,$$
$$(\hat{L}_x - i\hat{L}_y)\Phi_5 = \sqrt{(6)}\,\Phi_{12} + \sqrt{(6)}\,\Phi_9,$$
$$(\hat{S}_x - i\hat{S}_y)\Phi_{15} = \Phi_{11} + \Phi_{10} + \Phi_9,$$
$$(\hat{L}_x - i\hat{L}_y)\Phi_6 = 2\Phi_{13} - \sqrt{(6)}\,\Phi_{12} + \sqrt{(6)}\,\Phi_{10},$$
$$(\hat{L}_x - i\hat{L}_y)\Phi_7 = -2\Phi_{13} + \sqrt{(6)}\,\Phi_{11}.$$

The wave function of the state ${}^2H, \frac{1}{2}, 5\,(M_S = \frac{1}{2}, M_L = 5)$ is equal to Φ_1, or $\Phi({}^2H, \frac{1}{2}, 5) = \Phi_1$. If we apply the operator $\hat{L}_x - i\hat{L}_y$, we get

$$(\hat{L}_x - i\hat{L}_y)\,\Phi({}^2H, \tfrac{1}{2}, 5) = \sqrt{(10)}\,\Phi({}^2H, \tfrac{1}{2}, 4).$$

Since

$$(\hat{L}_x - i\hat{L}_y)\,\Phi_1 = -2\Phi_3 + \sqrt{(6)}\,\Phi_2,$$

$$\Phi({}^2H, \tfrac{1}{2}, 4) = 1/\sqrt{(10)}\,[\sqrt{(6)}\,\Phi_2 - 2\Phi_3].$$

The other states of the ${}^2H$ term which are necessary for the solution of our problem can be found similarly:

$$\Phi({}^2H, \tfrac{1}{2}, 3) = \frac{1}{\sqrt{(30)}}\,[\sqrt{(6)}\,\Phi_4 - 2\Phi_5 + 4\Phi_6 - 2\Phi_7],$$

$$\Phi({}^2H, \tfrac{1}{2}, 2) = \frac{1}{\sqrt{(30)}}\,[\Phi_8 - \Phi_9 + 3\Phi_{10} - 2\Phi_{11} - 3\Phi_{12} + \sqrt{(6)}\,\Phi_{13}].$$

The state $\Phi({}^2G, \frac{1}{2}, 4)$ is a linear combination of the states Φ_2 and Φ_3, which is orthogonal to the state $\Phi({}^2H, \frac{1}{2}, 4)$. From these conditions we can determine the wave function $\Phi({}^2G, \frac{1}{2}, 4)$,

$$\Phi({}^2G, \tfrac{1}{2}, 4) = \frac{\exp(i\alpha)}{\sqrt{(10)}}\,[2\Phi_2 + \sqrt{(6)}\,\Phi_3].$$

Since there are no phase relations between states of different terms we can put $\alpha = 0$.

The other states of the ${}^2G$ term can directly be obtained by consecutive applications of the operator $\hat{L}_x - i\hat{L}_y$. We have in that way

$$\Phi({}^2G, \tfrac{1}{2}, 4) = \frac{1}{\sqrt{(10)}}\,[2\Phi_2 + \sqrt{(6)}\,\Phi_3],$$

$$\Phi({}^2G, \tfrac{1}{2}, 3) = \frac{1}{\sqrt{(20)}}\,[\sqrt{(6)}\,\Phi_4 + 3\Phi_5 - \Phi_6 - 2\Phi_7],$$

$$\Phi({}^2G, \tfrac{1}{2}, 2) = \sqrt{\left(\frac{3}{140}\right)}\,[2\Phi_8 + 3\Phi_9 + \Phi_{10} - 4\Phi_{11} + 4\Phi_{12} + \sqrt{(\tfrac{2}{3})}\,\Phi_{13}].$$

It is still necessary to determine the functions $\Phi({}^4F, \frac{1}{2}, 2)$ and $\Phi({}^2F, \frac{1}{2}, 2)$ in order to solve our problem.

Since $\Phi({}^4F, \frac{3}{2}, 3) = \Phi_{14}$, we get by acting on it with the operator $\hat{S}_x - i\hat{S}_y$ the state $\Phi({}^4F, \frac{1}{2}, 3)$:

$$(\hat{S}_x - i\hat{S}_y)\,\Phi({}^4F, \tfrac{3}{2}, 3) = \sqrt{(3)}\,\Phi({}^4F, \tfrac{1}{2}, 3) = (\hat{S}_x - i\hat{S}_y)\,\Phi_{14},$$

$$\Phi({}^4F, \tfrac{1}{2}, 3) = \frac{1}{\sqrt{3}}\,\{\Phi_5 + \Phi_6 + \Phi_7\}.$$

From the state $\Phi(^4F, \frac{1}{2}, 3)$ we get the state $\Phi(^4F, \frac{1}{2}, 2)$ by operating with $\hat{L}_x - i\hat{L}_y$

$$\Phi(^4F, \tfrac{1}{2}, 2) = \frac{1}{\sqrt{3}}\{\Phi_9 + \Phi_{10} + \Phi_{11}\}.$$

The wave function $\Phi(^2F, \frac{1}{2}, 3)$ is determined by the condition of orthogonality to the three functions $\Phi(^2H, \frac{1}{2}, 3)$, $\Phi(^2G, \frac{1}{2}, 3)$, $\Phi(^4F, \frac{1}{2}, 3)$. The normalised function $\Phi(^2F, \frac{1}{2}, 3)$ which we get in this way is

$$\Phi(^2F, \tfrac{1}{2}, 3) = \frac{1}{\sqrt{(12)}}[-\sqrt{(6)}\,\Phi_4 + \Phi_5 + \Phi_6 - 2\Phi_7].$$

Finally we get

$$\Phi(^2F, \tfrac{1}{2}, 2) = \frac{1}{\sqrt{(12)}}[-2\Phi_8 + \Phi_9 - \Phi_{10} + \sqrt{(6)}\,\Phi_{13}].$$

We have now four states with $M_S = \frac{1}{2}$ and $M_L = 2$,

$$\Phi(^2H, \tfrac{1}{2}, 2) = \frac{1}{\sqrt{(30)}}[\Phi_8 - \Phi_9 + 3\Phi_{10} - 2\Phi_{11} - 3\Phi_{12} + \sqrt{(6)}\,\Phi_{13}],$$

$$\Phi(^2G, \tfrac{1}{2}, 2) = \sqrt{\left(\frac{3}{140}\right)}[2\Phi_8 + 3\Phi_9 + \Phi_{10} - 4\Phi_{11} + 4\Phi_{12} + \sqrt{(\tfrac{2}{3})}\,\Phi_{13}],$$

$$\Phi(^4F, \tfrac{1}{2}, 2) = \frac{1}{\sqrt{3}}(\Phi_9 + \Phi_{10} + \Phi_{11}),$$

$$\Phi(^2F, \tfrac{1}{2}, 2) = \frac{1}{\sqrt{(12)}}[-2\Phi_8 + \Phi_9 - \Phi_{10} + \sqrt{(6)}\,\Phi_{13}].$$

In the same group there are still the two ^2D states. These two states are mutually orthogonal and also orthogonal to the four states we have just written down.

From the condition of orthogonality and normalisation we get the following orthonormal functions:

$$\Phi(a\,^2D, \tfrac{1}{2}, 2) = \tfrac{1}{2}(-\Phi_8 - \Phi_9 + \Phi_{10} + \Phi_{12}),$$

$$\Phi(b\,^2D, \tfrac{1}{2}, 2) = \frac{1}{\sqrt{(84)}}[-5\Phi_8 + 3\Phi_9 + \Phi_{10} - 4\Phi_{11} - 3\Phi_{12} - 2\sqrt{(6)}\,\Phi_{13}].$$

The other wave functions of the ^2D states corresponding to other values of the two components of the angular momenta are easily obtained by applying the operators $\hat{L}_x - i\hat{L}_y$ and $\hat{S}_x - i\hat{S}_y$.

23*. We first of all write down the scheme corresponding to the configuration $npn'p$. We restrict ourselves to non-negative values of M_S and M_L.

pp		M_S = 1	M_S = 0
M_L	2	$\Phi_1(1^+, 1^+)$	$\Phi_2(1^+, 1^-)$ $\Phi_3(1^-, 1^+)$
	1	$\Phi_4(1^+, 0^+)$ $\Phi_5(0^+, 1^+)$	$\Phi_6(1^+, 0^-)$ $\Phi_7(0^+, 1^-)$ $\Phi_8(1^-, 0^+)$ $\Phi_9(0^-, 1^+)$
	0	$\Phi_{10}(1^+, -1^+)$ $\Phi_{11}(0^+, 0^+)$ $\Phi_{12}(-1^+, 1^+)$	$\Phi_{13}(1^+, -1^-)$ $\Phi_{14}(0^+, 0^-)$ $\Phi_{15}(-1^+, 1^-)$ $\Phi_{16}(1^-, -1^+)$ $\Phi_{17}(0^-, 0^+)$ $\Phi_{18}(-1^-, 1^+)$

From this scheme we see that this configuration corresponds to the following terms: ^1S, ^3S, ^1P, ^3P, ^1D and ^3D.

We shall use as zeroth order wave function the functions given in the table.

Since the energy does not depend on the values of M_S and M_L, the perturbation matrix will be of the form:

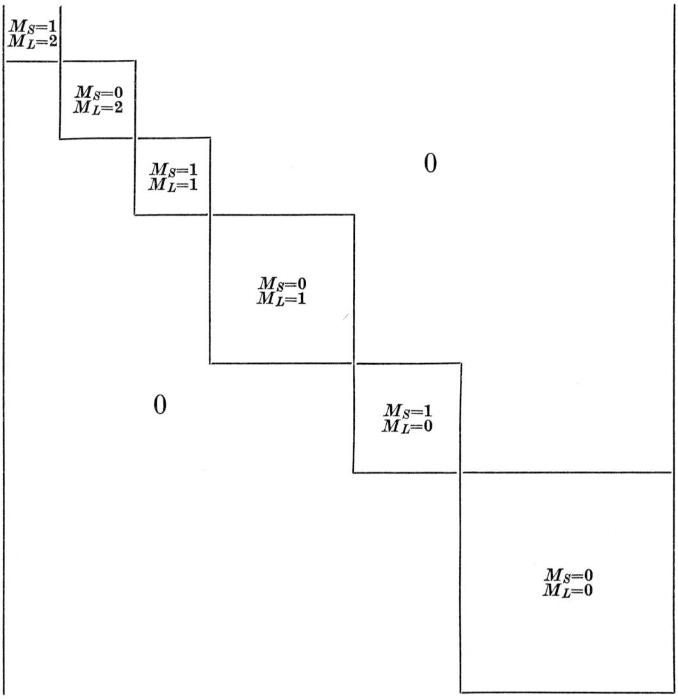

We shall denote by $\hat{V}$ the perturbation operator. In the first submatrix there is only one ^3D term. Hence

$$E^{(1)}(^3D) = V_{11}.$$

In the second submatrix ($M_L = 2, M_S = 0$) we have a combination of two terms, ^3D and ^1D,

$$E^{(1)}(^3D)+E^{(1)}(^1D) = V_{22}+V_{33}.$$

In the third submatrix ($M_L = 1, M_S = 1$) we have a combination of two terms, ^3D and ^3P,

$$E^{(1)}(^3D)+E^{(1)}(^3P) = V_{44}+V_{55}.$$

In the fourth submatrix ($M_L = 1, M_S = 0$) we have four terms, $^3D, {}^1D, {}^3P, {}^1P$,

$$E^{(1)}(^3D)+E^{(1)}(^1D)+E^{(1)}(^3P)+E^{(1)}(^1P) = V_{66}+V_{77}+V_{88}+V_{99}.$$

In the fifth submatrix ($M_L = 0, M_S = 1$) there are three terms, ^3D, ^3P, and ^3S,

$$E^{(1)}(^3D)+E^{(1)}(^3P)+E^{(1)}(^3S) = V_{10,10}+V_{11,11}+V_{12,12}.$$

Finally in the sixth submatrix ($M_L = 0, M_S = 0$) there are six terms, ^3D, ^3P, ^3S, ^1D, ^1P, and ^1S,

$$E^{(1)}(^3D)+E^{(1)}(^1D)+E^{(1)}(^3P)+E^{(1)}(^1P)+E^{(1)}(^3S)+E^{(1)}(^1S)$$
$$= V_{13,13}+V_{14,14}+V_{15,15}+V_{16,16}+V_{17,17}+V_{18,18}.$$

From these equations one obtains easily an expression for the term values in terms of the diagonal matrix elements:

$$E^{(1)}(^3D) = V_{11},$$
$$E^{(1)}(^1D) = V_{22}+V_{33}-V_{11},$$
$$E^{(1)}(^3P) = V_{44}+V_{55}-V_{11},$$
$$E^{(1)}(^1P) = V_{66}+V_{77}+V_{88}+V_{99}+V_{11}-V_{22}-V_{33}-V_{44}-V_{55},$$

and so on.

24. (a) $Z^{-\frac{1}{3}}\dfrac{\hbar^2}{\mu e^2}$, (b) $Z^{\frac{1}{3}}\dfrac{\mu e^4}{\hbar^2}$, (c) $Z^{\frac{4}{3}}\dfrac{\mu e^4}{\hbar^2}$, (d) $Z^{\frac{7}{3}}\dfrac{\mu e^4}{\hbar^2}$,

(e) $Z^{\frac{2}{3}}\dfrac{e^2}{\hbar}$, (f) $Z^{\frac{1}{3}}\hbar$, (g) $Z^{\frac{1}{3}}$.

25. The total energy consists of three parts, the kinetic energy of the electrons T, the interaction energy of the electrons and the nucleus

V_{ne}, and the energy of the interaction of the electrons with one another, V_{ee}. The last two terms have the following form,

$$V_{ne} = -\int \frac{Z}{r} \rho \, d\tau,$$

$$V_{ee} = \frac{1}{2} \int \frac{\rho(\boldsymbol{r})\, \rho(\boldsymbol{r}')}{|\boldsymbol{r}-\boldsymbol{r}'|} \, d\tau \, d\tau'.$$

To evaluate the kinetic energy let us consider an infinitesimal volume element of the atom, $d\tau$. The number of electrons with momenta between $\boldsymbol{p}$ and $\boldsymbol{p}+\boldsymbol{dp}$ is proportional to the phase volume and equal to:

$$dn = \frac{8\pi p^2 \, dp \, d\tau}{(2\pi)^3} = \frac{p^2 \, dp \, d\tau}{\pi^2}.$$

The electron density is obtained by integrating over p from 0 to some maximum value $p = p_0$,

$$\rho = \frac{p_0^3}{3\pi^2}.$$

The electronic kinetic energy in the volume $d\tau$ is equal to:

$$dT = \int_0^{p_0} \frac{p^2}{2} \, dn = \frac{p_0^5}{10\pi^2} \, d\tau.$$

If we express in this equation p_0 in terms of ρ and integrate over the atomic volume we find the kinetic energy of the electrons

$$T = \frac{1}{10\pi^2} (3\pi^2)^{\frac{5}{3}} \int \rho^{\frac{5}{3}} \, d\tau.$$

We find finally for the total energy

$$E = T + V_{ne} + V_{ee} = \frac{(3\pi^2)^{\frac{5}{3}}}{10\pi^2} \int \rho^{\frac{5}{3}} \, d\tau - Z \int \frac{\rho}{r} \, d\tau + \frac{1}{2} \iint \frac{\rho(\boldsymbol{r})\, \rho(\boldsymbol{r}')}{|\boldsymbol{r}-\boldsymbol{r}'|} \, d\tau \, d\tau'.$$

26. The total energy of the atom is given by the equation (compare the preceding problem)

$$E = -\int \frac{Z}{r} \rho d^3\boldsymbol{r} + \frac{1}{2} \int \frac{\rho(\boldsymbol{r})\, \rho(\boldsymbol{r}')}{|\boldsymbol{r}-\boldsymbol{r}'|} \, d^3\boldsymbol{r} d^3\boldsymbol{r}' + \frac{(3\pi^2)^{\frac{5}{3}}}{10\pi^2} \int \rho^{\frac{5}{3}} \, d^3\boldsymbol{r}, \qquad (1)$$

and we get thus

$$\delta E = -\int \frac{Z}{r} \delta\rho \, d^3\boldsymbol{r} + \int \frac{\rho(\boldsymbol{r})\, \delta\rho(\boldsymbol{r}')}{|\boldsymbol{r}-\boldsymbol{r}'|} \, d^3\boldsymbol{r} \, d^3\boldsymbol{r}' + \frac{(3\pi^2)^{\frac{5}{3}}}{6\pi^2} \int \rho^{\frac{2}{3}} \, \delta\rho \, d^3\boldsymbol{r},$$

with the subsidiary condition

$$\int \delta\rho \, d^3\boldsymbol{r} = 0. \qquad (2)$$

Using a Lagrangian multiplier we find thus

$$\rho(\boldsymbol{r}) = \left\{\frac{6\pi^2}{(3\pi^2)^{\frac{2}{3}}}[V(\boldsymbol{r})-\lambda]\right\}^{\frac{3}{2}}, \tag{3}$$

where

$$V(\boldsymbol{r}) = \frac{Z}{r} - \int \frac{\rho(\boldsymbol{r}')\,d^3\boldsymbol{r}'}{|\boldsymbol{r}-\boldsymbol{r}'|}. \tag{4}$$

Since Poisson's equation must hold, we get ($V_0 \equiv \lambda$)

$$\nabla^2(V-V_0) = 4\pi Z\rho, \tag{5}$$

and from (3) and (5) we get the Thomas–Fermi equation

$$\nabla^2(V-V_0) = \frac{2^{\frac{7}{2}}}{3\pi}[V-V_0]^{\frac{3}{2}}.$$

27. To solve this problem, we consider a distribution of the electrons in momentum space corresponding to a point $\boldsymbol{r}$ in coordinate space which is displaced by an amount $\boldsymbol{D}(\boldsymbol{r})$. The total kinetic energy is then given by the equation

$$E_{\text{kin}} = \frac{1}{4\pi^3}\int d^3\boldsymbol{r}\int d^3\boldsymbol{p}\,\frac{p^2}{2} = \int d^3\boldsymbol{r}\left(\frac{P_0^5}{10\pi^2}+\frac{P_0^3 D^2}{6\pi^2}\right), \tag{1}$$

where $P_0(\boldsymbol{r})$ is the radius of the undisplaced Fermi sphere which is connected with the density $\rho(\boldsymbol{r})$ by the equation

$$\rho(\boldsymbol{r}) = \frac{1}{4\pi^3}\int d^3\boldsymbol{p} = \frac{1}{3\pi^2}P_0^3, \tag{2}$$

where the integration over $d^3\boldsymbol{p}$ in both (1) and (2) is over the whole of the Fermi-sphere.

Instead of (1) of the preceding problem we now get using (1) and (2)

$$E = -\int \frac{Z}{r}\rho\,d^3\boldsymbol{r} + \frac{1}{2}\int \frac{\rho(\boldsymbol{r})\,\rho(\boldsymbol{r}')}{|\boldsymbol{r}-\boldsymbol{r}'|}\,d^3\boldsymbol{r}\,d^3\boldsymbol{r}' + \frac{(3\pi^2)^{\frac{5}{3}}}{10\pi^2}\int \rho^{\frac{5}{3}}\,d^3\boldsymbol{r} + \frac{1}{2}\int d^3\boldsymbol{r}D^2\rho.$$

Apart from the subsidiary condition (2) we now have also

$$\delta J = 0,$$

where $\boldsymbol{J}$ is the total angular momentum,

$$\boldsymbol{J} = \int \rho(\boldsymbol{r})\,[\boldsymbol{r}\wedge\boldsymbol{D}(\boldsymbol{r})]\,d^3\boldsymbol{r} = \int \boldsymbol{D}(\boldsymbol{r})\,r\sin\theta\,\rho(\boldsymbol{r})\,d^3\boldsymbol{r},$$

where θ is a polar angle.

Using Lagrangian multipliers and taking the variation with respect to $\boldsymbol{D}$, we find easily

$$\boldsymbol{D} = \lambda r\sin\theta,$$

which shows that the model corresponds to a rigid sphere rotating uniformly with angular velocity λ.

28*. The volume element expressed in terms of x is of the form

$$d\tau = 4\pi r^2\,dr = 8\pi\lambda^3 x^5\,dx.$$

We can evaluate the kinetic energy

$$T = \frac{12(3\pi^2)^{\frac{2}{3}}}{25\pi}\,\lambda^3 A^{\frac{5}{3}}$$

and the energy of interaction between the electrons and the nucleus,

$$V_{ne} = -8\pi A\lambda^2 Z.$$

To evaluate $V_{ee} = \frac{1}{2}\iint\frac{\rho(\boldsymbol{r})\,\rho(\boldsymbol{r}')}{|\boldsymbol{r}-\boldsymbol{r}'|}\,d\tau\,d\tau'$, we find first of all the potential produced by the electrons, φ_e. By solving the Poisson equation

$$\nabla^2\varphi_e = 4\pi\rho,$$

we get

$$\varphi_e = -\frac{16\pi A\lambda^2}{x^2}\,[1-e^{-x}(x+1)].$$

If we now use Green's theorem to evaluate V_{ee} we find

$$V_{ee} = -\frac{1}{2}\int\varphi_e\,\rho\,d\tau = 16\pi^2 A^2\lambda^5.$$

The normalisation condition determines A,

$$\int\rho\,d\tau = 16\pi A\lambda^3 = N.$$

Substituting this value for A in the expressions for T, V_{ne}, V_{ee}, we get

$$T = \frac{12}{25\pi}\left(\frac{3\pi N}{16}\right)^{\frac{5}{3}}\frac{1}{\lambda^2},$$

$$V_{ne} = -\frac{ZN}{2\lambda},$$

$$V_{ee} = \frac{N^2}{16\lambda}.$$

The minimum of $E = T+V_{ne}+V_{ee}$ is reached for

$$\lambda = \frac{9}{25}\left(\frac{3\pi N}{16}\right)^{\frac{2}{3}}\frac{1}{Z-N/8}$$

and is equal to

$$E = \frac{25}{36}\left(\frac{16}{3\pi}\right)^{\frac{2}{3}}N^{\frac{1}{3}}\left(Z-\frac{N}{8}\right)^2 \text{ at un.}$$

For a neutral atom we get

$$E = \frac{25.49}{36.64}\left(\frac{16}{3\pi}\right)^{\frac{2}{3}} Z^{\frac{7}{3}} = 0{\cdot}758Z^{\frac{7}{3}} \text{ at un.}$$

29. Let $\rho(r)$ be the expression for the electronic density in the Thomas–Fermi model. In that case $\rho(\boldsymbol{r})$ will lead to a minimum of the energy of the atom,

$$E = \frac{(3\pi^2)^{\frac{5}{3}}}{10\pi^2}\int \rho^{\frac{5}{3}}\,d\tau - Z\int\frac{\rho}{r}\,d\tau + \frac{1}{2}\iint\frac{\rho(\boldsymbol{r})\,\rho(\boldsymbol{r}')}{|\boldsymbol{r}-\boldsymbol{r}'|}\,d\tau\,d\tau'.$$

If we substitute in that expression instead of ρ the function $\lambda^3\,\rho(\lambda\boldsymbol{r})$, which satisfies the same normalisation condition as $\rho(\boldsymbol{r})$, we find

$$E(\lambda) = \lambda^2\,T + \lambda V,$$

where T is the kinetic and V the potential energy of the electrons in the atom. Since $E(\lambda)$ must have a minimum for $\lambda = 1$ we find that the following equation must hold:

$$2T + V = 0,$$

which is the virial theorem.

30. The energy of the interaction of the electrons can be written in the form

$$V_{ee} = -\frac{1}{2}\int\varphi_e\,\rho\,d\tau = \frac{Z}{2}\int\frac{\rho}{r}\,d\tau - \frac{1}{2}\int\varphi\rho\,d\tau, \tag{1}$$

where φ_e is the potential produced by the electrons, and φ is the potential of the self-consistent field, including the field of the nucleus

$$\varphi = \varphi_e + \frac{Z}{r}.$$

In the Thomas–Fermi model the following relations are satisfied:

$$\frac{p_0^2}{2} = \varphi - \varphi_0, \quad \rho = \frac{p_0^3}{3\pi^2},$$

where p_0 is the limiting value of the momentum, and φ_0 the potential at the boundary of the atom. If we eliminate p_0 and express φ in terms of ρ, we find

$$V_{ee} = \frac{Z}{2}\int\frac{\rho}{r}\,d\tau - \frac{(3\pi^2)^{\frac{2}{3}}}{4}\int\rho^{\frac{5}{3}}\,d\tau - \frac{\varphi_0 N}{2}.$$

The first two terms differ only by a numerical factor from the energy of the interaction of the electrons with the nucleus,

$$V_{ne} = -Z\int\frac{\rho}{r}\,d\tau$$

and the kinetic energy,

$$T = \frac{3(3\pi^2)^{\frac{2}{3}}}{10}\int \rho^{\frac{5}{3}}\,d\tau.$$

We have thus

$$V_{ee} = -\tfrac{1}{2}V_{ne} - \tfrac{5}{6}T - \frac{\varphi_0 N}{2}.$$

If we now take the value of T from the virial theorem,

$$2T = -V_{ne} - V_{ee},$$

we find finally

$$V_{ee} = -\tfrac{1}{7}V_{ne} - \tfrac{6}{7}\varphi_0 N.$$

For a neutral atom $(N = Z)$ $\varphi_0 = 0$ and

$$V_{ee} = -\tfrac{1}{7}V_{ne}.$$

31*. The energy of total ionisation is equal to the total energy of the electrons, but with opposite sign. Using the virial theorem we find (see preceding problem)

$$E_{\text{ion}} = -\tfrac{3}{7}V_{ne} + \tfrac{3}{7}\varphi_0 N.$$

We shall transform the expression

$$V_{ne} = -Z\int \frac{\rho}{r}\,d\tau$$

as follows: we introduce the potential φ_e, which is produced by the electrons

$$\nabla^2 \varphi_e = 4\pi\rho$$

and use Green's theorem

$$V_{ne} = -\frac{Z}{4\pi}\int \frac{\nabla^2 \varphi_e}{r}\,d\tau = Z\varphi_e(0)$$

[the surface integral over the boundary of the atom is equal to zero and $\nabla^2(1/r) = -4\pi\delta(r)$].

We have thus

$$E_{\text{ion}} = -\tfrac{3}{7}Z\varphi_e(0) + \tfrac{3}{7}N\varphi_0.$$

If we go over to Thomas–Fermi units,

$$r = xbZ^{-\frac{1}{3}}, \quad b = \frac{1}{2}\left(\frac{3\pi}{4}\right)^{\frac{2}{3}}, \quad \varphi - \varphi_0 = \frac{Z^{\frac{4}{3}}}{b}\frac{\chi(x)}{x},$$

we find

$$\varphi_e(r) = \varphi - \frac{Z}{r} = \varphi_0 - \frac{Z^{\frac{4}{3}}}{b}[1 - \chi(x)],$$

and since we have for small value of x

$$\chi(x) = 1 - ax + \tfrac{4}{3}x^{\frac{3}{2}},$$

with $a = a_0 = 1{\cdot}58$ for a neutral atom (for a positive ion $a > a_0$), we have finally

$$E_{\text{ion}} = \frac{3}{7}\frac{Z^{\frac{7}{3}}}{b}a - \frac{3}{7}\frac{(Z-N)^2 Z^{\frac{1}{3}}}{bx_0},$$

where x_0 is the radius of the $(Z-N)$ times ionised atom.

32*. Not only the total angular momentum but also separately both the total orbital angular momentum and the total spin angular momentum of the electrons are equal to zero in diamagnetic atoms. Because of its precession the electron acquires an extra velocity,

$$v' = \frac{e}{2\mu c}[\mathscr{H}\wedge \boldsymbol{r}].$$

If we denote by $\boldsymbol{A}$ the vector potential of the external magnetic field $\boldsymbol{A} = \frac{1}{2}[\mathscr{H}\wedge \boldsymbol{r}]$, we can write this last equation in the form

$$v' = \frac{e}{\mu c}\boldsymbol{A}.$$

The current density produced by the precession of the electron shells is equal to:

$$\boldsymbol{J} = \frac{e}{\mu c}\boldsymbol{A}\rho(\boldsymbol{r}),$$

where $\rho(\boldsymbol{r})$ is the charge density at $\boldsymbol{r}$ (the electronic charge is equal to $-e$). We first of all evaluate the vector potential $\boldsymbol{A}'$ produced by the magnetic field,

$$\boldsymbol{A}'(r) = \frac{e}{2\mu c^2}\int [\mathscr{H}\wedge \boldsymbol{r}']\frac{\rho(r')}{|\boldsymbol{r}-\boldsymbol{r}'|}\,d\tau'.$$

If we use the equation

$$\frac{1}{r_{12}} = \frac{1}{\sqrt{\{R^2+r^2-2Rr[\cos\theta\cos\vartheta+\sin\theta\sin\vartheta\cos(\Phi-\varphi)]\}}}$$

$$= \sum_{l,m}\sqrt{\left(\frac{4\pi}{2l+1}\right)}\frac{Y^*_{lm}(\theta,\Phi)\,Y_{lm}(\vartheta,\varphi)}{Y_{l0}(0)}\begin{cases}\dfrac{r^l}{R^{l+1}}, & r<R,\\[2ex] \dfrac{R^l}{r^{l+1}}, & R<r,\end{cases}$$

we get the following expression for $\boldsymbol{A}'$:

$$\boldsymbol{A}'(\boldsymbol{r}) = \frac{e}{6\mu c^2}[\mathscr{H}\wedge \boldsymbol{r}]\cdot\left[\frac{1}{r^3}\int_{r'<r} r'^2\rho(r')\,d\tau' + \int_{r'>r}\frac{\rho(r')}{r'}\,d\tau'\right].$$

Using this expression one can easily evaluate the magnetic strength along the z-axis.

This field strength is

$$\mathscr{H}'_z = \left(\frac{e\mathscr{H}}{2\mu c^2 r^5}\right)(z^2 - \tfrac{1}{3}r^2)\int_{r'<r} r'^2 \rho(r')\, d\tau' + \frac{e\mathscr{H}}{3\mu c^2}\int_{r'>r} \frac{\rho(r')}{r'}\, d\tau'.$$

The field at the centre of the atom, that is the field acting on the nucleus,

$$\mathscr{H}'_z(0) = \frac{e\mathscr{H}}{3\mu c^2}\int \frac{\rho(r')}{r'}\, d\tau' = \frac{e\mathscr{H}}{3\mu c^2}\varphi(0),$$

does depend on the electrostatic potential $\varphi(0)$, produced by the electrons.

In the Thomas–Fermi model we have

$$\varphi(0) = -1{\cdot}588\frac{Ze}{b}, \quad \text{where} \quad b = 0{\cdot}858\frac{a}{Z^{\frac{1}{3}}}, \quad a = \frac{\hbar^2}{\mu e^2},$$

and thus

$$\mathscr{H}'_z(0) = -0{\cdot}319.10^{-4}\, Z^{\frac{4}{3}}\, \mathscr{H}.$$

33*. $$\mathscr{H}'_z(0) = -\frac{27}{24}\frac{e^2\mathscr{H}}{\mu c^2 a} = -0{\cdot}599.10^{-4}\,\mathscr{H}.$$

34*. To derive the quantum rule we want, it is necessary to proceed in the same way as in the derivation of the usual Bohr–Sommerfeld quantisation rule. Especially it is necessary to find the exact functions near the points where the semi-classical approximation breaks down. Comparing these functions with the semi-classical ones in the "overlap" region (that is, in the region where both the exact and the semi-classical functions can both be applied) we can find the quantisation rule.

It is easily understood that in the semi-classical case which interests us ($n \gg 1$) the exact functions are Coulomb functions for $n \to \infty\,(E \to 0)$, and the overlap region is the interval $Z^{-1} \ll r \ll Z^{-\frac{1}{3}}$ (we use atomic units). Indeed, the self-consistent field $\varphi(r)$, in which each electron of the Thomas–Fermi distribution moves, agrees for distances $r \ll Z^{-\frac{1}{3}}$ with the Coulomb field of the nucleus $\varphi(r) \approx Z/r$. The semi-classical approximation can be applied for distances $r \gg Z^{-1}$.

The Coulomb functions for $l = 0, n \to \infty$ (that is, $E \to 0$) are of the form

$$\psi = c\frac{J_1[\sqrt{(8r)}]}{\sqrt{r}} \quad \text{(in Coulomb units)}$$

or

$$\psi = Z^{\frac{3}{2}}\, c\frac{J_1[\sqrt{(8Zr)}]}{\sqrt{(Zr)}} \quad \text{(in atomic units).} \tag{1}$$

For $r \gg Z^{-1}$, or $Zr \gg 1$, we can use the asymptotic expression for the Bessel function J_1, which is

$$\psi = Z^{\frac{3}{2}} c \frac{\sin\left[\sqrt{(8Zr)} - \frac{\pi}{4}\right]}{(2\pi^2 Z^3 r^3)^{\frac{1}{4}}} = c_1 \frac{\sin\left[\sqrt{(8Zr)} - \frac{\pi}{4}\right]}{r^{\frac{3}{4}}}. \tag{2}$$

The semi-classical function can be written in the form

$$\psi_{\text{semicl}} = c_{\text{semicl}} \frac{\sin\left(\int_0^r p_r\, dr + \alpha\right)}{r\sqrt{(p_r)}}. \tag{3}$$

Our problem is to determine the phases α. In the region which interests us, the radial momentum $p_r = \sqrt{\{2[E - V(r)]\}}$ (for the case $l = 0$ which we are considering there is no centrifugal potential) is equal to $\sqrt{(2Z/r)}$, since $|V(r)| \sim Z/r \gg |E|$; the last inequality is true by virtue of the fact that the energy of an electron in the Thomas–Fermi distribution is of the order $Z^{\frac{4}{3}}$. We have thus

$$\int_0^r p_r\, dr = \sqrt{(2Z)} \int_0^r \frac{dr}{\sqrt{r}} = \sqrt{(8Zr)},$$

and the wave function is equal to

$$\psi_{\text{semicl}} \approx c_{\text{semicl}} \frac{\sin[\sqrt{(8Zr)} + \alpha]}{r(2Z/r)^{\frac{1}{4}}} = c'_{\text{semicl}} \frac{\sin[\sqrt{(8Zr)} + \alpha]}{r^{\frac{3}{4}}}.$$

Comparing the two functions we find that $\alpha = -\pi/4$. The semi-classical function is thus of the form

$$\psi_{\text{semicl}} = c_{\text{semicl}} \frac{\sin\left(\int_0^r p_r\, dr - \frac{\pi}{4}\right)}{r\sqrt{p_r}}. \tag{4}$$

On the other hand, the function satisfying the boundary condition near the turning point r_0, which limits the motion of the particle at large values of r, is of the form

$$\psi_{\text{semicl}} = c''_{\text{semicl}} \frac{\sin\left(\int_r^{r_0} p_r\, dr + \frac{\pi}{4}\right)}{r\sqrt{p_r}}. \tag{5}$$

In order that expressions (4) and (5) are identical it is necessary that the sum of the phases of the sines is equal to a multiple of π which gives us the quantisation rule we are looking for

$$\int_0^{r_0} p_r\, dr = n\pi \quad \text{for} \quad \int_0^{r_0} \sqrt{\{2\mu[E - V(r)]\}}\, dr = \pi\hbar n. \tag{6}$$

The quantisation rule we have obtained is thus different from the Bohr–Sommerfeld quantisation rule.

One sees easily that in the particular case of a Coulomb field $V(r) = -Ze^2/r$ in the whole of space the quantisation rule (6) leads to the usual form of the spectrum

$$E_n = -\frac{Z^2 \mu e^4}{2\hbar^2}\frac{1}{n^2}.$$

(This can be checked by substituting into equation (6) for r_0 from the equation $E - V(r_0) = E + Ze^2/r_0 = 0$, or $r_0 = -Ze^2/E$, and performing the elementary integration.)

35. When the applied electrical field is of the same order as the Thomas–Fermi field, that is, the electrical field in the atom in the region where there are electrons of the Thomas–Fermi distribution, the relative change in the distribution will be of the order of unity. The dipole moment of the atom will thus in that case be of the order of the Thomas–Fermi (TF) value: Thomas–Fermi radius for one electron, and Zr_{TF} for the whole atom.

If we assume that the dipole moment is proportional to the field, we get for any, not too large, field $\mathscr{E}$:

$$d = Zr_{\text{TF}}\frac{\mathscr{E}}{\mathscr{E}_{\text{TF}}},$$

where $\mathscr{E}_{\text{TF}}$ is the Thomas–Fermi electrical field strength. Its order of magnitude can be estimated.

$$|\mathscr{E}_{\text{TF}}| = \left|\frac{d\varphi(r)}{dr}\right|_{r\sim r_{\text{TF}}} = \frac{Z^{\frac{1}{3}}}{b}\frac{Z^{\frac{4}{3}}}{b}\left|\frac{d}{dx}\left[\frac{\chi(x)}{x}\right]\right|_{x\sim 1}.$$

We have introduced here Thomas–Fermi units and expressed the potential in terms of the universal function $\chi(x)$. The derivative $d[\chi(x)/x]/dx$ is at $x\sim 1$ also of the order of unity. We find thus that $\mathscr{E}_{\text{TF}} \sim Z^{\frac{5}{3}}$ while the Thomas–Fermi radius is of the order $Z^{-\frac{1}{3}}$. We get thus

$$d \sim Z.Z^{-\frac{1}{3}}\frac{\mathscr{E}}{Z^{\frac{5}{3}}} = \frac{\mathscr{E}}{Z},$$

or

$$\frac{d}{\mathscr{E}} \sim \frac{1}{Z}$$

or, in usual units,

$$\frac{d}{\mathscr{E}} \sim \frac{\hbar^6}{Z\mu^3 e^6} = \left(\frac{\hbar^2}{Z^{\frac{1}{3}}\mu e^2}\right)^3,$$

that is, the polarisability is of the order of the cube of the Thomas–Fermi radius of the atom. (This result is completely analogous to the result of the classical theory of electrons for different models of the atom, such as the Thomson model, or the charged sphere model.)

We note in conclusion that the same quantity for a valence electron is of the order of unity in atomic units, since such an electron moves in a field $V(r) \approx -e^2/r$. Since $Z \gg 1$, the polarisability of an atom will thus be determined by the valence electrons and not by the Thomas–Fermi electrons.

36. $\dfrac{L-2S+1}{L-S+1} \leqslant g \leqslant \dfrac{L+2S}{L+S}$, if $L \geqslant S$;

$$\frac{2S+2-L}{S-L+1} \leqslant g \leqslant \frac{L+2S}{L+S}, \quad \text{if } L \leqslant S.$$

38*. The energy of an atom in a magnetic field is of the same order of magnitude as the spin–orbit interaction. The spin–orbit interaction operator which is equal to $\varphi(r)\,(\hat{\boldsymbol{l}}\,.\,\hat{\boldsymbol{s}})$ (see problem 23 of section 11) is of the same order as the operator $(e\hbar/2\mu c)(\hat{l}_z + 2\hat{s}_z)$, and their sum

$$V = \varphi(r)\,(\hat{\boldsymbol{l}}\,.\,\hat{\boldsymbol{s}}) + \frac{e\hbar}{2\mu c}\mathscr{H}(\hat{l}_z + 2\hat{s}_z)$$

can be considered to be a small perturbation. The square and the z-component of the orbital angular momentum, and also the square and the z-component of the spin, will be constants of motion in the unperturbed state. It is convenient for us to take another set of constants of motion. We shall characterise the stationary unperturbed states by the quantum numbers n, l, j, m_j ($\hat{\boldsymbol{l}}^2, \hat{\boldsymbol{j}}^2, \hat{j}_z$ commute with H_0). The degree of the degeneracy in the case of a Coulomb field is equal to $2n^2$, and in the case of arbitrary central field of force $2(2l+1)$. It is unnecessary for us to solve a secular equation of such a high order. We note that in the perturbed state the square of the orbital angular momentum and the z-component of the total angular momentum are integrals of motion. The wave function of the perturbed problem can therefore be constructed from the functions $\psi^{(0)}_{nljm_j}$, corresponding to the same value of n, l, m_j, or

$$\psi = c_1\psi^{(0)}(n, l, j = l-\tfrac{1}{2}, m_j) + c_2\psi^{(0)}(n, l, j = l+\tfrac{1}{2}, m_j)$$

or

$$\psi = c_1\frac{R^{(0)}_{nl}}{\sqrt{(2l+1)}}\begin{pmatrix}\sqrt{(l+m_j+\frac{1}{2})}\,Y_{l,m_j-\frac{1}{2}}\\ \sqrt{(l-m_j+\frac{1}{2})}\,Y_{l,m_j+\frac{1}{2}}\end{pmatrix}$$

$$+c_2\frac{R^{(0)}_{nl}}{\sqrt{(2l+1)}}\begin{pmatrix}\sqrt{(l+m_j+\frac{1}{2})}\,Y_{l,m_j-\frac{1}{2}}\\ -\sqrt{(l+m_j+\frac{1}{2})}\,Y_{l,m\,+\frac{1}{2}}\end{pmatrix}.$$

The matrix elements of the operator $\hat{V}$ are equal to

$$(\hat{V})_{nlj=l+\frac{1}{2}m_j}^{nlj=l+\frac{1}{2}m_j} = A\frac{l}{2} + \mathscr{H}\mu_0 m_j\left(1+\frac{1}{2l+1}\right),$$

$$(\hat{V})_{nlj=l-\frac{1}{2}m_j}^{nlj=l-\frac{1}{2}m_j} = -A\frac{l+1}{2} + \mathscr{H}\mu_0 m_j\left(1-\frac{1}{2l+1}\right),$$

$$(\hat{V})_{nlj=l+\frac{1}{2}m_j}^{nlj=l-\frac{1}{2}m_j} = (\hat{V})_{nlj=l-\frac{1}{2}m_j}^{nlj=l+\frac{1}{2}m_j} = \frac{\mathscr{H}\mu_0}{2l+1}\sqrt{[(l+\tfrac{1}{2})^2 - m_j^2]},$$

where

$$A = \int_0^\infty R_{nl}(r)\,\varphi(r)\,R_{nl}(r)\,r^2\,dr \quad \text{and} \quad \mu_0 = \frac{e\hbar}{2\mu c}.$$

The energy value E follows from the solution of the secular equation

$$\begin{vmatrix} E_{nl}^0 + A\frac{l}{2} + \mathscr{H}\mu_0 m_j\left(1+\frac{1}{2l+1}\right) - E & \frac{\mathscr{H}\mu_0}{2l+1}\sqrt{[(l+\frac{1}{2})^2 - m_j^2]} \\ \frac{\mathscr{H}\mu_0}{2l+1}\sqrt{[(l+\frac{1}{2})^2 - m_j^2]} & E_{nl}^{(0)} - A\frac{l+1}{2} + \mathscr{H}\mu_0 m_j\left(1-\frac{1}{2l+1}\right) - E \end{vmatrix} = 0.$$

We shall denote by E_+ and E_- the energies of the one-electron atom, taking spin–orbit interaction into account where E_+ refers to the state $j = l+\frac{1}{2}$, and E_- to $j = l-\frac{1}{2}$.

From problem 23 of section 11, it follows that

$$E_+ = E_{nl}^{(0)} + A\frac{l}{2}, \quad E_- = E_{nl}^{(0)} - A\frac{l+1}{2}.$$

The solution of the secular equation is

$$E = \tfrac{1}{2}(E_+ + E_-) + \mathscr{H}\mu_0 m_j \pm \sqrt{\left[\tfrac{1}{4}(E_+ - E_-)^2 + \mathscr{H}\mu_0 \frac{m_j}{2l+1}(E_+ - E_-) + \tfrac{1}{4}\mathscr{H}^2\mu_0^2\right]}.$$

Let us consider some limiting cases.

(a) In the case of weak fields, that is, when $\mu_0\mathscr{H} \ll E_+ - E_-$, we get the following expressions for the energy:

$$E = E_+ + \mathscr{H}\mu_0 m_j \frac{2l+2}{2l+1},$$

$$E = E_- - \mathscr{H}\mu_0 m_j \frac{2l}{2l+1}.$$

The first energy value corresponds to the energy of the nth level of the state with $j = l+\frac{1}{2}$, and the second to $j = l-\frac{1}{2}$ (see problem 44 of section 7).

(b) In the case of strong fields, that is, for $\mu_0 \mathscr{H} \gg E_+ - E_-$ we have

$$E = \tfrac{1}{2}(E_+ + E_-) + \mathscr{H}\mu_0 m_j \pm \tfrac{1}{2}\mathscr{H}\mu_0 \pm \frac{m_j}{2l+1}(E_+ - E_-).$$

Let E_c be the energy of the centre of gravity of the level, when there is no field present, that is,

$$E_c = \frac{E_+(l+1) + E_- l}{2l+1}$$

[The statistical weights of the states E_+ and E_- have a ratio $(l+1)/l$] and let ΔE be the difference $E_+ - E_-$. With this notation we get for E

$$E = E_c + \mathscr{H}\mu_0(m_j \pm \tfrac{1}{2}) \pm \frac{\Delta E}{2l+1}(m_j \mp \tfrac{1}{2}).$$

This expression is identical, as can easily be checked, with the expression of problem 40 of section 7.

The upper sign corresponds to the state with $m_l = m_j - \frac{1}{2}$, $m_s = \frac{1}{2}$, and the lower one to the state with $m_l = m_j + \frac{1}{2}$, $m_s = -\frac{1}{2}$.

39*. $c_1 = \sqrt{[\frac{1}{2}(1+\gamma)]}$, $c_2 = \sqrt{[\frac{1}{2}(1-\gamma)]}$ for the upper state,

$c_1 = \sqrt{[\frac{1}{2}(1-\gamma)]}$, $c_2 = -\sqrt{[\frac{1}{2}(1+\gamma)]}$ for the lower state,

where

$$\gamma = \frac{\frac{1}{2}\Delta E + \dfrac{m_j}{2l+1}\mathscr{H}\mu_0}{\sqrt{\left[\frac{1}{4}(\Delta E)^2 + \dfrac{m_j}{2l+1}\Delta E \mathscr{H}\mu_0 + \frac{1}{4}\mathscr{H}^2\mu_0^2\right]}}.$$

We shall consider some limiting cases.

(a) Let the magnetic field be vanishingly small, that is, $\Delta E \gg \mathscr{H}\mu_0$, with $\gamma \approx 1$. We shall get for the upper level $c_1 = 1, c_2 = 0$, and for the lower level $c_1 = 0, c_2 = 1$.

(b) Strong magnetic field: $\Delta E \ll \mathscr{H}\mu_0$. In that case $\gamma = m_j/(l+\frac{1}{2})$ and we have for the upper level

$$c_1 = \sqrt{\left(\frac{l+m_j+\frac{1}{2}}{2l+1}\right)}, \quad c_2 = \sqrt{\left(\frac{l-m_j+\frac{1}{2}}{2l+1}\right)}$$

and for the lower level

$$c_1 = \sqrt{\left(\frac{l-m_j+\frac{1}{2}}{2l+1}\right)} \quad c_2 = -\sqrt{\left(\frac{l+m_j+\frac{1}{2}}{2l+1}\right)}.$$

If we substitute these values of c_1 and c_2 into the expression for the wave functions (see problem 38 of section 7) we get for the upper level

$$\psi = R_{nl}^{(0)}\begin{pmatrix} Y_{l,m-\frac{1}{2}}(\vartheta,\varphi) \\ 0 \end{pmatrix}$$

and for the lower level

$$\psi = R_{nl}^{(0)}\begin{pmatrix} 0 \\ Y_{l,m-\frac{1}{2}}(\vartheta,\varphi) \end{pmatrix}.$$

40*. Since the energy in the magnetic field is considerably larger than the spin–orbit interaction energy, we can neglect the latter to a first approximation.

In this case $\hat{l}_z$ and $\hat{s}_z$ are constants of motion and the energy splitting follows from the equation

$$E^{(1)} = \frac{e\hbar}{2\mu c}\mathscr{H}(m_l + 2m_s).$$

In the second approximation we must take the spin–orbit interaction into account. The multiplet splitting which is added to the splitting in the magnetic field is equal to the average value of the operator

$$\frac{e^2}{2\mu^2 c^2}\frac{1}{r^3}(\hat{\boldsymbol{l}}\,.\,\hat{\boldsymbol{s}})$$

(see problem 23 of section 11) over the state with given values of m_l and m_s. For a given value of one of the components of the angular momentum the average value of the other two is equal to zero so that

$$(\hat{\boldsymbol{l}}\,.\,\hat{\boldsymbol{s}}) = m_l m_s.$$

The energy splitting of the level, when spin–orbit interaction is taken into account, is thus of the form

$$E^{(1)} = \frac{e\hbar}{2\mu c}\mathscr{H}(m_l + 2m_s) + \frac{e^2\hbar^2}{2\mu^2 c^2}\overline{r^{-3}}\, m_l m_s.$$

We can substitute in this equation the value of $\overline{r^{-3}}$, expressed in terms of the fine structure splitting when no field is present. One can easily show that (see problem 23 of section 11)

$$\frac{e^2\hbar^2}{2\mu^2 c^2}\overline{r^{-3}} = \frac{E_{n,l,j=l+\frac{1}{2}} - E_{n,l,j=l-\frac{1}{2}}}{l+\frac{1}{2}} = \frac{\Delta E}{l+\frac{1}{2}}.$$

We finally get for $E^{(1)}$ the expression

$$E^{(1)} = \frac{e\hbar}{2\mu c}\mathscr{H}(m_l + 2m_s) + \frac{\Delta E}{l+\frac{1}{2}}\, m_l m_s.$$

In fig. 41 we have given the scheme of the splitting of the $1s$ and $2p$ terms for an alkali metal atom in a strong magnetic field.

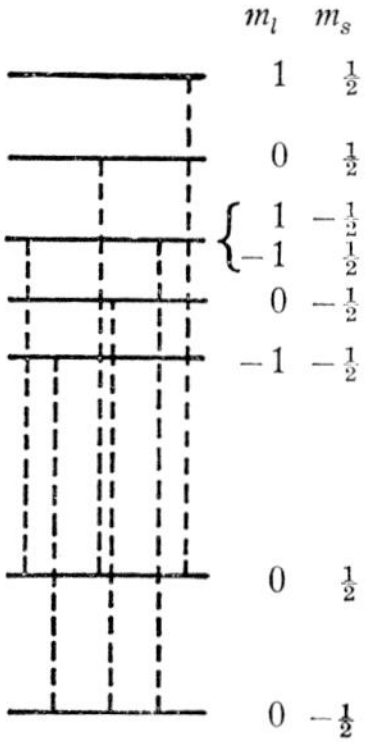

Fig. 41.

41. The perturbation energy in our case is equal to

$$\hat{V} = \frac{e\hbar}{\mu c}\beta\frac{1}{r^3}\hat{\boldsymbol{i}}\left[\frac{3(\boldsymbol{r}\,.\,\hat{\boldsymbol{s}})\,\boldsymbol{r}}{r^2} - \hat{\boldsymbol{s}}\right] + \frac{e\hbar}{\mu c}\mathscr{H}\hat{s}_z + \beta\mathscr{H}\hat{\imath}_z.$$

In this expression we shall neglect the last term since the magnetic moment of the nucleus is small compared to the magnetic moment of the electron ($\beta < e\hbar/2\mu c$).

If we proceed as in problem 38 of section 7, we find the following secular equation:

$$\begin{vmatrix} E_+ + \dfrac{\mu_0\mathscr{H}m_f}{i+\frac{1}{2}} - E & \dfrac{\mathscr{H}\mu_0}{i+\frac{1}{2}}\sqrt{[(i+\frac{1}{2})^2 - m_f^2]} \\ \dfrac{\mathscr{H}\mu_0}{i+\frac{1}{2}}\sqrt{[(i+\frac{1}{2})^2 - m_f^2]} & E_- - \dfrac{\mathscr{H}\mu_0}{i+\frac{1}{2}}m_f - E \end{vmatrix} = 0,$$

where E_+ and E_- are the energy terms taking hyperfine structure into account, while E_+ refers to the state with $f = i+\frac{1}{2}$, and E_- to $f = i-\frac{1}{2}$, m_f is the z-component of the total angular moment ($m_f = f, f-1, \ldots, -f$), and $\mu_0 = e\hbar/2\mu c$. Solving the secular equation we find

$$E = \frac{E_+ + E_-}{2} \pm \frac{\Delta E}{2}\sqrt{\left(1 + \frac{2\xi}{i+\frac{1}{2}}m_f + \xi^2\right)},$$

where $\Delta E = E_+ - E_-$, and $\xi = 2\mu_0\mathscr{H}/\Delta E$.

We can find the order of magnitude of the magnetic field strength which produces the splitting from the formula obtained. We have assumed that $\mathscr{H} \sim |\Delta E_{jj'}|/\mu_0$. In the case of the sodium atom we have

$$\Delta E_{jj'} = 0{\cdot}0583 \text{ cm}^{-1} = 1{\cdot}962 \times 10^{-16} \text{ erg};$$

since

$$\mu_0 = 0{\cdot}922 \times 10^{-20} \text{ Oe cm}^3$$

the field strength $\mathscr{H}$ will be of the order of magnitude of 600 Oe.

We consider two limiting cases.

(a) In the case of weak fields, that is when $\mu_0 \mathscr{H} \ll \Delta E$, we find the following expressions for the energy:

$$E = E_+ + \frac{\mathscr{H}\mu_0}{i+\frac{1}{2}} m_j,$$

$$E = E_- - \frac{\mathscr{H}\mu_0}{i+\frac{1}{2}} m_j.$$

(b) In the case of strong fields, that is when $\mathscr{H}\mu_0 \gg \Delta E$, we have

$$E = \tfrac{1}{2}(E_+ + E_-) \pm \mathscr{H}\mu_0.$$

44. The expression $(e\hbar/2\mu c)(\hat{L}_z + 2\hat{S}_z)\mathscr{H}$ is a small perturbation. We consider Russell–Saunders coupling. In that case $\hat{H}_0$ commutes with operators $\hat{\boldsymbol{J}}^2, \hat{J}_z, \hat{\boldsymbol{L}}^2, \hat{\boldsymbol{S}}^2$. The unperturbed energy levels are characterised by the quantum numbers J, L, S. Each of these states is degenerate with respect to the direction of the vector $\hat{\boldsymbol{J}}$, the degree of degeneracy being equal to $2J+1$.

Since the matrix elements of the operator $\hat{L}_z + 2\hat{S}_z$ which are non-diagonal with respect to $\hat{J}_z$ are equal to zero, the energy correction is simply equal to the average value of the operator $\hat{L}_z + 2\hat{S}_z$ in the state characterised by the quantum numbers J, J_z, L, S. To evaluate this average value we substitute in the equation of problem 44 of section 4, $g_1 = 1, g_2 = 2, \hat{\boldsymbol{J}}_1 = \hat{\boldsymbol{L}}, \hat{\boldsymbol{J}}_2 = \hat{\boldsymbol{S}}$. The result is

$$\overline{\hat{L}_z + 2\hat{S}_z} = J_z\left[\frac{3}{2} + \frac{S(S+1) - L(L+1)}{2J(J+1)}\right] = gJ_z.$$

According to what we have said a moment ago we find for the required splitting

$$E^{(1)}_{JLS} = \frac{e\hbar\mathscr{H}}{2\mu c} gJ_z.$$

46*. The normalised eigenfunctions of the hydrogen atom in the unperturbed state are of the form (see problem 36 of section 4)

$$\left.\begin{aligned} \psi_{nlj=l+\frac{1}{2}m_j} &= \frac{R_{n,j-\frac{1}{2}}}{\sqrt{(2j)}}\begin{pmatrix}\sqrt{(j+m_j)}\,Y_{j-\frac{1}{2},m_j-\frac{1}{2}}\\ \sqrt{(j-m_j)}\,Y_{j-\frac{1}{2},m_j+\frac{1}{2}}\end{pmatrix} = u_-,\\ \psi_{nlj=l-\frac{1}{2}m_j} &= \frac{R_{n,j+\frac{1}{2}}}{\sqrt{(2j+2)}}\begin{pmatrix}\sqrt{(j+1-m_j)}\,Y_{j+\frac{1}{2},m_j-\frac{1}{2}}\\ -\sqrt{(j+1+m_j)}\,Y_{j+\frac{1}{2},m_j+\frac{1}{2}}\end{pmatrix} = u_+.\end{aligned}\right\} \qquad (1)$$

The energy is determined by the two quantum numbers n, j. When a uniform electric field ($\mathscr{E}_x = \mathscr{E}_y = 0$, $\mathscr{E}_z = \mathscr{E}$) is present, $\hat{j}_z$ remains a constant of motion, but the orbital angular momentum ceases to be an integral of motion. The matrix elements of the perturbation operator, $\hat{V} = e\mathscr{E}z$, for transitions between states with different values of m_j are equal to zero. The diagonal elements of the operator $\hat{V}$ are also equal to zero, that is,

$$\sum_\sigma \int u_+^* z u_+ \, d\tau = \sum_\sigma \int u_-^* z u_- \, d\tau. \qquad (2)$$

We must therefore evaluate the matrix elements of $\hat{V}$ corresponding to a transition from the state $n, j, m_j, l = j+\frac{1}{2}$ to the state $n, j, m_j, l = j-\frac{1}{2}$ in order to find the energy splitting. These matrix elements are equal to:

$$\begin{aligned} V_{21} = V_{12} &= e\mathscr{E}\sum_\sigma \int u_-^* z u_+ \, d\tau \\ &= \int_0^\infty r^3 R_{n,j-\frac{1}{2}}(r)\, R_{n,j+\frac{1}{2}}(r)\, dr \frac{1}{2\sqrt{[j(j+1)]}} \\ &\times \Big\{\sqrt{[(j+m_j)(j-m_j+1)]} \int Y^*_{j-\frac{1}{2},m-\frac{1}{2}} Y_{j+\frac{1}{2},m-\frac{1}{2}} \cos\vartheta \, d\Omega \\ &\quad -\sqrt{[(j-m_j)(j+m_j+1)]} \int Y^*_{j-\frac{1}{2},m+\frac{1}{2}} Y_{j+\frac{1}{2},m_j+\frac{1}{2}} \cos\vartheta \, d\Omega\Big\}. \qquad (3)\end{aligned}$$

We shall first integrate over the angles. Since

$$\begin{aligned}\cos\vartheta\, Y_{lm}(\vartheta,\varphi) &= \sqrt{\left[\frac{(l+m+1)(l-m+1)}{(2l+1)(2l+3)}\right]}\, Y_{l+1,m}(\vartheta,\varphi) \\ &\quad + \sqrt{\left[\frac{(l+m)(l-m)}{(2l+1)(2l-1)}\right]}\, Y_{l-1,m}(\vartheta,\varphi),\end{aligned}$$

we find that the expression within braces in equation (3) is equal to

$$\frac{1}{2\sqrt{[j(j+1)]}}[(j+m_j)(j-m_j+1)-(j+m_j+1)(j-m_j)] = \frac{m_j}{\sqrt{[j(j+1)]}}.$$

If we then integrate over r, we get

$$-\tfrac{3}{2}n\sqrt{[n^2-(j+\tfrac{1}{2})^2]}.$$

We thus find finally for the perturbation matrix elements (3):

$$V_{12} = V_{21} = -\tfrac{3}{4}n\frac{\sqrt{[n^2-(j+\tfrac{1}{2})^2]}}{j(j+1)}m_j e\mathscr{E}.$$

We find the required energy correction by solving the secular equation

$$\begin{vmatrix} -\epsilon & V_{12} \\ V_{21} & -\epsilon \end{vmatrix} = 0, \quad \epsilon = \pm V_{12},$$

$$\epsilon = \pm\tfrac{3}{4}\sqrt{[n^2-(j+\tfrac{1}{2})^2]}\frac{nm_j}{j(j+1)}e\mathscr{E}.$$

For a given value of n the term with $j = n-\frac{1}{2}$ is not split in the electrical field since it is not degenerate with respect to the quantum number l (l has a fixed value, $l = j-\frac{1}{2} = n-1$). All the other terms of the fine structure are split into $2j+1$ equidistant levels ($m_j = -j, \ldots, +j$).

47. $\alpha = \frac{9}{2}a_{\rm B}^3$, where $a_{\rm B}$ is the Bohr radius.

48.

$$\mu_0 = \frac{e\hbar}{2\mu c}\frac{(j+\frac{1}{2})^2}{j(j+1)}m_j.$$

49*. In the case under consideration the spin-orbit interaction V_1, the relativistic correction due to the change in mass V_2, and the energy of the electron in the external homogeneous electrical field $V_3 = -e\mathscr{E}z$ are all of the same order of magnitude. We shall therefore consider their sum as a small perturbation of the original system. For our calculations we shall start from a state where the orbital angular momentum L, its z-component m_l, and the z-component of the spin m_s (the electrical field is along the z-axis) have well-defined values. If we evaluate the matrix elements of V_1, V_2, and V_3, we have (in atomic units)

$$(V_1)^{l'm_l'}_{lm_l} = \begin{cases} \frac{1}{2}\alpha^2\dfrac{m_l(m-m_l)}{n^3\, l(l+\frac{1}{2})(l+1)}\delta_{ll'} & \text{for} \quad m_l' = m_l, \\ \frac{1}{4}\alpha^2\dfrac{\sqrt{[(l+\frac{1}{2})^2-m^2]}}{n^3\, l(l+\frac{1}{2})(l+1)}\delta_{ll'} & \text{for} \quad m_l' = m+\frac{1}{2}, \\ & \quad m_l = m-\frac{1}{2} \text{ or the other way round,} \\ 0 \quad \text{otherwise,} & \end{cases}$$

$$(V_2)^{l'm_l'}_{lm_l} = -\frac{1}{2}\frac{\alpha^2}{n^3}\left(\frac{1}{l+\frac{1}{2}}-\frac{3}{4n}\right)\delta_{ll'}\,\delta_{m_l m_l'},$$

$$(V_3)^{l'm_l'}_{lm_l} = -\frac{3n}{2}\sqrt{\left[\frac{(n^2-l^2)(l^2-m_l^2)}{4l^2-1}\right]}e\mathscr{E}\,\delta_{l',l-1}\,\delta_{m_l' m_l}.$$

If $n = 2$, the energy of the state with quantum numbers

$$l = 1, \qquad m = m_l + m_s = \pm\tfrac{3}{2} \quad (j = \tfrac{3}{2})$$

will not change in the electrical field. The shift of this level due to V_1 and V_2 is equal to $\alpha^2/128$ at un (see problem 23 of section 11).

The splitting of the level with quantum numbers $n = 2, m = \pm\frac{1}{2}$, can be found from the solution of the secular equation

$$\begin{vmatrix} -\frac{11}{4}\delta - E^{(1)} & \delta\sqrt{2} & 0 \\ \delta\sqrt{2} & -\frac{7}{4}\delta - E^{(1)} & -3e\mathscr{E} \\ 0 & -3e\mathscr{E} & -\frac{15}{4}\delta - E^{(1)} \end{vmatrix} = 0,$$

where $3\delta = \alpha^2/32$ at un is the fine structure splitting of the level $n = 2$ when there is no external field. If we introduce in that equation ϵ which is connected with $E^{(1)}$ by the relation

$$E^{(1)} = \epsilon - \tfrac{11}{4}\delta$$

[$-\frac{11}{4}\delta$ is the energy of the centre of gravity of the three energy levels: $(E_1^{(1)} + E_2^{(1)} + E_3^{(1)})/3 = -\frac{11}{4}\delta$], we get

$$\begin{vmatrix} -\epsilon & \delta\sqrt{2} & 0 \\ \delta\sqrt{2} & \delta - \epsilon & -3e\mathscr{E} \\ 0 & -3e\mathscr{E} & -\delta - \epsilon \end{vmatrix} = 0,$$

or

$$\epsilon^3 - \epsilon[3\delta^2 + 9(e\mathscr{E})^2] - 2\delta^3 = 0.$$

We shall solve this equation both for weak fields ($e\mathscr{E} \ll \delta$) and for strong fields ($e\mathscr{E} \gg \delta$). In the first case we find

$$\epsilon_1 = -\delta - \sqrt{(3)}\, e\mathscr{E} - \frac{(e\mathscr{E})^2}{\delta},$$

$$\epsilon_2 = -\delta + \sqrt{(3)}\, e\mathscr{E} - \frac{(e\mathscr{E})^2}{\delta},$$

$$\epsilon_3 = 2\delta + 2\frac{(e\mathscr{E})^2}{\delta}.$$

In the second case the result is

$$\epsilon_1 = -3e\mathscr{E} - \frac{1}{2}\frac{\delta^2}{e\mathscr{E}} + \frac{1}{9}\frac{\delta^3}{(e\mathscr{E})^2},$$

$$\epsilon_2 = -\frac{2}{9}\frac{\delta^3}{(e\mathscr{E})^2},$$

$$\epsilon_3 = 3e\mathscr{E} + \frac{1}{2}\frac{\delta^3}{e\mathscr{E}} + \frac{1}{9}\frac{\delta^3}{(e\mathscr{E})^2}.$$

50. We must add to the Hamiltonian of the unperturbed problem (see problem 32 of section 1) the perturbation energy operator

$$\hat{H}' = -(\boldsymbol{d}.\mathscr{E}) = -d\mathscr{E}\cos\varphi,$$

so that the Schrödinger equation will be of the form

$$\frac{d^2\psi}{d\varphi^2} + \frac{2I}{\hbar^2}(E + d\mathscr{E}\cos\varphi)\psi = 0,$$

where φ is the angle of rotation, and E the energy of the rotator. The energy levels of the unperturbed problem

$$E_m^{(0)} = \frac{\hbar^2 m^2}{2I} \quad (m = 0, \pm 1, \pm 2, \ldots)$$

are, apart from the lowest level, twofold degenerate with respect to the direction of the angular momentum $M_z = m\hbar$. However, since this degeneracy is not lifted until we take the second-order perturbation theory into account we can up to that order apply the theory of perturbation of non-degenerate levels. (An exception is the level with $m = \pm 1$.) In other words the unperturbed wave functions

$$\psi_m^{(0)} = \frac{1}{\sqrt{(2\pi)}}\exp(im\varphi)$$

are still good functions for the zeroth approximation.

The perturbation operator matrix elements are equal to

$$H'_{mm'} \equiv \int_0^{2\pi} \psi_m^{(0)*}\hat{H}'\psi_{m'}^{(0)}\,d\varphi = -\frac{d\mathscr{E}}{2\pi}\int_0^{2\pi}\exp[i(m'-m)\varphi]\cos\varphi\,d\varphi$$

$$= \begin{cases} 0, & \text{if } m' \neq m \pm 1, \\ -\dfrac{d\mathscr{E}}{2}, & \text{if } m' = m \pm 1. \end{cases}$$

It follows from this that the first-order energy correction of all levels is equal to zero.

$$E_m^{(1)} = H'_{mm} = 0.$$

The second-order correction of the mth level is clearly given by

$$E_m^{(2)} = \frac{|H'_{m,m-1}|^2}{E_m^{(0)} - E_{m-1}^{(0)}} + \frac{|H'_{m,m+1}|^2}{E_m^{(0)} - E_{m+1}^{(0)}} = \frac{Id^2 \mathscr{E}^2}{\hbar^2(4m^2-1)},$$

so that the energy levels of the plane rotator in a weak electrical field are up to second-order terms given by

$$E_m = E_m^{(0)} + E_m^{(1)} + E_m^{(2)} = \frac{\hbar^2 m^2}{2I} + \frac{Id^2 \mathscr{E}^2}{\hbar^2(4m^2-1)}.$$

This result can be interpreted very simply. If we introduce in the usual manner the polarisibility, α, of the rotator as the ratio of the induced dipole moment to the external field strength, we find that the energy of the induced dipole in the field $\mathscr{E}$, which is equal to the work of polarisation, $-\alpha \int_0^{\mathscr{E}} \mathscr{E} d\mathscr{E}$, is equal to $-\frac{1}{2}\alpha\mathscr{E}^2$. If we compare this additional energy with the first non-vanishing energy correction, which we have just obtained, we find the following formula for the polarisation:

$$\alpha_m = -\frac{2Id^2}{\hbar^2(4m^2-1)}.$$

We see that for $m \neq 0$, the polarisibility is negative, that is, the electrical moment of the rotator, $\boldsymbol{d}$, is oriented antiparallel to the field $\boldsymbol{\mathscr{E}}$. The opposite is true for $m = 0$.

This result is completely in accordance with the classical effect of anti-parallel polarisation for fast rotation of a plane rotator, and parallel polarisation for slow rotation. Fast is meant here to be sufficiently fast for the rotator to be completely turned over in the field.

51. If the well has a width a, the polarisability α is given by the expression

$$\alpha = \frac{4096}{\pi^4 \hbar^2} \mu e^2 a^4 \sum_{p=1}^{\infty} \frac{p^2}{(4p^2-1)^5},$$

which follows from the expression $\Delta E = -\frac{1}{2}\alpha\mathscr{E}^2$ for the energy shift in an electric field $\mathscr{E}$, when we use second-order perturbation theory.

52. We shall choose the direction of $\mathscr{E}$ as the polar axis of a spherical system of coordinates, so that the perturbation operator will be

$$\hat{V} = V = -(\boldsymbol{d}.\boldsymbol{\mathscr{E}}) = -d\mathscr{E}\cos\theta, \tag{1}$$

where θ is the angle between the axis of the rotator (which is parallel to $\boldsymbol{d}$) and the field $\boldsymbol{\mathscr{E}}$.

The wave functions of the unperturbed stationary states of the rotator are of the form

$$\psi_{lm}^{(0)}(\theta,\varphi) = Y_{lm}(\theta,\varphi) = \sqrt{\left[\frac{(2l+1)(l+m)!}{4\pi(l-m)!}\right]}\, P_l^m(\cos\theta)\exp(im\varphi), \quad (2)$$

where θ is the polar angle introduced a moment ago, and φ the azimuthal angle of the axis of the rotator.

The functions (2) are orthonormal,

$$\int_0^{2\pi}\int_0^{\pi} Y_{lm}^*(\theta,\varphi)Y_{l'm'}(\theta,\varphi)\sin\theta\,d\theta\,d\varphi = \delta_{ll'}\,\delta_{mm'}. \quad (3)$$

In particular, the ground state wave function, for whose perturbation we are looking, is of the form

$$\psi_{00}^{(0)}(\theta,\varphi) = Y_{00} = \frac{1}{\sqrt{(4\pi)}}. \quad (4)$$

The unperturbed energy levels of the stationary states are equal to

$$E_l^{(0)} = \frac{l(l+1)\hbar^2}{2I}; \quad E_0^{(0)} = 0. \quad (5)$$

We shall write down the perturbation operator matrix element for a transition from the ground state $(0,0)$ to the excited state (l,m),

$$V_{lm}^{00} = (V_{00}^{lm})^* \equiv \int_0^{2\pi}\int_0^{\pi}[\psi_{00}^{(0)}(\theta,\varphi)]^*\,\hat{V}\psi_{lm}^{(0)}(\theta,\varphi)\sin\theta\,d\theta\,d\varphi.$$

If we use equations (1), (2), (4), the orthonormality condition (3), and replace $\cos\theta$ by $\sqrt{(4\pi/3)}\,Y_{10}(\theta,\varphi)$ we get

$$(V_{lm}^{00})^* = V_{00}^{lm} = -(\boldsymbol{d}.\mathscr{E})\sqrt{\left(\frac{4\pi}{3}\right)}\frac{1}{\sqrt{(4\pi)}}\int_0^{2\pi}\int_0^{\pi} Y_{lm}^*(\theta,\varphi)Y_{10}(\theta,\varphi)\sin\theta\,d\theta\,d\varphi$$

$$= -\frac{(\boldsymbol{d}.\mathscr{E})}{\sqrt{3}}\,\delta_{l1}\,\delta_{m0}. \quad (6)$$

There is thus only one non-vanishing matrix element, namely, the one from the ground state $l = m = 0$ to the state $l = 1$, $m = 0$. Since this matrix element is non-diagonal the first-order energy level correction vanishes.

The second-order correction is according to the general equations of perturbation theory and using equations (5) and (6) equal to

$$E_0^{(2)} = \sum_l{}' \sum_m \frac{|V_{lm}^{00}|^2}{E_0^{(0)}-E_l^{(0)}} = \frac{|V_{10}^{00}|^2}{E_0^{(0)}-E_1^{(0)}} = -\frac{(\boldsymbol{d}.\mathscr{E})^2 I}{3\hbar^2}. \quad (7)$$

The shift of the ground state level of a three-dimensional rotator is thus negative; the same would be true for the second-order ground state energy correction of any quantum mechanical system.

The condition of applicability of the perturbation theory is that the shift (7) is small compared to the difference $|E_0^{(0)} - E_1^{(0)}|$ between the ground state and the first excited state.

If we solve this condition for $\mathscr{E}$, we get

$$\mathscr{E} \ll \sqrt{(3)}\frac{\hbar^2}{dI}.$$

As we should have expected, this inequality also means that the perturbation energy ($\sim \mathscr{E}d$) is small compared to the distance between consecutive levels ($\sim \hbar^2/I$).

We use the relation between the normalised spherical harmonics

$$\cos\theta\, Y_{l,m} = \sqrt{\left[\frac{(l+1)^2 - m^2}{(2l+3)(2l+1)}\right]}\, Y_{l+1,m} + \sqrt{\left(\frac{l^2-m^2}{4l^2-1}\right)}\, Y_{l-1,m},$$

and we find for the polarisation in the lm-state

$$\alpha_{lm} = \frac{2I\mu^2}{\hbar^2}\left[\frac{(l+1)^2 - m^2}{(2l+3)(2l+1)(l+1)} - \frac{l^2 - m^2}{(2l+1)(2l-1)l}\right],$$

where the value for $l = 0$ is found by putting $m = 0$ and letting $l \to 0$.

It is rather paradoxical that a rigid dipole is "polarisable". Consider the case when $l = 1$. If $m = 0$, the dipole behaves diamagnetically, and if $m = \pm 1$, it behaves paramagnetically (or "normal").

The degeneracy is only partly lifted, as the energy shift depends only on $|m|$.

Finally we note that

$$\sum_{m=-l}^{+l} \alpha_{lm} = 0.$$

53. (a) $$E^{(1)} = \frac{m_j}{8j(j+1)}\frac{e\hbar}{\mu c}\left((2j+1)^2\mathscr{H} \pm \sqrt{\left\{(2j+1)^2\mathscr{H}^2 + \frac{36n^2}{\alpha^2}[n^2 - (j+\tfrac{1}{2})^2]\mathscr{E}^2\right\}}\right)$$

$$\left(\alpha = \frac{e^2}{\hbar c}\right)$$

(see problem 46 and 49 of section 7).

(b) $$E^{(1)}(n = 2, m_j = \pm\tfrac{3}{2}) = -\frac{\alpha^2}{128}\left(\frac{\mu e^4}{\hbar^2}\right) \pm \frac{e\hbar}{\mu c}\mathscr{H},$$

$$E^{(1)}(n = 2, m_j = \pm\tfrac{1}{2}) = \epsilon - \frac{11}{384}\alpha^2\left(\frac{\mu e^4}{\hbar^2}\right),$$

where ϵ follows from the solution of a cubic equation

$$\epsilon^3 \mp 2\beta\epsilon^2 - \epsilon(3\delta^2 + 9F^2 - \beta^2) \pm 2\delta^2\beta - 2\delta^3 = 0,$$

where

$$\beta = \frac{e\hbar}{2\mu c}\mathscr{H}, \quad F = e\mathscr{E}\left(\frac{\hbar^2}{\mu e^2}\right), \quad \delta = \frac{\alpha^2}{96}\left(\frac{\mu e^4}{\hbar^2}\right).$$

For a strong field ($\delta \ll F, \delta \ll \beta$) we have (see problem 49 of section 7)

$$\epsilon_1 = \pm\beta - 3F - \frac{1}{2}\frac{9F \mp \beta}{9F^2 \mp 3F\beta}\delta^2,$$

$$\epsilon_2 = \pm\frac{2\beta}{9F^2 - \beta^2}\delta^2,$$

$$\epsilon_3 = \pm\beta + 3F + \frac{1}{2}\frac{9F \pm \beta}{9F \pm 3F\beta}\delta^2.$$

54. Let the z-axis be along a magnetic field direction and the x-axis along the electrical direction. The potential energy operator of an electron in these fields will then be given by

$$w = \frac{e\hbar}{2\mu c}(\hat{l}_z + 2\hat{s}_z)\mathscr{H} + e\mathscr{E}x.$$

We shall consider this expression to be a small perturbation, and we shall characterise the unperturbed stationary states by the quantum numbers n, l, m, σ (m and σ are the z-components of the orbital angular momentum and the spin). The non-vanishing matrix elements of x are of the form:

$$(x)_{l,m-1}^{l-1,m} = (x)_{l-1,m}^{l,m-1} = \tfrac{3}{4}n\sqrt{\left[\frac{(n^2-l^2)(l-m+1)(l-m)}{(2l+1)(2l-1)}\right]}\,a,$$

$$(x)_{l-1,m-1}^{l,m} = (x)_{l,m}^{l-1,m-1} = -\tfrac{3}{4}n\sqrt{\left[\frac{(n^2-l^2)(l+m-1)(l+m)}{(2l+1)(2l-1)}\right]}\,a$$

$$\left(a = \frac{\hbar^2}{\mu e^2}\right).$$

We shall consider the case $n = 2$. Let us fix our ideas and assume that the z-component of the spin is equal to $+\frac{1}{2}$. We shall use the notation $\beta = (e\hbar/2\mu c)\mathscr{H}$, $\gamma = (3/\sqrt{2})e\mathscr{E}a$. In this notation the perturbation operator matrix will be of the form

$$\begin{pmatrix} 2\beta & 0 & 0 & -\gamma \\ 0 & \beta & 0 & 0 \\ 0 & 0 & 0 & \gamma \\ -\gamma & 0 & \gamma & \beta \end{pmatrix}.$$

We have numbered the states in order of decreasing values l and m. The state with quantum numbers $l = 1$, $m = 0$ does not combine with the other states. The energy of that state is equal to $E_1^{(1)} = \beta$. The other three eigenvalues of the perturbation matrix are determined from the solution of the secular equation

$$E^3 - 3\beta E^2 + 2(\beta^2 - \gamma^2) E + 2\gamma^2 \beta = 0.$$

If we solve it we find

$$E_2^{(1)} = \beta, \quad E_{3,4}^{(1)} = \beta \pm \sqrt{(\beta^2 + 2\gamma^2)}$$

or, using the definitions of β and γ,

$$E_2^{(1)} = \frac{e\hbar}{2\mu c}\mathscr{H}, \quad E_{3,4}^{(1)} = \frac{e\hbar}{2\mu c}\left[\mathscr{H} \pm \sqrt{\left(\mathscr{H}^2 + \frac{36}{\alpha^2}\mathscr{E}^2\right)}\right].$$

55. $E \leqslant -4e^2/3\pi a_B$, where $a_B = \hbar^2/\mu e^2$ is the Bohr radius.

56. $E_0 \leqslant \frac{1}{2}\hbar\omega$.

57. In this case we must use a trial wavefunction which is orthogonal to the wavefunction found in the preceding problem. The simplest form is $\psi = Bx \exp(-Cx^2)$, and we find $E_1 \leqslant \frac{3}{2}\hbar\omega$.

58. The Hamiltonian is of the following form (we shall use atomic units $e = \hbar = \mu = 1$)

$$\hat{H} = -\tfrac{1}{2}\nabla_1^2 - \tfrac{1}{2}\nabla_2^2 - \frac{Z}{r_1} - \frac{Z}{r_2} + \frac{1}{r_{12}}.$$

According to the variational principle it is necessary to evaluate the integral

$$E(Z') = \int \psi^*(r_1, r_2) \hat{H} \psi(r_1, r_2)\, d\tau_1\, d\tau_2,$$

and to determine the quantity Z' from the condition $dE/dZ' = 0$.

In our case we have

$$\psi(r_1, r_2) = c \exp[-Z'(r_1 + r_2)],$$

where $c = Z'^3/\pi$.

The integrals of the first four terms can easily be evaluated,

$$\int \psi(\boldsymbol{r}_1, \boldsymbol{r}_2)\left(-\tfrac{1}{2}\nabla_1^2 - \tfrac{1}{2}\nabla_2^2 - \frac{Z}{r_1} - \frac{Z}{r_2}\right)\psi(\boldsymbol{r}_1, \boldsymbol{r}_2)\, d\tau_1\, d\tau_2 = Z'^2 - 2ZZ'.$$

As to the integral involving $1/r_{12}$, it is convenient to evaluate it in elliptical coordinates,

$$s = r_1 + r_2, \quad t = r_1 - r_2, \quad u = r_{12},$$

$$d\tau_1\, d\tau_2 = \pi^2(s^2 - t^2)\, u\, ds\, dt\, du,$$

$$-u \leqslant t \leqslant u, \quad 0 \leqslant u \leqslant s \leqslant \infty.$$

The result of this calculation is

$$\int \psi^2(r_1, r_2)\frac{1}{r_{12}}\, d\tau_1\, d\tau_2 = \pi^2 c^2 \int_0^\infty ds \int_0^s du \int_{-u}^{+u} dt \exp(-2Z's)\frac{s^2 - t^2}{u}u = \tfrac{5}{8}Z'.$$

We get finally

$$E(Z') = Z'^2 - 2ZZ' + \tfrac{5}{8}Z'.$$

From minimising $E(Z')$ we find

$$Z' = Z - \tfrac{5}{16}.$$

For this value of Z' we get for the ground state energy

$$E = -(Z - \tfrac{5}{16})^2.$$

To find out how accurate our result is we calculate the ionisation potential of helium ($Z = 2$) and compare the result with the experimental value. The ionisation potential of helium I_{He} is equal to the difference of energy of the once ionised helium atom and the neutral helium atom in its ground state. The value of I_{He} is equal to (1 at un = 2 Ry = 27 eV)

$$I_{He} = 0{\cdot}8476 \text{ at un} = 1{\cdot}695 \text{ Ry}.$$

The experimental value is 1·810 Ry.

The ionisation potentials of other systems with two electrons, Li^+, Be^{++}, and so on are also known experimentally. Comparing them with the theoretical value we get

Element	He ($Z = 2$)	Li^+ ($Z = 3$)	Be^{++} ($Z = 4$)	B^{+++} ($Z = 5$)	C^{++++} ($Z = 6$)
$I(\text{Ry})_{th}$	1·6952	5·445	11·195	18·945	28·695
$I(\text{Ry})_{exp}$	1·810	5·560	11·307	19·061	28·816

The results of the evaluation of the ground state energy is in satisfactory agreement with the experimental values.

60. Both the orbital and the spin angular momentum are equal to zero in the ground state of the helium atom. As a consequence helium is diamagnetic. The diamagnetic susceptibility evaluated for one mole is given by the expression

$$\chi = -\frac{e^2 N_A}{6\mu c^2}(\overline{r_1^2}+\overline{r_2^2}),$$

where

$$\overline{r_1^2}+\overline{r_2^2} = \int (r_1^2+r_2^2)\,\psi^2\,d\tau_1\,d\tau_2$$

and where N_A is the Avogadro number. The approximate expression for the ground state wave function of the helium atom is

$$\psi(r_1, r_2) = \frac{Z'^3}{\pi a^3}\exp[-Z'(r_1+r_2)/a].$$

If we evaluate the average value $\overline{r_1^2}+\overline{r_2^2}$ using that wave function we get

$$\overline{r_1^2}+\overline{r_2^2} = \frac{6a^2}{Z'^2}.$$

Substituting this value into the expression for the diamagnetic susceptibility we find

$$\chi = -1{\cdot}67\,.\,10^{-6}.$$

The experimental value for the diamagnetic susceptibility is given by

$$\chi = -(1{\cdot}90 \pm 0{\cdot}02)\,.\,10^{-6}.$$

61*. It is convenient to introduce the following notation:

$$Z_1 = \alpha, \quad Z_2 = 2\beta, \quad 2Z_1^{\frac{3}{2}} = a, \quad cZ_2^{\frac{3}{2}} = b.$$

From the fact that the functions are orthonormal we get

$$\gamma Z_2 = \tfrac{1}{2}(\alpha+\beta), \quad b^2 = \frac{12\beta^5}{\alpha^2-\alpha\beta+\beta^2}.$$

In the new variables ψ_1 and ψ_2 have the following form:

$$\psi_1 = \psi_{100} = a\,e^{-\alpha r}\,Y_{00}, \quad \psi_2 = \psi_{200} = b[1-\tfrac{1}{3}(\alpha+\beta)\,r]\,e^{-\beta r}\,Y_{00}.$$

The approximate ground state wave function of the lithium atom can be written as follows:

$$\Phi = \frac{1}{\sqrt{(3!)}}\begin{vmatrix} \psi_1(1)\,\eta_+(\sigma_1) & \psi_1(2)\,\eta_+(\sigma_2) & \psi_1(3)\,\eta_+(\sigma_3) \\ \psi_1(1)\,\eta_-(\sigma_1) & \psi_1(2)\,\eta_-(\sigma_2) & \psi_1(3)\,\eta_-(\sigma_3) \\ \psi_2(1)\,\eta_+(\sigma_1) & \psi_2(2)\,\eta_+(\sigma_2) & \psi_2(3)\,\eta_+(\sigma_3) \end{vmatrix},$$

where $\eta_+(\frac{1}{2}) = 1, \eta_+(-\frac{1}{2}) = 0, \eta_-(\frac{1}{2}) = 0, \eta_-(-\frac{1}{2}) = 1$. In this state $S = \frac{1}{2}$, $M = \frac{1}{2}$.

The Hamiltonian operator (in units for which $e = \hbar = \mu = 1$) is for our problem of the form

$$\hat{H} = \sum_{i=1}^{3} \left(-\tfrac{1}{2}\nabla_i^2 - \frac{Z}{r_i} \right) + \frac{1}{r_{12}} + \frac{1}{r_{23}} + \frac{1}{r_{31}}.$$

We shall evaluate the energy in the state Φ.

The kinetic energy of an electron in the $1s$-state is equal to

$$T_1 = \int (-\psi_1 \tfrac{1}{2}\nabla^2 \psi_1)\, d\tau = \int_0^\infty \left(\frac{d\psi_1}{dr}\right)^2 r^2\, dr = \tfrac{1}{2}\alpha^2,$$

and in the $2s$-state

$$T_2 = \frac{\beta^2}{6} + \frac{\beta^4}{\alpha^2 - \alpha\beta + \beta^2}.$$

The interaction energy of the inner electrons with the nucleus is

$$V_1 = -\int \frac{Z\psi_1^2}{r}\, d\tau = Za^2 \int_0^\infty \frac{e^{-2\alpha r}}{r} r^2\, dr = -Z\alpha,$$

and of an outer electron with the nucleus (for lithium $Z = 3$)

$$V_2 = -\frac{Z\beta}{2} + \frac{Z\beta^2}{2} \frac{\alpha - 2\beta}{\alpha^2 - \alpha\beta + \beta^2}.$$

The energy of the Coulomb interaction of the inner electrons is equal to

$$K_{11} = \iint \frac{1}{r_{12}} |\psi_1(r_1)|^2 |\psi_1(r_2)|^2\, d\tau_1\, d\tau_2 = \tfrac{5}{8}\alpha.$$

The interaction energy of the outer electron with the inner electrons is equal to

$$2K_{12} = 2\iint \frac{1}{r_{12}} |\psi_1(r_1)|^2 |\psi_2(r_2)|^2\, d\tau_1\, d\tau_2$$

$$= 2\alpha - \frac{2\alpha^3}{(\alpha+\beta)^2} - \frac{\alpha^4 \beta(3\alpha+\beta)}{(\alpha+\beta)^3 (\alpha^2 - \alpha\beta + \beta^2)}.$$

The exchange interaction energy of the two electrons with parallel spin is

$$A = 2a^2 b^2 \int_0^\infty \exp[-(\alpha+\beta) r_2] [1 - \tfrac{1}{3}(\alpha+\beta) r_2] r_2^2\, dr_2$$

$$\times \int_{r_2}^\infty \exp[-(\alpha+\beta) r_1] [1 - \tfrac{1}{3}(\alpha+\beta) r_1] r_1\, dr_1$$

$$= \frac{4\alpha^3 \beta^5}{(\alpha+\beta)^5 (\alpha^2 - \alpha\beta + \beta^2)}.$$

If we put $\beta = \lambda\alpha$, we get

$$2T_1+T_2 = T = \alpha^2\varphi_1(\lambda),$$
$$2V_1+V_2+K_{11}+2K_{12}-A = -\alpha\varphi_2(\lambda),$$
$$E = \alpha^2\varphi_1(\lambda)-\alpha\varphi_2(\lambda).$$

The minimum value of the energy is realised for values of α and λ which satisfy the equations

$$\frac{\partial E}{\partial \alpha} = 0, \quad \frac{\partial E}{\partial \lambda} = 0,$$

or

$$2\alpha\varphi_1(\lambda)-\varphi_2(\lambda) = 0, \quad \alpha\varphi_1'(\lambda)-\varphi_2'(\lambda) = 0.$$

Eliminating α, we get

$$\frac{\varphi_1'(\lambda)}{\varphi_1(\lambda)}-\frac{2\varphi_2'(\lambda)}{\varphi_2(\lambda)} = 0, \quad \lambda = 0{\cdot}2846.$$

The corresponding values of α and β are

$$\alpha = 2{\cdot}694, \quad \beta = 0{\cdot}767.$$

If we substitute these values of the variational parameters into E, we find for the ground state energy of the lithium atom

$$E = -7{\cdot}414 \text{ at un}, \quad \text{or} \quad E = -200{\cdot}8 \text{ eV}.$$

Experimentally $E_{\text{exp}} = -202{\cdot}54$ eV. Applying perturbation theory, that is putting $\alpha = 3, \beta = \frac{3}{2}$, we get for the ground state energy a less accurate value than from the variational method, namely $E = -7{\cdot}05$, or in eV:

$$E = -190{\cdot}84 \text{ eV}.$$

62. The average value of the total energy is equal to

$$\overline{\hat{H}} = \frac{\int \psi_0^*(1+\lambda u)\,\hat{H}\psi_0(1+\lambda u)\,d\tau}{\int \psi_0^*\,\psi_0(1+\lambda u)^2\,d\tau}. \tag{1}$$

If we integrate the numerator by parts the equation is simplified.

The kinetic energy operator of the electrons is of the form

$$\hat{T} = -\frac{1}{2}\sum_{i=1}^{n}\nabla_i^2,$$

where n is the number of electrons and ∇_i^2 is the Laplace operator of the

ith electron (we use atomic units). We shall write down an expression for the average value of the kinetic energy in a form which is symmetric in ψ_0^* and ψ_0,

$$\bar{\hat{T}} = -\frac{1}{2}\sum_{i=1}^{n}\frac{1}{2}\int\{\psi_0^*(1+\lambda u)\nabla_i^2(1+\lambda u)\psi_0+\psi_0(1+\lambda u)\nabla_i^2(1+\lambda u)\psi_0^*\}\,d\tau.$$

If we differentiate under the integral sign we get

$$\bar{\hat{T}} = -\frac{1}{2}\sum_{i=1}^{n}\frac{1}{2}\int[\psi_0^*(1+\lambda u)^2\nabla_i^2\psi_0+\psi_0(1+\lambda u)^2\nabla_i^2\psi_0^*$$
$$+2\lambda\psi_0\psi_0^*(1+\lambda u)\nabla_i^2 u+2\lambda(1+\lambda u)(\nabla_i(\psi_0^*\psi_0)\,.\,\nabla_i u)]\,d\tau. \quad (2)$$

We shall transform the last two terms, using the identity

$$\{\nabla_i\cdot[\psi_0^*\psi(1+\lambda u)\nabla_i u]\}$$
$$=\psi_0^*\psi_0(1+\lambda u)\nabla_i^2 u+(1+\lambda u)(\nabla_i(\psi_0^*\psi_0)\,.\,\nabla_i u)+\lambda\psi_0^*\psi(\nabla_i u\,.\,\nabla_i u). \quad (3)$$

If we integrate this identity over the whole of configuration space we get

$$\int\{\psi_0^*\psi_0(1+\lambda u)\nabla_i^2 u+(1+\lambda u)(\nabla_i(\psi_0^*\psi_0)\,.\,\nabla_i u)\}\,d\tau$$
$$=-\lambda\int\psi_0^*\psi(\nabla_i u\,.\,\nabla_i u)\,d\tau. \quad (4)$$

Substituting expression (4) into equation (2) we get

$$\bar{\hat{T}} = -\frac{1}{2}\sum_{i=1}^{n}\frac{1}{2}\int[\psi_0^*(1+\lambda u)^2\nabla_i^2\psi_0+\psi_0(1+\lambda u)^2\nabla_i^2\psi_0^*]\,d\tau$$
$$+\frac{\lambda^2}{2}\sum_{i=1}^{n}\int\psi_0^*\psi_0(\nabla_i u\,.\,\nabla_i u)\,d\tau.$$

The Hamiltonian operator $\hat{H}$ is equal to $\hat{H}=\hat{H}_0+u=\hat{T}+\hat{V}+u$. Since $\hat{V}$ commutes with $(1+\lambda u)$, expression (1) can be written as follows:

$$\bar{\hat{H}} = E_0+\frac{\frac{1}{2}\int(1+\lambda u)^2(\psi_0^*\hat{H}\psi_0+\psi_0\hat{H}\psi_0^*)\,d\tau+\frac{1}{2}\lambda^2\sum_{i=1}^{n}\int\psi_0^*\psi_0(\nabla_i u\,.\,\nabla_i u)\,d\tau}{\int(1+\lambda u)^2\psi_0^*\psi_0\,d\tau}.$$

Since $\hat{H}_0\psi_0=E_0\psi_0$, we have

$$\bar{\hat{H}} = E_0+\frac{\int\psi_0^*u(1+\lambda u)^2\psi_0\,d\tau+\frac{1}{2}\lambda^2\sum_{i=1}^{n}\int\psi_0^*\psi_0(\nabla_i u\,.\,\nabla_i u)\,d\tau}{\int(1+\lambda u)^2\psi_0^*\psi_0\,d\tau}$$

or

$$\hat{H} = E_0 + \frac{(u)_{00} + 2\lambda(u^2)_{00} + \lambda^2(u^3)_{00} + \frac{1}{2}\lambda^2 \sum_{i=1}^{n} [(\nabla_i u . \nabla_i u)]_{00}}{1 + 2\lambda(u)_{00} + \lambda^2(u^2)_{00}}, \tag{5}$$

where $(u)_{00} = \int \psi_0^* u \psi_0 \, d\tau$, $(u^2)_{00} = \int \psi_0^* u^2 \psi_0 \, d\tau$ and so on. We shall expand the second term in equation (5) in a power series, neglecting terms like $(u^3)_{00}, (u)_{00}^3, \ldots$.

For the energy correction ΔE we find an approximate expression of the form

$$\Delta E \approx (u)_{00} + 2\lambda(u^2)_{00} - 2\lambda(u)_{00}^2 + \frac{1}{2}\lambda^2 \sum_{i=1}^{n} [(\nabla_i u . \nabla_i u)]_{00}. \tag{6}$$

The variational parameter λ is determined from the condition

$$\frac{d\Delta E}{d\lambda} = 2(u^2)_{00} - 2(u)_{00}^2 + \lambda \sum_{i=1}^{n} [(\nabla_i u . \nabla_i u)]_{00} = 0,$$

so that we find for λ the expression

$$\lambda = 2 \frac{(u)_{00}^2 - (u^2)_{00}}{\sum_{i=1}^{n} [(\nabla_i u . \nabla_i u)]_{00}}.$$

Substituting this value into equation (6) we get the required relation

$$\Delta E \approx (u)_{00} - 2 \frac{[(u)_{00}^2 - (u^2)_{00}]^2}{\sum_{i=1}^{n} [(\nabla_i u . \nabla_i u)]_{00}}. \tag{7}$$

63. If the atom is in a uniform electrical field of strength $\mathscr{E}$, along the z-axis the perturbation operator is

$$u = -\mathscr{E} \sum_{i=1}^{n} z_i = -\mathscr{E} z,$$

whose matrix element $(u)_{00}$ is equal to zero (we use atomic units).

From equation (7) of the preceding problem we have

$$\Delta E \approx -2\mathscr{E}^2 \frac{[(z^2)_{00}]}{n},$$

where n is the number of electrons.

It follows thus that the polarisibility is equal to:

$$\alpha = \frac{4[(z^2)_{00}]^2}{n}.$$

It is necessary to note that this equation which is obtained by using one variational parameter λ is a reasonable approximation only for hydrogen and helium. For the case of atoms with several electron shells the deformation of the shells will not be the same. To obtain a better result from the variational method we must thus introduce for each electron shell its own variational parameter.

For the hydrogen atom we have

$$(z^2)_{00} = \tfrac{1}{3}(r^2)_{00} = \frac{1}{3}\frac{4\pi}{\pi}\int_0^\infty \exp(-2r)\,r^4\,dr = 1$$

and thus

$$\alpha = 4 \text{ at un.}$$

In the cgs system we have

$$\alpha = 4\left(\frac{\hbar^2}{\mu e^2}\right)\text{cm}^3.$$

For the helium atom we put the ground state wave function in the form (see problem 59 of section 7)

$$\psi_0 = \frac{Z_{\text{eff}}^3}{\pi}\exp[-Z_{\text{eff}}(r_1+r_2)], \quad \text{where} \quad Z_{\text{eff}} = \tfrac{27}{16}.$$

The final result is

$$\alpha = 0{\cdot}98 \text{ at un} \quad \text{or} \quad \alpha = 0{\cdot}98\left(\frac{\hbar^2}{\mu e^2}\right)^3\text{cm}^3.$$

From this value of α we get for the dielectric constant of helium under normal conditions

$$\epsilon = 1{\cdot}00049,$$

while its experimental value is 1·00074.

The relatively large difference between the theoretical and experimental values is mainly due to the fact that we have used a rather rough approximation for the unperturbed wave function.

64. The potential of a point charge is the same as the potential of a sphere charged on the surface outside that sphere, if the total charge is the same in both cases. Inside this sphere the difference between the two potentials is given by

$$\Delta\varphi = -Ze\left(\frac{1}{r}-\frac{1}{a}\right).$$

The change in the potential energy of the atomic electrons is equal to:

$$\Delta V = -Ze^2\sum_{i=1}^{N}\left(\frac{1}{r_i}-\frac{1}{a}\right)\epsilon(r_i),$$

where we have introduced the function

$$\epsilon(r) = \begin{cases} 1 & r < a, \\ 0 & r > a. \end{cases}$$

According to first-order perturbation theory the energy shift is

$$\Delta E = -Ze^2 \int |\psi|^2 \sum_{i=1}^{N} \left(\frac{1}{r_i} - \frac{1}{a}\right) \epsilon(r_i)\, d\tau_1 \ldots d\tau_N.$$

If we integrate over all variables except one, we get

$$\int |\psi(r_1 \ldots r_N)|^2\, d\tau_2 \ldots d\tau_N = \frac{1}{N}\rho(r),$$

where $\rho(r)$ is the electron density.

In this way we get

$$\Delta E = -Ze^2 \int \rho(r) \left(\frac{1}{r} - \frac{1}{a}\right) \epsilon(r)\, d\tau.$$

We can use the fact that $\rho(r)$ changes little in the region $r < a$. If we take this quantity from under the integral sign and put it equal to its value at the origin we get

$$\Delta E = -Ze^2\, \rho(0) \frac{2\pi}{3} a^2. \tag{1}$$

65. We use for the shift the equation (compare preceding problem)

$$\Delta E = \int |\psi_0|^2\, \Delta V d^3 \boldsymbol{r}, \tag{1}$$

where ΔV is the difference between the actual potential energy and the Coulomb potential and ψ_0 the ground state wave function. If $\rho(\boldsymbol{r})$ is the nuclear charge density which is normalised so that

$$\int \rho(\boldsymbol{r})\, d^3 \boldsymbol{r} = Ze, \tag{2}$$

while its mean square radius $\langle r^2 \rangle$ satisfies the equation

$$\int \rho(\boldsymbol{r})\, r^2\, d^3 \boldsymbol{r} = Ze\langle r^2 \rangle, \tag{3}$$

we have for ΔV the equation

$$\Delta V(r) = -\frac{Ze}{r} + \frac{1}{r}\int_0^r \rho(\boldsymbol{r}')\, d^3 \boldsymbol{r}' - \int_r^\infty \frac{\rho(\boldsymbol{r}')\, d^3 \boldsymbol{r}'}{r'}. \tag{4}$$

Substituting (4) into (1), integrating by parts, using the fact that ψ_0 changes little over the nuclear charge density, we find, using (2) and (3), that

$$\Delta E = -\frac{2\pi Ze^2}{3} |\psi_0(0)|^2 \langle r^2 \rangle. \tag{5}$$

If the charge distribution is a spherical shell as in the preceding problem, (5) reduces to equation (1) of that problem.

66*. The wave function of an s-electron is

$$\psi_n(r) = \frac{1}{\sqrt{(4\pi)}} \frac{\chi_n(r)}{r},$$

where χ_n satisfies the equation

$$\chi_n'' + \frac{2\mu}{\hbar^2}[E_n - V(r)]\chi_n = 0$$

and the normalising condition $\int_0^\infty \chi_n^2 \, dr = 1$.

In the semi-classical approximation the solution of the equation for χ_n is of the form

$$\chi_n = \frac{A_n}{\sqrt{(p_n)}} \cos\left(\frac{1}{\hbar}\int_0^r p_n \, dr + \varphi\right), \tag{1}$$

where

$$p_n = \sqrt{\{2\mu[E_n - V(r)]\}}.$$

This solution is, however, useless for small values of r. Indeed, if r is small ($r \ll \hbar^2/Z^{\frac{1}{3}} \mu e^2$), then first of all we can neglect the screening of the field of the nucleus and put $V(r) = -Ze^2/r$; secondly, we can also neglect E_n compared to $V(r)$. If we put $p = \sqrt{(2\mu Ze^2/r)}$ into the condition of applicability of the semi-classical approximation

$$\frac{d(\hbar/p)}{dr} \ll 1,$$

we get

$$r \gg \frac{\hbar^2}{Z\mu e^2}.$$

To obtain an expression for χ_n which can be used for small values of r, we start again from the original equation and substitute for $V(r)$

$$-Ze^2/r$$

and also neglect E_n:

$$\chi_n'' + \frac{2\mu}{\hbar^2}\frac{Ze^2}{r}\chi_n = 0.$$

This equation has the following solution:

$$\chi_n = C_n \sqrt{(r)}\, J_1\left[2\sqrt{\left(\frac{2\mu Z e^2 r}{\hbar^2}\right)}\right]. \tag{2}$$

To find the connection between the constants C_n and A_n, we note that the region of applicability of the semi-classical solution (1),

$$r \gg \frac{\hbar^2}{Z\mu e^2},$$

and the region of applicability of solution (2), where the screening of the nuclear field has been neglected,

$$r \ll \frac{\hbar^2}{Z^{\frac{1}{3}}\mu e^2},$$

overlap for large values of Z so that solutions (1) and (2) must coincide in the region

$$\frac{\hbar^2}{Z\mu e^2} \ll r \ll \frac{\hbar^2}{Z^{\frac{1}{3}}\mu e^2}.$$

We shall show that solutions (1) and (2) are the same in the entire region of applicability. To do that we put in solution (1) $p = \sqrt{(2\mu Z e^2/r)}$. We then get

$$\chi_n = \frac{A_n \sqrt[4]{r}}{\sqrt[4]{(2\mu Z e^2)}} \cos\left[\frac{2\sqrt{(2\mu Z e^2 r)}}{\hbar} + \varphi\right] \quad \frac{\hbar^2}{Z\mu e^2} \ll r \ll \frac{\hbar^2}{Z^{\frac{1}{3}}\mu e^2}. \tag{3}$$

The condition $r \gg \hbar^2/Z\mu e^2$ means that the argument of the Bessel function in expression (2) is large. But for large values of the argument ($x \gg 1$) we have

$$J_1(x) \approx \sqrt{\left(\frac{2}{\pi x}\right)} \cos\left(x - \frac{3\pi}{4}\right).$$

The solution (2) is thus of the form

$$\chi_n = C_n \sqrt{(r)} \sqrt{\left[\frac{\hbar}{\pi\sqrt{(2\mu Z e^2 r)}}\right]} \cos\left[\frac{2\sqrt{(2\mu Z e^2 r)}}{\hbar^2} - \frac{3\pi}{4}\right] \quad \frac{\hbar^2}{Z\mu e^2} \ll r \ll \frac{\hbar^2}{Z^{\frac{1}{3}}\mu e^2}.$$

Comparing this expression with expression (3) we find

$$\varphi = -\frac{3\pi}{4}, \quad A_n = C_n \sqrt{\left(\frac{\hbar}{\pi}\right)}.$$

We must now find $\psi^2(0)$. For $x \ll 1$ we have $J_1(x) = x/2$ and using equation (2) we find

$$\left.\frac{\chi_n}{r}\right|_{r\to 0} = C_n \sqrt{\left(\frac{2\mu Z e^2}{\hbar^2}\right)} = \sqrt{\left(\frac{2\pi\mu Z e^2}{\hbar^3}\right)} A_n.$$

Hence,

$$\psi_n^2(0) = \frac{\mu Z e^2}{2\hbar^3} A_n^2.$$

The constant A_n can be determined from the normalisation condition

$$\int_0^\infty \chi_n^2\, dr = A_n^2 \int_0^\infty \frac{\cos^2\left(\frac{1}{\hbar}\int_0^r p_n\, dr - \frac{3\pi}{4}\right)}{\sqrt{\{2\mu[E_n - V(r)]\}}}\, dr \approx \frac{A_n^2}{2} \int \frac{dr}{\sqrt{\{2\mu[E_n - V(r)]\}}} = 1.$$

If we differentiate the quantum condition

$$\int \sqrt{\{2\mu[E_n - V(r)]\}}\, dr = \pi(n+\gamma)\hbar,$$

which determines E_n as a function of n, with respect to n we find

$$\mu \frac{dE_n}{dn} \int \frac{dr}{\sqrt{\{2\mu[E_n - V(r)]\}}} = \pi\hbar.$$

Comparing this last expression with the normalisation condition we get

$$A_n^2 = \frac{2\mu}{\pi\hbar} \frac{dE_n}{dn},$$

and finally

$$\psi_n^2(0) = \frac{Ze^2\,\mu^2}{\pi\hbar^4} \frac{dE_n}{dn}. \tag{4}$$

For an unscreened Coulomb field we have

$$E_n = -\frac{Z^2}{2n^2} \frac{\mu e^4}{\hbar^2}$$

and we would then get from equation (4) $\psi_n^2(0) = \frac{Z^3}{\pi n^3}\left(\frac{\mu e^2}{\hbar^2}\right)^3$ which is the same as the result obtained from a rigorous calculation. In atomic spectroscopy the following equation is often used for the energy levels of the excited states of the valence electron

$$E_n = -\frac{\mu e^4}{2\hbar^2} \frac{1}{(n-\sigma)^2},$$

where σ is the so-called quantum defect which depends only weakly on n. If we use this formula for E_n we get

$$\psi^2(0) = \frac{Z}{\pi} \frac{\mu^3 e^6}{\hbar^6} \frac{(1 - d\sigma/dn)}{(n-\sigma)^3}.$$

67*. We shall denote the radius vectors of the electrons with respect to the nucleus by $\mathbf{r}_1$ and $\mathbf{r}_2$. We have the following Schrödinger equation for the stationary states of the helium-like two-electron system:

$$\hat{H}\Psi(\mathbf{r}_1, \mathbf{r}_2) = E\Psi(\mathbf{r}_1, \mathbf{r}_2), \tag{1}$$

where the Hamiltonian $\hat{H}$ is of the form

$$\hat{H} = -\frac{\hbar^2}{2\mu}(\nabla^2_{r_1} + \nabla^2_{r_2}) - Ze^2\left(\frac{1}{r_1} + \frac{1}{r_2}\right) + \frac{e^2}{r_{12}}. \tag{2}$$

We consider the last term as a perturbation, and denote it by $\hat{H}'$. We shall perform our calculations in atomic units ($\hbar = 1, e = 1, \mu = 1$).

Since the unperturbed Hamiltonian, $\hat{H}^{(0)} \equiv \hat{H} - \hat{H}'$ is the sum of two one-electron Hamiltonians, the unperturbed wave function of the ground state of our system will be the product of two hydrogen-like functions in the field of the nucleus Z (since the ground state of hydrogen-like atom lies far below the excited states, it is clear that the ground state of a helium-like system can be constructed out of the ground states of the corresponding hydrogen-like systems),

$$\Psi^{(0)}(\mathbf{r}_1, \mathbf{r}_2) = \psi_0(\mathbf{r}_1)\psi_0(\mathbf{r}_2) = \frac{Z^{\frac{3}{2}}}{\sqrt{\pi}} \exp(-Zr_1) . \frac{Z^{\frac{3}{2}}}{\sqrt{\pi}} \exp(-Zr_2)$$

$$= \frac{Z^3}{\pi} \exp[-Z(r_1 + r_2)], \tag{3}$$

and the unperturbed ground state energy is equal to the sum of the ground state energy of two hydrogen-like atoms,

$$E^{(0)} = -\frac{Z^2}{2} - \frac{Z^2}{2} = -Z^2. \tag{4}$$

The required first-order correction to the ground state energy is equal to the average value of the perturbing energy $\hat{H}'$ in the unperturbed state (3),

$$E^{(1)} = \overline{H'} = \iint \Psi_0^*(\mathbf{r}_1, \mathbf{r}_2)\hat{H}'\Psi_0(\mathbf{r}_1, \mathbf{r}_2)\, d^3\mathbf{r}_1\, d^3\mathbf{r}_2$$

$$= \frac{Z^6}{\pi^2} \iint \exp[-2Z(r_1 + r_2)] \frac{1}{r_{12}}\, d^3\mathbf{r}_1\, d^3\mathbf{r}_2. \tag{5}$$

The integral we have just obtained can be evaluated in different ways. We shall use one of the simpler ways.

The quantity $1/r_{12} = 1/|\boldsymbol{r}_1 - \boldsymbol{r}_2|$ is the generating function for the Legendre polynomials,

$$\frac{1}{|\boldsymbol{r}_1-\boldsymbol{r}_2|} = \begin{cases} \dfrac{1}{r_1}\sum\limits_{l=0}^{\infty}\left(\dfrac{r_2}{r_1}\right)^l P_l(\cos\theta), & \text{if} \quad r_1 \geqslant r_2, & (6') \\[2ex] \dfrac{1}{r_2}\sum\limits_{l=0}^{\infty}\left(\dfrac{r_1}{r_2}\right)^l P_l(\cos\theta), & \text{if} \quad r_1 \leqslant r_2, & (6'') \end{cases}$$

where θ is the angle between r_1 and r_2. If we integrate, for instance, first over r_2 and split the domain of integration according to equations (6′) and (6″) we get

$$E^{(1)} = \frac{Z^6}{\pi^2}\int d^3\boldsymbol{r}_1 \exp(-2Zr_1)$$
$$\times\left\{\int_{r_2=0}^{r_1} d^3\boldsymbol{r}_2 \exp(-2Zr_2)\frac{1}{r_1}\sum_{l=0}^{\infty}\left(\frac{r_2}{r_1}\right)^l P_l(\cos\theta)\right.$$
$$\left.+\int_{r_2=r_1}^{\infty} d^3\boldsymbol{r}_2 \exp(-2Zr_2)\frac{1}{r_2}\sum_{l=0}^{\infty}\left(\frac{r_1}{r_2}\right)^l P_l(\cos\theta)\right\}.$$

If we take the polar axis of a spherical system of coordinates along r_1 and use the orthogonality relation of the Legendre polynomials,

$$\int_0^{\pi} P_l(\cos\theta)\sin\theta\, d\theta = \int_0^{\pi} P_l(\cos\theta)\, P_0(\cos\theta)\sin\theta\, d\theta = \frac{2}{2l+1}\,\delta_{l0},$$

we find that after integrating over the angle only two terms are left in the expression within braces,

$$E^{(1)} = \frac{Z^6}{\pi^2}\int d^3\boldsymbol{r}_1 \exp(-2Zr_1)\,.\,4\pi\left\{\frac{1}{r_1}\int_0^{r_1}\exp(-2Zr_2)\, r_2^2\, dr_2\right.$$
$$\left.+\int_{r_1}^{\infty}\exp(-2Zr_2)\, r_2\, dr_2\right\}.$$

If we now perform the remaining elementary integration we find finally

$$E^{(1)} = \tfrac{5}{8}Z. \tag{7}$$

The positive sign of this correction corresponds, as should be, to the mutual repulsion of the electrons.

From equations (4) and (7) we get for the ground state energy of a helium-like atom or ion to a first approximation,

$$E \approx E^{(0)} + E^{(1)} = -(Z^2 - \tfrac{5}{8}Z), \tag{8}$$

or, in the usual units,

$$E \approx -(Z^2 - \tfrac{5}{8}Z)\frac{\mu e^4}{\hbar^2} = -(Z^2 - \tfrac{5}{8}Z)\,27{\cdot}1\ \text{eV}. \tag{8'}$$

The single ionisation potential V_1 is the energy necessary to take one of the electrons just into a state of the continuous spectrum.

It is clear that this quantity is equal to the difference between the total ionisation energy, $-E$, given by equation (8), and the energy of ionisation of a one-electron atom or ion, which is equal to $\frac{1}{2}Z^2$ at un,

$$V_1 = (Z^2 - \tfrac{5}{8}Z) - \tfrac{1}{2}Z^2 = (\tfrac{1}{2}Z^2 - \tfrac{5}{8}Z)\ \text{at un} = (\tfrac{1}{2}Z^2 - \tfrac{5}{8}Z)\,27{\cdot}1\ \text{eV}. \tag{9}$$

Perturbation theory is applicable to the present problem provided $Z \gg 1$, as can be seen from equation (8). (In the given simple example we do not have to consider very strictly the requirement of the applicability of perturbation theory; this is necessary if we want to find the first-order correction to the wave function or the second-order correction to the energy.)

It is of interest to note, however, that our evaluation, which is relatively gross, leads already for $Z = 2$ (helium atom) to a value of the binding energy which agrees with the experimental value within about 5%. For larger values of Z ($Z = 3$: Li^+, $Z = 4$: Be^{++}, ...) the agreement is naturally even better.

The agreement with experimental values is not so good for the first ionisation potential V_1 evaluated from equation (9); for helium, for instance ($Z = 2$), the difference is about 15%.

68. To first order the cubic term does not contribute to the change in energy. The perturbed eigenvalues are

$$E_n = (n+\tfrac{1}{2})\hbar\omega + \tfrac{3}{4}\beta(2n^2+2n+1)\left(\frac{\hbar}{\mu\omega}\right)^2 - \tfrac{1}{8}\frac{\alpha^2}{\hbar\omega}(30n^2+30n+11)\left(\frac{\hbar}{\mu\omega}\right)^3$$

$$-\tfrac{1}{8}\frac{\beta^2}{\hbar\omega}(34n^3+51n^2+59n+21)\left(\frac{\hbar}{\mu\omega}\right)^4 .$$

69. The lowest three energy levels of the unperturbed system are a non-degenerate singlet state with $E = E_0 = \hbar\omega$, a four-fold degenerate level with $E = E_1 = 2\hbar\omega$, corresponding to a three-fold degenerate triplet state and a non-degenerate singlet state, and a five-fold degenerate level with $E = E_2 = 3\hbar\omega$, corresponding to one three-fold degenerate triplet state and two non-degenerate singlet states. To first order in V_0 the energies of the triplet states remain unchanged. The final energy eigenvalues with their remaining degrees of degeneracy are (note that the two singlet states corresponding to E_2 still have the same energy value):

$$E'_0 = \hbar\omega + V_0/\sqrt{(2\pi)},\ g_0 = 1;\ E'_{1t} = 2\hbar\omega,\ g_{1t} = 3;$$

$$E'_{1s} = 2\hbar\omega + V_0/\sqrt{(2\pi)},\ g_{1s} = 1;\ E'_{2t} = 3\hbar\omega,\ g_{2t} = 3;$$

$$E'_{2s} = 3\hbar\omega + 3V_0/4\sqrt{(2\pi)},\ g_{2s} = 2.$$

70*. First of all we show that the characteristic time parameter τ enters into the perturbing field in such a way that the total pulse P, which is classically transferred to the oscillator by the electrical field over the total duration of the perturbation, does not depend on τ,

$$P = \int_{-\infty}^{\infty} e\mathscr{E}(t)\,dt = \frac{eA}{\sqrt{(\pi)}\,\tau}\int_{-\infty}^{\infty} \exp\left[-\left(\frac{t}{\tau}\right)^2\right]dt = eA = \text{const.} \quad (1)$$

This means graphically that the area under the curve is the same for all values of τ (see fig. 42).

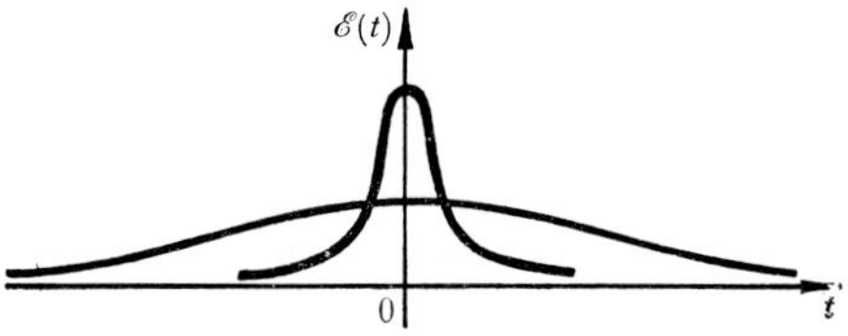

Fig. 42.

The probability for a transition from the nth stationary state of the discrete spectrum to the kth is equal to

$$w_{nk} = \frac{1}{\hbar^2}\left|\int_{-\infty}^{\infty} V_{kn}\exp(i\omega_{kn}t)\,dt\right|^2 \quad (2)$$

where $V_{kn} = \int_{-\infty}^{\infty} \psi_k^{(0)*}\,\hat{V}\psi_n^{(0)}\,dx$ is the matrix element of the perturbation $\hat{V}$, $\omega_{kn} = (1/\hbar)\,[E_k^{(0)} - E_n^{(0)}]$, where $\psi_k^{(0)}, \psi_n^{(0)}, E_k^{(0)}, E_n^{(0)}$ are the wave functions and energy levels of the corresponding (unperturbed) stationary states.

If we denote by e, μ, and ω the charge, the mass, and the eigenfrequency of the oscillator, and by x its displacement from its equilibrium position we get in the case under consideration of a uniform field for the perturbation operator,

$$\hat{V}(x, t) = -ex\mathscr{E}(t) \sim x.$$

It is well known that in the matrix of the coordinate of the oscillator in the energy representation only the following matrix elements are different from zero, $x_{n,n+1} = x_{n+1,n} = \sqrt{[(n+1)\hbar/2\mu\omega]}$. Since we have assumed that the oscillator was original in its ground state ($n = 0$), we are dealing with only the following non-vanishing matrix elements of the perturbation:

$$V_{01} = V_{10} = -\frac{P}{\sqrt{(\pi)}\,\tau}\sqrt{\left(\frac{\hbar}{2\mu\omega}\right)}\exp\left[-(t/\tau)^2\right]. \tag{3}$$

In first-order perturbation theory the uniform field can thus produce a transition of the oscillator only to the first excited state ($k = n+1 = 1$). (To evaluate the probability for a transition to the second excited state we must use second-order perturbation theory—as this transition, $n = 0 \rightarrow n = 2$, goes through the intermediate state $n = 1$, and so on; each excited energy state corresponds to a certain order of perturbation theory and a correspondingly increasing number of intermediate states. From the character of the coordinate matrix it follows that each of the intermediate states is next to the previous one—for instance, $0 \rightarrow 1 \rightarrow 2 \rightarrow 3$.)

If we substitute expression (3) and also

$$\omega_{kn} = \omega_{10} \equiv (1/\hbar)\left[E_1^{(0)} - E_0^{(0)}\right] = \omega$$

into equation (2) we get the probability for excitation,

$$w_{01} = \frac{P^2}{2\pi\tau^2\,\mu\hbar\omega}\left|\int_{-\infty}^{\infty}\exp\left[i\omega t - (t/\tau)^2\right]dt\right|^2. \tag{4}$$

If we evaluate the well-known integral

$$\int_{-\infty}^{\infty}\exp(i\beta x - \alpha x^2)\,dx = \sqrt{\left(\frac{\pi}{\alpha}\right)}\exp(-\beta^2/4\alpha),$$

we get finally

$$w_{01} = \frac{P^2}{2\mu\hbar\omega}\exp\left[-\tfrac{1}{2}(\omega\tau)^2\right]. \tag{5}$$

We see that for a given total classical transferred pulse P the probability for excitation decreases steeply with increasing effective duration of the perturbation τ and if $\tau \gg 1/\omega$ (duration of perturbation much less than the clasşical period of the oscillator), this probability is very small; and we are dealing with a so-called adiabatic perturbation.

On the other hand for a fast perturbation ($\tau \ll 1/\omega$) the probability of excitation (5) is practically constant. In the limit where $\tau \rightarrow 0$ we can use a well-known expression for the δ-function,

$$\lim_{\tau\rightarrow 0}\mathscr{E}(t) = A\delta(t) = \frac{P}{e}\delta(t);$$

we are dealing with so-called instantaneous perturbation and the probability will be equal to

$$\lim_{\tau\to 0} w_{01} = \frac{P^2}{2\mu\hbar\omega}, \tag{5'}$$

which is equal to the ratio of the classical imparted energy, $(1/2\mu)P^2$, to the distance between the oscillator energy levels $\hbar\omega$.

The criterion of applicability of perturbation theory is that the probability for excitation should be small compared to the probability that the oscillator stays in the ground state,

$$w_{01} \ll (1-w_{01}), \quad \text{or} \quad w_{01} \ll 1. \tag{6}$$

We see from equation (5) that a sufficient condition for satisfying conditions (6) is that

$$\frac{P^2}{2\mu} = \frac{(eA)^2}{2\mu} \ll \hbar\omega. \tag{6'}$$

It is clear, however, that if the degree of adiabatic change is sufficiently large, that is, if $\tau \gg \omega^{-1}$, condition (6′) is much too strong, and perturbation theory is also applicable if $P^2/2\mu$ is of the order of $\hbar\omega$.

We emphasise that the excitation of the oscillator when condition (6′) is satisfied is a quantum effect. Indeed, in classical mechanics excitation of the oscillator by an amount $\hbar\omega$ would in general not be possible because of the conservation law for the energy. In quantum mechanics, however, the excitation is possible even though we can see from equations (5) and (5′) that its probability is small. This does not violate the law of conservation of energy, since the quantity $P^2/2\mu$ in quantum mechanics cannot, in general, be interpreted in the classical way as the energy given to the oscillator by the field.

71. The perturbation in the present case is of a form which is qualitatively the same as the Gaussian perturbation of the preceding problem. The condition $P = \int_{-\infty}^{\infty} e\mathscr{E}(t)\,dt = \text{const}$ leads, as can easily be verified, to the following law of change of the field:

$$\mathscr{E}(t) = \frac{P}{e}\frac{\tau}{\pi}\frac{1}{\tau^2+t^2}. \tag{1}$$

We can use all equations of the preceding problem without any change up to equation (3) which now will be

$$V_{01} = V_{10} = -P\frac{\tau}{\pi}\sqrt{\left(\frac{\hbar}{2\mu\omega}\right)}\frac{1}{\tau^2+t^2}, \tag{2}$$

so that we get for the probability of a transition to the first excited state instead of equation (4)

$$w_{01} = \frac{P^2 \tau^2}{2\pi^2 \mu\hbar\omega} \left| \int_{-\infty}^{\infty} \frac{\exp(i\omega t)\,dt}{t^2+\tau^2} \right|^2. \tag{3}$$

The integral in expression (3) can be most simply evaluated using the theory of residues. We use a variable t in the complex plane

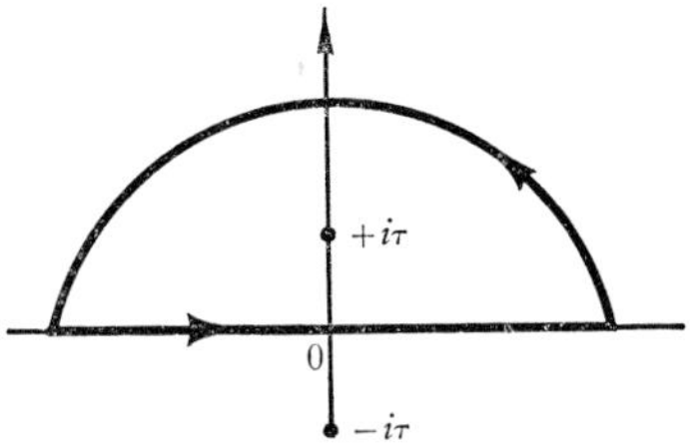

Fig. 43.

and we close the circuit, which starts along the real axis corresponding to the integration in expression (3), by a semi-circle of infinitely large radius in the upper half plane; the result of integration over this contour is the same as the result of the integral in expression (3), since the integrand contains a factor $\exp(i\omega t)$. (It is of course necessary that the rest of the integrand also tends to zero if $|t| \to \infty$.) The only singularity (in our case a pole of the first order) of the integrand inside the contour is at the point $t = +i\tau$. The integral over t is equal to $2\pi i$ times the residue at that point,

$$\int_{-\infty}^{+\infty} \frac{\exp(i\omega t)\,dt}{t^2+\tau^2} = 2\pi i \,\mathrm{Res} \frac{\exp(i\omega t)}{t^2+\tau^2}\bigg|_{t=i\tau} = 2\pi i \frac{\exp(-\omega\tau)}{2i\tau} = \frac{\pi}{\tau}\exp(-\omega\tau).$$

Substituting this result into equation (3) we find finally

$$w_{01} = \frac{P^2}{2\mu\hbar\omega}\exp(-2\omega\tau). \tag{4}$$

The qualitative dependence of this transition probability on the effect of duration of the perturbation τ and also the discussion of the applicability of perturbation theory can proceed exactly as in the preceding problem. We note that in the limiting case of an instantaneous perturbation ($\tau \to 0$) it follows from equation (1) and one possible expression for the δ-function, $\lim_{\tau \to 0} \mathscr{E}(t) = (1/e) P\delta(t)$, that the probability of excitation tends to $P^2/2\mu\hbar\omega$, which is exactly the same as the result (5′) of the preceding problem. We should have expected this result since for a given value of $\int_{-\infty}^{\infty} \mathscr{E}(t)\,dt$ the transition under the action of a sudden perturbation should not depend on the actual form of the "pulse" function $\mathscr{E}(t)$. This can be formally expressed also by noting that there are a number of different "pulse" functions which lead all to the same function $\delta(t)$.

72. Instead of equations (1) to (4) of the preceding problem we now have

$$\mathscr{E}(t) = \frac{P}{e\tau}e^{-t/\tau}; \quad V_{10} = V_{01} = -\frac{P}{\tau}\sqrt{\left(\frac{\hbar}{2\mu\omega}\right)}e^{-t/\tau};$$

$$w_{10} = \frac{P^2}{2\tau^2\mu\hbar\omega}\left|\int_0^\infty e^{i\omega t - t/\tau}\,dt\right|^2 = \frac{P^2}{2\mu\hbar\omega(1+\omega^2\tau^2)}.$$

73*. The general equation for the probability of a transition per unit time from a state of the discrete spectrum into a state corresponding to an infinitesimal energy interval of the continuous spectrum through the action of a periodic perturbation of frequency ω is of the form

$$dw_{n\nu} = \frac{2\pi}{\hbar}|F_{\nu n}|^2\,\delta(E_\nu - E_n^{(0)} - \hbar\omega)\,d\nu. \tag{1}$$

In this expression n is a set of quantum numbers characterising the state of the discrete spectrum, ν a set of quantities for the states of the continuous spectrum, $d\nu$ the corresponding infinitesimal interval, $E_n^{(0)}$ and E_ν are the unperturbed energy levels of the discrete spectrum and the continuous spectrum, while $F_{\nu n}$ is the matrix element of the perturbation operator for the transition under consideration, which will be defined more exactly in a moment. The wave functions of the discrete spectrum are normalised to unity and those of the continuous spectrum to $\delta(\nu-\nu')$.

Let r and $-e$ be the radius vector and charge of the atomic electron and let $\mathscr{E}_0$ be the amplitude of the intensity of the uniform electrical field. In our case the perturbation operator $\hat{V}$ is of the form

$$\hat{V} = e(\mathscr{E}(t)\,.\,\boldsymbol{r}) = e(\mathscr{E}_0\,.\,\boldsymbol{r})\sin\omega t = \hat{F}\exp(-i\omega t) + \hat{F}^*\exp(i\omega t), \tag{2}$$

where

$$\hat{F} = \frac{i}{2}e(\mathscr{E}_0\,.\,\boldsymbol{r}). \tag{2'}$$

The δ-function which occurs in equation (1) shows the resonance character of possible transitions; the energy can make a transition to a state of the continuous spectrum only if (strictly speaking, equation (3) is only approximately taken into account as we shall discuss at the end of this problem)

$$E_\nu = E_n^{(0)} + \hbar\omega \tag{3}$$

(absorption of a "quantum" of frequency ω). Since for the hydrogen atom in the ground state $(E_\nu - E_n^{(0)})_{\min} = \mu e^4/2\hbar^2$, it follows that the minimum frequency which is necessary for ionisation is given by the equation

$$\omega_{\min} = \frac{\mu e^4}{2\hbar^3} \equiv \omega_0.$$

We shall now evaluate the matrix element

$$F_{\nu n} = \int \psi_\nu^* \hat{F} \psi_n^{(0)} \, d\tau. \tag{4}$$

We shall substitute in this expression equation (2′) and also

$$\psi_n^{(0)} = \psi_{100} = (\pi a^3)^{-\frac{1}{2}} \exp(-r/a), \quad \psi_\nu \approx \psi_k = (2\pi)^{-\frac{3}{2}} \exp[i(\boldsymbol{k}.\boldsymbol{r})],$$

where we have used the normalisation mentioned earlier.

We shall choose for our set of quantum numbers ν, the wave vector of the free electron $\boldsymbol{k}$; it is clear that the use of plane waves for ψ_ν is, strictly speaking, only correct if $\omega \gg \omega_0$, that is, when the electron is moving fast.

We get in this way from equation (4)

$$F_{\nu n} = \frac{ie}{2}(2\pi)^{-\frac{3}{2}}(\pi a^3)^{-\frac{1}{2}} \int \exp[-i(\boldsymbol{k}.\boldsymbol{r}) - r/a] (\mathscr{E}_0.\boldsymbol{r}) \, d^3 r. \tag{4′}$$

To evaluate the integral we introduce a spherical system of coordinates (r, ϑ, φ) with the polar axis along $\boldsymbol{k}$ and we denote by Θ the angle between $\boldsymbol{k}$ and $\mathscr{E}_0$. We get then

$$(\mathscr{E}_0.\boldsymbol{r}) = \mathscr{E}_0 r[\cos\Theta \cos\vartheta + \sin\Theta \sin\vartheta \cos(\varphi - \varphi_0)],$$

where φ_0 is the azimuth of $\mathscr{E}_0$ in this system of coordinates. Substituting this last expression into equation (4′) the second term will clearly give zero, when integrated over φ, and we get $(\cos\vartheta \equiv x)$:

$$F_{\nu n} = \frac{ie}{2^{\frac{5}{2}} \pi^2 a^{\frac{3}{2}}} 2\pi \mathscr{E}_0 \cos\Theta \int_{-1}^{1} \left[\int_0^\infty \exp(-ikrx - r/a) r^3 \, dr \right] x \, dx$$

$$= \frac{ie\mathscr{E}_0 \cos\Theta}{\pi (2a)^{\frac{3}{2}}} \int_{-1}^{1} \frac{3!\, x \, dx}{(1/a + ikx)^4} = -\frac{e\mathscr{E}_0 \cos\Theta}{\pi (2a)^{\frac{3}{2}}} \frac{16 k a^5}{(1 + k^2 a^2)^3}.$$

Substituting this expression for $|F_{\nu n}|^2$ into equation (1) and writing

$$d\nu = d^3 \boldsymbol{k} = k^2 \, dk \, d\Omega_k = k^2 \frac{dk}{dE_\nu} d\Omega_k \, dE_\nu = \frac{\mu k}{\hbar^2} d\Omega_k \, dE_\nu$$

(we use the equation $E_\nu = \hbar^2 k^2 / 2\mu$), we get

$$dw_{n\nu} = \frac{2^6}{\pi} \cdot \frac{\mu a^7 e^2}{\hbar^3} \cdot \frac{\mathscr{E}_0^2 k^3 \cos^2\Theta}{(1 + k^2 a^2)^6} \delta(E_\nu - E_n^{(0)} - \hbar\omega) \, d\Omega_k \, dE_\nu,$$

where $d\Omega_k$ is an element of solid angle with its axis along $\boldsymbol{k}$. If we finally integrate $dw_{n\nu}$ over E_ν, we find the probability of ionisation while the electron has a final wave vector $\boldsymbol{k}$ within the element $d\Omega_k$. For this integration we need clearly only consider the point $E_\nu = E_n^{(0)} + \hbar\omega$, that is $k^2 = 2\mu E_\nu/\hbar^2 = 2\mu(\omega - \omega_0)/\hbar$, so that $(1 + k^2 a^2) = \omega/\omega_0$ (we have put $E_n^{(0)} = -\mu e^4/2\hbar^2 = -\hbar\omega_0$). If we also take into account the fact that $a = \hbar^2/\mu e^2$, we find finally

$$dw_k = \frac{64a^3}{\pi\hbar}\mathscr{E}_0^2\left(\frac{\omega_0}{\omega}\right)^6\left(\frac{\omega}{\omega_0} - 1\right)^{\frac{3}{2}}\cos^2\Theta\, d\Omega_k. \tag{5}$$

The angular distribution of the electrons which are expelled from the atom by a high frequency electrical field has axial symmetry around the only preferred direction, namely, that of the field $\mathscr{E}$, as we should have expected since the initial state of the atom was spherically symmetric. The angular distribution is also symmetrical with respect to the plane $\Theta = \pi/2$, where it is equal to zero (no electrons fly away in a direction perpendicular to the field $\mathscr{E}$); in other words forward ($\Theta \approx 0$), and backward ($\Theta \approx \pi$) expulsion have the same probability.

This last result is also natural because the field $\mathscr{E}(t)$ is oscillating with a frequency which is much larger than the eigenfrequency ω_0 of the electron in the atom. It is useful, however, to note that this result (which is qualitatively clear also in classical mechanics) follows automatically from the quantum mechanical calculation. The appearance in expression (5) of the factor $\cos^2\Theta$, which is symmetric with respect to the directions $\mathscr{E}_0$ and $-\mathscr{E}_0$, is caused simply by the structure of the matrix element $F_{\nu n}$ which in general does not depend on the time and which thus cannot be connected with a field of sufficiently high frequency. In actual fact, however, this high frequency behaviour is contained in the energy δ-function, and for frequencies of the field which are less than ω_0, the probability (1) is strictly equal to zero. (One can see in other words that if $\omega < \omega_0$ the δ-function in expression (1) does in general exclude the appearance of the matrix element (4′) with real values of k, which means that the transition of an electron to the continuous spectrum is impossible in that case.)

If we integrate expression (5) over all possible angles, and use the fact that $4\pi\overline{\cos^2\Theta} = 4\pi/3$ we get the total probability w_i of the ionisation of an atom per unit time as a function of the frequency of the field

$$w_i(\omega) = \frac{256}{3}\frac{a^3}{\hbar}\mathscr{E}_0^2\left(\frac{\omega_0}{\omega}\right)^6\left(\frac{\omega}{\omega_0} - 1\right)^{\frac{3}{2}}. \tag{6}$$

If $\omega \gtrsim \omega_0$, that is, if we are near the threshold for ionisation this probability increases from zero as $(\omega - \omega_0)^{\frac{3}{2}}$. If $\omega \gg \omega_0$ it decreases steeply as $\omega^{-\frac{8}{3}}$, with increasing ω. By differentiating expression (6) one sees easily that it has a maximum for $\omega = \frac{4}{3}\omega_0$.

In conclusion we want to discuss briefly the limit of applicability of the perturbation theory which we have used. First of all it follows from equation (1) that the time t reckoned from $t = 0$ must be sufficiently large. The uncertainty ΔE_ν of the energy of the perturbed final states of the electron is connected to the value of t by the uncertainty relation $\Delta E_\nu . t \sim \hbar$.

On the other hand the time t cannot be too large since the distortion of the initial wave function of the atom must still be relatively small. This is equivalent to requiring that for all intermediate times t the probability for ionisation, which is equal to $w_i t$, must be less than unity. One can use equation (6) to obtain the limits of applicability of the perturbation theory when we change one of the three quantities t, $\mathscr{E}_0$, or ω while keeping the other two fixed.

74*. The wave function of the initial state of the electron, in atomic units, is of the form

$$\psi_0 = \frac{Z^{\frac{3}{2}}}{\sqrt{\pi}} \exp(-Zr).$$

The final state of the electron can be described by a plane wave since its velocity is large,

$$\psi_1 = \frac{1}{(2\pi)^{\frac{3}{2}}} \exp[i(\boldsymbol{k}.\boldsymbol{r})].$$

The perturbing interaction, also in atomic units, is of the form

$$V = -\sum_p \frac{1}{|\boldsymbol{r}-\boldsymbol{r}_p|},$$

where the sum extends over all protons in the nucleus.

We shall denote by Ψ_0 and Ψ_1 the initial and final states of the nucleus, so that we can write the matrix element of the perturbation in the form

$$V_{01} = -\iint dV\, d\tau\, \psi_1^* \frac{1}{(2\pi)^{\frac{3}{2}}} \exp[-i(\boldsymbol{k}.\boldsymbol{r})] \sum_p \frac{1}{|\boldsymbol{r}-\boldsymbol{r}_p|} \Psi_0 \frac{Z^{\frac{3}{2}}}{\sqrt{\pi}} \exp(-Zr), \tag{1}$$

where $d\tau$ is the nuclear volume element, and dV the electron volume element.

We shall expand the potential of the Coulomb interaction in a Fourier integral,

$$\sum_p \frac{1}{|\boldsymbol{r}-\boldsymbol{r}_p|} = \frac{1}{2\pi^2} \int \frac{d^3\boldsymbol{q}}{q^2} \sum_p \exp[i(\boldsymbol{q}.\boldsymbol{r}-\boldsymbol{r}_p)]$$

$$= \frac{1}{2\pi^2} \int \frac{d^3\boldsymbol{q}}{q^2} \exp[i(\boldsymbol{q}.\boldsymbol{r})] \sum_p \exp[-i(\boldsymbol{q}.\boldsymbol{r}_p)]. \tag{2}$$

If we substitute this expansion into expression (1) and change the order of integration, we get

$$V_{01} = -\frac{Z^{\frac{3}{2}}}{2^{\frac{3}{2}}\pi^4}\int\frac{d^3\boldsymbol{q}}{q^2}\int d\tau\,\Psi_1^*\sum_p \exp[-i(\boldsymbol{q}\,.\,\boldsymbol{r}_p)]\,\Psi_0$$

$$\times\int dV \exp[-i(\boldsymbol{k}\,.\,\boldsymbol{r})+i(\boldsymbol{q}\,.\,\boldsymbol{r})-Zr].$$

From equation (2) it follows that $q_{\text{eff}}r \sim 1$. Since, moreover, $r \gg r_p$ (the dimensions of the K-orbit are much larger than the dimensions of the nucleus) we have $|(\boldsymbol{q}\,.\,\boldsymbol{r}_p)| \ll 1$. We can thus expand the exponent in the integral in powers of $(\boldsymbol{q}\,.\,\boldsymbol{r}_p)$ and discard all terms from the third onwards,

$$\int d\tau\,\Psi_1^*\sum_p \exp[-i(\boldsymbol{q}\,.\,\boldsymbol{r}_p)]\,\Psi_0 \approx \int d\tau\,\Psi_1^*\left[Z-i\left(\boldsymbol{q}\,.\sum_p \boldsymbol{r}_p\right)\right]\Psi_0$$

$$= -i\left(\boldsymbol{q}\int d\tau\,\Psi_1^*\,.\sum_p \boldsymbol{r}_p\right)\Psi_0 = -i(\boldsymbol{q}\,.\,\boldsymbol{d}_{01}),$$

where we have used the mutual orthogonality of the nuclear wave functions Ψ_1^* and Ψ_0 and where we have denoted by $\boldsymbol{d}_{01}$ the matrix element of the nuclear dipole moment.

If we integrate over the electron coordinates, we get

$$\int dV \exp[i(\boldsymbol{q}-\boldsymbol{k}\,.\,\boldsymbol{r})-Zr] = \frac{8\pi Z}{[Z^2+(\boldsymbol{q}-\boldsymbol{k})^2]^2}.$$

We get thus

$$V_{01} = i\frac{Z^{\frac{5}{2}}\sqrt{2}}{\pi^3}\left(\boldsymbol{d}_{01}\,.\int\frac{d^3\boldsymbol{q}}{q^2}\,\frac{\boldsymbol{q}}{[Z^2+(\boldsymbol{q}-\boldsymbol{k})^2]^2}\right).$$

In the integral over $\boldsymbol{q}$ we introduce a new variable $\boldsymbol{q}' = \boldsymbol{q}-\boldsymbol{k}$:

$$\int\frac{\boldsymbol{q}'+\boldsymbol{k}}{(\boldsymbol{q}'+\boldsymbol{k})^2}\,.\,\frac{d^3\boldsymbol{q}'}{[Z^2+q'^2]^2}.$$

The integrand has a maximum near $q' = 0$. The width of this maximum is of the order Z. Since the final velocity of the electron is much larger than its velocity in the atom, we have $k \gg Z$. We can to a first approximation neglect therefore the terms in $\boldsymbol{q}'$, and we find

$$\frac{\boldsymbol{k}}{k^2}\int\frac{d^3\boldsymbol{q}'}{(Z^2+q'^2)^2} = \frac{\pi^2}{Z}\frac{\boldsymbol{k}}{k^2}.$$

We get thus finally

$$V_{01} = i\frac{\sqrt{2}}{\pi}Z^{\frac{3}{2}}\frac{(\boldsymbol{k}\,.\,\boldsymbol{d}_{01})}{k^2}.$$

The probability that the electron leaves within a solid angle $d\Omega$ is equal to the following expression:

$$dw = 2\pi |V_{01}|^2 k^2 \frac{dk}{dE} d\Omega = 2\pi \frac{2}{\pi^2} Z^3 \frac{|(\boldsymbol{k}.\boldsymbol{d}_{01})|^2}{k^4} k\, d\Omega = \frac{4Z^3}{\pi} \frac{|(\boldsymbol{k}.\boldsymbol{d}_{01})|^2}{k^3} d\Omega$$

where we have used the general formula for the transition probability under the action of a constant perturbation.

If we integrate this expression over all angles, we get the total probability of ejecting an electron per unit time

$$w = \frac{16}{3} \frac{Z^3}{k} |\boldsymbol{d}_{01}|^2,$$

or, using ordinary units,

$$w = \frac{16}{3} \frac{\mu^3 e^6}{\hbar^7} \frac{Z^3 e^2}{\hbar v} |\boldsymbol{d}_{01}|^2.$$

On the other hand the probability for dipole radiation is

$$w_{\text{rad}} = \frac{4}{3} \frac{\omega^3}{\hbar c^3} |\boldsymbol{d}_{01}|^2,$$

where ω is the frequency of the radiation.

We find thus that the internal conversion coefficient, which is defined as

$$\alpha = \frac{w}{w + w_{\text{rad}}} = \frac{w/w_{\text{rad}}}{1 + w/w_{\text{rad}}},$$

is equal to

$$\alpha = \frac{4\left(\frac{Ze^2}{\hbar c}\right)^3 \frac{e^2}{\hbar v} \left(\frac{\mu c^2}{\hbar \omega}\right)^3}{1 + 4\left(\frac{Ze^2}{\hbar c}\right)^3 \frac{e^2}{\hbar v} \left(\frac{\mu c^2}{\hbar \omega}\right)^3}.$$

75. The first inequality shows that the wave function of the atom, that is of the electron shell, does not have time to change appreciably during the duration of the kick τ. The second inequality shows that the nucleus can be assumed to have stayed practically fixed during the kick. To find the required transition probability we must expand the original wave function of the electron shell ψ_0 (see first condition!) in terms of the eigenfunctions of the shell with respect to the moving nucleus; each of these functions describes a stationary state of the moving atom. (The set of these functions includes, of course, also the functions corresponding to the continuous spectrum, that is, describing states of the ionised atom.) The coefficients of this expansion determine also according to the general rules of quantum mechanics the probability that the atom is in the corresponding states.

We can obtain this expansion either in a system of coordinates in which the nucleus is originally at rest or, more conveniently, in a system of coordinates which is moving with the nucleus after the kick. In this last system the eigenfunctions of the possible final states are the set of the ordinary stationary wave functions of an atom at rest,

$$\psi_n(\boldsymbol{r}_1, \boldsymbol{r}_2, \ldots, \boldsymbol{r}_i, \ldots, \boldsymbol{r}_N), \tag{1}$$

where $\boldsymbol{r}_i$ is the coordinate of the ith electron of the shell with respect to the nucleus, and where the index n denotes the totality of quantum numbers characterising the stationary states of the atom. The initial wave function can be transformed, if we go over to our present system of coordinates, to the form

$$\psi_0' = \exp\left(-i\mu \boldsymbol{v} \,.\, \sum_i \boldsymbol{r}_i/\hbar\right) \psi_0(\boldsymbol{r}_1, \boldsymbol{r}_2, \ldots, \boldsymbol{r}_i, \ldots, \boldsymbol{r}_N), \tag{2}$$

where μ is the mass of an electron.

Indeed, the exponential factor is the wave function of the centre of gravity of the electron shell which in our chosen system of coordinates moves clearly with a velocity $-\boldsymbol{v}$ while $\psi_0(\boldsymbol{r}_1, \ldots, \boldsymbol{r}_N)$ is the wave function of the shell in its own centre of mass system. (We have

$$\psi_{\text{c.o.m.}} = \exp\left[\frac{i}{\hbar}(\boldsymbol{P}_{\text{c.o.m.}} \cdot \boldsymbol{R}_{\text{c.o.m.}})\right] = \exp\left\{\frac{i}{\hbar}\left(\sum_i -\mu_i \boldsymbol{v} \,.\, \frac{\sum_i \mu_i \boldsymbol{r}_i}{\sum_i \mu_i}\right)\right\},$$

from which we get the required expression for $\exp[i(\boldsymbol{q} . \boldsymbol{r})]$ because $\mu_1 = \mu_2 = \ldots = \mu_i = \ldots \equiv \mu$.) Because of our second condition the coordinates $\boldsymbol{r}_i$ in expression (2) can be taken from the same origin as in expression (1).

The required expansion of ψ_0' is of the form

$$\psi_0'(\boldsymbol{r}_1, \ldots, \boldsymbol{r}_i, \ldots, \boldsymbol{r}_N) = \sum_n c_n \psi_n(\boldsymbol{r}_1, \ldots, \boldsymbol{r}_i, \ldots, \boldsymbol{r}_N).$$

If we multiply both sides of this equation by

$$\psi_n^* \, d\tau_1 \ldots d\tau_i \ldots d\tau_N \quad (d\tau_i = dx_i \, dy_i \, dz_i),$$

integrate over all electron coordinates, and sum over all their spins, we get, assuming the orthonormality of the functions $\psi_n, \psi_m, \ldots$,

$$c_n = \int \ldots \int \psi_n^*(\boldsymbol{r}_1, \ldots, \boldsymbol{r}_i, \ldots, \boldsymbol{r}_N)$$

$$\times \exp\left[i\left(\boldsymbol{q} \cdot \sum_i \boldsymbol{r}_i\right)\right] \psi_0(\boldsymbol{r}_1, \ldots, \boldsymbol{r}_i, \ldots, \boldsymbol{r}_N) \, d\tau_1 \ldots d\tau_i \ldots d\tau_N.$$

The required probability for a transition to the state n is equal to

$$w_n = |c_n|^2 = \left| \int \psi_n^* \exp\left[i\left(\boldsymbol{q} \cdot \sum_{i=1}^{N} \boldsymbol{r}_i \right) \right] \psi_0 \, d\tau \right|^2, \tag{3}$$

where we have introduced the notation

$$\boldsymbol{q} = -\frac{\mu \boldsymbol{v}}{\hbar}, \quad d\tau = \prod_{i=1}^{N} d\tau_i.$$

We can easily generalise our result to the case of a transition to the continuous spectrum, that is, to the case of ionisation of the atom.

We note that provided $qa \ll 1$ the probability (3) to excite an atom "by a kick" is proportional to the probability for the corresponding optical transition. Indeed, in that case one can expand the exponential factor and retain only the first two terms,

$$\exp\left[i\left(\boldsymbol{q} \cdot \sum_i \boldsymbol{r}_i \right) \right] \approx 1 + i\left(\boldsymbol{q} \cdot \sum_i \boldsymbol{r}_i \right).$$

The first term gives zero since ψ_0 and ψ_n are orthogonal to one another. If we choose the z-axis in the direction of $\boldsymbol{q}$ we get

$$w_n \approx q^2 \left| \int \psi_n^* \left(\sum_i z_i \right) \psi_0 \, d\tau \right|^2,$$

which is proportional to the probability for the optical transition $0 \to n$.

76. According to equation (3) of the preceding problem the probability for a transition of the hydrogen atom to the nth stationary state is equal to

$$w_n = \left| \int \psi_n^* \exp\left[i(\boldsymbol{q} \cdot \boldsymbol{r}) \right] \psi_0 \, d^3 \boldsymbol{r} \right|^2, \tag{1}$$

where $\psi_0 = (\pi a^3)^{-\frac{1}{2}} \exp(-r/a)$ ($a = \hbar^2/\mu e^2$ is the first Bohr radius), and $\boldsymbol{q} = -\mu \boldsymbol{v}/\hbar$ ($\boldsymbol{v}$ is the recoil velocity of the proton). Since $\boldsymbol{v} = \boldsymbol{p}/M$, where M is the proton mass, we have

$$\boldsymbol{q} = -\frac{\mu}{M} \frac{\boldsymbol{p}}{\hbar}. \tag{2}$$

The required total probability of excitation and ionisation of the atom is clearly equal to $1 - w_0$, where w_0 is the probability that the atom stays in the ground state which according to equation (1) is equal to

$$w_0 = \left| \int \psi_0^2 \exp\left[i(\boldsymbol{q} \cdot \boldsymbol{r}) \right] d^3 \boldsymbol{r} \right|^2,$$

which after substitution for ψ_0 leads to

$$w_0 = \frac{1}{(\pi a^3)}\left|\int \exp\left[-\frac{2r}{a}+i(\boldsymbol{q}.\boldsymbol{r})\right]d^3\boldsymbol{r}\right|^2. \tag{3}$$

The integral occurring in this equation has been evaluated before (see problem 74 of section 7); it is equal to

$$\int \exp\left[-\frac{2r}{a}+i(\boldsymbol{q}.\boldsymbol{r})\right]d^3\boldsymbol{r} = \frac{16\pi a^3}{(4+q^2a^2)^2}.$$

The probability that the atom stays in the ground state is thus equal to

$$w_0 = \frac{1}{(1+\frac{1}{4}q^2a^2)^4}, \tag{4}$$

and the total probability for excitation and ionisation of the atom is equal to

$$1-w_0 = 1-\frac{1}{(1+\frac{1}{4}q^2a^2)^4}. \tag{5}$$

As should be expected, in the limiting cases of weak ($\frac{1}{2}qa \ll 1$) and strong ($\frac{1}{2}qa \gg 1$) kicks the probability (5) tends respectively to zero or unity. The corresponding limiting expressions are of the form

$$\tfrac{1}{2}qa \ll 1:\quad 1-w_0 \approx q^2a^2,$$

$$\tfrac{1}{2}qa \gg 1:\quad 1-w_0 \approx 1-\left(\frac{2}{qa}\right)^8.$$

We note that these results can be seen qualitatively without explicit calculations from equation (3) for w_0: this quantity is small or nearly equal to unity according to whether or not the factor $\exp[i(\boldsymbol{q}.\boldsymbol{r})]$ oscillates fast over the range of the factor $\exp(-2r/a)$, and the measure of whether the oscillation is fast is the quantity $qa/2$.

We formulated in the preceding problem the criterion for the applicability of the approximation we have used. Since the ionisation of the atom plays an essential role in the problem under consideration we must slightly alter the first part of the criterion, namely, the criterion of the suddenness of the impact. The length of the impact τ must, namely, be small compared to the Bohr period of the most important transitions,

$$\tau \ll \frac{\hbar}{|E_k-E_0|}, \tag{6}$$

where E_0 and E_k are the energies of the initial and the final state. In our case $E_0 = -\mu e^4/2\hbar^2$, and E_k is equal to the kinetic energy of the electron which is flying away.

The effective value of E_k can be estimated from equation (1).

For $qa \gg 1$ the integral in equation (1) is appreciably different from zero only for those states whose wave functions, ψ_n, contain a factor of the kind $\exp[i(\boldsymbol{k}.\boldsymbol{r})]$ with $\boldsymbol{k} \approx \boldsymbol{q}$, because only under those circumstances can the fast oscillating factors in the function under the integral sign cancel each other's effect. In that case we have thus

$$(E_k)_{\text{eff}} \sim \frac{\hbar^2 q^2}{\mu} \sim \frac{\mu}{M} E_p, \tag{7}$$

where E_p is the recoil energy of the proton.

If $qa \ll 1$ the value of the integral in equation (1) is effectively determined, as can easily be understood, by the degree of overlap of the functions ψ_0 and ψ_n^*, so that an appreciable transition probability is obtained only to final states with $k \lesssim 1/a$, so that

$$(E_k)_{\text{eff}} \sim \frac{\hbar^2}{\mu a^2} \sim E_0. \tag{8}$$

Either equations (6) and (7) or equations (6) and (8), depending on the value of qa, that is, of p, give thus the required condition for the applicability of the "ballistic" approximation.

77. Since the protons, which capture the negative mesons, are approximately uniformly distributed over the nuclear volume, we find that the required probability is clearly equal to the ratio

$$\frac{\int_{V_{\text{nucl}}} |\psi(\boldsymbol{r})|^2 \, d^3\boldsymbol{r}}{\int |\psi(\boldsymbol{r})|^2 \, d^3\boldsymbol{r}}, \tag{1}$$

where ψ is the wave function of the meson in its K-orbit.

The integration in the denominator is over the whole space, and in the enumerator over the nuclear volume V_{nucl} which is approximately proportional to Z ($V_{\text{nucl}} = \frac{4}{3}\pi R^3 = \frac{4}{3}\pi(r_0 A^{\frac{1}{3}})^3 = \frac{4}{3}\pi r_0^3 A \approx \frac{4}{3}\pi r_0^3 . 2Z$, where $r_0 = \text{const} \approx 1{\cdot}2 . 10^{-13}$ cm).

We shall assume that the nuclear radius R is small compared to the radius of the K-orbit of the meson $a = \hbar^2/Z\mu e^2$, where μ is the meson mass. Taking into account the fact that the mesonic wave function varies appreciably only over distances of the order of magnitude of a, we get for the enumerator $\approx |\psi(0)|^2 . V_{\text{nucl}}$, where $\psi(0)$ is the value of the mesonic wave function at the origin (that is, inside the nucleus).

If we use the normalised wave function (which makes the denominator equal to unity),

$$\psi(\boldsymbol{r}) = \frac{1}{\sqrt{(\pi a^3)}} \exp(-r/a),$$

it is immediately clear (since $a \sim 1/Z$) that the ratio (1), and thus also the required probability, is proportional to

$$Z^3 . Z = Z^4.$$

We note that our approximation $a \gg R$ is not valid for heavy nuclei. Indeed, since

$$R \approx r_0 A^{\frac{1}{3}} \approx r_0 (2Z)^{\frac{1}{3}},$$

where

$$r_0 \approx \tfrac{1}{2}(\tfrac{1}{137})^2 \text{ at un}$$

and $\mu \approx 200$ at un the inequality $a \gg R$ is equivalent to (compare problem 15 of section 5)

$$Z \ll 45.$$

78. The potential φ to be derived is the sum of the potential $\varphi_p = e/r$, due to the proton, and the potential $\overline{\varphi_e}$, due to the electron {the averaging is over the wave function of the hydrogen atom ground state $\psi_0 = [1/\sqrt{(\pi a^3)}] \exp(-r/a)$, with $a = \hbar^2/\mu e^2$},

$$\varphi(\boldsymbol{r}) = \varphi_p(\boldsymbol{r}) + \overline{\varphi_e}(\boldsymbol{r}) = \frac{e}{r} - e \int \frac{\psi_0^2(\boldsymbol{r}')}{|\boldsymbol{r} - \boldsymbol{r}'|} d^3 \boldsymbol{r}'. \tag{1}$$

The potential $\overline{\varphi_e}$, is, of course, the same as the potential of a static electron "cloud" of density $\rho(\boldsymbol{r}') = -e\psi_0^2(\boldsymbol{r}')$. We can thus simply find $\overline{\varphi_e}$ as the spherically symmetric solution of Poisson's equation

$$\frac{1}{r}\frac{d^2(r\overline{\varphi_e})}{dr^2} = -4\pi\rho(r) = 4\pi e \frac{1}{\pi a^3} \exp(-2r/a),$$

from which $\overline{\varphi_e}$ is easily obtained by integrating twice from r to ∞. The integral in equation (1) is most easily evaluated by using for $1/|\boldsymbol{r} - \boldsymbol{r}'|$ an expansion in Legendre functions:

if $r' < r$,

$$\frac{1}{|\boldsymbol{r} - \boldsymbol{r}'|} = \frac{1}{r} \sum_{l=0}^{\infty} P_l(\cos\theta') \left(\frac{r'}{r}\right)^l,$$

if $r' > r$,

$$\frac{1}{|\boldsymbol{r} - \boldsymbol{r}'|} = \frac{1}{r'} \sum_{l=0}^{\infty} P_l(\cos\theta') \left(\frac{r}{r'}\right)^l,$$

where θ' is the angle between $\boldsymbol{r}'$ and $\boldsymbol{r}$.

If we split the integration over r' into two parts ($r' \leqslant r$ and $r' \geqslant r$), we see that in both parts integration over the angle θ' reduces the sums $\sum_{l=0}^{\infty}$

to their first term ($l = 0$) by virtue of the orthogonality of the Legendre functions, and we get

$$\varphi(r) = \frac{e}{r} - \frac{4e}{a^3}\left(\frac{1}{r}\int_0^r \exp(-2r'/a)\, r'^2\, dr' + \int_r^\infty \exp(-2r'/a)\, r'\, dr'\right).$$

As was clear from the start, the potential $\overline{\varphi_e}$ (and thus also φ) is spherically symmetric. If we perform the integration we find finally

$$\varphi(r) = e\left(\frac{1}{r} + \frac{1}{a}\right)\exp(-2r/a).$$

In the limits $r \ll a$ and $r \gg a$ we get the translucent result $\varphi \approx e/r$ (Coulomb field of the proton) and $\varphi \approx (e/a)\exp(-2r/a)$ (proton practically completely screened by the electron).

79. Let $\psi(x, y, z)$ be a solution of the Schrödinger equation referring to the discrete energy spectrum. Let us consider the one-parameter set of normalised functions of the form $\lambda^{\frac{3}{2}}\psi(\lambda x, \lambda y, \lambda z)$.

The expression

$$I(\lambda) = \lambda^3 \int \left[\frac{\hbar^2}{2\mu}|\nabla\psi(\lambda x, \lambda y, \lambda z)|^2 + V(x, y, z)\,|\psi(\lambda x, \lambda y, \lambda z)|^2\right] dx\, dy\, dz,$$

considered as a function of λ, must have an extremum for $\lambda = 1$, that is

$$\left(\frac{dI}{d\lambda}\right)_{\lambda=1} = 0.$$

If we introduce new variables of integration $\lambda x, \lambda y, \lambda z$, we get

$$I(\lambda) = \lambda^2\,\overline{T} + \lambda^{-\nu}\,\overline{V}.$$

Hence we find

$$2\overline{T} - \nu\overline{V} = 0.$$

The virial theorem can easily be generalised to the case of a system of many particles.

80. Let μ be the electron mass, ξ_0, η_0, ζ_0, the coordinates of the nucleus, and ξ_i, η_i, ζ_i ($i = 1, 2, \ldots, n$) the coordinates of the ith electron.

We introduce centre of mass coordinates,

$$X = \frac{M\xi_0 + \mu\sum_{i=1}^{n}\xi_i}{M + n\mu} \quad \text{(similarly for } Y, Z\text{)} \tag{1}$$

and relative coordinates ($i = 1, 2, \ldots, n$)

$$x_i = \xi_i - \xi_0 \quad \text{(similarly for } y_i, z_i\text{)}, \tag{2}$$

and get then

$$\frac{\partial}{\partial \xi_i} = \frac{\mu}{M+n\mu}\frac{\partial}{\partial X} + \frac{\partial}{\partial x_i}, \quad \frac{\partial}{\partial \xi_0} = \frac{M}{M+n\mu}\frac{\partial}{\partial X} - \sum_{i=1}^{n}\frac{\partial}{\partial x_i}. \tag{3}$$

The Hamiltonian of the atom (in the fixed system of coordinates) is of the form

$$\hat{H} = -\frac{\hbar^2}{2M}\left(\frac{\partial^2}{\partial \xi_0^2} + \frac{\partial^2}{\partial \eta_0^2} + \frac{\partial^2}{\partial \zeta_0^2}\right) - \frac{\hbar^2}{2\mu}\sum_{i=1}^{n}\left(\frac{\partial^2}{\partial \xi_i^2} + \frac{\partial^2}{\partial \eta_i^2} + \frac{\partial^2}{\partial \zeta_i^2}\right) + V, \tag{4}$$

where V is the potential energy of the atom which depends only on the relative coordinates (2).

If we introduce into expression (4) the new variables (1), (2), and (3), we get for $\hat{H}$ two terms, one depending only on the centre of mass coordinates, and the other only on the relative coordinates. Separating in the usual way the centre of mass motion, which is of no interest, we get the following Schrödinger equation in the centre of mass system:

$$\left[-\frac{\hbar^2}{2\mu}\sum_{i=1}^{n}\nabla_i^2 - \frac{\hbar^2}{2M}\sum_{i=1}^{n}\sum_{k=1}^{n}\left(\frac{\partial^2}{\partial x_i \partial x_k} + \frac{\partial^2}{\partial y_i \partial y_k} + \frac{\partial^2}{\partial z_i \partial z_k}\right)\right. \tag{5}$$

$$\left. + V(x_1, \ldots, x_n)\right] \Psi(x_1, \ldots, x_n) = E\Psi(x_1, \ldots, x_n).$$

The required effect of the finite mass M (or, in other words, of the motion of the nucleus) is caused by the second term within the square brackets. Clearly this effect produces a correction (compared to the terms for which $M = \infty$) of relative order μ/M.

If we take together the terms with $i = k$ and $i \neq k$, and introduce the reduced mass

$$\mu_r = \frac{\mu M}{\mu + M}, \tag{6}$$

we get

$$\left[-\frac{\hbar^2}{2\mu_r}\sum_i \nabla_i^2 - \frac{\hbar^2}{M}\sum_{i<k}\sum\left(\frac{\partial^2}{\partial x_i \partial x_k} + \frac{\partial^2}{\partial y_i \partial y_k} + \frac{\partial^2}{\partial z_i \partial z_k}\right)\right.$$

$$\left. + V(x_1, \ldots, x_n)\right] \Psi(x_1, \ldots, x_n) = E\Psi(x_1, \ldots, x_n). \tag{7}$$

From this equation it follows immediately that the effect of a finite value of M leads first of all to replacing the electron μ by the reduced mass μ_r and secondly to an additional perturbing term

$$\hat{H}' = -\frac{\hbar^2}{M}\sum_{i<k}\sum\left(\frac{\partial^2}{\partial x_i \partial x_k} + \frac{\partial^2}{\partial y_i \partial y_k} + \frac{\partial^2}{\partial z_i \partial z_k}\right). \tag{8}$$

The first of these modifications of the Schrödinger equation expresses the so-called "elementary" or "normal" effect of the moving nucleus; this effect is also present in the hydrogen atom. It is obvious that the normal effect affects all terms in the same way, and reduces the frequency of all spectral lines of the atom in the ratio

$$\frac{\mu_r}{\mu} = \frac{M}{M+\mu} \approx 1 - \frac{\mu}{M}.$$

The second effect, however, is essentially different for different states of the atom. Indeed, first-order perturbation theory which is applicable here as $\mu/M \ll 1$ leads to an energy correction

$$\Delta E = \int \Psi^{(0)*} \hat{H}' \Psi^{(0)} \, d\tau, \tag{9}$$

where $\Psi^{(0)}$ is the wave function of the atom for $M = \infty$, and

$$d\tau = \prod_{i=1}^{n} d\tau_i = \prod_{i=1}^{n} dx_i \, dy_i \, dz_i$$

is a volume element of the configuration space of the atom.

Substituting expression (8) into equation (9), interchanging summation and integration, integrating by parts, and using the fact that $\Psi \to 0$ at infinity we get

$$\Delta E = \frac{\hbar^2}{M} \sum_{i<k}\sum \int (\nabla_i \Psi^{(0)*} . \nabla_k \Psi^{(0)}) \, d\tau, \tag{10}$$

which clearly depends strongly on the form of $\Psi^{(0)}$.

In particular, if the motions of the electrons were completely independent, that is, if the unperturbed wave function of the atom were simply a product of wave functions φ_i of separate electrons,

$$\Psi^{(0)} = \prod_{i=1}^{n} \varphi_i(x_i, y_i, z_i), \tag{11}$$

the effect would in general vanish. Indeed, from equations (10) and (11) and the normalisation relation $\int \varphi_i^* \varphi_i \, d\tau_i = 1$, it follows that

$$\Delta E = \frac{\hbar^2}{M} \sum_{i<k}\sum \left(\int \varphi_i \nabla \varphi_i^* \, d\tau_i . \int \varphi_k^* \nabla \varphi_k \, d\tau_k \right) = 0,$$

since both integrals in this expression are equal to zero (see problem 16 of section 5).

In actual fact, the electrons in an atom are not independent so that this effect—which in contradistinction to the "normal" effect is called the "specific" effect for a finite mass, and which is characteristic for atoms with $n \geqslant 2$—is, generally speaking, not only different from zero but in a number of cases exceeds appreciably the "normal" effect.

The correlation between the electronic motions is caused partly—trivially—by the electrostatic interaction between electrons, partly also, and mainly, by the fact that they are identical particles; this part of the specific effect, which is basically and essentially a quantum phenomenon, can thus be called an "exchange" effect.

Indeed, the total wave function of the atom (neglecting spin–orbit interaction, and thus written as a product of coordinate and spin functions) must be antisymmetric in any two electrons. To each value of the total spin S of an electron shell (and thus to each kind of symmetry with respect to the permutation group of the spin function) there corresponds thus a well-defined permutation symmetry of the coordinate function of the shell. Using a rough classical analogy we can characterise this last fact by a preferential motion of the electrons which is either parallel (when the coordinate function is symmetric) or antiparallel (when the coordinate function is antisymmetric). Since the centre of mass of the atom is at rest it is clear that in the first case the nucleus must move considerably, but only very little in the second case. In other words, in the first case the specific (or, rather, exchange) effect will add to the normal effect, while in the second case it will act in the opposite direction.

We note that both effects play an essential role for the so-called isotope shifts of spectral lines of very light elements (since the mass M is different for different isotopes, and the difference will be the more pronounced the lighter the element).

A quantitative consideration of the exchange effect of the moving nucleus will be given in the next problem for the case of an atom (or ion) with two electrons.

81. The neglect of the electrostatic interaction between the electrons is equivalent to choosing for the unperturbed coordinate function of the two-electron system a superposition of products of hydrogen-like functions. Since the total wave function must be anti-symmetric in the two electrons, it follows that the coordinate function must be of the form

$$\Psi^{(0)}(1; 2) = \frac{1}{\sqrt{2}}[\varphi_{100}(1)\,\varphi_{nlm}(2) \pm \varphi_{nlm}(1)\,\varphi_{100}(2)], \tag{1}$$

where 1 and 2 indicate simply all coordinates of the electrons, while $\varphi_{100} \equiv u$ and $\varphi_{nlm} \equiv v$ are hydrogen-like electron functions corresponding to the given quantum numbers; the first one corresponds to some effective charge Z_i, and the second one to an effective charge Z_o (in general $Z_o \neq Z_i$; the indices "i" and "o" correspond to "inner" and "outer" electron, in correspondence with the qualitative meaning of φ_{100} and φ_{nlm}, which describe the ground state and an excited state in an atom); the factor $2^{-\frac{1}{2}}$ is a normalisation factor.

The upper sign in equation (1) corresponds clearly to an anti-symmetric spin state of the two-electrons (para-terms, $S = 0$) and the lower sign to a symmetric spin state (ortho-terms, $S = 1$).

If we substitute expression (1) into equation (10) of the preceding problem (with $i, k = 1, 2$), we get

$$\Delta E = \frac{\hbar^2}{M}\iint v^*(2)\{\nabla u^*(1).[u(1)\nabla v(2) \pm v(1)\nabla u(2)]\}\, d\tau_1\, d\tau_2.$$

The first term gives zero since (see problem 16 of section 5)

$$\int v^*(i)\nabla v(i)\, d\tau_i = 0$$

and the same for $u(i)$. The second term gives

$$\Delta E = \pm\frac{\hbar^2}{M}\int\left(v(1)\,\nabla u^*(1)\, d\tau_1 . \int v^*(2)\,\nabla u(2)\right) d\tau_2$$

$$= \pm\frac{\hbar^2}{M}\left|\int v(i)\,\nabla u^*(i)\, d\tau_i\right|^2. \tag{2}$$

This is finally the general expression for the required first-order exchange correction for a finite nuclear mass.

Let us find the selection rule for expression (2). The functions u and v satisfy the following equations (in atomic units):

$$\left(-\tfrac{1}{2}\nabla^2 - \frac{Z_i}{r}\right) u = E_1 u, \tag{3}$$

$$\left(-\tfrac{1}{2}\nabla^2 - \frac{Z_o}{r}\right) v = E_n v, \tag{4}$$

where

$$E_n = -\frac{Z_o^2}{2n^2}, \quad E_1 = -\frac{Z_i^2}{2}. \tag{5}$$

Multiplying equation (3) by $\boldsymbol{r}v^*$, and the complex conjugate of equation (4) by $\boldsymbol{r}u$, subtracting the two results and integrating the resulting expression over the whole of space, we get

$$\frac{1}{2}\int \boldsymbol{r}(u\nabla^2 v^* - v^* \nabla^2 u)\, d\tau = (Z_i - Z_o)\int \frac{\boldsymbol{r}}{r} uv^*\, d\tau + (E_1 - E_n)\int \boldsymbol{r}uv^*\, d\tau. \tag{6}$$

Transforming the left-hand side of this equation, we get

$$\frac{1}{2}\int \boldsymbol{r}(u\nabla^2 v^* - v^* \nabla^2 u)\, d\tau = \frac{1}{2}\int \boldsymbol{r}(\nabla . (u\nabla v^* - v^* \nabla u))\, d\tau$$

$$= -\frac{1}{2}\int (u\nabla v^* - v^* \nabla u)\, d\tau = \int v^* \nabla u\, d\tau$$

where we have used the equations

$$\int \boldsymbol{r}(\nabla . \boldsymbol{A})\, d\tau = -\int \boldsymbol{A}\, d\tau \quad \text{and} \quad \int \operatorname{grad}\chi\, d\tau = \oint \chi\, d^2\boldsymbol{S}.$$

(One sees the validity of these easily by multiplying both sides by a constant vector integrating the equation so obtained, and using Gauss' theorem.) Equation (6) becomes then [if $Z_i = Z_o$, so that u and v are eigenfunctions of the same Hamiltonian, equation (7) is the same as equation (2) of problem 16 of section 5]

$$\int v\nabla u^*\, d\tau = (Z_i - Z_o)\int \frac{\boldsymbol{r}}{r} u^* v\, d\tau + (E_1 - E_n)\int \boldsymbol{r}u^* v\, d\tau. \tag{7}$$

The selection rule for both integrals on the right-hand side is the orthogonality relation of the spherical harmonics $Y_{lm}(\theta, \varphi)$ which occur in them. It is clear that both integrals vanish, unless

$$l_u - l_v = \pm 1.$$

Since we have assumed that one of the electrons is in a 1s-state (which is practically always the case), so that $l_u = 0$ we conclude that the left-hand side of equation (7), and thus also the exchange correction (2), vanishes unless $l_v = 1$, that is, unless the second electron is in a p-state. In other words, the specific (or rather the exchange) effect of a moving nucleus is only present for the P-terms of two-electron atoms or ions (He, Li^+Be^{++}, ...), at any rate, in first order in μ/M.

As far as the sign of the specific effect is concerned, we see from equation (2) that for the para-terms $\Delta E > 0$, that is, the specific effect has the same sign as the normal effect, while for ortho-terms their signs are opposite. This is in complete agreement with the qualitative conclusions reached at the end of the preceding problem, since in the para-states both electrons move, roughly speaking, mainly in the same direction, while in the ortho-states they are moving mainly in the opposite direction.

To find a numerical value for the specific effect, we use equations (2) and (7). From equation (7) it follows that

$$\left|\int v\nabla u^*\,d\tau\right|^2 = \left|(Z_i-Z_o)\int \frac{x}{r}u^*\,v\,d\tau+(E_1-E_n)\int xu^*\,v\,d\tau\right|^2$$
$$+\left|(Z_i-Z_o)\int \frac{y}{r}u^*\,v\,d\tau+(E_1-E_n)\int yu^*\,v\,d\tau\right|^2$$
$$+\left|(Z_i-Z_o)\int \frac{z}{r}u^*\,v\,d\tau+(E_1-E_n)\int zu^*\,v\,d\tau\right|^2. \tag{8}$$

The integrals in this equation can easily be evaluated. We express the hydrogen-like wave functions in spherical polars and use atomic instead of Coulomb units [this means changing $\boldsymbol{r}$(Coul un)$\to Z\boldsymbol{r}$(at un), R_{nl}(Coul un)$\to Z^{-\frac{3}{2}}R_{nl}$(at un)], and we get

$$u\equiv\psi_{100}(\boldsymbol{r}) = R_{10}(r)\,Y_{00}(\theta,\varphi), \tag{9}$$
$$v\equiv\psi_{nlm}(\boldsymbol{r}) = R_{nl}(r)\,Y_{lm}(\theta,\varphi), \tag{10}$$

where Y_{lm} is the normalised spherical harmonic and R_{nl} the normalised radial function; in v we shall put $l = 1$ in agreement with the selection rule,

$$R_{n1} = Z_o^{\frac{3}{2}}\frac{2}{3!\,n^3}\sqrt{\left[\frac{(n+1)!}{(n-2)!}\right]}\,2Z_o r\exp(-Z_o r/n)\,F\left(-n+2, 4, \frac{2Z_o\boldsymbol{r}}{n}\right)\text{(at un)}, \tag{11}$$

$$R_{10} = 2Z_i^{\frac{3}{2}}\exp(-Z_i r) \quad \text{(at un)}. \tag{12}$$

From equations (9) and (10) it follows that the integrals in equation (8) are equal to:

$$\int \frac{x}{r}u^*\,v\,d\tau = \int \sin\theta\cos\varphi Y_{00}Y_{1m}\,d\Omega\int_0^\infty R_{10}R_{n1}r^2\,dr,$$
$$\int zu^*\,v\,d\tau = \int \cos\theta Y_{00}Y_{1m}\,d\Omega\int_0^\infty R_{10}R_{n1}r^3\,dr, \quad \text{and so on.}$$

The integration over the angles is elementary, if we use the fact that $Y_{00} = 1/\sqrt{(4\pi)}$, $\sin\theta\cos\varphi = \frac{1}{2}\sqrt{(8\pi/3)}(Y_{11}+Y_{1,-1})$ and so on, and use the orthonormality of the spherical harmonics. As to the radial integrals they contain, according to equations (11) and (12), exponents, powers, and the confluent hypergeometric function. Looking these integrals up in the standard literature we get finally the following expression (in ordinary units) for the specific effect for P-terms:

$$\Delta E = \pm\frac{64}{3}\frac{\mu}{M}(Z_i Z_o)^5\frac{(Z_i n-Z_o)^{2n-4}}{(Z_i n+Z_o)^{2n+4}}n^3(n^2-1)\frac{\mu e^4}{\hbar^2},$$

where n is the principal quantum number of the p-electron, and Z_i and Z_o are the effective nuclear charges for the $1s$- and np-electrons. These quantities are usually determined by the variational method. For the $2\,^3P$ term of the Li^+-ion, for instance, $Z_i = 2{\cdot}98$, $Z_o = 2{\cdot}16$. [These values of the effective charges are pretty well the same as those to be expected from elementary considerations based on (i) a bare charge of the Li nucleus of 3; (ii) the fact that the dimensions of the outer electron orbit ($n = 2$) are roughly speaking about $2^2 = 4$ times larger than those of the inner electron; (iii) the outer electron has $l = 1$ and does not therefore penetrate very far inside the inner electron orbit. In other words, the inner electron screens the nucleus from the outer electron so that we should expect $Z_i \approx 3, Z_o \approx 2$.]

82. The average value of the dipole moment of a system of N particles is equal to

$$\langle \boldsymbol{d} \rangle = \int \dots \int \psi^*(\boldsymbol{r}_1, \dots, \boldsymbol{r}_N) \left(\sum_{i=1}^{N} e_i \boldsymbol{r}_i \right) \psi(\boldsymbol{r}_1, \dots, \boldsymbol{r}_N)\, d\tau_1 \dots d\tau_N,$$

where e_i is the charge of the ith particle, $\sum_{i-1}^{N} e_i \boldsymbol{r}_i = \hat{\boldsymbol{d}}$ the dipole moment operator, which in the coordinate representation is simply equivalent to multiplying by $\boldsymbol{d} = \sum_i e_i \boldsymbol{r}_i$, and the integration is over the configuration space of all particles. If the system is in a state of well-defined parity, $\psi^* \psi$ is an even function in the coordinates of all the particles, while the dipole moment $\boldsymbol{d}$ is odd in those coordinates; that is, the complete expression under the integral sign is thus odd. Since the integration goes to infinity for each of the coordinates, that is, is over a symmetrical domain, it is clear that $\langle \boldsymbol{d} \rangle = 0$.

This result can be obtained also in a different (less rigorous) way. Since the reflection operator commutes with the operators of the angular momentum components and the latter commute with one another in the semi-classical case, that is, if $M_i \gg \hbar$ any state with a well-defined parity has a well-defined value for the angular momentum vector $\boldsymbol{J}$ (in particular, the spin of the system $\boldsymbol{I}$).

A non-vanishing angular momentum defines a definite direction in space, which must also be the direction of the (average) dipole moment $\langle \boldsymbol{d} \rangle$:

$$\langle \boldsymbol{d} \rangle = \text{const.}\, \boldsymbol{J}. \tag{1}$$

Since $\langle \boldsymbol{d} \rangle$ is a polar vector and $\boldsymbol{J}$ an axial vector, the (physically necessary) invariance of equation (1) under a reflection means that the constant in equation (1) must be a pseudoscalar (which in our case is clearly not so). The constant must thus be equal to zero, or $\langle \boldsymbol{d} \rangle = 0$.

83. The operator of the x-component of the dipole moment of a system of particles of mass μ and charge e is of the form $\hat{d}_x = e\sum_{i=1}^{N} x_i$, and its matrix elements d_{mn} which occur in the sum rule are equal to

$$(d_x)_{mn} = \int \psi_m^*(\boldsymbol{r}_1, \ldots, \boldsymbol{r}_N)\, \hat{d}_x \psi_n(\boldsymbol{r}_1, \ldots, \boldsymbol{r}_N)\, d^3\boldsymbol{r}_1 \ldots d^3\boldsymbol{r}_N,$$

where $\psi_m, \psi_n, \ldots$ form a complete orthonormal set of stationary wave functions for the system of particles under consideration (without time factors).

We shall introduce the operator of the time derivative of the dipole moment component $\hat{\dot{d}}$ (we drop the index x from now on) and use the formula

$$\dot{d}_{nm} = i\frac{E_n - E_m}{\hbar} d_{nm},$$

where E_n are the energy levels of the system of particles; we find easily the following chain of equations:

$$\sum_n (E_n - E_m)|d_{mn}|^2 = \sum_n (E_n - E_m)\, d_{nm}\, d_{mn}$$

$$= \frac{1}{2}\frac{\hbar}{i}\sum_n \dot{d}_{nm}\, d_{mn} - \frac{1}{2}\frac{\hbar}{i}\sum_n \dot{d}_{mn}\, d_{nm} = \frac{1}{2}\frac{\hbar}{i}(d\dot{d} - \dot{d}d)_{mm},$$

where we have used the fact that $\hat{d}$ is a Hermitean operator ($d_{mn}^* = d_{nm}$) and the multiplication rule for matrices.

If we take the definition of $\hat{d}$ into account and the fact that the coordinate operators $\hat{x}_i$ commute with the velocity operators

$$\hat{\dot{x}}_k = \frac{\hbar}{i\mu}\frac{\partial}{\partial x_k}$$

if $i \neq k$, while

$$\hat{x}_i \hat{\dot{x}}_i - \hat{\dot{x}}_i \hat{x}_i = \frac{i\hbar}{\mu},$$

we find

$$\sum_n (E_n - E_m)|d_{mn}|^2 = \frac{1}{2}\frac{\hbar}{i} e^2 \sum_{i=1}^{N} (x_i \dot{x}_i - \dot{x}_i x_i)_{mm} = \frac{\hbar^2 e^2}{2\mu} N,$$

which had to be proved.

We emphasise that the sum is independent of m, that is, it is the same for all "initial" states.

84. The proof follows by taking the mm-matrix element of the double commutator

$$\left[\left[H, \sum_j \exp\{i(\boldsymbol{q} \cdot \boldsymbol{r}_j)\}\right]_-, \sum_k \exp\{i(\boldsymbol{q} \cdot \boldsymbol{r}_k)\}\right]_-.$$

Note that the result of the previous problem follows from the sum rule obtained here by taking the dipole approximation, that is, putting $\exp\{i(\boldsymbol{q}\cdot\boldsymbol{r}_j)\} \cong 1+i(\boldsymbol{q}\cdot\boldsymbol{r}_j)$.

85. We write (compare problem 82 of section 7)

$$\sum_n (E_n - E_m)|p_{nm}|^2 = \frac{\hbar}{2i}(p\dot{p} - \dot{p}p)_{mm}$$

$$= \frac{\hbar}{2i}([p, \dot{p}]_-)_{mm}$$

$$= -\frac{\hbar^2}{2} V''_{mm},$$

where we have used the relations

$$i\hbar\hat{\dot{p}} = [\hat{p}, \hat{H}]_- = [\hat{p}, V]_- = -i\hbar V',$$

$$[\hat{p}, V']_- = -i\hbar V''.$$

86. The Schrödinger equation for an electron in a central field with potential energy V is

$$\nabla^2\psi + \frac{2\mu}{\hbar^2}(E-V)\psi = 0.$$

If we write the wave function for an eigenstate with quantum numbers n and l in the form

$$\psi_l^n = F_l^{(n)}(r)(ax+by+cz)^l,$$

we find that

$$a^2+b^2+c^2 = 0,$$

and that the $F_l^{(n)}$ satisfy the equations (primes indicate differentiation with respect to r)

$$F_l'' + \frac{2l+2}{r}F_l' + \frac{2\mu}{\hbar^2}(E-V)F_l = 0, \tag{1}$$

and

$$\int_0^\infty F_l^{(n)} F_l^{(n')} r^{2l+2}\, dr = \delta_{nn'}. \tag{2}$$

One finds easily that the f_a^b given by equation (B) of the problem satisfies the equation

$$(2l+1) f_{n,l}^{n',l+1} = \frac{2\mu(E_{n',l+1} - E_{n,l})}{3\hbar^2}(l+1)\left[\int_0^\infty r^{2l+4} F_{l+1}^{(n')} F_l^{(n)}\, dr\right]^2. \tag{3}$$

Multiplying (1) for $F_l^{(n)}$ by $r^{2l+4}F_{l+1}^{(n')}$ and for $F_{l+1}^{(n')}$ by $r^{2l+4}F_l^{(n)}$ and integrating the difference of the two expressions thus obtained over r, we get

$$\frac{2\mu}{\hbar^2}(E_{n,l+1}-E_{n,l})\int_0^\infty F_l^{(n)}F_{l+1}^{(n')}r^{2l+4}\,dr$$

$$= \int_0^\infty (F_{l+1}^{(n')}F_l^{(n)\prime\prime}-F_l^{(n)}F_{l+1}^{(n')\prime\prime})\,r^{2l+4}\,dr$$

$$+\int_0^\infty [(2l+2)\,F_{l+1}^{(n')}F_l^{(n)\prime}-(2l+4)\,F_l^{(n)}F_{l+1}^{(n')\prime}]\,r^{2l+3}\,dr,$$

and hence for the first sum from (A), using (3),

$$S_1 = -\frac{2}{3}\frac{l+1}{2l+1}\sum_{n'}\left[\int_0^\infty r^{2l+4}F_{l+1}^{(n')}F_l^{(n)}\,dr\right]\cdot\left[\int_0^\infty r^{2l+3}F_{l+1}^{(n')}F_l^{(n)\prime}\,dr\right].$$

If P and Q are functions of r, one finds by expanding these functions in terms of the $F_k^{(n)}$ that

$$\int_0^\infty PQr^{2k+2}\,dr = \sum_{n'}\left[\int_0^\infty PF_k^{(n')}r^{2k+2}\,dr\right]\left[\int_0^\infty QF_k^{(n')}r^{2k+2}\,dr\right]. \qquad (4)$$

If we now put

$$P_1 = F_l^{(n)},\quad Q_1 = \frac{1}{r}F_l^{(n)},\quad k = l+1,$$

we find

$$S_1 = -\frac{2}{3}\frac{l+1}{2l+1}\int P_1 Q_1 r^{2l+4}\,dr = \frac{1}{3}\frac{(l+1)(2l+3)}{2l+1},$$

as had to be proved.

The expression for S_2 can be proved similarly; we then use (4) with

$$P_2 = r^2 F_l^{(n)},\quad Q_2 = \frac{1}{r^{2l}}\frac{d}{dr}(r^{2l+1}F_l^{(n)}),\quad k = l-1.$$

Note that $S_1+S_2 = 1$, as should be the case.

Molecules

1. If we neglect the difference between the centre of mass of the molecule and the centre of mass of the nuclei and take into account the fact that the centre of mass is fixed in the origin of the system of coordinates, the Schrödinger equation for a diatomic molecule will have the following form:

$$\left\{-\frac{\hbar^2}{2\mu}\sum_i\left(\frac{\partial^2}{\partial x_i^2}+\frac{\partial^2}{\partial y_i^2}+\frac{\partial^2}{\partial z_i^2}\right)\right.$$

$$-\frac{\hbar^2}{2M\rho^2}\left[\frac{\partial}{\partial\rho}\left(\rho^2\frac{\partial}{\partial\rho}\right)+\frac{1}{\sin\theta}\frac{\partial}{\partial\theta}\left(\sin\theta\frac{\partial}{\partial\theta}\right)+\frac{1}{\sin^2\theta}\frac{\partial^2}{\partial\varphi^2}\right]$$

$$\left.+V(x_i, y_i, z_i;\ \rho, \theta, \varphi)\right\}\psi(\ldots x_i, y_i, z_i, \ldots \rho, \theta, \varphi) = E\psi.$$

Here x_i, y_i, z_i are the coordinates of the ith electron with respect to a fixed system of reference, the angles θ, φ determine the position in space of the line which connects the two nuclei, ρ is the distance between the nuclei, and M the reduced mass of the two nuclei. This equation is difficult to handle in so far as in the potential energy V of the electrostatic interaction, the angles θ and φ occur. In order to put the Schrödinger equation in a more convenient form we introduce a new system of coordinates, ξ, η, ζ, which rotates with the nuclei. The ζ-axis is along the line connecting the nuclei and the ξ-axis lies in the xy-plane. We take the positive ξ-axis in such a way that the z-, ζ-, and ξ-axes form a right-handed system. The connection between the old and the new coordinates has the following form:

$$\xi_i = -x_i \sin\varphi + y_i \cos\varphi,$$

$$\eta_i = -x_i \cos\theta\cos\varphi - y_i \cos\theta\sin\varphi + z_i \sin\theta,$$

$$\zeta_i = \quad x_i \sin\theta\cos\varphi + y_i \sin\theta\sin\varphi + z_i \cos\theta.$$

If we indicate the difference between differentiation keeping x_i, y_i, and z_i constant and differentiation keeping ξ_i, η_i, and ζ_i constant by primes, we find:

$$\frac{\partial'}{\partial\theta} = \frac{\partial}{\partial\theta} + \sum_i \left(\zeta_i \frac{\partial}{\partial\eta_i} - \eta_i \frac{\partial}{\partial\zeta_i}\right),$$

$$\frac{\partial'}{\partial\varphi} = \frac{\partial}{\partial\varphi} - \sum_i \left[\sin\theta\left(\zeta_i \frac{\partial}{\partial\xi_i} - \xi_i \frac{\partial}{\partial\zeta_i}\right) + \cos\theta\left(\xi_i \frac{\partial}{\partial\eta_i} - \eta_i \frac{\partial}{\partial\xi_i}\right)\right].$$

One sees easily that

$$\sum_i \left(\frac{\partial^2}{\partial x_i^2} + \frac{\partial^2}{\partial y_i^2} + \frac{\partial^2}{\partial z_i^2}\right) = \sum_i \left(\frac{\partial^2}{\partial \xi_i^2} + \frac{\partial^2}{\partial \eta_i^2} + \frac{\partial^2}{\partial \zeta_i^2}\right).$$

The potential energy in the new system of coordinates will be of the form:

$$V = \frac{Z_1 Z_2 e^2}{\rho} + \sum_{i>k} \frac{e^2}{r_{ik}^*} - \sum_{k=1}^{n} \frac{Z_1 e^2}{r_{1k}^*} - \sum_{k=1}^{n} \frac{Z_2 e^2}{r_{2k}^*},$$

where

$$r_{ik}^* = \sqrt{[(\xi_i - \xi_k)^2 + (\eta_i - \eta_k)^2 + (\zeta_i - \zeta_k)^2]}$$

is the distance between the ith and kth electron in the new coordinates, while

$$r_{1k}^* = \sqrt{\left[\xi_k^2 + \eta_k^2 + \left(\zeta_k + \rho \frac{M_2}{M_1 + M_2}\right)^2\right]}$$

is the distance of the kth electron from the first nucleus, and

$$r_{2k}^* = \sqrt{\left[\xi_k^2 + \eta_k^2 + \left(\zeta_k - \rho \frac{M_1}{M_1 + M_2}\right)^2\right]}$$

the distance between the kth electron and the second nucleus. In this way the potential energy expressed in the new coordinates does not depend on the angles θ and φ. If we take into account all the relations which we have just written down, the Schrödinger equation will be of the following form:

$$\left\{-\frac{\hbar^2}{2\mu} \sum_i \left(\frac{\partial^2}{\partial \xi_i^2} + \frac{\partial^2}{\partial \eta_i^2} + \frac{\partial^2}{\partial \zeta_i^2}\right) - \frac{\hbar^2}{2M} \frac{1}{r^2} \left[\frac{\partial}{\partial r}\left(r^2 \frac{\partial}{\partial r}\right) + \cot\theta\left(\frac{\partial}{\partial\theta} - i\hat{L}_\xi\right)\right.\right.$$
$$\left. + \left(\frac{\partial}{\partial\theta} - i\hat{L}_\xi\right)^2 + \frac{1}{\sin^2\theta}\left(\frac{\partial}{\partial\varphi} - i\sin\theta\,\hat{L}_\eta - i\cos\theta\,\hat{L}_\zeta\right)^2\right]$$
$$\left. + V(\xi_i, \eta_i, \zeta_i; \rho) - E\right\} \psi(\xi_i, \eta_i, \zeta_i; \rho, \theta, \varphi) = 0.$$

In this equation $\hat{L}_\xi$, $\hat{L}_\eta$, and $\hat{L}_\zeta$ are the operators of the components of the orbital angular momentum of the electrons in the ξ, η, ζ system expressed in units $\hbar$.

2. We shall denote the spin variable of the ith electron with respect to a fixed space by s_i', and with respect to the moving system by s_i. The function $\psi(\ldots s_i \ldots)$ with the spins referring to the ξ, η, ζ system is related to the function $\psi(\ldots s_i' \ldots)$ with the spins referring to the xyz-system by a linear transformation:

$$\psi(\ldots s_i \ldots) = \sum_{s_1' \ldots s_i' \ldots} S(s_1, \ldots, s_i, \ldots, s_1', \ldots, s_i', \ldots)\,\psi(\ldots s_i' \ldots) = \hat{S}\psi(\ldots s_i' \ldots),$$

where

$$S(s_1, s_2, \ldots, s_i, \ldots, s_1', \ldots, s_i', \ldots) = S(s_1;\, s_1')\, S(s_2;\, s_2') \ldots S(s_i;\, s_i') \ldots,$$

and

$$S(\tfrac{1}{2};\, \tfrac{1}{2}) = \cos\frac{\theta}{2} \exp[\tfrac{1}{2}i(\varphi + \tfrac{1}{2}\pi)],$$

$$S(\tfrac{1}{2};\, -\tfrac{1}{2}) = i\sin\frac{\theta}{2} \exp[\tfrac{1}{2}i(\varphi + \tfrac{1}{2}\pi)],$$

$$S(-\tfrac{1}{2};\, \tfrac{1}{2}) = i\sin\frac{\theta}{2} \exp[\tfrac{1}{2}i(\varphi + \tfrac{1}{2}\pi)],$$

$$S(-\tfrac{1}{2};\, -\tfrac{1}{2}) = \cos\frac{\theta}{2} \exp[\tfrac{1}{2}i(\varphi + \tfrac{1}{2}\pi)].$$

The Schrödinger equation which we are looking for will have the form

$$[\hat{S}\hat{H}\hat{S}^{-1} - E]\,\psi(\ldots \xi_i, \eta_i, \zeta_i, s_i \ldots;\, \rho, \theta, \varphi) = 0,$$

where $\hat{H}$ is the Hamiltonian given in the solution of the preceding problem.

After some simple manipulations, we find finally for the Schrödinger equation the following expression:

$$\left\{-\frac{\hbar^2}{2\mu}\Sigma\left(\frac{\partial^2}{\partial\xi_i^2} + \frac{\partial^2}{\partial\eta_i^2} + \frac{\partial^2}{\partial\zeta_i^2}\right) - \frac{\hbar^2}{2M}\frac{1}{r^2}\left[\frac{\partial}{\partial r}\left(r^2\frac{\partial}{\partial r}\right) + \cot\theta\left(\frac{\partial}{\partial\theta} - i\hat{M}_\xi\right)\right.\right.$$
$$\left. + \left(\frac{\partial}{\partial\theta} - i\hat{M}_\xi\right)^2 + \frac{1}{\sin^2\theta}\left(\frac{\partial}{\partial\varphi} - i\sin\theta\,\hat{M}_\eta - i\cos\theta\,\hat{M}_\zeta\right)^2\right]$$
$$\left. + V - E\right\}\psi(\ldots \xi_i, \eta_i, \zeta_i, s_i \ldots;\, \rho, \theta, \varphi) = 0.$$

Here $\hat{M}_\xi$, $\hat{M}_\eta$, and $\hat{M}_\zeta$ are now in contra-distinction to the preceding problem, the operators of the components of the total angular momentum of the electrons (orbital and spin).

3. We shall assume that the problem with fixed centres is solved, that is that we know the electron terms $E^{\mathrm{el}}(\rho)$ and the wave functions Φ_{el}. To fix our ideas we consider the *a* case; let, in the state which corresponds to the wave function Φ_{el}, the momentum of the total angular momentum (orbital and spin) along the axis of the molecule be equal to Ω. We multiply the Schrödinger equation

$$\hat{H}\Phi_{\mathrm{el}}(\xi_i, \eta_i, \zeta_i; \sigma_i; \rho) f(\rho)\, \Theta(\theta, \varphi) = E\Phi_{\mathrm{el}}(\xi_i, \eta_i, \zeta_i; \sigma_i; \rho) f(\rho)\, \Theta(\theta, \varphi)$$

from the left by Φ_{el}^* and integrate over the coordinates in the system in which the centres are fixed and sum over σ_i. If we note that

$$\int \Phi_{\mathrm{el}}^* \hat{M}_\xi \Phi_{\mathrm{el}}\, d\tau = \int \Phi_{\mathrm{el}}^* \hat{M}_\eta \Phi_{\mathrm{el}}\, d\tau = 0,$$

we have

$$\left[B \frac{\partial}{\partial\rho}\left(\rho^2 \frac{\partial}{\partial\rho}\right) - E^{\mathrm{el}}(\rho) - V'(\rho) - E^{\mathrm{rot}} + E\right] f(\rho) = 0,$$

$$B\left[\frac{1}{\sin\theta}\frac{\partial}{\partial\theta}\left(\sin\theta \frac{\partial}{\partial\theta}\right) + \frac{1}{\sin^2\theta}\left(\frac{\partial}{\partial\varphi} - i\Omega\cos\theta\right)^2\right]\Theta(\theta, \varphi) + E^{\mathrm{rot}}\,\Theta(\theta, \varphi) = 0.$$

In the last two equations we have introduced the following abbreviated notation:

$$V'(\rho) = \frac{1}{M}\int \Phi_{\mathrm{el}}^* \left[\frac{1}{\rho^2}(\hat{M}_\xi^2 + \hat{M}_\eta^2) - \frac{\partial^2}{\partial\rho^2}\right]\Phi_{\mathrm{el}}\, d\tau,$$

$$B = \frac{\hbar^2}{2M}\frac{1}{\rho^2}.$$

The quantity B is called the rotational constant.

4. The basic physical fact in the quantum mechanics of molecules is the very large value of the ratio M/μ, where M and μ are respectively the nuclear and the electronic masses. Indeed, the presence of this large dimensionless parameter, of the order of magnitude of 1000 to 10,000, causes a considerable difference between the orders of magnitude of the quantities mentioned in the present problem.

Let a be of the order of magnitude of the linear dimension of a diatomic molecule. The distance between the nuclei will clearly be of the same order of magnitude. Indeed, it cannot be larger than a because of the meaning of this quantity but it can neither be considerably smaller than a because of the mutual electrostatic repulsion of the nuclei. (We can easily arrive similarly to the conclusion that a must be of the order of magnitude of the linear dimensions of an atom. This fact about the order of magnitude of a does not play any role in the estimates in which we are interested.)

1. We shall first of all estimate the order of magnitude of the energy of the valence electrons and at the same time the order of magnitude of the intervals between electronic levels. Since the valence electrons move, in contra-distinction to the electrons of the closed inner shells at each of the nuclei, in a region of space of linear dimensions $\sim a$, so that the uncertainty in momentum Δp will be of the order of magnitude $\hbar/a$, the zero-point energy of the electron, or the difference in energy of successive electronic levels, will be of the order of magnitude E_{el}:

$$E_{el} \sim \frac{(\Delta p)^2}{\mu} \sim \frac{\hbar^2}{\mu a^2}. \qquad (1)$$

We shall now consider the vibrations of the nuclei in the molecule. We can use as a model, at any rate for the ground state and the lower excited levels, the motion of a harmonic oscillator with a mass of the order of M (or, rather, with the reduced mass of the nuclei) and a stiffness coefficient K. We can estimate the latter from the fact that a change of the distance between the nuclei of order a must correspond to a change of order unity in the electronic wave function, that is, it must be connected to a change in energy of the order of magnitude

$$Ka^2 \sim E_{el},$$

so that we get in the usual way for the frequency of the vibrations of the nuclei in the molecule $\omega \sim \sqrt{(K/M)}$ and using equation (1) we get for the intervals between vibrational levels

$$E_{vib} \approx \hbar\omega \sim \hbar\left(\frac{E_{el}}{Ma^2}\right)^{\frac{1}{2}} \sim \frac{\hbar^2}{(\mu M)^{\frac{1}{2}} a^2} \sim \left(\frac{\mu}{M}\right)^{\frac{1}{2}} E_{el}. \qquad (2)$$

Finally the rotational levels of the molecule can clearly be treated as the levels of a rotator with a moment of inertia $I \sim Ma^2$, so that we get for the intervals between rotational levels

$$E_{rot} \sim \frac{\hbar^2}{I} \sim \frac{\hbar^2}{Ma^2} \sim \frac{\mu}{M} E_{el}. \qquad (3)$$

It is clear from equations (1), (2), and (3) that the quantities E_{el}, E_{vib}, and E_{rot} form a geometric progression with the factor $(\mu/M)^{\frac{1}{2}} \sim 10^{-2}$.

2. Let b be the amplitude of the zero point vibrations of the nuclei in the molecule. It will be of the order of magnitude of $\sqrt{(\hbar/M\omega)}$ [we can obtain this estimate in different ways, for instance, by equating the order of magnitude of the oscillator energy ($\sim \hbar\omega$) and the potential

energy for a displacement over a distance b ($\sim M\omega^2 b^2$) which from the expression (2) for ω gives us

$$b \sim \left(\frac{\mu}{M}\right)^{\frac{1}{4}} a]. \tag{4}$$

The ratio of the vibrational amplitude b to the equilibrium distance a of the nuclei in the molecule is thus of the order of magnitude of $(\mu/M)^{\frac{1}{4}} \ll 1$. We can consider this quantity to be a small expansion parameter in the theory of molecules. According to the results obtained earlier E_{el}, E_{vib}, and E_{rot} are quantities of respectively the zeroth, second, and fourth order in this small parameter.

3. The periods of the electronic motion and of the nuclear vibrations in the molecule are of the following order of magnitude:

$$T_{\mathrm{el}} \sim \frac{1}{\omega_{\mathrm{el}}} \sim \frac{\hbar}{E_{\mathrm{el}}} \sim \frac{\mu a^2}{\hbar}, \tag{5}$$

$$T_{\mathrm{vib}} \sim \frac{1}{\omega_{\mathrm{vib}}} \sim \frac{(\mu M)^{\frac{1}{2}} a^2}{\hbar}. \tag{6}$$

The corresponding characteristic velocities are clearly equal to

$$v_{\mathrm{el}} \sim \frac{a}{T_{\mathrm{el}}} \left[\text{or} \sim \sqrt{\left(\frac{E_{\mathrm{el}}}{\mu}\right)}\right] \sim \frac{\hbar}{\mu a}, \tag{7}$$

$$v_{\mathrm{vib}} \sim \frac{b}{T_{\mathrm{vib}}} \left[\text{or} \sim \sqrt{\left(\frac{E_{\mathrm{vib}}}{M}\right)}\right] \sim \frac{\hbar}{\mu^{\frac{1}{4}} M^{\frac{3}{4}} a}. \tag{8}$$

From equations (5) to (8) it follows that

$$\frac{T_{\mathrm{vib}}}{T_{\mathrm{el}}} \sim \left(\frac{M}{\mu}\right)^{\frac{1}{2}} \gg 1, \quad \frac{v_{\mathrm{vib}}}{v_{\mathrm{el}}} \sim \left(\frac{\mu}{M}\right)^{\frac{3}{4}} \ll 1. \tag{9}$$

The last inequality means that the nuclei are moving slowly compared to the electrons in the molecule. This fact makes it possible to consider the nuclear motion in the adiabatic approximation.

5. Nitrogen molecule (atoms in the ^4S state): $^1\Sigma_g^+$, $^3\Sigma_u^+$, $^5\Sigma_g^+$, $^7\Sigma_u^+$.

Bromine molecule (atoms in the ^2P state):

$$2\,{}^1\Sigma_g^+,\ {}^1\Sigma_u^-,\ {}^1\Pi_g,\ {}^1\Pi_u,\ {}^1\Delta_g;$$

$$2\,{}^3\Sigma_u^+,\ {}^3\Sigma_g^-,\ {}^3\Pi_g,\ {}^3\Pi_u,\ {}^3\Delta_u.$$

LiH molecule (Li atom in ^2S_g state, H atom in ^2S_g state): $^1\Sigma^+$, $^3\Sigma^+$.

HBr molecule (Br atom in ^2P_u state): $^1\Sigma^+$, $^3\Sigma^+$, $^1\Pi$, $^3\Pi$.

CN molecule (C atom in ^3P_g state, N atom in ^4S_u state): $^2\Sigma^+$, $^4\Sigma^+$, $^6\Sigma^+$, $^2\Pi$, $^4\Pi$, $^6\Pi$.

(The number which stands before the term symbols indicates the multiplicity of the terms.)

6. The helium atom in its ground state is characterised by the fact that both its electrons are in the lowest state (parahelium). The total eigenfunction of the ground state of the helium atom can be written approximately in the form

$$\frac{1}{\sqrt{2}}\psi_a(1)\,\psi_a(2)\,[\eta_+(\sigma_1)\,\eta_-(\sigma_2)-\eta_+(\sigma_2)\,\eta_-(\sigma_1)],$$

where ψ_a is a hydrogen function.

The hydrogen atom has the eigenfunctions

$$\psi_b(3)\,\eta_+(\sigma_3) \quad \text{or} \quad \psi_b(3)\,\eta_-(\sigma_3).$$

If both atoms are at a large distance apart the wave functions of the system can be written as a product

$$\frac{1}{\sqrt{2}}\psi_a(1)\,\psi_a(2)\,\psi_b(3)\,[\eta_+(\sigma_1)\,\eta_-(\sigma_2)-\eta_+(\sigma_2)\,\eta_-(\sigma_1)]\,\eta_+(\sigma_3).$$

Taking exchange of electrons into account the wave functions of the system must be anti-symmetric with respect to an interchange of any two electrons. There is only one anti-symmetric eigenfunction which is also the eigenfunction of our system in zeroth approximation

$$\begin{aligned}\Psi = \frac{1}{\sqrt{[6(1-S)]}}\{&\psi_a(1)\,\psi_a(2)\,\psi_b(3)\,[\eta_+(1)\,\eta_-(2)-\eta_+(2)\,\eta_-(1)]\,\eta_+(3)\\ &+\psi_a(3)\,\psi_a(1)\,\psi_b(2)\,[\eta_+(3)\,\eta_-(1)-\eta_+(1)\,\eta_-(3)]\,\eta_+(2)\\ &+\psi_a(2)\,\psi_a(3)\,\psi_b(1)\,[\eta_+(2)\,\eta_-(3)-\eta_+(3)\,\eta_-(2)]\,\eta_+(1)\}.\end{aligned} \tag{1}$$

In expression (1) $\frac{1}{\sqrt{[6(1-S)]}}$ is a normalising factor and

$$\begin{aligned}S &= \int\psi_a(1)\,\psi_a(2)\,\psi_b(3)\,\psi_a(2)\,\psi_a(3)\,\psi_b(1)\,d\tau_1\,d\tau_2\,d\tau_3\\ &= \int\psi_a(1)\,\psi_a(2)\,\psi_b(3)\,\psi_a(1)\,\psi_a(3)\,\psi_b(2)\,d\tau_1\,d\tau_2\,d\tau_3\\ &= \int\psi_a(1)\,\psi_a(3)\,\psi_b(2)\,\psi_a(2)\,\psi_a(3)\,\psi_b(1)\,d\tau_1\,d\tau_2\,d\tau_3.\end{aligned}$$

If we apply the usual perturbation theory we have

$$\epsilon = \sum_\sigma \int \Psi\hat{H}\Psi\,d\tau,$$

where Ψ is the eigenfunction in zeroth approximation and $\hat{H}$ the perturbation energy, while the summation is over the spin variables. We must take into account that $\hat{H}$ is different for different parts of Ψ, and indeed for $\psi_a(1)\psi_a(2)\psi_b(3)$ the perturbation energy is equal to

$$\hat{H} = e^2\left(\frac{2}{R} - \frac{1}{r_{a3}} - \frac{1}{r_{b1}} - \frac{1}{r_{b2}} + \frac{1}{r_{13}} + \frac{1}{r_{23}}\right),$$

and for $\psi_a(1)\psi_a(3)\psi_b(2)$ it is equal to

$$\hat{H} = e^2\left(\frac{2}{R} - \frac{1}{r_{a2}} - \frac{1}{r_{b1}} - \frac{1}{r_{b3}} + \frac{1}{r_{12}} + \frac{1}{r_{32}}\right).$$

If we take into account the fact that the integrals which differ only in the numbering of the electrons are the same, we have

$$\epsilon = \frac{K-A}{1-S}, \tag{2}$$

where

$$K = e^2\int\left(\frac{2}{R} + \frac{1}{r_{13}} + \frac{1}{r_{23}} - \frac{1}{r_{a3}} - \frac{1}{r_{b1}} - \frac{1}{r_{b2}}\right)\psi_a^2(1)\,\psi_a^2(2)\,\psi_b^2(3)\,d\tau_1\,d\tau_2\,d\tau_3,$$

$$A = e^2\int\left(\frac{2}{R} + \frac{1}{r_{13}} + \frac{1}{r_{23}} - \frac{1}{r_{a3}} - \frac{1}{r_{b1}} - \frac{1}{r_{b2}}\right)$$
$$\times\,\psi_a(1)\,\psi_a(2)\,\psi_b(3)\,\psi_a(1)\,\psi_a(3)\,\psi_b(2)\,d\tau_1\,d\tau_2\,d\tau_3.$$

The integrals K, A, S are in general of the same kind as the corresponding integrals in the hydrogen molecule problem. An evaluation of the integrals shows that equation (2) corresponds to a repulsion. This is correct not only for helium but for all inert gases.

7. If we split off the motion of the centre of mass we find for the wave function of the relative motion of the nuclei the following equation:

$$\nabla^2\psi + \frac{2M}{\hbar^2}\left[E + 2D\left(\frac{1}{\rho} - \frac{1}{2\rho^2}\right)\right]\psi = 0,$$

where M is the reduced mass of the nuclei. If we separate the variables in spherical coordinates and put

$$\psi = \frac{\chi(\rho)}{\rho}Y_{KM}(\theta, \varphi),$$

we find for χ the differential equation

$$\frac{d^2\chi}{d\rho^2} + \left[-\lambda^2 + \frac{2\gamma^2}{\rho} - \frac{\gamma^2 + K(K+1)}{\rho^2}\right]\chi = 0$$

$$\lambda = \sqrt{\left(-\frac{2Ma^2E}{\hbar^2}\right)}, \quad \gamma^2 = \frac{2Ma^2}{\hbar^2}D.$$

If we make the substitution $\chi(\rho) = \rho^s \exp(-\lambda\rho)\, u(\rho)$, where

$$s = \tfrac{1}{2} + \sqrt{[\gamma^2 + (K+\tfrac{1}{2})^2]},$$

the last equation goes over into the hypergeometric equation

$$\rho u'' + (2s - 2\lambda\rho)\, u' + (-2s\lambda + 2\gamma^2)\, u = 0.$$

The solution of that equation which is finite for $\rho = 0$ is of the following form:

$$u = cF\left(s - \frac{\gamma^2}{\lambda}, 2s, 2\lambda\rho\right).$$

In the states of the discrete spectrum the wave functions χ must tend to zero as $r \to \infty$. This means that the expression for u must become a polynomial,

$$s - \frac{\gamma^2}{\lambda} = -v,$$

where v is a non-negative integer. From this condition we get the energy levels

$$E_{vK} = -\frac{\hbar^2}{2Ma^2} \frac{\gamma^4}{\{v + \frac{1}{2} + \sqrt{[\gamma^2 + (K+\frac{1}{2})^2]}\}^2}.$$

The dimensionless parameter γ^2 is proportional to the reduced mass of the nuclei M, so that $\gamma^2 \gg 1$. If v and K are not very large,

$$v \ll \gamma, \quad K \ll \gamma,$$

the expression for E_{vK} is of the following form:

$$E_{vK} = -D + \hbar\omega_0(v + \tfrac{1}{2}) + \frac{\hbar^2}{2Ma^2}(K + \tfrac{1}{2})^2 - \frac{3\hbar^2}{2Ma^2}(v + \tfrac{1}{2})^2 - \frac{3\hbar^3 (K+\frac{1}{2})(v+\frac{1}{2})}{2M^2 a^4 \omega_0},$$

where

$$\omega_0 = \sqrt{\left(\frac{2D}{Ma^2}\right)}.$$

The dissociation energy is approximately equal to

$$E_0 = D - \frac{\hbar\omega_0}{2}.$$

The second and third term in the expression for E_{vK} give us the vibrational and rotational energy. The fourth term takes into account the anharmonicity of the vibrations and, finally, the fifth term gives the correction to the energy taking into account the interaction between the rotation and the vibration of the nuclei.

Since D is of the order of unity in atomic units ($e = m = \hbar = 1$), it follows from the expression which we have obtained that

$$D:\hbar\omega_0:\frac{\hbar^2}{2Ma^2} \sim 1:\sqrt{\left(\frac{\mu}{M}\right)}:\frac{\mu}{M},$$

where μ is the mass of an electron.

It is thus clear that the difference in energy between two quantum states with different electronic motion (a quantity of the order of D) is large compared to the difference in energy of different vibrational states which in turn is large compared to the difference between rotational states.

8. We can find the minimum of the effective potential

$$W = -2D\left(\frac{1}{\rho}-\frac{1}{2\rho^2}\right)+\frac{A^2}{\rho^2}, \quad \text{where} \quad A^2 = \frac{\hbar^2 K(K+1)}{2Ma^2},$$

by putting its derivative equal to zero

$$W' = -2D\left(-\frac{1}{\rho_0^2}+\frac{1}{\rho_0^3}\right)-\frac{2A^2}{\rho_0^3} = 0,$$

from which we get

$$\rho_0 = 1+\frac{A^2}{D}.$$

We shall now expand the effective potential near the equilibrium position which we have found

$$W(\rho)\approx -2D\left(\frac{1}{\rho_0}-\frac{1}{\rho_0^2}\right)+\frac{A^2}{\rho_0^2}+\frac{D+A^2}{\rho_0^4}(\rho-\rho_0)^2,$$

and retaining only terms up to the order A^2, we get

$$W(\rho)\approx -D+A^2+(D-3A^2)(\rho-\rho_0)^2.$$

We can substitute this expression into the equation for χ

$$\frac{d^2\chi}{d\rho^2}+\frac{2Ma^2}{\hbar^2}[E+D-A^2-(D-3A^2)(\rho-\rho_0)^2]\chi = 0.$$

We find thus for the energy levels

$$E_{vK} = -D+A^2+\hbar\omega(v+\tfrac{1}{2}).$$

In the same approximation we have

$$\omega = \omega_0\left(1-\frac{3}{2}\frac{A^2}{D}\right), \quad \omega_0 = \sqrt{\left(\frac{2D}{Ma^2}\right)}.$$

Finally, we have for the energy E_{vK}

$$E_{vK} = -D + \hbar\omega_0(v+\tfrac{1}{2}) + \frac{\hbar^2 K(K+1)}{2Ma^2} - \frac{3}{2}\frac{\hbar K(K+1)(v+\tfrac{1}{2})}{M^2 a^4 \omega_0}.$$

As we have not taken into account in these calculations the effect of anharmonicity, we have not got here a term $-(3\hbar^2/2Ma^2)(v+\tfrac{1}{2})^2$ which we found in the preceding problem.

9. In the infra-red band we are dealing with transitions in the electronic ground state where vibrational and rotational quantum numbers are changing. For the frequency of the transition between two states $v', J' \rightarrow v'', J''$ we have

$$\omega = \omega_0(v'-v'') + \frac{\hbar}{2Ma^2}[J'^2 + J' - J''^2 - J''].$$

From the selection rule for J it follows that $J'' = J' \pm 1$.

We get thus the following frequencies:

$$\omega = \omega_0(v'-v'') - \frac{\hbar}{2Ma^2}2(J'+1) \quad \begin{pmatrix} J'' = J'+1, \\ J' = 0, 1, 2, \ldots \end{pmatrix}$$

and

$$\omega = \omega_0(v'-v'') + \frac{\hbar}{2Ma^2}2J' \quad \begin{pmatrix} J'' = J'-1, \\ J' = 1, 2, \ldots \end{pmatrix}.$$

We note that these two sets of frequencies are in molecular spectroscopy called the P and R branches.

From the expressions which we have obtained it is clear that the difference in frequency of two successive lines for given values of v' and v'' is (in cm^{-1}) equal to

$$\Delta\nu = \frac{\Delta\omega}{2\pi c} = \frac{\hbar}{2\pi c M a^2}.$$

The moment of inertia of the $H^{35}Cl$ molecule is equal to

$$I = Ma^2 = \frac{\hbar}{2\pi c \Delta\nu} = 2{\cdot}65 \,.\, 10^{-40} \text{ g cm}^2.$$

If we use the value of the reduced mass,

$$M = M_{\text{H}}\frac{1.35}{1+35} = 0{\cdot}972 M_{\text{H}} = 1{\cdot}61 \,.\, 10^{-24} \text{ g}$$

we find for the distance between the nuclei in HCl,

$$a = \sqrt{\left(\frac{I}{M}\right)} = 1{\cdot}29 \,.\, 10^{-8} \text{ cm}.$$

The equilibrium distances a in the molecules DCl and HCl are the same since the form of the potential curves is determined by the electronic states. It follows thus that

$$\frac{\Delta\nu_{\text{DCl}}}{\Delta\nu_{\text{HCl}}} = \frac{M_{\text{HCl}}}{M_{\text{DCl}}}; \quad \Delta\nu_{\text{DCl}} = 10{\cdot}7 \text{ cm}^{-1}.$$

10. The distance between the first two rotational levels is given by

$$\Delta\nu_{\text{rot}} = \frac{\hbar}{2\pi c I} = 41{\cdot}5 \text{ cm}^{-1}.$$

We find thus

$$\frac{\Delta\nu_{\text{rot}}}{\Delta\nu_{\text{vib}}} = 0{\cdot}0104.$$

11. The dissociation energy of the D_2 molecule is equal to $4{\cdot}54$ eV.

12*. If we go over to a new variable $\xi = (r-a)/a$, we can write the equation for the radial function χ/r as follows:

$$\frac{d^2\chi}{d\xi^2} + \frac{2Ma^2}{\hbar^2}(E-V)\chi = 0.$$

If we put $z = \alpha \exp(-2\beta\xi)$, we find

$$\chi'' + \frac{1}{z}\chi' + \left(-\frac{s^2}{z^2} - \tfrac{1}{4} + \frac{v+s+\frac{1}{2}}{z}\right)\chi = 0,$$

where

$$s = \sqrt{\left[\frac{Ma^2(D-E)}{2\beta^2\hbar^2}\right]}, \quad v+s+\tfrac{1}{2} = \frac{Ma^2 D}{\alpha\beta^2\hbar^2}, \quad \alpha^2 = \frac{2Ma^2 D}{\beta^2\hbar^2}.$$

If we make the substitution $\chi = \exp(-z/2)\, z^s\, u(z)$ this equation will go over into the hypergeometric equation $zu'' + (2s+1-z)u' + vu = 0$, the solution of which is the confluent hypergeometric function

$$u = F(-v, 2s+1, z).$$

This function satisfies the condition that χ tends to zero as $r \to +\infty$ for positive values of s (discrete spectrum). As $r \to -\infty$ the wave function must tend to zero.† In order that that condition is satisfied it is necessary that F reduces to a polynomial, that is, that v is a non-negative integer. This condition determines the energy spectrum

$$E_n = \hbar\omega(v+\tfrac{1}{2}) - \frac{\hbar^2\omega^2}{4D}(v+\tfrac{1}{2})^2, \quad \text{where} \quad \omega = 4\beta\sqrt{\left(\frac{D}{2Ma^2}\right)}.$$

† Strictly speaking, the wave function should vanish at $r = 0$. This only introduces minor changes in the final result (see D. ter Haar, *Phys. Rev.* **70**, 222 (1946)).

In this way the difference between vibrational levels decreases with increasing quantum number v.

The dissociation energy is equal to

$$E_0 = D - \frac{\hbar\omega}{2} + \frac{\hbar^2\omega^2}{16D}.$$

14. As parameters to characterise the rotation, we shall use the Euler angles θ, ψ, and φ. In this case the coordinates of a point x, y, z in the fixed system are connected with the coordinates ξ, η, ζ in the moving system by the following equations:

$$\left.\begin{aligned} x &= \xi(\cos\psi\cos\varphi - \sin\psi\sin\varphi\cos\theta) \\ &\quad -\eta(\cos\psi\sin\varphi + \sin\psi\cos\varphi\cos\theta) + \zeta\sin\psi\sin\theta, \\ y &= \xi(\sin\psi\cos\varphi + \cos\psi\sin\varphi\cos\theta) \\ &\quad +\eta(-\sin\psi\sin\varphi + \cos\psi\cos\varphi\cos\theta) - \zeta\cos\psi\sin\theta, \\ z &= \xi\sin\varphi\sin\theta + \eta\cos\varphi\sin\theta + \zeta\cos\theta. \end{aligned}\right\} \quad (1)$$

To find the form of the operators $\hat{J}_\xi, \hat{J}_\eta, \hat{J}_\zeta$, we use the fact that the operator $\hat{J}_\xi$ is $\hat{J}_\xi = -i(\partial/\partial\alpha)$, where α is the angle in the plane perpendicular to the ξ-axis. Since under a rotation of the ξ, η, ζ-system with respect to, for instance, the ξ-axis over an infinitesimal angle $d\alpha$, the values of the angles change, we can write for $\hat{J}_\xi$, in units $\hbar$,

$$\hat{J}_\xi = -i\left(\frac{\partial\theta}{\partial\alpha}\frac{\partial}{\partial\theta} + \frac{\partial\varphi}{\partial\alpha}\frac{\partial}{\partial\varphi} + \frac{\partial\psi}{\partial\alpha}\frac{\partial}{\partial\psi}\right).$$

For an infinitesimal rotation around the ξ-axis over an angle $d\alpha$ we have

$$\left.\begin{aligned} \xi &= \xi', \\ \eta &= \eta' - \zeta' d\alpha, \\ \zeta &= \eta' d\alpha + \zeta', \end{aligned}\right\} \quad (2)$$

and

$$z = \xi'\sin(\varphi + d\varphi)\sin(\theta + d\theta) + \eta'\cos(\varphi + d\varphi)\sin(\theta + d\theta) + \zeta'\cos(\theta + d\theta). \quad (3)$$

On the other hand if we substitute expression (2) into equation (1) we get

$$z = \xi'\sin\varphi\sin\theta + \eta'(\cos\varphi\sin\theta + \cos\theta\, d\alpha) + \zeta'(\cos\theta - \cos\varphi\sin\theta\, d\alpha). \quad (4)$$

Comparing equations (3) and (4) we find

$$\frac{d\theta}{d\alpha} = \cos\varphi, \quad \frac{d\varphi}{d\alpha} = -\sin\varphi\cot\theta.$$

Similarly we have $d\psi/d\alpha = \sin\varphi/\sin\theta$.

We find finally for $\hat{J}_\xi$ the expression

$$\hat{J}_\xi = -i\left(\cos\varphi\frac{\partial}{\partial\theta} - \sin\varphi\cot\theta\frac{\partial}{\partial\varphi} + \frac{\sin\varphi}{\sin\theta}\frac{\partial}{\partial\psi}\right).$$

In the same way we find expressions for the two other operators

$$\hat{J}_\eta = -i\left(-\sin\varphi\frac{\partial}{\partial\theta} - \cos\varphi\cot\theta\frac{\partial}{\partial\varphi} + \frac{\cos\varphi}{\sin\theta}\frac{\partial}{\partial\psi}\right),$$

$$\hat{J}_\zeta = -i\frac{\partial}{\partial\varphi}.$$

17.
$$E_J = \frac{\hbar^2}{2A}J(J+1).$$

Each energy level is $(2J+1)$-fold degenerate with respect to the direction of the angular momentum referring to a fixed system of reference and equally degenerate with respect to the direction of the angular momentum referring to the body itself.

18. Since

$$\hat{H} = \frac{1}{2A}\hat{J}^2 + \frac{1}{2}\left(\frac{1}{C} - \frac{1}{A}\right)\hat{J}_\zeta^2,$$

we have

$$E = \frac{1}{2A}J(J+1) + \frac{1}{2}\left(\frac{1}{C} - \frac{1}{A}\right)k^2; \quad J_\zeta = k, \ |k| \leqslant J.$$

In this case the energy levels are $2(2J+1)$-fold degenerate. The degeneracy as far as the direction of the angular momentum in the fixed system of reference is concerned is $(2J+1)$-fold as before.

19.
$$-\frac{\hbar^2}{2A}\left\{\frac{1}{\sin\theta}\frac{\partial}{\partial\theta}\left(\sin\theta\frac{\partial u}{\partial\theta}\right) + \frac{1}{\sin^2\theta}\left(\frac{\partial^2 u}{\partial\varphi^2} + \frac{\partial^2 u}{\partial\psi^2}\right)\right.$$
$$\left. - 2\frac{\cos\theta}{\sin^2\theta}\frac{\partial^2 u}{\partial\psi\,\partial\varphi}\right\} - \frac{\hbar^2}{2}\left(\frac{1}{C} - \frac{1}{A}\right)\frac{\partial^2 u}{\partial\varphi^2} = Eu.$$

20*. Since $\hat{J}^2$ commutes with $\hat{J}_\zeta = -i(\partial/\partial\varphi)$ and $\hat{J}_z = -i(\partial/\partial\psi)$ we can write the eigenfunction in the form

$$\Phi_{kJM_J} = \Theta_{kJM_J}(\theta)\exp(iM_J\psi)\exp(ik\varphi),$$

where M_J and k are the components of the angular momentum with respect to the fixed z-axis and the moving ζ-axis respectively.

Since ψ and φ enter symmetrically in equation (14) and $|k|\leqslant J$, we have $|M_J|\leqslant J$.

Let us consider the operators

$$\hat{J}_\xi+i\hat{J}_\eta = -i\exp(-i\varphi)\left(\frac{\partial}{\partial\theta}-i\cot\theta\frac{\partial}{\partial\varphi}+\frac{i}{\sin\theta}\frac{\partial}{\partial\psi}\right), \tag{1}$$

$$\hat{J}_\xi-i\hat{J}_\eta = -i\exp(i\varphi)\left(\frac{\partial}{\partial\theta}+i\cot\theta\frac{\partial}{\partial\varphi}-\frac{i}{\sin\theta}\frac{\partial}{\partial\psi}\right). \tag{2}$$

One can easily show that

$$\hat{J}_\zeta(\hat{J}_\xi-i\hat{J}_\eta)\,\Phi_{kJM_J} = (k+1)\,(\hat{J}_\xi-i\hat{J}_\eta)\,\Phi_{kJM_J},$$

that is, the expression $(\hat{J}_\xi-i\hat{J}_\eta)\,\Phi_{kJM_J}$ is the eigenfunction corresponding to the value $k+1$ of the operator $\hat{J}_\zeta$.

If we put $k = J$, we have

$$(\hat{J}_\xi-i\hat{J}_\eta)\,\Phi_{JJM_J}\equiv 0.$$

This last relation can be put in the form

$$\left(\frac{\partial}{\partial\theta}+i\cot\theta\frac{\partial}{\partial\rho}-\frac{i}{\sin\theta}\frac{\partial}{\partial\psi}\right)\Theta_{JJM_J}(\theta)\exp(iM_J\psi)\exp(iJ\varphi)\equiv 0.$$

We find thus an ordinary differential equation of the first order to determine Θ_{JJM_J}

$$\frac{d\Theta_{JJM_J}}{d\theta}+\frac{M_J-J\cos\theta}{\sin\theta}\,\Theta_{JJM_J} = 0.$$

The general solution of this equation is of the form

$$\left.\begin{aligned} \Theta_{JJM_J} &= c\frac{(\sin\theta)^J}{(\tan\frac{1}{2}\theta)^{M_J}} \\ \text{or}\qquad & \\ \Theta(\theta) &= c(1-\cos\theta)^{\frac{1}{2}(J-M_J)}(1+\cos\theta)^{\frac{1}{2}(J+M_J)}. \end{aligned}\right\} \tag{3}$$

Since the function Θ must be finite, we have $|M^J|\leqslant J$.

To determine the function Θ_{kJM_J}, we consider the action of the operator $\hat{J}_\xi+i\hat{J}_\eta$ upon it. Since

$$\hat{J}_\zeta(\hat{J}_\xi+i\hat{J}_\eta)\,\Theta_{kJM_J} = (k-1)\,(\hat{J}_\xi+i\hat{J}_\eta)\,\Theta_{kJM_J},$$

we have

$$(\hat{J}_\xi+i\hat{J}_\eta)\,\Theta_{kJM_J} = \alpha_k\,\Theta_{k-1JM_J}. \tag{4}$$

Substituting into equation (4) the explicit form of the operator $\hat{J}_\xi+i\hat{J}_\eta$ from expression (1) we are led to the equation

$$\frac{d\Theta_{kJM_J}}{d\theta}+\frac{k\cos\theta-M_J}{\sin\theta}\,\Theta_{kJM_J} = i\alpha_k\,\Theta_{k-1JM_J},$$

which, if we introduce a new variable $x = \cos\theta$, becomes of the form

$$\sqrt{(1-x^2)}\frac{dP_{kJM_J}(x)}{dx}+\frac{M_J-kx}{\sqrt{(1-x^2)}}P_{kJM_J}(x) = -i\alpha_k P_{k-1JM_J}(x),$$

where

$$P_{kJM_J}(x) = \Theta_{kJM_J}(\arccos x).$$

If we put

$$P_{kJM_J} = (1-x)^{-\frac{1}{2}(k-M_J)}(1+x)^{-\frac{1}{2}(k+M_J)}v_{kJM_J}, \tag{5}$$

we find a simple relation to determine v_{kJM_J}

$$\frac{dv_{kJM_J}}{dx} = -i\alpha_k v_{k-1JM_J}. \tag{6}$$

The function P_{JJM_J} which we found earlier [see equation (3)] can be written in the form of expression (5)

$$P_{JJM_J}(x) = (1-x)^{-\frac{1}{2}(J-M_J)}(1+x)^{-\frac{1}{2}(J+M_J)}v_{JJM_J},$$

where we understand by v_{JJM_J} the expression

$$v_{JJM_J}(x) = c(1-x)^{J-M_J}(1+x)^{J+M_J}. \tag{7}$$

From equation (7) and the recurrence relation (6) it follows that

$$v_{kJM_J} = c\frac{d^{J-k}}{dx^{J-k}}[(1-x)^{J-M_J}(1+x)^{J+M_J}].$$

Hence

$$\Phi_{kJM_J}(\theta,\psi,\varphi)$$
$$= c\exp(ik\varphi)\exp(iM_J\psi)(1-\cos\theta)^{-\frac{1}{2}(k-M_J)}(1+\cos\theta)^{-\frac{1}{2}(k+M_J)}$$
$$\times\left(\frac{\partial}{\partial\cos\theta}\right)^{J-k}[(1-\cos\theta)^{J-M_J}(1+\cos\theta)^{J+M_J}];$$

for $M_J = 0$ these generalised spherical functions go over, as we should have expected, into the ordinary spherical functions and are the wave functions of the rotator

$$\Phi_{kJ0}(\theta,\varphi) = c\exp(ik\varphi)\frac{1}{\sin^k\theta}\frac{d^{J-k}}{(d\cos\theta)^{J-k}}(\sin^{2J}\theta).$$

21.

$$\hat{H}_{kk} = \frac{\hbar^2}{4}\left(\frac{1}{A}+\frac{1}{B}\right)[J(J+1)-k^2]+\frac{\hbar^2k^2}{2C},$$

$$\hat{H}_{kk+2} = \hat{H}_{k+2k} = \frac{\hbar^2}{8}\left(\frac{1}{A}-\frac{1}{B}\right)\sqrt{[(J-k)(J-k-1)(J+k+1)(J+k+2)]}.$$

22. For an asymmetric top the degeneracy referring to the direction of the angular momentum with respect to a fixed system of reference remains the same. The degeneracy with respect to the quantum number k disappears completely so that a given value of J corresponds to $2J+1$ different levels. In the case $J = 1$ the energy levels are determined from the solution of a secular equation of the form

$$\begin{vmatrix} H_{11}-E & H_{10} & H_{1,-1} \\ H_{10} & H_{00}-E & H_{0-1} \\ H_{-11} & H_{-10} & H_{-1,-1}-E \end{vmatrix} = \begin{vmatrix} H_{11}-E & 0 & H_{1,-1} \\ 0 & H_{00}-E & 0 \\ H_{-1,1} & 0 & H_{-1,-1}-E \end{vmatrix} = 0.$$

Since $H_{-1,-1} = H_{11}$, we have

$$(H_{00}-E)(H_{11}^2-E^2-2H_{11}E-H_{1,-1}^2) = 0,$$

and thus

$$E_1 = \frac{\hbar^2}{2}\left(\frac{1}{A}+\frac{1}{B}\right), \quad E_2 = \frac{\hbar^2}{2}\left(\frac{1}{C}+\frac{1}{A}\right), \quad E_3 = \frac{\hbar^2}{2}\left(\frac{1}{B}+\frac{1}{C}\right).$$

23*.

E		m	
E_1	$\Phi_{k=0,\,m}$	1	$[\sqrt{(3)/4\pi}]\exp(i\psi)\sin\theta$
		0	$[1/\pi]\sqrt{\frac{3}{8}}\cos\theta$
		-1	$[\sqrt{(3)/4\pi}]\exp(-i\psi)\sin\theta$
E_2	$c(\Phi_{k=1,\,m}+\Phi_{k=-1,\,m})$	1	$[1/\pi]\sqrt{(\frac{3}{16})}\exp(i\psi)(\cos\varphi+i\cos\theta\sin\varphi)$
		0	$[1/\pi]\sqrt{(\frac{3}{8})}\cos\varphi\sin\theta$
		-1	$[1/\pi]\sqrt{(\frac{3}{16})}\exp(-i\psi)(\cos\varphi-i\cos\theta\sin\varphi)$
E_3	$c(\Phi_{k=1,\,m}-\Phi_{k=-1,\,m})$	1	$[1/\pi]\sqrt{(\frac{3}{16})}\exp(i\psi)(\cos\theta\cos\varphi+i\sin\varphi)$
		0	$[1/\pi]\sqrt{(\frac{3}{8})}\sin\theta\sin\varphi$
		-1	$[1/\pi]\sqrt{(\frac{3}{16})}\exp(-i\psi)(-\cos\theta\cos\varphi+i\sin\varphi)$

25. The splitting of the terms is determined by the spin–spin interaction. To determine this splitting it is necessary to average the operator of the spin–spin interaction, $\alpha(\hat{\boldsymbol{S}}.\boldsymbol{n})^2$, over a rotational state.

For a given value of K the quantum number J takes on the values

$$J = K+1, K, K-1.$$

The matrix elements of $(\boldsymbol{n}\,.\,\hat{\boldsymbol{S}})$ which are different from zero have the form

$$(\boldsymbol{n}\,.\,\hat{\boldsymbol{S}})_{K\ \ \ J=K}^{K-1\ J=K} = (\boldsymbol{n}\,.\,\hat{\boldsymbol{S}})_{K-1\ J=K}^{K\ \ \ J=K} = \sqrt{\left(\frac{K+1}{2K+1}\right)},$$

$$(\boldsymbol{n}\,.\,\hat{\boldsymbol{S}})_{K\ \ \ J=K}^{K+1\ J=K} = (\boldsymbol{n}\,.\,\hat{\boldsymbol{S}})_{K+1\ J=K}^{K\ \ \ J=K} = \sqrt{\left(\frac{K}{2K+1}\right)},$$

$$(\boldsymbol{n}\,.\,\hat{\boldsymbol{S}})_{K+1\ J=K+1}^{K\ \ \ J=K+1} = (\boldsymbol{n}\,.\,\hat{\boldsymbol{S}})_{K\ \ \ J=K+1}^{K+1\ J=K+1} = \sqrt{\left(\frac{K+2}{2K+3}\right)},$$

$$(\boldsymbol{n}\,.\,\hat{\boldsymbol{S}})_{K-1\ J=K-1}^{K\ \ \ J=K-1} = (\boldsymbol{n}\,.\,\hat{\boldsymbol{S}})_{K\ \ \ J=K-1}^{K-1\ J=K-1} = \sqrt{\left(\frac{K-1}{2K-1}\right)}.$$

Using these last relations we find for the splitting of the components of the triplet

$$\Delta E_{J=K+1} = \frac{K+2}{2K+3}\alpha, \quad \Delta E_{J=K} = \alpha, \quad \Delta E_{J=K-1} = \frac{K-1}{2K-1}\alpha.$$

26*. We find first of all the off-diagonal elements of the operator w,

$$(w)_{n\Lambda\Omega'vJ'}^{n\Lambda\Omega vJ}.$$

One sees easily that the only matrix elements which are different from zero correspond to $\Omega' = \Omega \pm 1, J = J'$.

Since

$$(\hat{M}_\xi)_{n\Lambda\Omega\pm1}^{n\Lambda\Omega} = \mp i(\hat{M}_\eta)_{n\Lambda\Omega\pm1}^{n\Lambda\Omega},$$

we have

$$(w)_{n\Lambda\Omega\pm1vJ}^{n\Lambda\Omega vJ} = \frac{\hbar^2}{M}\left[\pm\frac{\partial}{\partial\theta}+\frac{i}{\sin\theta}\frac{\partial}{\partial\varphi}+(\Omega\pm1)\cot\theta\right]_{\Delta\Omega\pm1J}^{\Delta\Omega J}\left(\frac{1}{\rho^2}\hat{M}_\eta\right)_{n\Lambda\Omega\pm1v}^{n\Lambda\Omega v}.$$

To evaluate the matrix elements

$$\left[\pm\frac{\partial}{\partial\theta}+\frac{i}{\sin\theta}\frac{\partial}{\partial\varphi}+(\Omega\pm1)\cot\theta\right]_{\Delta\Omega\pm1J}^{\Delta\Omega J},$$

we note that if we put in the operator $J_\xi + iJ_\eta$ (see problem 13 of section 8) $\varphi = 0$ and $-i(\partial/\partial\varphi)\to\hat{M}_\zeta$, we have

$$(J_\xi \pm iJ_\eta)_{\substack{\varphi\to0\\-i(\partial/\partial\varphi)\to\hat{M}_\zeta}} = \mp i\left[\pm\frac{\partial}{\partial\theta}+\frac{i}{\sin\theta}\frac{\partial}{\partial\varphi}+(\Omega\pm1)\cot\theta\right].$$

If we evaluate this last equation we have

$$\left[\pm\frac{\partial}{\partial\theta}+\frac{i}{\sin\theta}\frac{\partial}{\partial\varphi}+(\Omega\pm 1)\cot\theta\right]^{\Lambda\Omega J}_{\Lambda\Omega\pm 1 J}=\pm i\sqrt{[(J\mp\Omega)(J\pm\Omega+1)]}.$$

In the case of small vibrations near the equilibrium position the matrix element

$$\left(\frac{1}{\rho^2}\hat{M}_\eta\right)^{n\Lambda\Omega v}_{n\Lambda\Omega\pm 1 v}$$

can be put approximately equal to

$$\frac{1}{\rho_0^2}(\hat{M}_\eta)^{n\Lambda\Omega}_{n\Lambda\Omega\pm 1},$$

where ρ_0 is the equilibrium distance between the nuclei. Since

$$(\hat{L}_\eta+\hat{S}_\eta)^{n\Lambda\Omega}_{n\Lambda\Omega\pm 1}=(\hat{S}_\eta)^{n\Lambda\Omega}_{n\Lambda\Omega\pm 1}=\pm\frac{i}{2}\sqrt{[S(S+1)-\Sigma(\Sigma\pm 1)]},$$

we have finally:

$$(w)^{n\Lambda\Omega vJ}_{n\Lambda\Omega\pm 1vJ}=B_0\sqrt{[S(S+1)-\Sigma(\Sigma\pm 1)]}\sqrt{[J(J+1)-(\Lambda+\Sigma)(\Lambda+\Sigma\pm 1)]}.$$

Here $B_0=\hbar^2/2M\rho_0^2$ is the value of the rotational constant in the equilibrium state which corresponds to $\rho=\rho_0$.

In the general case the doublet splitting may be of the same order of magnitude as the matrix elements we have evaluated. If we consider, therefore, the displaced levels of the doublet term we can apply perturbation theory in a somewhat different form. Instead of the functions

$$\psi_{n\Lambda\Lambda+\frac{1}{2}vJ},\quad \psi_{n\Lambda\Lambda-\frac{1}{2}vJ}$$

we use as zeroth approximation a linear combination of them

$$\psi=c_1\psi_{n\Lambda\Lambda+\frac{1}{2}vJ}+c_2\psi_{n\Lambda\Lambda-\frac{1}{2}vJ}.$$

If we substitute this expression in the perturbed equation and use standard methods we find the secular equation

$$\begin{vmatrix} E^{(0)}_{n\Lambda\Lambda+\frac{1}{2}vJ}-E & w^{n\Lambda\Lambda+\frac{1}{2}vJ}_{n\Lambda\Lambda-\frac{1}{2}vJ} \\ w^{n\Lambda\Lambda-\frac{1}{2}vJ}_{n\Lambda\Lambda+\frac{1}{2}vJ} & E^{(0)}_{n\Lambda\Lambda-\frac{1}{2}vJ}-E \end{vmatrix}=0.$$

From the solution of the secular equation it follows that

$$E=\tfrac{1}{2}E^{(0)}\pm\tfrac{1}{2}\sqrt{\{\Delta E^{(0)}+4B_0^2[(J+\tfrac{1}{2})^2-\Lambda^2]\}},\tag{1}$$

where

$$E^{(0)}=E^{(0)}_{n\Lambda\Lambda+\frac{1}{2}vJ}+E^{(0)}_{n\Lambda\Lambda-\frac{1}{2}vJ},$$

$$\Delta E^{(0)}=E^{(0)}_{n\Lambda\Lambda+\frac{1}{2}vJ}-E^{(0)}_{n\Lambda\Lambda-\frac{1}{2}vJ}.$$

In the *a* case when the multiplet splitting is large compared to the distance between rotational levels, it follows approximately from equation (1) that

$$E_1 = E^{(0)}_{n\Lambda\Lambda+\frac{1}{2}vJ} + \frac{B_0^2[(J+\frac{1}{2})^2-\Lambda^2]}{\Delta E^{(0)}},$$

$$E_2 = E^{(0)}_{n\Lambda\Lambda-\frac{1}{2}vJ} - \frac{B_0^2[(J+\frac{1}{2})^2-\Lambda^2]}{\Delta E^{(0)}}.$$

In the *b* case we get from equation (1)

$$E_{1,2} = \tfrac{1}{2}E^{(0)} \pm B_0[(J+\tfrac{1}{2})^2-\Lambda^2] \pm \frac{\Delta E^{(0)}}{8B_0[(J+\frac{1}{2})^2-\Lambda^2]}.$$

27. $K = 0, 2, 4, \ldots$, if the total spin $S = 2$, or $S = 0$; $K = 1, 3, 5, \ldots$, if the total spin $S = 1$.

28. The magnetic moment of the molecule is equal to

$$\frac{e\hbar}{2\mu c}(\Lambda+2\Sigma)\,\boldsymbol{n},$$

where $\boldsymbol{n}$ is a unit vector in the direction of the molecular axis.

To determine the splitting of the energy it is necessary to average the quantity

$$-\frac{e\hbar}{2\mu c}(\Lambda+2\Sigma)(\boldsymbol{n}\,.\,\mathscr{H})$$

over a rotational state, that is to determine the matrix element

$$(\boldsymbol{n})^{JM_j}_{JM_j}.$$

Since $\hat{\boldsymbol{J}}$ is a vector quantity which is conserved, it is clear that the matrix elements of the vector $\boldsymbol{n}$ must be proportional to the matrix elements of $\hat{\boldsymbol{J}}$, that is,

$$(\boldsymbol{n})^{JM_j'}_{JM_j} \sim (\hat{\boldsymbol{J}})^{JM_j'}_{JM}.$$

If we consider $\boldsymbol{n}$ as an operator we have

$$\hat{\boldsymbol{n}} = \text{const}\,\hat{\boldsymbol{J}}.$$

To determine the constant we multiply the last expression to the left and to the right by $\hat{\boldsymbol{J}}$. Since the eigenvalues of $\hat{\boldsymbol{J}}^2$ are equal to $J(J+1)$, and $(\hat{\boldsymbol{J}}\,.\,\hat{\boldsymbol{n}})$ is equal to Ω, we have

$$\hat{\boldsymbol{n}} = \frac{\Omega}{J(J+1)}\hat{\boldsymbol{J}}.$$

In this way the operator of the perturbation energy is equal to

$$-\frac{e\hbar}{2\mu c}(\Lambda+2\Sigma)\,.\,\frac{\Omega}{J(J+1)}(\mathscr{H}\,.\,\hat{\boldsymbol{J}}).$$

If we evaluate the diagonal matrix elements we get for the energy splitting the following expression:

$$\Delta E_{M_j} = -\frac{e\hbar\mathscr{H}}{2\mu c}(\Lambda+2\Sigma)\frac{\Omega}{J(J+1)}M_j.$$

29. The perturbation operator in the given case is, as one can easily verify, of the following form:

$$-\frac{e\hbar}{2\mu c}\left(\mathscr{H}\cdot\left[\frac{\Lambda^2}{K(K+1)}\hat{\boldsymbol{K}}+2\hat{\boldsymbol{S}}\right]\right).$$

It follows from this that the Zeeman splitting is equal to

$$\Delta E_M = -M_j\frac{e\hbar\mathscr{H}}{2\mu c}\left[\Lambda^2\frac{J(J+1)-S(S+1)+K(K+1)}{2K(K+1)J(J+1)}\right.$$
$$\left.+\frac{J(J+1)+S(S+1)-K(K+1)}{J(J+1)}\right].$$

30. The energy of the Zeeman splitting is equal to

$$\Delta E_{M_K M_S} = -\frac{e\hbar}{2\mu c}\mathscr{H}\cdot\left[\frac{\Lambda^2}{K(K+1)}M_K+2M_S\right].$$

31*. Since the interaction energy of the magnetic moment with the external magnetic field is of the same order of magnitude as the spin–dipole interaction it is necessary to consider the two simultaneously. The perturbation operator has the form

$$\hat{V} = A(\boldsymbol{n}\,.\,\hat{\boldsymbol{S}})-\mu_0\Lambda(\boldsymbol{n}\,.\,\boldsymbol{\mathscr{H}})-2\mu_0(\hat{\boldsymbol{S}}\,.\,\boldsymbol{\mathscr{H}}).$$

As zeroth approximation wave functions we use the wave functions of the states in which the angular momentum K and the components of $\boldsymbol{K}$ and $\boldsymbol{S}$ along the direction of the magnetic field have well-defined values. The z-axis is taken along the magnetic field. Since the component of the total angular momentum along the direction of the magnetic field is conserved we can, for the case of a doublet term, apply the perturbation theory for two-fold degeneracy. If we evaluate the matrix elements of the perturbation operator, we have

$$V^{M_K,-\frac12}_{M_K,-\frac12} = -M_K\frac{\Lambda}{2K(K+1)}A-M_K\frac{\Lambda^2}{K(K+1)}\mu_0\mathscr{H}+\mu_0\mathscr{H},$$

$$V^{M_K-1,\frac12}_{M_K-1,\frac12} = (M_K-1)\frac{\Lambda}{2K(K+1)}A-(M_K-1)\frac{\Lambda^2}{K(K+1)}\mu_0\mathscr{H}-\mu_0\mathscr{H},$$

$$V^{M_K,-\frac12}_{M_K-1,\frac12} = \tfrac12 A\frac{\Lambda}{K(K+1)}\sqrt{[(K-M+1)(K+M)]} = \tfrac12 A(n_x+in_y)^{M_K}_{M_K-1},$$

$$V^{M_K-1,\frac12}_{M_K,-\frac12} = \tfrac12 A(n_x-in_y)^{M_K-1}_{M_K} = \tfrac12 A\frac{\Lambda}{K(K+1)}\sqrt{[(K-M+1)(K+M)]}.$$

Substituting those expressions into the secular equation and solving it we get

$$E_{1,2}^{(1)} = \frac{\Lambda^2}{K(K+1)}(M_K - \tfrac{1}{2})\,\mu_0\mathscr{H} - \frac{A\Lambda}{4K(K+1)}$$

$$\pm \frac{1}{2K(K+1)}\sqrt{\{[A\Lambda(M-\tfrac{1}{2}) - \Lambda^2\mu_0\mathscr{H} + 2\mu_0\mathscr{H}(K+1)K]^2}$$

$$+ A^2\Lambda^2(K+M_K)(K-M_K+1)\}.$$

We consider two limiting cases.

If $\mu_0\mathscr{H} \gg A$, we get for $E_{1,2}^{(1)}$ the expression

$$E_{1,2}^{(1)} = -\frac{\Lambda^2}{K(K+1)}(M_K - \tfrac{1}{2} \mp \tfrac{1}{2})\,\mu_0\mathscr{H} \mp \mu_0\mathscr{H},$$

which coincides with the equation obtained in the preceding problem.

For $A \gg \mu_0\mathscr{H}$ we get

$$E_1^{(1)} = \frac{A\Lambda}{2(K+1)} - (M_K - \tfrac{1}{2})\left[\frac{\Lambda^2}{(K+1)(K+\tfrac{1}{2})} + \frac{1}{K+\tfrac{1}{2}}\right]\mu_0\mathscr{H},$$

$$E_2^{(1)} = -\frac{A\Lambda}{2K} - (M_K - \tfrac{1}{2})\left[\frac{\Lambda^2}{K(K+\tfrac{1}{2})} - \frac{1}{K+\tfrac{1}{2}}\right]\mu_0\mathscr{H}.$$

The second terms of these equations, that is, the terms which are linear in $\mathscr{H}$, agree with the corresponding expressions which we obtain if we substitute in the equation found in problem 28 of section 8

$$J = K \pm \tfrac{1}{2}, \quad S = \tfrac{1}{2}, \quad \text{and} \quad M_j = M_K - \tfrac{1}{2}.$$

32. Because of the axial symmetry the dipole moment of the molecule is directed along the line connecting the nuclei, that is

$$\boldsymbol{p} = p\boldsymbol{n}.$$

If we proceed as in the solution of problem 27 of section 8, we find

$$\Delta E_{M_j} = -\mathscr{E}p\frac{\Omega}{J(J+1)}M_j.$$

33. $$\Delta E_{M_j} = -\mathscr{E}pM_j\Lambda\frac{J(J+1) - S(S+1) + K(K+1)}{2K(K+1)J(J+1)}.$$

34. The potential energy $V(x_1, x_2)$ of the system is to lowest order in R^{-1}

$$V(x_1, x_2) = \tfrac{1}{2}kx_1^2 + \tfrac{1}{2}kx_2^2 + \frac{e^2x_1x_2}{R^3},$$

where x_1 and x_2 are the displacements of the two electrons from their respective centres.

The Schrödinger equation for the two electrons involved can thus be reduced to one for two linear harmonic oscillators by bringing the kinetic and the potential energies simultaneously in the form of squares only, and the eigenvalues of the energy are

$$\frac{\hbar}{2\sqrt{\mu}}\left[(2n_1+1)\sqrt{\left(k+\frac{e^2}{R^3}\right)}+(2n_2+1)\sqrt{\left(k-\frac{e^2}{R^3}\right)}\right], \quad n_{1,2}=0,1,\ldots.$$

In the ground state we have $n_1 = n_2 = 0$, and expanding the radicals, we find for the dispersion energy of attraction, again to lowest order in R^{-1},

$$\frac{e^4}{8k^2R^6}\hbar\sqrt{\left(\frac{k}{\mu}\right)}.$$

35*. The interaction energy consists of two parts. First of all, there is a contribution due to the polarisability α of the hydrogen atom (see problem 47 of section 7). The field $\mathcal{E}$ of the H^+ ion at the position of the hydrogen atom is e/R^2 so that that part of the interaction energy is given by the relation

$$E_{\text{pol}} = -\tfrac{1}{2}\alpha\mathcal{E}^2 = -\tfrac{9}{4}\frac{\hbar^2}{\mu a_B^2}\left(\frac{a_B}{R}\right)^4. \tag{1}$$

The second contribution comes from the symmetry properties of the problem (compare problems 6 and 12 of section 2). Let $\psi_0(x, y, z)$ be the wavefunction of the electron corresponding to it moving in the vicinity of proton 1 which is situated at $x = \frac{1}{2}R, y = z = 0$ (see fig. 43a),

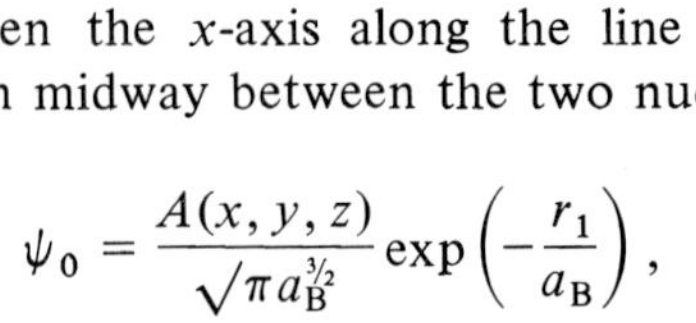

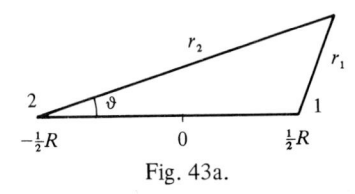

Fig. 43a.

where we have chosen the x-axis along the line connecting the two protons and the origin midway between the two nuclei. In that case ψ_0 will be of the form

$$\psi_0 = \frac{A(x,y,z)}{\sqrt{\pi a_B^{3/2}}}\exp\left(-\frac{r_1}{a_B}\right), \tag{2}$$

with $A(x, y, z)$ a slowly varying function which satisfies the boundary condition

$$A(x,y,z) \to 1 \quad \text{as} \quad x,y,z \to \tfrac{1}{2}R, 0, 0. \tag{3}$$

The function ψ_0 must satisfy the equation

$$-\frac{\hbar^2}{2\mu}\nabla^2\psi_0+(U-E_0)\psi_0 = 0\,, \tag{4}$$

where

$$U = -\frac{e^2}{r_1}-\frac{e^2}{r_2}+\frac{e^2}{R}\,, \tag{5}$$

while E_0 is the binding energy of the hydrogen atom. If we denote by ψ_{00} the function ψ_0 with $A(x, y, z) \equiv 1$, ψ_{00} satisfies the equation

$$-\frac{\hbar^2}{2\mu}\nabla^2\psi_{00}+\left(-\frac{e^2}{r_1}-E_0\right)\psi_{00} = 0\,, \tag{6}$$

so that we get for $A(x, y, z)$ the equation

$$\frac{\partial A}{\partial x}-\frac{A}{R}+\frac{A}{\frac{1}{2}R+x} = 0\,, \tag{7}$$

where we have neglected all second derivatives and all derivatives with respect to y and z of A and where we have put $r_2 = \frac{1}{2}R+x$. The solution of (7) satisfying the boundary condition (3) is

$$A = \frac{R}{\frac{1}{2}R+x}\exp\frac{x-\frac{1}{2}R}{R}\,. \tag{8}$$

There is, of course, also a solution of (4) which is centred around proton 2. It will be $\psi_0(-x, y, z)$ (which in accordance with the assumptions of our problem can be assumed not to overlap with $\psi_0(x, y, z)$) and also corresponds to an energy E_0. The true states of the system will be the symmetric and antisymmetric combinations of $\psi_0(x, y, z)$ and $\psi_0(-x, y, z)$:

$$\psi_{s,a} = \frac{1}{\sqrt{2}}\{\psi_0(x, y, z) \mp \psi_0(-x, y, z)\}\,, \tag{9}$$

and these combinations also satisfy equation (4) with energies $E_{s,a}$. If we multiply equation (4) for $\psi_0(x, y, z)$ by $\psi_{s,a}$ and equation (4) for $\psi_{s,a}$ by $\psi_0(x, y, z)$ and integrate the difference of the two resulting equations over the half-space $x \geqslant 0$, we find, using the fact that for $x = 0$ we have $\psi_s = \sqrt{2}\,\psi_0$, $\psi_a = 0$, $\partial\psi_s/\partial x = 0$, $\partial\psi_a/\partial x = \sqrt{2}\,\partial\psi_0/\partial x$, that

$$E_{s,a}-E_0 = \mp\frac{\hbar^2}{\mu}\iint\psi_0\frac{\partial\psi_0}{\partial x}\,dy\,dz\,, \tag{10}$$

where the integration is over the plane $x = 0$.

Using expressions (2) and (8) for ψ_0 we then find

$$E_{s,a} - E_0 = \frac{\hbar^2}{\mu a_B^2}\left[\mp\frac{2}{e}\frac{R}{a_B}\exp\left(-\frac{R}{a_B}\right) - \tfrac{9}{4}\left(\frac{a_B}{R}\right)^4\right]. \tag{11}$$

We note that this result is valid, provided $R/a_B \gg 1$. The symmetric state has the lowest energy. The exact E_s has a minimum for $R \approx 2a_B$, equal to $-16{\cdot}3$ eV (compare next problem), while the minimum for E_a—which follows from (11)—occurs at $R \approx 13a_B$ and equals $-0{\cdot}002$ eV.

36*. We start from equation (4) of the preceding problem and use as a trial wavefunction

$$\psi = a\psi_1 + b\psi_2 , \tag{1}$$

where

$$\psi_1 = \frac{\alpha^{3/2}}{\sqrt{\pi}}\exp(-\alpha r_1), \quad \psi_2 = \frac{\alpha^{3/2}}{\sqrt{\pi}}\exp(-\alpha r_2), \tag{2}$$

so that we have two parameters, α and R, in our problem. In contrast to the discussion in the preceding problem we now must take into account the overlap integral

$$S = \int \psi_1 \psi_2 d^3r . \tag{3}$$

From symmetry considerations it follows that the lowest energy will correspond to the case where $a = b = \{2(1+S)\}^{-1/2}$, where we have used the requirement that ψ must be normalized. The energy of the system is now found from the equation

$$E = \frac{1}{1+S}\left\{\int \psi_1 \hat{H}\psi_1 d^3r + \int \psi_2 \hat{H}\psi_1 d^3r\right\}, \tag{4}$$

where

$$\hat{H} = -\frac{\hbar^2}{2\mu}\nabla^2 + \frac{e^2}{R} - \frac{e^2}{r_1} - \frac{e^2}{r_2}. \tag{5}$$

Hence we find

$$E = \frac{1}{1+S}\frac{\hbar^2}{\mu a_B^2}\left[-\tfrac{1}{2}\alpha^2 + \frac{\alpha(\alpha-1) - C + (\alpha-2)I}{1+S} + \frac{1}{R}\right], \tag{6}$$

where C is the dimensionless Coulomb interaction integral

$$C = a_B\frac{\alpha^3}{\pi}\int \frac{\exp(-2\alpha r_1)}{r_2} d^3\boldsymbol{r} , \tag{7}$$

and I the dimensionless exchange integral,

$$I = a_B\frac{\alpha^3}{\pi}\int \frac{\exp(-\alpha r_1 - \alpha r_2)}{r_2} d^3\boldsymbol{r} . \tag{8}$$

Integral (7) can be evaluated by expanding r_2^{-1} in terms of r_1/R (if $r_1 < R$) or R/r_1 (if $r_1 > R$) (see fig. 43a)

$$\frac{1}{r_2} = \begin{cases} \dfrac{1}{R}\sum_{l=0}^{\infty}\left(\dfrac{r_1}{R}\right)^l P_l(\cos\vartheta), & \text{if } r_1 < R, \\ \dfrac{1}{r_1}\sum_{l=0}^{\infty}\left(\dfrac{R}{r_1}\right)^l P_l(\cos\vartheta), & \text{if } r_1 > R, \end{cases} \tag{9}$$

where the P_l are Legendre polynomials. Substituting (9) into (7) one finds

$$C = \frac{a_B}{R}[1-(1+\alpha R)e^{-2\alpha R}]. \tag{10}$$

To evaluate I we introduce prolate ellipsoidal coordinates:

$$\xi = \frac{r_1+r_2}{R}, \quad \eta = \frac{r_1-r_2}{R}, \tag{11}$$

and the angle φ around the molecular axis. The integral then becomes

$$\begin{aligned} I &= a_B \frac{\alpha^3}{\pi}\int_1^{\infty} d\xi \int_{-1}^{+1} d\eta \int_0^{2\pi} d\varphi \left(\frac{R}{2}\right)^2 (\xi^2-\eta^2)\frac{e^{-\alpha R\xi}}{\xi-\eta} \\ &= \alpha a_B(1+\alpha R)e^{-\alpha R}. \end{aligned} \tag{12}$$

The overlap integral can be evaluated similarly, and we find

$$S = [1+\alpha R+\tfrac{1}{3}(\alpha R)^2]e^{-\alpha R}. \tag{13}$$

Substituting (10), (12), and (13) into (6) and using as variational parameters α and $\beta \equiv \alpha R$, we can write the energy in the form

$$E = F(\beta)\alpha^2 + G(\beta)\alpha, \tag{14}$$

so that minimization with respect to α leads to

$$E = -\frac{\{G(\beta)\}^2}{4F(\beta)}. \tag{15}$$

To obtain the minimum with respect to β one best proceeds graphically. The result is that the minimum occurs for $\alpha a_B \sim 1{\cdot}25$, $\beta \sim 2{\cdot}5$, or $R \sim 2a_B$, and $E \sim 0{\cdot}6\,(\hbar^2/\mu a_B^2)$.

To find the dissociation energy D we must deduct from this value of E the binding energy of the hydrogen atom, $-\hbar^2/2\mu a_B^2$, so that we find

$$D \sim 0{\cdot}1\,\hbar^2/\mu a_B^2.$$

Strictly speaking, one should correct for the energy of the zero-point vibrations of the molecule, but for the rough estimate which we have just given this makes no significant difference. If one performs the calculations with greater accuracy, one can estimate the magnitude of the zero-point energy by looking at the curve $E(R)$ in the neighbourhood of the minimum. The results then are $\alpha a_B = 1\cdot 24$, $R = 2\cdot 08 a_B$, $E = -0\cdot 587\ (\hbar^2/\mu a_B^2)$, and for the zero-point energy $\frac{1}{2}\hbar\omega = 0\cdot 0046\ (\hbar^2/\mu a_B^2)$. The experimental value of D lies, in fact, very close to $0\cdot 1\ \hbar^2/\mu a_B^2$, so that the rough estimates we made give a slightly better value than, if we had taken greater care.

37*. For large values of R we can neglect exchange effects, that is, assume that the first electron is near the nucleus a and the second electron near the nucleus b (see fig. 44).

The interaction between the two atoms which is of the form

$$V = \frac{1}{R} - \frac{1}{r_{a2}} - \frac{1}{r_{b1}} + \frac{1}{r_{12}} \tag{1}$$

must be considered as a small perturbation.

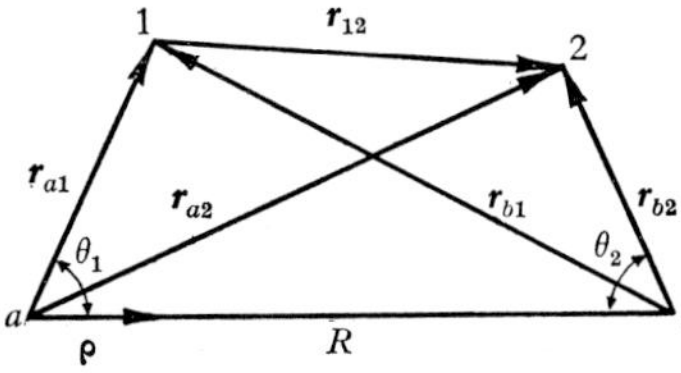

Fig. 44.

In the first approximation the interaction energy of the two atoms is equal to the diagonal matrix element of V, that is

$$\int \psi_0(r_{1a})\,\psi_0(r_{2b})\,V\psi_0(r_{1a})\,\psi_0(r_{2b})\,d\tau_1\,d\tau_2,$$

where

$$\psi_0(r_{1a}) = 2\exp(-r_{1a}); \quad \psi_0(r_{2b}) = 2\exp(-r_{2b}).$$

In the S-state the diagonal matrix elements, that is, the average values of the dipole, quadrupole, ... moments, are equal to zero so that for an evaluation of the interaction energy it is necessary to go over to second-order perturbation theory.

In the perturbation operator (1) we restrict ourselves to the dipole–dipole interaction as it decreases most slowly with distance. To obtain the operator of the dipole–dipole interaction we expand the potential V

in decreasing powers to R. An expansion in spherical harmonics gives

$$\frac{1}{r_{b1}} = \frac{1}{|R\boldsymbol{\rho} - \boldsymbol{r}_{1a}|} = \sum_{\lambda=0} \frac{r_{a1}^{\lambda}}{R^{\lambda+1}} P_\lambda(\cos\theta) = \frac{1}{R} + \frac{(\boldsymbol{r}_{a1}\cdot\boldsymbol{\rho})}{R^2} + \frac{3(\boldsymbol{r}_{a1}\cdot\boldsymbol{\rho})^2 - \boldsymbol{r}_{a1}^2}{2R^3} + \ldots,$$

$$\frac{1}{r_{12}} = \frac{1}{|R\boldsymbol{\rho} + \boldsymbol{r}_{b2} - \boldsymbol{r}_{a1}|} = \frac{1}{R} + \frac{(\boldsymbol{r}_{a1} - \boldsymbol{r}_{b2}\cdot\boldsymbol{\rho})}{R^2} + \frac{3(\boldsymbol{r}_{a1} - \boldsymbol{r}_{b2}\cdot\boldsymbol{\rho})^2 - (\boldsymbol{r}_{a1} - \boldsymbol{r}_{b2})^2}{2R^3} + \ldots,$$

$$\frac{1}{r_{a2}} = \frac{1}{R} + \frac{(\boldsymbol{r}_{b2}\cdot\boldsymbol{\rho})}{R^2} + \frac{3(\boldsymbol{r}_{b2}\cdot\boldsymbol{\rho})^2 - \boldsymbol{r}_{b2}^2}{2R^3} + \ldots.$$

Substituting this expansion into V we find an expression for the dipole–dipole interaction

$$V = -\frac{2z_1 z_2 - x_1 x_2 - y_1 y_2}{R^3}, \tag{2}$$

where the z-axis is taken along the line connecting the nuclei. We have already shown that the average value (2) over the unperturbed eigenfunction $\psi = \psi_0(r_{a1})\psi_0(r_{b2})$ is equal to zero. The non-diagonal elements of (2) which correspond to a transfer from the ground state to excited states can be written in the form

$$V_{00}^{mn} = -\frac{2z_{0m} z_{0n} - x_{0m} x_{0n} - y_{0m} y_{0n}}{R^3}.$$

From the selection rules we see that the matrix elements z_{0n}, x_{0n}, y_{0n} are different from zero only for transitions from the ground state to the states

$$\psi_n(r)\cos\theta, \quad \psi_n(r)\sin\theta\cos\varphi,$$

where these three matrix elements are equal to each other. The interaction energy in second approximation is equal to

$$E^{(2)} = \sum_{mn} \frac{(V_{00}^{mn})^2}{2E_0 - E_m - E_n} = \frac{1}{R^6} \sum_{mn} \frac{4z_{0m}^2 z_{0n}^2 + x_{0m}^2 x_{0n}^2 + y_{0m}^2 y_{0n}^2}{2E_0 - E_m - E_n}$$

or

$$E^{(2)} = \frac{6}{R^6} \sum_{mn} \frac{z_{0m}^2 z_{0n}^2}{2E_0 - E_m - E_n}. \tag{3}$$

Since $E_0 < E_m$ and $E_0 < E_n$, $E^{(2)}$ is negative, and we find thus that two atoms in a non-excited state, which are at a large distance apart, attract each other with a force which is inversely proportional to the sixth power of the distance. To evaluate the sum in (3) approximately, we note that the difference in energy between different excited levels is small compared to the difference of energy between the ground state and the excited levels. We can, therefore, write expression (3) approximately in the form

$$E^{(2)} = \frac{3}{R^6} \sum_m z_{0m}^2 \sum_n \frac{z_{0n}^2}{E_0 - E_n}.$$

From the theory of the quadratic Stark effect it follows that

$$\sum \frac{z_{0n}^2}{E_0 - E_n} = -\tfrac{1}{2}\alpha,$$

where α is the polarisation of the atom. For the hydrogen atom ground state $\alpha = 4{\cdot}5$ atomic units. To evaluate the sum $\sum\limits_{m \neq 0} z_{0m}^2$ we use the matrix sum rule

$$(AB)_{nk} = \sum_m A_{nm} B_{mk},$$

or

$$\int \psi_n^* (AB) \psi_k \, d\tau = \sum_m \left(\int \psi_n^* A \psi_m \, d\tau \, . \int \psi_m^* B \psi_k \, d\tau \right).$$

Putting in the last relation $A = B = z$, $n = k = 0$, we get

$$(z^2)_{00} = \sum_m z_{0m} z_{m0} = \sum z_{0m}^2$$

or

$$\sum_{m \neq 0} z_{0m}^2 = \sum_m z_{0m}^2 - z_{00}^2 = (z^2)_{00} - z_{00}^2.$$

Since in an S-state $z_{00} = 0$ and $(z^2)_{00} = \frac{1}{3}(r^2)_{00}$, because of symmetry, we get

$$\sum_{m \neq 0} z_{0m}^2 = \tfrac{1}{3}(r^2)_{00} = 1.$$

We find thus finally for $V(r)$ the expression

$$V(R) = -\frac{6{\cdot}75}{R^6}.$$

To ascertain how the interaction forces between two neutral spherical symmetric hydrogen atoms arise, we consider the wave function of the system. For the wave function of the system we find in first approximation

$$\psi = \psi_0(r_{a1}) \psi_0(r_{b2}) \left[1 + \frac{1}{2E_0 R^3} (x_1 x_2 + y_1 y_2 - 2 z_1 z_2) \right].$$

The probability density $w(1, 2)$ has, if we neglect terms multiplied by R^{-6}, the form

$$w(r_{a1}, r_{b2}) = w_0(r_{a1}) w_0(r_{b2}) \left[1 + \frac{1}{E_0 R^3} (x_1 x_2 + y_1 y_2 - 2 z_1 z_2) \right].$$

If there is no interaction between the atoms the probability density is simply equal to the product of $w(1)$ and $w(2)$, that is, there is in that case no correlation whatever between the positions of the electrons. In the case of an interaction the position of the first electron is not independent of the position of the second one. The electrons occupy statistically more often positions where the mutual potential energy has the least possible value.

The interaction force can thus in first approximation be explained not by the deformation of the electron clouds but by the correlation between the electron positions.

38. The procedure is very similar to that of the preceding problem, but now we get a result in first-order perturbation theory. The perturbing energy is again of the form

$$V = \frac{e^2}{R^3}(x_1x_2+y_1y_2-2z_1z_2)',$$

and we must use degenerate-level perturbation theory. The p-states are three-fold degenerate and as we have two electrons, we have a six-fold degeneracy altogether. If r_0 is the quantity given by the equation

$$r_0 = \int_0^\infty F_0(r)F_1(r)r^3 dr,$$

where $F_0(r)$ and $F_1(r)$ are the radial parts of the s- and p-state wave-functions,

$$F_0(r) \propto \exp(-r/a_B), \quad F_1(r) \propto r\exp(-r/2a_B),$$

and if v is given by the equation

$$v = \frac{e^2 r_0}{R^3},$$

the secular equation for the energy correction E_1 becomes

$$\begin{vmatrix} -E_1 & v & 0 & 0 & 0 & 0 \\ v & -E_1 & 0 & 0 & 0 & 0 \\ 0 & 0 & -E_1 & -2v & 0 & 0 \\ 0 & 0 & -2v & -E_1 & 0 & 0 \\ 0 & 0 & 0 & 0 & -E_1 & v \\ 0 & 0 & 0 & 0 & v & -E_1 \end{vmatrix} = 0,$$

where the first two rows (columns) correspond to one or other of the two hydrogen atoms being in a p-state with $m = 1$, the next two to states with $m = 0$, and the final two to states with $m = -1$. We see that now the interaction energy is proportional to R^{-3}.

39*. Let us prove the additivity for a system consisting of three atoms. From our calculations it will be clear that it can be applied to an arbitrary number of atoms. We write the interaction energy in the following form

$$V = V(1,2)+V(2,3)+V(3,1),$$

where we have denoted by 1, 2, and 3 all the coordinates of the first, second, and third atoms. We consider the interaction of atoms which are at large distances apart. In that case exchange forces play no role whatever. If we ignore exchange forces, we can put the wave function of three atoms in zeroth approximation in the form

$$\psi = \psi_{ai}(1)\,\psi_{bk}(2)\,\psi_{cl}(3),$$

where i, k, l indicate the quantum states of the atoms a, b, c. The functions $\psi_{ai}(1)$, belonging to different values of i, are orthogonal. The same can be said also about the functions $\psi_{bk}(2)$ and $\psi_{cl}(3)$.

The perturbation energy in second approximation is of the form

$$\epsilon = V_{000}^{000}+\Sigma' \frac{|V_{ikl}^{000}|^2}{E_{a0}+E_{b0}+E_{c0}-E_{ai}-E_{bk}-E_{cl}}. \tag{1}$$

The prime on the summation sign indicates that i, k, l cannot be simultaneously equal to zero. The first term is the classical multipole interaction. In our case it is equal to zero. In expression (1) all terms where at the same time $i\neq 0$, $k\neq 0$, $l\neq 0$ disappear because of the orthogonality of the functions.

Three partial sums with

$$i=k=0,\ l\neq 0; \quad i=l=0,\ k\neq 0; \quad k=l=0,\ i\neq 0$$

refer to the polarisation interaction of the lth, kth, and ith atoms in the resulting field of the two remaining atoms. In the case where the distribution of changes in the atoms has spherical symmetry these sums are also equal to zero. It is necessary to remember that these sums cannot be obtained by taking additively into account the interaction energy of each pair of atoms. We must finally consider those terms where two indices are different from zero. Thus, if we make again the same assumption about the charge distribution in the atoms the interaction energy can be expanded in three partial sums

$$\epsilon = \sum_{i\neq 0\, k\neq 0}' \frac{|V_{ik0}^{000}|^2}{E_{a0}+E_{b0}-E_{ai}-E_{bk}} + \sum_{k\neq 0\, l\neq 0}' \frac{|V_{0kl}^{000}|^2}{E_{b0}+E_{c0}-E_{bk}-E_{cl}}$$
$$+\sum_{i\neq 0\, l\neq 0} \frac{(V_{i0l}^{000})^2}{E_{a0}+E_{c0}-E_{ai}-E_{cl}}. \tag{2}$$

Because of the orthonormality of the eigenfunctions of the atom we get for the matrix elements

$$V_{ik0}^{000} = \int \psi_{a0}^*(1)\,\psi_{b0}^*(2)\,\psi_{c0}^*(3)\,\{V(1,2)+V(2,3)+V(3,1)\}$$
$$\times\,\psi_{ai}(1)\,\psi_{bk}(2)\,\psi_{c0}(3)\,d\tau_1\,d\tau_2$$
$$= \int \psi_{a0}^*(1)\,\psi_{b0}^*(2)\,V(1,2)\,\psi_{ai}(1)\,\psi_{bk}(2)\,d\tau_1\,d\tau_2 = [V(1,2)]_{ik}^{00}.$$

Expression (2) consists thus of three terms, each of which is the dispersion interaction of a pair of atoms. One sees easily that these considerations can be extended to an arbitrary number of atoms.

In the case when the distance between the atoms is not very large, one must take into account the transition of electrons from one atom to another, that is, exchange forces.

40. We are dealing with a system with cylindrical symmetry, and can write the Hamiltonian in the form

$$\hat{H} = \hat{\Omega}(\rho, z) - \frac{\hbar^2}{2\mu}\frac{\partial^2}{\partial\chi^2}, \tag{1}$$

where ρ, z, and χ are cylindrical coordinates, and where (V: potential energy)

$$\hat{\Omega} = -\frac{\hbar^2}{2\mu}\left(\frac{\partial^2}{\partial z^2}+\frac{\partial^2}{\partial \rho^2}+\frac{1}{\rho}\frac{\partial}{\partial\rho}\right)+V(\rho, z). \tag{2}$$

The eigenfunctions of the molecule which are characterised by the two quantum numbers λ and n are of the form

$$\phi_{\lambda,n} = \frac{1}{\sqrt{(2\pi)}}\psi_{\lambda,n}(\rho, z)\exp(i\lambda\chi), \tag{3}$$

while the $\psi_{\lambda,n}$ satisfy the relation

$$\sum_n \psi_{\lambda,n}^*(\rho, z)\,\psi_{\lambda,n}(\rho', z') = \frac{1}{\sqrt{(\rho\rho')}}\,\delta(\rho-\rho')\,\delta(z-z'). \tag{4}$$

We shall prove (A) for the case $\lambda' = \lambda$. The other two cases follow in a similar way. We have

$$S_0 = \sum_{n'} f_{\lambda n;\,\lambda n'} = -\frac{2\mu}{3\hbar^2}\sum_{n'}(E_{n'}-E_n)\left|\int \phi_{\lambda,n}^*\, z\phi_{\lambda,n'}\,d\tau\right|^2,$$

where we have used the fact that for $\lambda' = \lambda$: $P_{\alpha\beta} = (ez)_{\alpha\beta}$ and where $d\tau = \rho\,d\rho\,dz\,d\chi = d\omega\,d\chi$.

Using the fact that the E_n are the energy eigenvalues, using (3) for the eigenfunctions and (1) and (4), and integrating by parts, we find

(primes on a function indicate that its arguments are primed)

$$S_0 = -\frac{2\mu}{3\hbar^2}\sum_{n'}\iint d\tau\, d\tau'\, \phi^*_{\lambda,n}\,\phi'_{\lambda,n}\, zz'\phi'^*_{\lambda,n'}(\hat{H}-E_n)\,\phi_{\lambda,n'}$$

$$= -\frac{2\mu}{3\hbar^2}\iint d\omega\, d\omega'\, \psi^*_{\lambda,n}\psi_{\lambda,n}\, zz'\left[\sum_{n'}\psi'^*_{\lambda,n'}\left(\hat{\Omega}+\frac{\hbar^2}{2\mu}\frac{\lambda^2}{\rho^2}-E_n\right)\psi_{\lambda,n'}\right]$$

$$= -\frac{2\mu}{3\hbar^2}\iint d\omega\, d\omega'\, \psi^*_{\lambda,n}\psi'_{\lambda,n}\, zz'\left(\hat{\Omega}+\frac{\hbar^2}{2\mu}\frac{\lambda^2}{\rho^2}-E_n\right)\frac{1}{\sqrt{(\rho\rho')}}\,\delta(\rho-\rho')\,\delta(z-z')$$

$$= -\frac{2\mu}{3\hbar^2}\int \rho\, d\rho\, dz\, z\psi_{\lambda,n}\left(\hat{\Omega}+\frac{\hbar^2}{2\mu}\frac{\lambda^2}{\rho^2}-E_n\right) z\psi^*_{\lambda,n}.$$

As

$$\left(\hat{\Omega}+\frac{\hbar^2}{2\mu}\frac{\lambda^2}{\rho^2}-E_n\right)\psi_{\lambda,n} = 0$$

we find

$$S_0 = -\frac{1}{3}\int \frac{\partial\psi^*_{\lambda,n}}{\partial z}\, z\psi_{\lambda,n}\,\rho\, d\rho\, dz,$$

and after integration by parts, we find

$$S_0 = \tfrac{1}{3}.$$

Scattering

1*. The potential energy of the particles is

$$V(r) = -V_0 \quad (r < a),$$
$$V(r) = 0 \quad (r > a).$$

It is necessary to find the phase shifts, that is, the asymptotic form of the radial functions satisfying the equations

$$r > a \quad \chi_l'' + \left[k^2 - \frac{l(l+1)}{r^2}\right]\chi_l = 0, \quad k^2 = \frac{2\mu E}{\hbar^2},$$

and

$$r < a \quad \chi_l'' + \left[k'^2 - \frac{l(l+1)}{r^2}\right]\chi_l = 0, \quad k'^2 = \frac{2\mu(E+V_0)}{\hbar^2}$$

with the boundary condition $\chi_l(0) = 0$.

When the de Broglie wavelength is considerably larger than the dimensions of the well, the main contribution to the scattering arises from the S-wave. The solution χ_0, which satisfies the boundary condition, is of the form

$$\chi_0 = A \sin k'r \qquad (r < a),$$
$$\chi_0 = \sin(kr + \delta_0) \quad (r > a).$$

The phase δ_0 and the coefficient A follow from the condition that both the wave function and its derivative are continuous at $r = a$. We get in this way

$$\delta_0 = \arctan\left(\frac{k}{k'}\tan k'a\right) - ka.$$

The partial cross-section for $l = 0$ is thus

$$\sigma_0 = \frac{4\pi}{k^2}\sin^2\delta_0 = \frac{4\pi}{k^2}\sin^2\left[\arctan\left(\frac{k}{k'}\tan k'a\right) - ka\right]. \tag{1}$$

For small velocities of the incident particles ($k \to 0$) δ_0 will be proportional to k

$$\delta_0 \approx ka\left(\frac{\tan k_0 a}{k_0 a} - 1\right), \quad k_0^2 = \frac{2\mu V_0}{\hbar^2}. \tag{2}$$

Because of the factor $1/k^2$ the cross-section σ_0 will be

$$\sigma \approx 4\pi a^2 \left(\frac{\tan k_0 a}{k_0 a} - 1\right)^2 \quad \text{(for small } k\text{)}. \tag{3}$$

We consider the cross-section σ_0 as a function of the well depth, which determines k_0. If the well is shallow ($k_0 a \ll 1$), we have

$$\sigma_0 = 4\pi a^2 \frac{k_0^4 a^4}{9} = \frac{16\pi}{9} \frac{a^6 V_0^2 \mu^2}{\hbar^4}.$$

We note that we get from perturbation theory

$$f(\vartheta) = -\frac{1}{4\pi} \frac{2\mu}{\hbar^2} \int V(r)\, d\tau = \frac{2\mu}{\hbar^2} V_0 \frac{a^3}{3}$$

and thus

$$\sigma = 4\pi |f(\vartheta)|^2 = \frac{16\pi}{9} \frac{a^6 V_0^2 \mu^2}{\hbar^4}.$$

The cross-section increases with increasing V_0 and diverges for $k_0 a = \pi/2$. The condition $k_0 a = \pi/2$ is the condition for the appearance of the first level in the well. If we deepen the well even further, the cross-section starts to decrease again and tends to zero for $\tan k_0 a = k_0 a$. When V_0 is further increased the cross-section continues to oscillate between 0 and ∞, becoming infinite whenever a new level appears in the well. The sharp oscillation of the cross-section for the scattering of slow particles explains why the cross-section for the scattering of slow electrons by an atom can differ appreciably from the geometric cross-section.

We note that if $k_0 a$ is near to an integer times $\frac{1}{2}\pi$, we must alter equations (2) and (3). Indeed, in that case $\tan k'a$ is a large number and we cannot use the expansion which leads from equation (1) to equation (2). In that case we can still neglect the term with $ka \ll 1$ in the square brackets of equation (1). We have thus

$$\delta_0 = \arctan\left(\frac{k}{k'} \tan k'a\right)$$

and we get for the cross-section σ_0

$$\sigma_0 = \frac{4\pi[1 + O(\kappa a)]}{\kappa^2 + k^2},$$

where

$$\kappa = \frac{k'}{\tan k'a} \ll \frac{1}{a}.$$

This equation for resonance scattering gives the dependence of the cross-section on k for small values of k, for the case where the potential of the well is such that a small change in its depth or width will make a discrete level appear or disappear.

2*. $$\sigma = 4\pi a^2\left(\frac{\tanh \kappa a}{\kappa a}-1\right)^2, \quad \text{where} \quad \kappa = \frac{\sqrt{(2\mu V_0)}}{\hbar}.$$

If $V_0 \to \infty$, $\sigma = 4\pi a^2$, that is, it is four times as large as the elastic scattering cross-section for an impenetrable sphere in classical mechanics.

3*. $$\frac{d\sigma}{d\Omega} = \frac{1}{k^2}\sum_{l=0}^{\infty}(2l+1)\sin^2\delta_l$$
$$+\frac{6\cos\vartheta}{k^2}\sum_{l=0}^{\infty}(l+1)\sin\delta_l\sin\delta_{l+1}\cos(\delta_{l+1}-\delta_l)$$
$$+\frac{5}{k^2}\frac{3\cos^2\vartheta-1}{2}\sum_{l=0}^{\infty}\left[\frac{l(l+1)(2l+1)}{(2l-1)(2l+3)}\sin^2\delta_l\right.$$
$$\left.+\frac{3(l+1)(l+2)}{2l+3}\sin\delta_l\sin\delta_{l+2}\cos(\delta_{l+2}-\delta_l)\right]+\dots$$

or

$$\int_0^\pi d\sigma = \frac{4\pi}{k^2}\sum_{l=0}^{\infty}(2l+1)\sin^2\delta_l,$$

$$\int_0^\pi \cos\vartheta\, d\sigma = \frac{8\pi}{k^2}\sum_{l=0}^{\infty}(l+1)\sin\delta_l\sin\delta_{l+1}\cos(\delta_{l+1}-\delta_l),$$

$$\int_0^\pi \frac{3\cos^2\vartheta-1}{2}\,d\sigma = \frac{4\pi}{k^2}\sum_{l=0}^{\infty}\frac{l(l+1)(2l+1)}{(2l+1)(2l+3)}\sin^2\delta_l$$
$$+\frac{12\pi}{k^2}\sum_{l=0}^{\infty}\frac{(l+1)(l+2)}{2l+3}\sin\delta_l\sin\delta_{l+2}\cos(\delta_{l+2}-\delta_l).$$

4*. The radial function satisfies the equation

$$\chi_l'' + \left[k^2-\frac{l(l+1)}{r^2}-\frac{2\mu A}{\hbar^2 r^2}\right]\chi = 0$$

and the boundary conditions $\chi_l(0) = 0$ and $\chi =$ finite as $r \to \infty$. The solution satisfying these conditions is

$$\chi_l = \sqrt{(r)}\,J_\lambda(kr),$$

where

$$\lambda = \sqrt{\left[(l+\tfrac{1}{2})^2+\frac{2\mu A}{\hbar^2}\right]}.$$

From the asymptotic behaviour of $J_\lambda(kr)$ we get the phase shifts,

$$\delta_l = -\frac{\pi}{2}(\lambda-l-\tfrac{1}{2}) = -\frac{\pi}{2}\left\{\sqrt{\left[(l+\tfrac{1}{2})^2+\frac{2\mu A}{\hbar^2}\right]}-(l+\tfrac{1}{2})\right\}.$$

The independence of δ_l on k means that we get for the scattering amplitude

$$f(\vartheta, k) = \frac{1}{k} f_0(\vartheta),$$

where $f_0(\vartheta)$ is independent of the energy of the scattered particles. The scattering cross-section

$$d\sigma = \frac{1}{k^2} |f_0(\vartheta)|^2 \, d\Omega$$

is inversely proportional to the energy and is characterised by a universal angular distribution.

Since the sum

$$f(\vartheta) = \frac{1}{2ik} \sum_{l=0}^{\infty} (2l+1) P_l(\cos\vartheta) [\exp(2i\delta_l) - 1],$$

which determines the scattering amplitude diverges for $\vartheta \to 0$, it is clear that large values of l are essential for the evaluation of $f(\vartheta)$ at small values of ϑ. For large values of l we have

$$-\delta_l \approx -\frac{\pi\mu A}{(2l+1)\hbar^2} \ll 1, \qquad (1)$$

so that

$$f(\vartheta) \approx \frac{1}{k} \sum_{l=0}^{\infty} (2l+1) P_l(\cos\vartheta)\, \delta_l$$

$$\approx -\frac{\pi\mu A}{\hbar^2 k} \sum_{l=0}^{\infty} P_l(\cos\vartheta) = -\frac{\pi\mu A}{k\hbar^2} \frac{1}{2\sin\frac{1}{2}\vartheta}.$$

If

$$\frac{8\mu A}{\hbar^2} \ll 1,$$

expression (1) for δ_l is valid for all l and thus for all values of ϑ

$$f(\vartheta) \approx -\frac{\pi\mu A}{k\hbar^2} \frac{1}{2\sin\frac{1}{2}\vartheta},$$

$$d\sigma = \frac{\pi^3 \mu A^2}{2\hbar^2 E} \cot\tfrac{1}{2}\vartheta \, d\vartheta.$$

5. The Schrödinger equation of our problem is of the form

$$-\frac{\hbar^2}{2\mu} \nabla^2 \psi + V\psi = E\psi.$$

Introducing, as usual, $k^2 = 2\mu E/\hbar^2$ we get this equation in the form

$$\nabla^2 \psi + k^2 \psi = \frac{2\mu}{\hbar^2} V\psi. \tag{1}$$

We shall consider the right-hand side as an inhomogeneous term. We must thus find a solution of an inhomogeneous equation, satisfying a given boundary condition,

$$\psi \underset{r\to\infty}{\approx} \exp[i(\boldsymbol{k}_0 . \boldsymbol{r})] + f\frac{\exp(ikr)}{r}.$$

Using the expression for the Green function of equation (1),

$$G(\boldsymbol{r}-\boldsymbol{r}') = -\frac{1}{4\pi}\frac{\exp(ik|\boldsymbol{r}-\boldsymbol{r}'|)}{|\boldsymbol{r}-\boldsymbol{r}'|},$$

we easily find the required solution

$$\psi = \exp[i(\boldsymbol{k}_0 . \boldsymbol{r})] - \frac{\mu}{2\pi\hbar^2}\int V(\boldsymbol{r}')\psi(\boldsymbol{r}')\frac{\exp(ik|\boldsymbol{r}-\boldsymbol{r}'|)}{|\boldsymbol{r}-\boldsymbol{r}'|}\,d^3\boldsymbol{r}'.$$

For large values of r we have

$$\frac{\exp(ik|\boldsymbol{r}-\boldsymbol{r}'|)}{|\boldsymbol{r}-\boldsymbol{r}'|} \approx \frac{\exp(ikr)}{r}.\exp[-i(\boldsymbol{k}.\boldsymbol{r}')] \quad (r\to\infty),$$

where

$$\boldsymbol{k} = k\frac{\boldsymbol{r}}{r}.$$

Hence,

$$\psi \approx \exp[i(\boldsymbol{k}_0 . \boldsymbol{r})] - \frac{\mu}{2\pi\hbar^2}\frac{\exp(ikr)}{r}\int d^3\boldsymbol{r}' \exp[-i(\boldsymbol{k}.\boldsymbol{r}')]\,V(\boldsymbol{r}')\psi(\boldsymbol{r}'),$$

so that

$$f(\boldsymbol{k}) = -\frac{\mu}{2\pi\hbar^2}\int d^3\boldsymbol{r} \exp[-i(\boldsymbol{k}.\boldsymbol{r})]\,V(\boldsymbol{r})\psi(\boldsymbol{r}).$$

This equation is convenient for various approximate calculations. For instance, by substituting $\psi(\boldsymbol{r}) \approx \exp[i(\boldsymbol{k}_0 . \boldsymbol{r})]$ we get the scattering amplitude in the first Born approximation.

6. The scattering amplitude is in the Born approximation given by the equation

$$f_{\rm Born}(\vartheta) = -\frac{\mu}{2\pi\hbar^2}\int \exp[i(\boldsymbol{q}.\boldsymbol{r})]\,V(r)\,d\tau = -\frac{\pi\mu A}{\hbar^2 q},$$

where

$$\boldsymbol{q} = \boldsymbol{k}' - \boldsymbol{k}, \quad q = 2k\sin\tfrac{1}{2}\vartheta.$$

Hence

$$d\sigma_{\text{Born}} = |f(\vartheta)|^2 \, d\Omega = \frac{\pi^3}{2\hbar^2} \frac{\mu A^2}{E} \cot \tfrac{1}{2}\vartheta \, d\vartheta.$$

In classical mechanics we have the following connection between the angle of scattering and the impact parameter ρ:

$$\int_{r_0}^{\infty} \frac{\mu v \rho \, dr}{r^2 \sqrt{[2\mu(E-V)-(\mu v \rho/r)^2]}} = \frac{\pi-\vartheta}{2},$$

where r_0 is the zero of the expression under the square root sign. If we integrate, we get

$$\rho^2 = \frac{A}{E} \frac{1}{\vartheta} \frac{(\pi-\vartheta)^2}{2\pi-\vartheta},$$

and thus

$$d\sigma = -2\pi\rho \frac{d\rho}{d\vartheta} d\vartheta = \frac{2\pi^3 A}{E} \frac{\pi-\vartheta}{\vartheta^2(2\pi-\vartheta)^2} d\vartheta.$$

If

$$\frac{8\mu A}{\hbar^2} \ll 1$$

we can apply the Born approximation for all angles (see problem 4 of section 9).

In the opposite limiting case, where $8\mu A/\hbar^2 \gg 1$ the classical result holds for not too small angles,

$$\vartheta \gtrsim \frac{\hbar^2}{8\mu A},$$

while for smaller angles,

$$\vartheta \lesssim \frac{\hbar^2}{8\mu A},$$

the Born approximation result is valid.

7*. We have the following equation for the radial function,

$$\chi'' + \frac{2\mu}{\hbar^2} [E + V_0 \exp(-r/a)] \chi = 0.$$

Using the notation $k^2 = 2\mu E/\hbar^2$, $\kappa^2 = 2\mu V_0/\hbar^2$ and the independent variable $\xi = \exp(-r/2a)$, we must solve the equation

$$\chi'' + \frac{1}{\xi}\chi' + 4a^2\left(\frac{k^2}{\xi^2} + \kappa^2\right)\chi = 0;$$

its solution is a Bessel function of imaginary order $\chi = J_{\pm 2aki}(2a\kappa\xi)$. The function χ must vanish at $r = 0$, that is, at $\xi = 1$, so that we get, apart from a normalising constant,

$$\chi = J_{-2aki}(2a\kappa) J_{2aki}(2a\kappa\xi) - J_{2aki}(2a\kappa) J_{-2aki}(2a\kappa\xi). \tag{1}$$

The asymptotic form of χ as $r\to\infty$ ($\xi\to 0$) is

$$\chi = J_{-2aki}(2a\kappa)\frac{\exp(2aki\ln a\kappa)}{\Gamma(2aki+1)}\exp(-ikr)$$
$$-J_{2aki}(2a\kappa)\frac{\exp(-2aki\ln a\kappa)}{\Gamma(-2aki+1)}\exp(ikr).$$

The coefficients of $\exp(-ikr)$ and $\exp(ikr)$ can be considered to be functions of a complex k. If we denote them by $a(k)$ and $b(k)$ one can easily show that they satisfy the relations

$$a(-k) = -b(k),$$
$$a^*(k) = -b(k)$$

(in taking the conjugate complex one must not change k to k^*).

We write the asymptotic form of χ in the form

$$\chi = A[\exp(-ikr-i\delta_0)-\exp(ikr+\delta_0)] = -2iA\sin(kr+\delta_0).$$

The phase shift δ_0 follows from the equation

$$\exp(2i\delta_0) = \frac{J_{2aki}(2a\kappa)}{J_{-2aki}(2a\kappa)}\frac{\Gamma(2aki+1)}{\Gamma(-2aki+1)}\exp(-4aki\ln a\kappa).$$

A bound state corresponds to an imaginary value of $k = ik_n$ and a negative value of the energy.

For $k_n > 0$ the coefficient of $\exp(-ikr) = \exp(k_n r)$ in the first term of the asymptotic expression for χ must vanish. That means that either $J_{2ak_n}(2a\kappa) = 0$, or $1/\Gamma(2ak_n+1) = 0$. From the second condition we get

$$2ak_n+1 = -n \quad (n = 0, 1, 2, \ldots),$$

or

$$E = -\frac{\hbar^2 k_n^2}{2\mu} = -\frac{\hbar^2(n+1)^2}{8\mu a^2}.$$

However, the order of the Bessel functions now becomes integral and since $J_n(x) = (-1)^n J_{-n}(x)$, the two solutions are linearly dependent and the wave function (1) vanishes identically. The energy levels are thus spurious. The first condition gives us the true discrete spectrum,

$$J_{2ak_n}(2a\kappa) = 0, \quad E_n = -\frac{\hbar^2 k_n^2}{2\mu}. \tag{2}$$

The zeros of the expression $\exp[2i\delta_0(k)]$ lie thus on the imaginary axes, and apart from the values ik_n, corresponding to the discrete spectrum (2), contain also the redundant zeros.

9. (a) $$d\sigma = \frac{\pi\mu^2 V_0}{4\hbar^4\alpha^6}\exp\left(-\frac{4\mu E}{\hbar^2\alpha^2}\sin^2\frac{\vartheta}{2}\right)d\Omega,$$

$$\sigma = \frac{\pi^2 V_0^2 \mu}{4\hbar^2\alpha^4 E}\left[1-\exp\left(-\frac{4\mu E}{\hbar^2\alpha^2}\right)\right];$$

(b) $$d\sigma = \frac{16\mu^2 V_0^2}{\hbar^4}\frac{\alpha^2}{(\alpha^2+q^2)^4}d\Omega,$$

$$\sigma = \frac{64\pi}{3}\frac{\mu^2 V_0}{\hbar^4}\frac{16k^4+12k^2\alpha^2+3\alpha^4}{\alpha^4(\alpha^2+4k^2)^3}.$$

10. (a) We evaluate the atomic form factor for hydrogen,

$$F(q) = \int \exp[i(\boldsymbol{q}.\boldsymbol{r})]\,n(\boldsymbol{r})\,d\tau = \frac{1}{\pi a^3}\int \exp[i(\boldsymbol{q}.\boldsymbol{r})-2r/a]\,d\tau = \frac{1}{(1+q^2a^2/4)^2}$$

where $a = \hbar^2/\mu e^2$ is the Bohr radius.

We find thus for the differential cross-section

$$d\sigma = \frac{4a^2(8+q^2a^2)^2}{(4+q^2a^2)^4}d\Omega \quad (q = 2k\sin\tfrac{1}{2}\theta) \tag{1}$$

and for the total cross-section

$$\sigma = \frac{\pi a^2}{3}\frac{7k^4a^4+18k^2a^2+12}{(k^2a^2+1)^3}.$$

In this case the Born approximation can be applied, provided

$$ka \gg 1,$$

so that we can simplify the last expression to

$$\sigma = \frac{7\pi}{3k^2}. \tag{2}$$

(b) For the helium atom we get from the variational method the following expression for the electron density distribution:

$$n(r) = \frac{2}{\pi b^3}\exp(-2/rb), \quad b = \tfrac{16}{27}a.$$

The differential and total cross-sections for elastic scattering by a helium atom have in this approximation the same form as for hydrogen. We must only replace a by b in equations (1) and (2), and introduce a factor $Z^2 = 4$. In particular,

$$\sigma = \frac{28\pi}{3k^2}.$$

11. It is well known that the solution of the problem of elastic scattering by a potential $V(\boldsymbol{r})$ in the Born approximation leads to the following expression for the wave function:

$$\psi \approx \psi^{(0)} + \psi^{(1)}, \tag{1}$$

where

$$\left.\begin{aligned} \psi^{(0)}(\boldsymbol{r}) &= \exp[i(\boldsymbol{k}.\boldsymbol{r})], \\ \psi^{(1)}(\boldsymbol{r}) &= -\frac{\mu}{2\pi\hbar^2}\int V(\boldsymbol{r}')\psi^{(0)}(\boldsymbol{r}')\frac{\exp(ik|\boldsymbol{r}-\boldsymbol{r}'|)}{|\boldsymbol{r}-\boldsymbol{r}'|}\,d^3\boldsymbol{r}', \end{aligned}\right\} \tag{2}$$

so that asymptotically we have

$$\psi^{(1)}(\boldsymbol{r}) \underset{r\to\infty}{\approx} \left\{-\frac{\mu}{2\pi\hbar^2}\int V(\boldsymbol{r}')\exp[i(\boldsymbol{q}.\boldsymbol{r}')]\,d^3\boldsymbol{r}'\right\}\frac{\exp(ikr)}{r} \equiv f(\theta)\frac{\exp(ikr)}{r}, \tag{2'}$$

where $\boldsymbol{q} = \boldsymbol{k} - \boldsymbol{k}'$ is the change in the wave vector of the particle during the scattering, and where θ is the angle of scattering.

The differential cross-section for scattering into a solid angle $d\Omega$ over an angle θ with the direction $\boldsymbol{k}$ of the incoming particle is equal to

$$d\sigma = |f(\theta)|^2\,d\Omega = \frac{\mu^2}{4\pi^2\hbar^4}\left|\int V(\boldsymbol{r}')\exp[i(\boldsymbol{q}.\boldsymbol{r}')]\,d^3\boldsymbol{r}'\right|^2 d\Omega. \tag{3}$$

We shall evaluate the integral in equation (3). Taking the polar axis of our system of spherical polars along the vector $\boldsymbol{q}$ and substituting for $V(r')$, we get

$$\int \exp[-\kappa r' + i(\boldsymbol{q}.\boldsymbol{r}')]\frac{d^3\boldsymbol{r}'}{r'} = 2\pi\int_0^\infty \exp(-\kappa r')r'\,dr'\int_{-1}^{1}\exp(iqr'x)\,dx$$

$$= \frac{4\pi}{\kappa^2+q^2},$$

and thus

$$d\sigma = \left(\frac{2A\mu}{\hbar^2}\right)^2\frac{d\Omega}{(\kappa^2+q^2)^2},$$

$$q = 2k\sin\tfrac{1}{2}\theta. \tag{4}$$

Integrating expression (4) over the angles we get the total scattering cross-section

$$\sigma = \frac{4\pi}{\kappa^2+4k^2}\left(\frac{2A\mu}{\kappa\hbar^2}\right)^2. \tag{5}$$

Let us consider some limiting cases of equation (4).

(1) $k \ll \kappa$, so that also $q \ll \kappa$:

$$d\sigma \approx \left(\frac{2A\mu}{\hbar^2\kappa^2}\right)^2 d\Omega, \tag{4'}$$

that is, the scattering cross-section, in complete accordance with the general theory of scattering of slow particles, depends neither on the scattering angle nor on the particle energy.

(2) $k \gg \kappa$ (fast particles).

There are now two characteristic ranges of scattering angles,

(a) $\theta \ll \kappa/k$, or $q \ll \kappa$; the cross-section for these small angles of scattering tends for $\theta \to 0$ to the constant limit (4′);

(b) $\theta \gg \kappa/k$, or $q \gg \kappa$ (large deflection). In this case we have, introducing the particle energy $E = \hbar^2 k^2/2\mu$,

$$d\sigma \approx \left(\frac{A}{4E}\right)^2 \frac{d\Omega}{\sin^4 \frac{1}{2}\theta}. \tag{4''}$$

This is essentially the Rutherford scattering formula, as it should be, since in the essential region of distances, $r \sim 1/q$ the field is practically a Coulomb field.

The condition of applicability of the Born formulae (3) and (4) is that the correction term $\psi^{(1)}$, which describes the influence of the scattering field, is small compared to the main term $\psi^{(0)}$. This condition is always satisfied, if $|\psi^{(1)}|_{r=0} \ll |\psi^{(0)}| = 1$. This condition of applicability is sufficient; it can, however, be shown to be too strong, especially for small scattering angles.

We have

$$\psi^{(1)}|_{r=0} = -\frac{\mu A}{2\pi\hbar^2} \int \exp[i(\boldsymbol{k}.\boldsymbol{r}')] \frac{\exp(-\kappa r' + ikr')}{r'^2} d^3\boldsymbol{r}'$$

$$= -\frac{2\mu A}{\hbar^2} \int_0^\infty \exp(-\kappa r' + ikr') \frac{\sin kr'}{kr'} dr'. \tag{2''}$$

For slow particles ($k \lesssim \kappa$) the important region in the integral is $r' \sim 1/\kappa$, where $\exp(ikr')(\sin kr'/kr') \sim 1$, so that the integral is of the order of magnitude of $1/\kappa$ and the required condition of applicability of equation (4) is of the form

$$\frac{|A|\mu}{\kappa\hbar^2} \ll 1. \tag{6'}$$

For fast particles ($k \gg \kappa$) the presence of the fast oscillating function $[\exp(ikr')\sin kr']$ means that the important region is $r' \sim 1/k$, so that the integral is of the order of magnitude of $1/k$, and the condition of applicability of the Born equations will be (introducing the particle velocity $v = \hbar k/\mu$):

$$\frac{|A|}{\hbar v} \ll 1. \tag{6''}$$

The meaning of inequalities (6′) and (6″) becomes very obvious if we consider them in conjunction with the limiting cases of expression (5) for the total cross-section. Dropping numerical coefficients we see then that both inequalities are equivalent to the condition

$$\sigma \ll \frac{1}{\kappa^2} \quad \left(\text{or } \sqrt{\sigma} \ll \frac{1}{\kappa}\right).$$

The condition of applicability of the Born approximation is thus that the scattering cross-section is small compared to the square of the effective range of the scattering field, or, in other words, that the scattering amplitude $\sqrt{\sigma}$ is small compared to the range of the field $1/\kappa$.

Using the results of this problem we can consider as a particular case the scattering of electrons by neutral atoms (that is, by a screened Coulomb field). We must then replace A by $-Ze^2$ and $1/\kappa$ by the Thomas–Fermi radius of the atom which is of the order of magnitude of

$$\frac{\hbar^2}{Z^{\frac{1}{3}}\mu e^2}.$$

12. The scattering potential, or rather potential energy, is equal to

$$V(\boldsymbol{r}) = B\delta(\boldsymbol{r}), \tag{1}$$

where we have taken the force centre as the origin. The constant B is clearly equal to the volume integral of the potential,

$$B = \int V(\boldsymbol{r})\, d^3\boldsymbol{r} = \text{const.} \tag{2}$$

The differential cross-section for elastic scattering into a unit solid angle (in the centre of mass system) is equal to

$$\frac{d\sigma}{do} = \frac{\mu^2}{4\pi^2\hbar^4}\left|\int V(\boldsymbol{r}) \exp[i(\boldsymbol{q}\,.\,\boldsymbol{r})]\, d^3\boldsymbol{r}\right|^2, \tag{3}$$

where μ is the reduced mass of the colliding particles and $\hbar\boldsymbol{q} \equiv \boldsymbol{p} - \boldsymbol{p}'$ is the change in momentum of their relative motion.

From equations (1) and (3) and the properties of $\delta(r)$, we get

$$\frac{d\sigma}{do} = \frac{\mu^2 B^2}{4\pi^2\hbar^4}. \tag{4}$$

The scattering by a delta-function potential is thus isotropic and does not depend on the velocity. It is well known that the same properties characterise the scattering of sufficiently slow particles by a potential well of finite dimensions. We shall elucidate the connection between these facts shortly.

The total scattering cross-section is equal to

$$\sigma = 4\pi \frac{d\sigma}{do} = \frac{\mu^2 B^2}{\pi \hbar^4}. \tag{5}$$

If we bear in mind the meaning of the constant B ($B = \int V\, d^3\boldsymbol{r}$), we see that equations (4) and (5) are obtained from the general Born equation (3) always when $\exp[i(\boldsymbol{q}.\boldsymbol{r}_{\text{eff}})] \approx 1$, or $|(\boldsymbol{q}.\boldsymbol{r}_{\text{eff}})| \ll 1$. In order that this will be the case for all scattering angles, it is necessary (since $q_{\max} \sim k$) that $kr_{\text{eff}} \ll 1$ or $r_{\text{eff}} \ll \lambda\!\!\!^{-}$, where r_{eff} is of the order of the dimensions of the region where $V(\boldsymbol{r})$ is appreciably different from zero ("range of the interaction").

The delta-function potential considered by us is thus an idealised interaction potential with a very small range or, rather, with a range which is much less than the de Broglie wavelength of the relative motion of the colliding particles. The delta-function potential can thus, for instance, serve to describe formally the interaction of sufficiently slow neutrons with protons or very heavy nuclei.

In those cases the formal application of the Born approximation leads to the correct result, even though perturbation theory is essentially inapplicable to a delta-function potential of the form considered by us. Indeed, the condition of applicability of perturbation theory for slow particles ($ka \lesssim 1$) is of the form

$$|V| \ll \frac{\hbar^2}{\mu a^2}.$$

If we write this condition as $|V| a^3 \ll (\hbar^2/\mu) a$ and bear in mind that for our delta-function potential $|V| a^3$ has a fixed, finite value, we see that for sufficiently small values of a (and always if $a = 0$) the perturbation theory, and thus also the Born approximation, is inapplicable.

From equations (4) and (5) it is also clear that the formal application of the Born approximation leads to the correct result in the case where the true cross-section does not depend on the velocity. In this case we are, roughly speaking, not interested in evaluating the scattering cross-section for the given potential, but rather in "adjusting" the constant of the potential to the known value of the cross-section.

13. $$\delta_l = \int^{\infty} \left[k^2 - \frac{2\mu}{\hbar^2} V(r) - \frac{l(l+1)}{r^2}\right]^{\frac{1}{2}} dr - \int^{\infty} \left[k^2 - \frac{l(l+1)}{r^2}\right]^{\frac{1}{2}} dr.$$

14. If α is the ratio of the masses of the scattered and the scattering particle, we have

$$|f_{\text{lab}}(\vartheta', \varphi')|^2 = \frac{\{1 + 2\alpha \cos\vartheta + \alpha^2\}^{3/2}}{|1 + \alpha \cos\vartheta|} |f_{\text{com}}(\vartheta, \varphi)|^2 .$$

15. (ii)

$$f(k) = \frac{4\pi\lambda\,|v(k)|^2}{1+\dfrac{2\lambda}{\pi}\displaystyle\int d^3k'\,\frac{|v(k')|^2}{k^2-k'^2}},$$

where

$$v(k) = \int_0^\infty \frac{\sin kr}{kr}\,v(r)r^2\,dr\,.$$

(iii) $$k\cot\delta = \frac{(k^2b^2+1)^2+\xi(k^2b^2-1)}{2\xi b}, \quad \text{where } \xi = 2\pi\lambda b^3\,.$$

16*. We take the origin in the centre of rotation and write the potential energy of the interaction of the incident particle with the rotator in the form

$$V = \frac{e^2}{|\boldsymbol{r}-\boldsymbol{n}a|}, \tag{1}$$

where e is the charge of both particles, a the distance of the rotating particle from the origin, $\boldsymbol{n}$ a unit vector along the "axis" of the rotator, and $\boldsymbol{r}$ the radius vector of the scattered particle.

The probability per unit time for the transition under consideration is in first-order perturbation theory equal to

$$dw_{0l} = \frac{2\pi}{\hbar}|V_{0l}|^2\,\rho_E\,d\Omega, \tag{2}$$

where V_{0l} is the matrix element of the perturbation (1) for the transition considered, taken between normalised wave functions; $\rho_E\,d\Omega$ is the number of final states per unit energy, where $d\Omega$ is an element of solid angle around the scattering direction.

The initial and final wave functions of the system are of the form

$$\psi_{\text{in}} = \frac{1}{\sqrt{\Omega}}\exp\,[i(\boldsymbol{k}_0\cdot\boldsymbol{r})]\,Y_{00}(\boldsymbol{n}),$$

$$\psi_{\text{fin}} = \frac{1}{\sqrt{\Omega}}\exp\,[i(\boldsymbol{k}_l\cdot\boldsymbol{r})]\,Y_{lm}(\boldsymbol{n}), \tag{3}$$

where the Y_{lm} are normalised spherical harmonics describing stationary states of the rotator, $\boldsymbol{k}_0$ and $\boldsymbol{k}_l$ are the wave vectors of the scattered particle before and after the scattering, and Ω is a normalising volume.

From equations (1) and (3) we get

$$V_{l0} = V_{0l}^* = \iint \psi_{\text{fin}}^* V \psi_{\text{in}}\, d^3\boldsymbol{r}\, do$$

$$= \frac{e^2}{\Omega} \iint Y_{lm}^*(\boldsymbol{n}) Y_{00}(\boldsymbol{n}) \exp[i(\boldsymbol{k}_0 - \boldsymbol{k}_l . \boldsymbol{r})] \frac{d^3\boldsymbol{r}\, do}{|\boldsymbol{r} - \boldsymbol{n}a|}. \tag{4}$$

where do is an element of solid angle around $\boldsymbol{n}$.

From the calculations it will become clear that for any given l only one state of the rotator (with $m = 0$) can be excited. The density of final states $\rho_E\, d\Omega$ is thus solely determined by the number of states of the scattered particle and hence equal to

$$\rho_E\, d\Omega = \frac{\Omega p_l^2 \dfrac{dp_l}{dE} d\Omega}{(2\pi\hbar)^3}, \tag{5}$$

where $p_l = \hbar k_l$ is the momentum of the scattered particle which is related to the initial momentum through the law of conservation of energy

$$E = \frac{p_0^2}{2\mu} = \frac{p_l^2}{2\mu} + \frac{\hbar^2 l(l+1)}{2I}, \tag{6}$$

where μ is the mass of the scattered particle and I the moment of inertia of the rotator.

If we write $\boldsymbol{k}_0 - \boldsymbol{k}_l = \boldsymbol{q}$ and use the fact that $Y_{00} = 1/\sqrt{(4\pi)}$ and change the variables to $\boldsymbol{r} - \boldsymbol{n}a = \boldsymbol{r}'$ we get instead of equation (4)

$$V_{l0} = \frac{e^2}{\sqrt{(4\pi)}\Omega} \int do\, Y_{lm}^*(\boldsymbol{n}) \exp[ia(\boldsymbol{q} . \boldsymbol{n})] \int \exp[i(\boldsymbol{q} . \boldsymbol{r}')] \frac{d^3\boldsymbol{r}'}{r'}. \tag{7}$$

The Fourier component of the Coulomb potential $\int \exp[i(\boldsymbol{q} . \boldsymbol{r})] (d^3\boldsymbol{r}'/r')$ which occurs here is equal to $4\pi/q^2$ (see, for instance, problem 42 of section 3). If we choose $\boldsymbol{q}$ as the axis for quantising the rotator and expand $\exp[ia(\boldsymbol{q} . \boldsymbol{n})]$ in a series of Legendre polynomials we get

$$V_{l0} = \frac{\sqrt{(4\pi)}\, e^2}{\Omega q^2} \sum_{l'=0}^{\infty} (2l'+1) i^{l'} \sqrt{\left(\frac{\pi}{2qa}\right)} J_{l'+\frac{1}{2}}(qa)$$

$$\times \int_0^{2\pi} \int_0^{\pi} Y_{lm}^*(\vartheta, \varphi) P_{l'}(\cos\vartheta) \sin\vartheta\, d\vartheta\, d\varphi.$$

Writing

$$Y_{lm}^* = \sqrt{\left[\frac{2l+1}{4\pi} . \frac{(l-m)!}{(l+m)!}\right]} P_l^m(\cos\vartheta) \exp(-im\varphi),$$

we see that only those matrix elements are different from zero for which the transition is to a state with $m = 0$. Using the relation

$$\int_0^\pi P_l P_{l'} \sin\vartheta \, d\vartheta = \frac{2}{2l+1}\,\delta_{ll'}$$

and summing over l', we get for those matrix elements the expression

$$V^*_{0l} = V_{l0} = 4\pi i^l \frac{e^2}{\Omega q^2}\sqrt{\left[(2l+1)\frac{\pi}{2qa}\right]}\, J_{l+\frac{1}{2}}(qa). \tag{8}$$

The differential cross-section $d\sigma_l(\theta)$ is equal to the probability (2) divided by the current density of the incident particles. The latter is according to equation (3) equal to v_0/Ω. Moreover, the quantity

$$p_l^2 \frac{dp_l}{dE} = \tfrac{1}{2} p_l \frac{d(p_l^2)}{dE}$$

occurring in expression (5) is according to equation (6) equal to $\mu p_l = \mu^2 v_l$, where v_0 and v_l are the velocities of the scattered particle before and after the scattering process. Combining this result with equations (2), (5), and (8) we get finally

$$d\sigma_l(\theta) = \frac{\Omega}{v_0}\, dw_{0l} = 2\pi(2l+1)\frac{v_l}{v_0}\frac{\mu^2 e^4}{\hbar^4 q^4}\frac{J^2_{l+\frac{1}{2}}(qa)}{qa}\, d\Omega. \tag{9}$$

The cross-section does not contain the normalisation volume Ω, which is as should be since it has no physical meaning.

Equation (9) determines thus the required differential cross-section for the scattering of a particle into the element of solid angle $d\Omega$ with simultaneous excitation of the lth level of the rotator. The quantity q occurring in equation (9) depends both on the scattering angle θ and on l, since $\boldsymbol{p}_0 = \boldsymbol{p}_l + \hbar\boldsymbol{q}$, where $\hbar\boldsymbol{q}$ is the momentum transferred from the particle to the rotator during the collision. From the definition of $\boldsymbol{q}$ and equation (6) we get

$$q^2 = k_0^2 + k_l^2 - 2k_0 k_l \cos\theta = 2k_0^2 - \frac{\mu}{I} l(l+1) - 2k_0 \sqrt{\left[k_0^2 - \frac{\mu}{I} l(l+1)\right]} \cos\theta. \tag{10}$$

To find the angular distribution of the scattered particles irrespective of their energy, one must sum expression (9) over all values for l for a given value of θ. To find the total scattering cross-section for a given energy transfer, we must integrate expression (9) over the angles for a given value of l. Because the cross-section (9) depends in a rather complicated way on l and θ, both operations can in general be performed only numerically. However, for the most essential cases one can approximately give also an analytical evaluation, which we shall consider in the following.

Let us now discuss equation (9); we note first of all that it is valid for the case where $l = 0$, that is, for elastic scattering. In this case $q = 2k_0 \sin \frac{1}{2}\theta$, so that the elastic scattering is equal to

$$d\sigma_{\text{el}}(\theta) = d\sigma_0(\theta) = 2\pi \frac{\mu^2 e^4}{\hbar^4 q^4} \frac{J_{\frac{1}{2}}^2(qa)}{qa} d\Omega = \frac{4\mu^2 e^4 a^4}{\hbar^4} \frac{\sin^2(2k_0 a \sin \frac{1}{2}\theta)}{(2k_0 a \sin \frac{1}{2}\theta)^6} d\Omega, \tag{11}$$

where we have used the relation $J_{\frac{1}{2}}(qa) = \sqrt{(2/\pi qa)} \sin qa$. The total elastic scattering cross-section diverges clearly for small angles θ.

In the limiting case where $2k_0 a \ll 1$ expression (11) leads to

$$d\sigma_{\text{el}}(\theta) \approx \frac{\mu^2 e^4}{4\hbar^4 k_0^4 \sin^4 \frac{1}{2}\theta} d\Omega = \left(\frac{e^2}{2\mu v_0^2}\right)^2 \frac{d\Omega}{\sin^4 \frac{1}{2}\theta}, \tag{12}$$

that is, the Rutherford formula. This we should have expected since $2k_0 a \ll 1$ means that the dimension of the region in which the scattering Coulomb centre (the rotator) moves, $2a$, is small compared to the de Broglie wavelength of the scattered particle $\lambda\!\!\!^{-} = 1/k_0$ and this in turn is practically equivalent to having a fixed scattering centre, which is necessary for the validity of the Rutherford formula (12).

For our next consideration it is convenient to transform expression (9) slightly, going over from $d\Omega$ to dq. From equation (10) we have for a given value of l

$$q\,dq = k_0 k_l \sin\theta\, d\theta = \frac{k_0 k_l}{2\pi} d\Omega, \tag{13}$$

so that equation (9) leads to

$$d\sigma_l = \left(\frac{2\pi e^2}{\hbar v_0}\right)^2 (2l+1) \frac{J_{l+\frac{1}{2}}^2(qa)}{qa} \frac{dq}{q^3}. \tag{14}$$

We shall limit our considerations to the excitation of the rotator by a fast particle, whose initial energy $E = \mu v_0^2/2$ is large compared to a "rotational quantum" $\hbar^2/2l$ so that it follows from equation (6) that $0 \leqslant l \leqslant l_{\text{max}}$ with $l_{\text{max}} \gg 1$. (One verifies easily that for values of μ and I corresponding to the collisions of electrons—and *a fortiori* mesons or protons—with molecules the condition $E \gg \hbar^2/2I$ ensures the validity of the necessary condition for the applicability of the perturbation theory used by us to scattering in a Coulomb field, $e^2/\hbar v \ll 1$.) Furthermore, we make the approximation (confirmed by the final result) that the deciding part is played by collisions corresponding to small scattering angles ($\theta \ll 1$) and to a transfer of energy which is small compared to the energy of the incident particle ($\Delta E \ll E$). In this case we have clearly $k_0 - k_l \ll k_0$, so that

$$\Delta E \equiv \frac{\hbar^2 l(l+1)}{2I} = \frac{\hbar^2}{2\mu}(k_0^2 - k_l^2) \approx \frac{\hbar^2 k_0}{\mu}(k_0 - k_l), \tag{15}$$

$$q^2 = (k_0 - k_l)^2 + 2k_0 k_l (1 - \cos\theta) \approx (k_0 - k_l)^2 + k_0^2 \theta^2. \tag{16}$$

Combining equations (15) and (16) we get

$$q \approx \sqrt{\left\{\left[\frac{\mu l(l+1)}{2Ik_0}\right]^2 + (k_0\theta)^2\right\}} = \sqrt{\left\{\left(\frac{\Delta E}{\hbar v_0}\right)^2 + (k_0\theta)^2\right\}}. \tag{17}$$

For $\theta = 0$ we have $q \approx \Delta E/\hbar v_0$, and for $\Delta E = 0$ we have

$$q \approx k_0 \theta = \frac{p_0 \theta}{\hbar},$$

so that, roughly speaking, $\Delta E/v_0$ and $p_0\theta$ are the longitudinally and transversely transferred momenta.

It is clear from equation (17) that for not too small angles of scattering, for which $k_0\theta \gg \Delta E/\hbar v_0$, q depends only on θ, and not on l. In that case we can sum expression (14) in its general form over l and thus obtain the angular distribution of the (both elastically and inelastically) scattered particles. We use the identity

$$\sum_{l=0}^{\infty} (2l+1) J_{l+\frac{1}{2}}^2(x) = \frac{2x}{\pi}. \tag{18}$$

(This identity can be obtained by integrating the absolute square of the expansion

$$\exp(ix\cos\theta) = \sum_{l=0}^{\infty} (2l+1) i^l \sqrt{\left(\frac{\pi}{2x}\right)} J_{l+\frac{1}{2}}(x) P_l(\cos\theta)$$

over $\sin\theta\, d\theta$ and using the orthonormality relation

$$\int_0^{\pi} P_l P_{l'} \sin\theta\, d\theta = \frac{2}{2l+1}\delta_{ll'}.)$$

From equations (14) and (18) it follows that

$$d\sigma(q) = \sum_{l=0}^{\infty} d\sigma_l(q) = \left(\frac{2\pi e^2}{\hbar v_0}\right)^2 \frac{dq}{q^3}\frac{1}{qa}\frac{2qa}{\pi} = 8\pi\left(\frac{e^2}{\hbar v_0}\right)^2 \frac{dq}{q^3}. \tag{19}$$

Since for the angles considered at this moment $q \approx k_0\theta$, equation (19) can be written in the form

$$d\sigma(\theta) = 8\pi\left(\frac{e^2}{\mu v_0^2}\right)^2 \frac{d\theta}{\theta^3} \approx \left(\frac{2e^2}{\mu v_0^2}\right)^2 \frac{d\Omega}{\theta^4}, \tag{19'}$$

which agrees with the Rutherford formula (12) (we remind ourselves that $\theta \ll 1$). This result is connected with the fact that for the angles considered $p_0\theta \gg \Delta E/v_0$, that is, the scattering is accompanied by the transfer of predominantly "transverse" momentum and is practically elastic.

The range of scattering angles for which the angular distribution (19) or (19′) is approximately valid is

$$\frac{\Delta E}{2E} \ll \theta \ll 1 \tag{20}$$

(where we have used the relation $p_0 v_0 = 2E$). Since we have assumed $(\Delta E)_{\text{eff}} \ll E$, there can be a substantial range of angles satisfying condition (20). The character of the angular distribution (19′)—a fast increase for decreasing θ—confirms our earlier assumption that $\theta_{\text{eff}} \ll 1$.

We must make the following observation. We have obtained equation (19) by summing over all values of l. However, inequality (20) and equation (17) impose on l the limit $\mu l^2/Ik_0 \ll k_0\theta$, or $l \ll l_1 \equiv k_0\sqrt{(I\theta/\mu)}$, which at first sight seems to make equation (19) incorrect. In actual fact, however, it turns out that the condition $l \ll l_1$ does practically not limit the applicability of that equation in general. Indeed, one can easily check that for a given value of x the dominant role in equation (18) is played by the terms with $l \leqslant x$. It follows that in order to satisfy the condition that equation (19) is applicable, $l_{\text{eff}} \ll l_1$, it is sufficient that $qa \ll l_1$, or $\theta \ll I/\mu a^2 = \mu_{\text{rot}}/\mu$ ($\mu_{\text{rot}} = I/a^2$ is the mass of the rotating particle). In practically all cases of interest $\mu_{\text{rot}} \gtrsim \mu$, so that the required inequality is automatically satisfied for $\theta \ll 1$.

We shall now evaluate the total scattering cross-section with simultaneous excitation of the lth level of the oscillator, σ_l. We integrate thereto expression (14) over q between limits corresponding to $\theta = 0$ and $\theta = \pi$. Since the main contribution to this cross-section corresponds to small scattering angles, we can use equation (17) to find the limits of q and put $\theta = 0$ and $\theta \sim 1$. We have thus

$$\sigma_l = \int_{q_{\min}}^{q_{\max}} d\sigma_l(q) = \left(\frac{2\pi e^2}{\hbar v_0}\right)^2 (2l+1)\,a^2 \int_{aq_{\min}}^{aq_{\max}} J^2_{l+\frac{1}{2}}(x)\,\frac{dx}{x^4}, \tag{21}$$

with

$$q_{\min} = \frac{\Delta E}{\hbar v_0}, \quad q_{\max} \sim k_0 \tag{22}$$

(to estimate the value of $q_{\max}$ one may neglect the first term under the square root sign since $\Delta E \ll E$).

While for large values of x the integrand in expression (21) decreases steeply with increasing values of x for any l its behaviour for small values of x depends strongly on the value of l. Since $J_{l+\frac{1}{2}}(x) \sim x^{l+\frac{1}{2}}$ as $x \to 0$, the integrand will be proportional to x^{2l-3} as $x \to 0$ and thus tend to zero for $l \geqslant 2$ and to infinity for $l = 0, 1$. We shall consider these two cases separately.

(1) $l = 2, 3, 4, \ldots$.

In this case we have

$$\int_{aq_{\min}}^{aq_{\max}} J^2_{l+\frac{1}{2}}(x)\frac{dx}{x^4} \approx \int_0^{\infty} J^2_{l+\frac{1}{2}}(x)\frac{dx}{x^4} = \frac{2}{3\pi}\frac{(l-2)!}{(l+2)!}. \tag{23}$$

Let us first discuss the validity of replacing the limits of integration by 0 and ∞. Since the first (largest) maximum, and also the first root, of the function $J_{l+\frac{1}{2}}(x)$ (where $l \gg 1$) is near $x = l$, while the function for $x \ll l$ behaves like $x^{l+\frac{1}{2}}$, and for $x > l$ is heavily damped and oscillating, it is clear that the change in limits is allowed provided

$$aq_{\min} \ll l, \quad aq_{\max} \gg l. \tag{24}$$

Using equations (22) and (15) (to eliminate l) and again the relation $\mu_{\text{rot}} = I/a^2$, we can write these conditions in the form

$$\frac{\Delta E}{E} \ll \frac{\mu_{\text{rot}}}{\mu}, \quad \frac{\Delta E}{E} \ll \frac{\mu}{\mu_{\text{rot}}}. \tag{24'}$$

Since, as was mentioned earlier, usually $\mu_{\text{rot}} \gtrsim \mu$, the first condition is automatically satisfied for $\Delta E \ll E$ but the second condition is more restrictive. For the case $\mu \sim \mu_{\text{rot}}$ at least we have thus proved the validity of the change in limits in expression (23).

From equations (21) and (23) we have finally

$$\sigma_l = \frac{8}{3}\left(\frac{e^2}{\hbar v_0}\right)^2 (2l+1)\frac{(l-2)!}{(l+2)!}\pi a^2 \quad (l = 2, 3, \ldots). \tag{25}$$

This total cross-section decreases steeply for increasing l (for $l \gg 2$ as $1/l^3$, that is, as $1/(\Delta E)^{\frac{3}{2}}$), which also confirms our earlier assumption that $(\Delta E)_{\text{eff}} \ll E$.

(2) $l = 0, 1$.

The case $l = 0$ corresponds to elastic scattering and was considered earlier.

For $l = 1$ the integral (21) diverges logarithmically as $x \to 0$, so that the main contribution to it comes from values of $x \ll 1$. If we therefore replace $J_{\frac{3}{2}}(x)$ by

$$\left(\frac{x}{2}\right)^{\frac{3}{2}}\left[\frac{1}{\Gamma(\frac{5}{2})}\right] = \frac{2x^{\frac{3}{2}}}{3\sqrt{(2\pi)}}$$

and integrate from

$$aq_{\min} = \frac{a\Delta E}{\hbar v_0} = \frac{a\hbar}{Iv_0} \ll 1$$

to some value $aq \sim 1$, we get

$$\sigma_1 = \frac{8}{3}\left(\frac{e^2}{\hbar v_0}\right)^2 \ln\left(c\frac{Iv_0}{\hbar a}\right)\pi a^2 \quad (l = 1), \tag{26}$$

where c is a dimensionless quantity of order unity (its exact value is not very important since the argument of the logarithm is much larger than unity).

The total inelastic scattering cross-section for all angles and all possible excitations of the rotator is finally equal to

$$\sigma_{\text{inel}} = \sigma_1 + \sum_{l=2}^{\infty} \sigma_l = \frac{8}{3}\left(\frac{e^2}{\hbar v_0}\right)^2 \pi a^2 \left[\ln\left(c\frac{Iv_0}{\hbar a}\right) + \sum_{l=2}^{\infty}(2l+1)\frac{(l-2)!}{(l+2)!}\right],$$

where we have used equations (25) and (26). Using the relation

$$\sum_{k=1}^{\infty} \frac{k!}{(k+n-1)!} = \frac{1}{(n-2)(n-1)!},$$

we get after some simple manipulations

$$\sum_{l=2}^{\infty}(2l+1)\frac{(l-2)!}{(l+2)!} = \tfrac{1}{3},$$

so that we finally get

$$\sigma_{\text{inel}} = \frac{8}{3}\left(\frac{e^2}{\hbar v_0}\right)^2 \ln\left(c_1\frac{Iv_0}{\hbar a}\right)\pi a^2, \tag{27}$$

where $c_1 = ce^{\frac{1}{3}}$ ($e = 2{\cdot}718\ldots$). Since $e^2/\hbar v_0 \ll 1$, σ_{inel} is small compared to πa^2 which is the geometric cross-section of the region in which the rotator moves. The cross-section for excitation of the $l = 1$ level is much larger than the total cross-section for the excitation of all the higher levels.

17*. The radial functions which satisfy the boundary condition $\chi_l(a) = 0$ can be expressed as follows in terms of Bessel functions:

$$\chi_l = \sqrt{(r)}\,[J_{-l-\frac{1}{2}}(ka)\,J_{l+\frac{1}{2}}(kr) - J_{l+\frac{1}{2}}(ka)\,J_{-l-\frac{1}{2}}(kr)].$$

From the asymptotic behaviour of the Bessel functions we find for the phase shifts

$$\cot\delta_l = (-1)^{l+1}\frac{J_{-l-\frac{1}{2}}(ka)}{J_{l+\frac{1}{2}}(ka)}.$$

We get thus for the total electrostatic scattering cross-section

$$\sigma = \frac{4\pi}{k^2}\sum_{l=0}^{\infty}(2l+1)\frac{J^2_{l+\frac{1}{2}}(ka)}{J^2_{l+\frac{1}{2}}(ka) + J^2_{-l-\frac{1}{2}}(ka)} \approx 2\pi a^2.$$

20. Since $R \gg \lambda$, we can apply semi-classical considerations. All particles will be incident with $l \leqslant R/\lambda\!\!\!^-$, so that

$$\eta_l = 0 \quad \text{if} \quad l < R/\lambda\!\!\!^-, \qquad \eta_l = 1 \quad \text{if} \quad l > R/\lambda\!\!\!^-.$$

If we substitute these values of η_l into the expression for the total cross-sections,

$$\sigma_r = \pi\lambda\!\!\!^{-2}\,\Sigma(2l+1)\,(1-|\eta_l|^2), \quad \sigma_s = \pi\lambda\!\!\!^{-2}\,\Sigma(2l+1)\,|1-\eta_l|^2,$$

we get

$$\sigma_r = \sigma_s = \pi\lambda\!\!\!^- \sum_{l=0}^{R/\lambda\!\!\!^-} (2l+1) \approx \pi R^2.$$

We see that the total cross-section $\sigma = \sigma_r + \sigma_s$ is equal to twice the geometric cross-section of the nucleus.

21*. The radial Schrödinger equation can be written in the form

$$\chi_l'' + \left\{k^2 - \tilde{V}(r) - \frac{l(l+1)}{r^2}\right\}\chi_l = 0, \tag{1}$$

or, equivalently,

$$\chi_l(r) = j_l(kr) + \int_0^\infty dr' G(r, r')\tilde{V}(r')\chi_l(r'). \tag{2}$$

If we put

$$\omega(r) = \sqrt{\{\tilde{V}(r)\}}\chi_l(r),$$
$$g(r) = \sqrt{\{\tilde{V}(r)\}}j_l(kr),$$
$$K(r, r') = \sqrt{\{\tilde{V}(r)\tilde{V}(r')\}}G(r, r'),$$

we can write equation (2) in the form of an integral equation with a symmetric kernel:

$$\omega(r) = g(r) + \int_0^\infty dr' K(r, r')\omega(r'). \tag{3}$$

In terms of the new functions the quantity F_l has the form

$$F_l = I_1/I_2^2, \tag{4}$$

with

$$I_1 = \int_0^\infty dr\omega(r)\left\{\omega(r) - \int_0^\infty dr'\omega(r')K(r, r')\right\}, \tag{5}$$

$$I_2 = \int_0^\infty dr\omega(r)g(r). \tag{6}$$

If $\omega(r)$ satisfies equation (3), $I_1 = I_2$. Using the definition of the phase shifts, taking the Green function in the form

$$G(r, r') = \frac{1}{k} j_l(kr) n_l(kr'), \qquad \text{when } r < r',$$
$$= \frac{1}{k} n_l(kr) j_l(kr'), \qquad \text{when } r > r', \tag{7}$$

where n_l is a spherical Neumann function, and using the asymptotic form of the j_l and n_l, we find

$$I_2 = -k \tan \delta_l, \tag{8}$$

so that, if $\omega(r)$ satisfies equation (3), we have

$$F_l = -\frac{1}{k} \cot \delta_l. \tag{9}$$

Considering now a solution $\omega(r) + \delta\omega(r)$, where $\omega(r)$ again satisfies equation (3) one finds that

$$\delta I_1 = 2\delta I_2, \tag{10}$$

whence follows

$$\delta F_l = 0.$$

22. The wave function of a system of two identical particles will be the product of an orbital and a spin function. Independent of whether the spin of the particles is integer or half-integer, an even total spin corresponds to a symmetric orbital wave function, and an odd total spin to an antisymmetric one.

We can introduce the centre of mass system and separate the centre of mass variable; the orbital wave function will then be of the form

$$\Psi(\boldsymbol{r}_1, \boldsymbol{r}_2) = \varphi(\boldsymbol{R}) \psi(\boldsymbol{\rho}),$$

with

$$\boldsymbol{R} = \frac{\boldsymbol{r}_1 + \boldsymbol{r}_2}{2}, \quad \boldsymbol{\rho} = \boldsymbol{r}_1 - \boldsymbol{r}_2.$$

If we interchange $\boldsymbol{r}_1$ and $\boldsymbol{r}_2$ the function $\varphi(\boldsymbol{R})$, which describes the centre of mass motion, clearly does not change. The wave function of the relative motion of the two particles will thus be even,

$$\psi(\boldsymbol{\rho}) = \psi(-\boldsymbol{\rho})$$

if the total spin S is even, and odd,

$$\psi(\boldsymbol{\rho}) = -\psi(-\boldsymbol{\rho}),$$

if S is odd.

The unperturbed wave function can be written as

$$\psi(\boldsymbol{\rho}) = \exp[i(\boldsymbol{k}_0 \cdot \boldsymbol{\rho})] + \exp[-i(\boldsymbol{k}_0 \cdot \boldsymbol{\rho})], \tag{1}$$

for even S and as

$$\psi(\boldsymbol{\rho}) = \exp[i(\boldsymbol{k}_0 \cdot \boldsymbol{\rho})] - \exp[-i(\boldsymbol{k}_0 \cdot \boldsymbol{\rho})], \tag{2}$$

for odd S. The interaction between the particles produces a scattered wave $[F(\vartheta)/\rho]\exp(ik\rho)$, where ϑ is the angle between $\boldsymbol{k}_0$ and the direction in which the particles disappear in the centre of mass system. The scattering amplitude can be expressed in terms of the scattering amplitude for a particle with a mass equal to the reduced mass of the two particles in the field $V(r)$. Indeed, $\psi(\boldsymbol{\rho})$ satisfies the equation:

$$\left[-\frac{\hbar^2}{\mu}(\nabla_\rho^2 + k^2) - V(\rho)\right]\psi(\boldsymbol{\rho}) = 0.$$

If the incident wave $\exp[i(\boldsymbol{k}_0 \cdot \boldsymbol{\rho})]$ corresponds to a scattered wave $[f(\vartheta)/r]\exp(ik\rho)$, we get for the incident wave (1)

$$F_0(\vartheta) = f(\vartheta) + f(\pi - \vartheta) \quad \text{(even spin)},$$

and for the wave (2)

$$F_1(\vartheta) = f(\vartheta) - f(\pi - \vartheta) \quad \text{(odd spin)}.$$

The probability that one of the particles is scattered into a solid angle $d\Omega$ (the other particle moves in the opposite direction) is related to the density of the incident beam,

$$d\sigma_0 = |f(\vartheta) + f(\pi - \vartheta)|^2,$$

$$d\sigma_1 = |f(\vartheta) - f(\pi - \vartheta)|^2.$$

The amplitude $f(\vartheta)$ can be expressed in terms of the phase shifts δ_l as follows:

$$f(\vartheta) = \frac{1}{2ik}\sum_{l=0}^{\infty}(2l+1)P_l(\cos\vartheta)[\exp(2i\delta_l) - 1].$$

Taking into account that

$$P_l[\cos(\pi - \vartheta)] = P_l(-\cos\vartheta) = (-1)^l P_l(\cos\vartheta),$$

we get

$$F_0(\vartheta) = \frac{1}{ik}\sum_{\text{even } l}(2l+1)P_l(\cos\vartheta)[\exp(2i\delta_l) - 1],$$

$$F_1(\vartheta) = \frac{1}{ik}\sum_{\text{odd } l}(2l+1)P_l(\cos\vartheta)[\exp(2i\delta_l) - 1].$$

For slow particles, small values of l will give the major contribution to the scattering. If the total spin is even, the cross-section is spherically symmetric (as is the case for the scattering of different particles) and does not vanish as $k \to 0$.

If the total spin is odd, the scattering is determined by the term with $l = 1$. Since $\delta_l \sim k^{2l+1}$ for small wave numbers k, the cross-section vanishes as E^2 (as $E \to 0$) and has an angular dependence $\sim \cos^2 \vartheta$.

23. In the case of the Coulomb field the scattering amplitude, in Coulomb units, will be

$$f(\vartheta) = -\frac{1}{2k^2 \sin^2 \frac{1}{2}\vartheta} \exp\left(-\frac{2i}{k} \ln \sin \tfrac{1}{2}\vartheta\right) . \frac{\Gamma(1+i/k)}{\Gamma(1-i/k)} .$$

Using the results of the preceding problem we find for the differential cross-section for the case of even spin

$$d\sigma_0 = |f(\vartheta) + f(\pi - \vartheta)|^2 \, d\Omega$$

$$= \frac{1}{4k^4} \left\{ \frac{1}{\sin^4 \frac{1}{2}\vartheta} + \frac{1}{\cos^4 \frac{1}{2}\vartheta} + \frac{2 \cos [(2/k) \ln \tan \frac{1}{2}\vartheta]}{\sin^2 \frac{1}{2}\vartheta \cos^2 \frac{1}{2}\vartheta} \right\} d\Omega.$$

This formula gives the cross-section for the scattering of α-particles whose spin is zero.

In the case of two electrons we can have a state of total spin 1. The differential cross-section is then

$$d\sigma_1 = |f(\vartheta) - f(\pi - \vartheta)|^2$$

$$= \frac{1}{4k^4} \left\{ \frac{1}{\sin^4 \frac{1}{2}\vartheta} + \frac{1}{\cos^4 \frac{1}{2}\vartheta} - \frac{2 \cos [(2/k) \ln \tan \frac{1}{2}\vartheta]}{\sin^2 \frac{1}{2}\vartheta \cos^2 \frac{1}{2}\vartheta} \right\} d\Omega.$$

If the scattered electrons are not polarised there are three possible values for the z-component of the total spin, $S_z = 0$, $S_z = \pm 1$, and two possible values for the total spin, $S = 0$ and $S = 1$. The probabilities for each of the possible values of the z-component are $W_{-1} = \frac{1}{4}$, $W_0 = \frac{1}{2}$, $W_{+1} = \frac{1}{4}$. The values $S_z = \pm 1$ correspond certainly to a total spin $S = 1$. Since the different values of the z-component for $S = 1$ are equally probable, the probability for $S_z = 0$, $S = 1$ is the same as for $S_z = \pm 1$, $S = 1$, that is, $\frac{1}{4}$. We find thus that the probability for total spin $S = 0$ is $w_0 - \frac{1}{4} = \frac{1}{4}$, and for total spin $S = 1$ equal to $\frac{3}{4}$. For an unpolarised electron beam we get thus

$$d\sigma = \tfrac{1}{4} d\sigma_0 + \tfrac{3}{4} d\sigma_1$$

$$= \frac{1}{4k^4} \left\{ \frac{1}{\sin^4 \frac{1}{2}\vartheta} + \frac{1}{\cos^4 \frac{1}{2}\theta} - \frac{\cos [(2/k) \ln \tan \frac{1}{2}\vartheta]}{\sin^2 \frac{1}{2}\vartheta \cos^2 \frac{1}{2}\vartheta} \right\} d\Omega.$$

The last, interference, term inside the braces is characteristic for the scattering of identical particles. As $\hbar \to 0$ the equation for $d\sigma$ must go over to the classical Rutherford formula which in the centre of mass system is of the form

$$d\sigma = \frac{1}{4k^4}\left(\frac{1}{\sin^4 \frac{1}{2}\vartheta} + \frac{1}{\cos^4 \frac{1}{2}\vartheta}\right) d\Omega.$$

The transition to this formula is somewhat unusual. If $e^2/\hbar v \gg 1$, so that we can apply classical considerations, the interference term in the usual units is of the form

$$\frac{\cos\left[(2e^2/\hbar v)\ln\tan \frac{1}{2}\vartheta\right]}{\sin^2 \frac{1}{2}\vartheta \cos^2 \frac{1}{2}\vartheta},$$

and oscillates fast. The quantum mechanical differential cross-section is thus essentially different from the classical one even for large values of $e^2/\hbar v$ if ϑ is strictly fixed. However, if we average over a small range of angles $\Delta\vartheta \sim \hbar v/e^2$ the interference term vanishes, and the quantum mechanical equation goes over into the classical one.

24. $$\hat{f}^2 = \tfrac{3}{4}f_3^2 + \tfrac{1}{4}f_1^2 + \tfrac{1}{4}(f_3^2 - f_1^2)(\hat{\boldsymbol{\sigma}}_n \cdot \hat{\boldsymbol{\sigma}}_p).$$

The average value of the operator $(\hat{\boldsymbol{\sigma}}_n \cdot \hat{\boldsymbol{\sigma}}_p)$ in the state characterised by the spin function

$$\begin{bmatrix} \exp(-i\alpha)\cos\beta \\ \exp(i\alpha)\sin\beta \end{bmatrix}_n \begin{pmatrix} 1 \\ 0 \end{pmatrix}_p$$

is equal to

$$\overline{(\hat{\boldsymbol{\sigma}}_n \cdot \hat{\boldsymbol{\sigma}}_p)} = \cos^2\beta - \sin^2\beta.$$

We find thus for the scattering cross-section the expression

$$\sigma = \pi[3f_3^2 + f_1^2 + (f_3^2 - f_1^2)\cos 2\beta]$$

or

$$\sigma = \left[\tfrac{3}{4}\sigma_3 + \tfrac{1}{4}\sigma_1 + \frac{\cos 2\beta}{4}(\sigma_3 - \sigma_1)\right].$$

In the case where the neutron beam is unpolarised $\overline{\cos 2\beta} = 0$ and we get for the cross-section

$$\sigma = \tfrac{3}{4}\sigma_3 + \tfrac{1}{4}\sigma_1.$$

25. The spin states of the neutron and the proton before the interaction are described by the function

$$\begin{pmatrix} 1 \\ 0 \end{pmatrix}_n \begin{pmatrix} 0 \\ 1 \end{pmatrix}_p.$$

This function can be expanded in terms of the spin functions of the singlet and triplet states,

$$\begin{pmatrix}1\\0\end{pmatrix}_n\begin{pmatrix}0\\1\end{pmatrix}_p = \frac{1}{\sqrt{2}}\left\{\frac{1}{\sqrt{2}}\left[\begin{pmatrix}1\\0\end{pmatrix}_n\begin{pmatrix}0\\1\end{pmatrix}_p+\begin{pmatrix}0\\1\end{pmatrix}_n\begin{pmatrix}1\\0\end{pmatrix}_p\right]+\frac{1}{\sqrt{2}}\left[\begin{pmatrix}1\\0\end{pmatrix}_n\begin{pmatrix}0\\1\end{pmatrix}_p-\begin{pmatrix}0\\1\end{pmatrix}_n\begin{pmatrix}1\\0\end{pmatrix}_p\right]\right\}.$$

The scattered wave is of the form

$$\frac{\exp(ikr)}{r}\frac{1}{\sqrt{2}}\left\{f_3\frac{1}{\sqrt{2}}\left[\begin{pmatrix}1\\0\end{pmatrix}_n\begin{pmatrix}0\\1\end{pmatrix}_p+\begin{pmatrix}0\\1\end{pmatrix}_n\begin{pmatrix}1\\0\end{pmatrix}_p\right]+f_1\frac{1}{\sqrt{2}}\left[\begin{pmatrix}1\\0\end{pmatrix}_n\begin{pmatrix}0\\1\end{pmatrix}_p-\begin{pmatrix}0\\1\end{pmatrix}_n\begin{pmatrix}1\\0\end{pmatrix}_p\right]\right\}$$

or

$$\frac{\exp(ikr)}{r}\left[\frac{f_3+f_1}{2}\begin{pmatrix}1\\0\end{pmatrix}_n\begin{pmatrix}0\\1\end{pmatrix}_p+\frac{f_3-f_1}{2}\begin{pmatrix}0\\1\end{pmatrix}_n\begin{pmatrix}1\\0\end{pmatrix}_p\right].$$

It follows thus that the probability for reorientation is equal to

$$\frac{1}{2}\frac{(f_3-f_1)^2}{f_3^2+f_1^2}.$$

26. We introduce the operator of the total spin of two protons,

$$\tfrac{1}{2}(\hat{\boldsymbol{\sigma}}_{p_1}+\hat{\boldsymbol{\sigma}}_{p_2}) = \hat{\boldsymbol{S}}.$$

One can easily show that

$$(\hat{\boldsymbol{\sigma}}_n\cdot\hat{\boldsymbol{S}})^2 = \hat{\boldsymbol{S}}^2-(\hat{\boldsymbol{\sigma}}_n\cdot\hat{\boldsymbol{S}}).$$

Using this expression we get for $\hat{f}^2$

$$\hat{f}^2 = \tfrac{1}{4}[(f_1+3f_3)^2+(5f_3^2-2f_3f_1-3f_1^2)(\hat{\boldsymbol{\sigma}}_n\cdot\hat{\boldsymbol{S}})+(f_3-f_1)^2\hat{\boldsymbol{S}}^2].$$

For the case where the scattering takes place in the para-state the cross-section is equal to

$$\sigma^{\text{para}} = \pi(f_1+3f_3)^2.$$

This cross-section does not, of course, depend on the polarisation of the incident neutrons since there is no preferential direction in space.

The scattering cross-section for ortho-hydrogen is equal to

$$\sigma^{\text{ortho}} = \pi[(f_1+3f_3)^2+(5f_3^2-2f_3f_1-3f_1^2)\cos 2\beta+2(f_3-f_1)^2],$$

where 2β is the angle between the direction of the total spin of the two protons and the spin of the neutron.

If the neutron beam is unpolarised the average value of $\cos 2\beta$ taken over a mixed ensemble is equal to zero and σ is of the form

$$\sigma^{\text{ortho}} = \pi[(f_1+3f_3)^2+2(f_3-f_1)^2],$$

while the ratio of the cross-section is equal to

$$\frac{\sigma^{\text{ortho}}}{\sigma^{\text{para}}} = 1+2\left(\frac{f_3-f_1}{f_1+3f_3}\right)^2.$$

27. $$(2I+1)\frac{d\sigma}{d\Omega} = (I+1)|a_+|^2+I|a_-|^2.$$

28. $$\left.\frac{d\sigma}{d\Omega}\right|_{\sigma\parallel n} = \frac{\mu^2}{4\pi^2\hbar^4}[A-\tfrac{4}{3}\pi a^3 B]^2,$$

$$\left.\frac{d\sigma}{d\Omega}\right|_{\sigma\parallel -n} = \frac{\mu^2}{4\pi\hbar^4}[A+\tfrac{4}{3}\pi a^3 B]^2;$$

$$P = \frac{\dfrac{8\mu^2}{3\hbar^4}a^3 ABNL}{1-\dfrac{\mu^2}{\pi\hbar^4}NL\left(A^2+\dfrac{16\pi^2}{9}a^6 B^2\right)}.$$

29*. At large distances from the target (the target consists of scalar particles) the wave function of the incident particles will be of the form

$$\exp(ikz)\begin{pmatrix}1\\0\end{pmatrix} \approx \frac{1}{2kr}\sum_{l=0}^{\infty} i^{l+1}(2l+1)\begin{pmatrix}1\\0\end{pmatrix}P_l(\cos\vartheta)$$
$$\times\{\exp[-i(kr-\tfrac{1}{2}\pi l)]-\exp[i(kr-\tfrac{1}{2}\pi l)]\}. \quad (1)$$

We shall expand the function $\begin{pmatrix}1\\0\end{pmatrix}P_l(\cos\vartheta)$ in terms of the eigenfunctions of the operator $\hat{j}^2$. The result is

$$\begin{pmatrix}1\\0\end{pmatrix}P_l(\cos\theta) = \frac{2\sqrt{\pi}}{\sqrt{(2l+1)}}\begin{pmatrix}1\\0\end{pmatrix}Y_{l0}(\vartheta) = \frac{2\sqrt{\pi}}{2l+1}[\sqrt{(l+1)}\,\Psi_l^+ + \sqrt{(l)}\,\Psi_l^-]. \quad (2)$$

In this equation Ψ_l^+ and Ψ_l^- are the Pauli functions (see problem 36 of section 4),

$$\Psi_l^+ = \frac{1}{\sqrt{(2l+1)}}\begin{bmatrix}\sqrt{(l+1)}\,Y_{l0}\\ \sqrt{(l)}\,Y_{l1}\end{bmatrix} \quad (j=l+\tfrac{1}{2}, l, j_z=\tfrac{1}{2}),$$

$$\Psi_l^- = \frac{1}{\sqrt{(2l+1)}}\begin{bmatrix}\sqrt{(l)}\,Y_{l0}\\ -\sqrt{(l+1)}\,Y_{l1}\end{bmatrix} \quad (j=l-\tfrac{1}{2}, l, j_z=\tfrac{1}{2}).$$

Substituting expression (2) into equation (1) we get

$$\exp(ikz)\begin{pmatrix}1\\0\end{pmatrix} \approx \frac{\sqrt{\pi}}{kr}\sum_{l=0}^{\infty} i^{l+1}\{\sqrt{(l+1)}\,\Psi_l^+ + \sqrt{(l)}\,\Psi_l^-\}$$
$$\times\{\exp[-i(kr-\tfrac{1}{2}\pi l)]-\exp[i(kr-\tfrac{1}{2}\pi l)]\}. \quad (3)$$

The interaction changes only the outgoing wave $\exp(ikr)/r$. Since for any interaction law $\hat{j}^2$, $\hat{l}^2$ and $\hat{j}_z$ will be integrals of motion (see problem 40 of section 4) this change will in general be different for states with different quantum numbers j, l. We get the following expression for the scattered wave:

$$\Psi_s \approx \frac{\exp(ikr)}{r}\frac{\sqrt{\pi}}{ik}\sum_{l=0}^{\infty}[\sqrt{(l+1)}(\eta_l^+-1)\Psi_l^+ + \sqrt{(l)}(\eta_l^- - 1)\Psi_l^-],$$

$$\eta_l^+ = \eta(j = l+\tfrac{1}{2}, l), \quad \eta_l^- = \eta(j = l-\tfrac{1}{2}, l)$$

or

$$\Psi_s \approx \frac{\exp(ikr)}{r}\frac{\sqrt{\pi}}{ik}\sum_{l=0}^{\infty}\frac{1}{\sqrt{(2l+1)}}$$

$$\times\left\{\begin{pmatrix}1\\0\end{pmatrix} Y_{l0}[(l+1)(\eta_l^+-1)+l(\eta_l^--1)]+\begin{pmatrix}0\\1\end{pmatrix} Y_{l1}(\eta_l^+-\eta_l^-)\right\}.$$

It is thus clear that the re-orientation of the particle spin can take place when $\eta_l^+ \neq \eta_l^-$. We can express the scattering cross-section in terms of η_l^+ and η_l^-.

The differential cross-section for scattering with a simultaneous change in polarisation, $d\sigma_1$, is equal to

$$d\sigma_1 = \frac{\pi}{k^2}\left|\sum_{l=0}^{\infty}\sqrt{\left[\frac{l(l+1)}{2l+1}\right]}(\eta_l^+-\eta_l^-)Y_{l1}\right|^2 d\Omega,$$

while the cross-section, $d\sigma_2$, without change in polarisation is equal to

$$d\sigma_2 = \frac{\pi}{k^2}\left|\sum_{l=0}^{\infty}\frac{1}{\sqrt{(2l+1)}}Y_{l0}[(l+1)(\eta_l^+-1)+l(\eta_l^--1)]\right|^2 d\Omega.$$

If the relative velocity of the particles is not large we need take into account only scattering of the S- and P-waves,

$$(|\eta_l^+-1| \ll 1, \quad |\eta_l^- - 1| \ll 1, \quad \text{if} \quad l > 1).$$

In that case we have

$$d\sigma_1 \approx \frac{1}{4k^2}(\eta_1^+-\eta_1^-)\sin^2\vartheta\, d\Omega,$$

$$d\sigma_2 \approx \frac{1}{4k^2}|\cos\vartheta\,(2\eta_1^+ + \eta_1^- - 1) + \eta_0^+ - 1|^2\, d\Omega.$$

From the expression for $d\sigma_1$ it is clear that particles whose spin orientation is changed are only weakly scattered in the direction perpendicular to the z-axis.

30. In all three cases the distribution is isotropic.

31*. The eigenfunctions of the operator I_z for the nucleon–pion system can be written in the form of all possible products of the functions φ and ψ. The functions φ correspond to the different charge states of the pion $(\varphi_+, \varphi_0, \varphi_-)$ and the functions ψ to those of the nucleon (ψ_p, ψ_n). We have given all possible combinations in the following table:

$(p^+) = \varphi_+ \psi_p$ $I_z = \frac{3}{2}$	$(p^0) = \varphi_0 \psi_p$ $I_z = \frac{1}{2}$	$(p^-) = \varphi_- \psi_p$ $I_z = -\frac{1}{2}$
$(n^+) = \varphi_+ \psi_n$ $I_z = \frac{1}{2}$	$(n^0) = \varphi_0 \psi_n$ $I_z = -\frac{1}{2}$	$(n^-) = \varphi_- \psi_n$ $I_z = -\frac{3}{2}$

In this table (p^+) indicates the function of the system consisting of a positive pion and a proton, (n^0) that of a system consisting of a neutral pion and a neutron, and so on. These functions will, generally speaking, not be eigenfunctions of the operator of the square of the total isotopic spin of the system $\boldsymbol{I}^2$.

The eigenfunctions of $\boldsymbol{I}^2$, which are at the same time eigenfunctions of I_z, will be linear combinations of the functions from the table, multiplied by Clebsch–Gordan coefficients. The Clebsch–Gordan coefficients for $m' = \pm\frac{1}{2}$ are of the form (see problem 36 of section 4)

	$m' = \frac{1}{2}$	$m' = -\frac{1}{2}$
$I = j+\frac{1}{2}$	$\sqrt{\left(\frac{j+M+\frac{1}{2}}{2j+1}\right)}$	$\sqrt{\left(\frac{j-M+\frac{1}{2}}{2j+1}\right)}$
$I = j-\frac{1}{2}$	$-\sqrt{\left(\frac{j-M+\frac{1}{2}}{2j+1}\right)}$	$\sqrt{\left(\frac{j+M+\frac{1}{2}}{2j+1}\right)}$

In our case $M = I_z$, $j = 1$, $m' = \tau_z$.

If we use this table we get the eigenfunctions $\Phi^I_{I_z}$ of the operators $\boldsymbol{I}^2$ and I_z:

$$\Phi^{\frac{3}{2}}_{\frac{3}{2}} = (p^+),$$

$$\Phi^{\frac{3}{2}}_{\frac{1}{2}} = \sqrt{(\tfrac{2}{3})}\,(p^0) + \sqrt{(\tfrac{1}{3})}\,(n^+),$$

$$\Phi^{\frac{3}{2}}_{-\frac{1}{2}} = \sqrt{(\tfrac{1}{3})}\,(p^-) + \sqrt{(\tfrac{2}{3})}\,(n^0),$$

$$\Phi^{\frac{3}{2}}_{-\frac{3}{2}} = (n^-),$$

$$\Phi^{\frac{1}{2}}_{\frac{1}{2}} = -\sqrt{(\tfrac{1}{3})}\,(p^0) + \sqrt{(\tfrac{2}{3})}\,(n^+),$$

$$\Phi^{\frac{1}{2}}_{-\frac{1}{2}} = -\sqrt{(\tfrac{2}{3})}\,(p^-) + \sqrt{(\tfrac{1}{3})}\,(n^0).$$

One can easily express the eigenfunctions of the nucleon–pion system in terms of these eigenfunctions:

$$(p^+) = \Phi^{\frac{3}{2}}_{\frac{3}{2}}, \quad (n^+) = \sqrt{(\tfrac{1}{3})}\,\Phi^{\frac{3}{2}}_{\frac{1}{2}} + \sqrt{(\tfrac{2}{3})}\,\Phi^{\frac{1}{2}}_{\frac{1}{2}},$$

$$(p^0) = \sqrt{(\tfrac{2}{3})}\,\Phi^{\frac{3}{2}}_{\frac{1}{2}} - \sqrt{(\tfrac{1}{3})}\,\Phi^{\frac{1}{2}}_{\frac{1}{2}}, \quad (n^0) = \sqrt{(\tfrac{2}{3})}\,\Phi^{\frac{3}{2}}_{-\frac{1}{2}} + \sqrt{(\tfrac{1}{3})}\,\Phi^{\frac{1}{2}}_{-\frac{1}{2}},$$

$$(p^-) = \sqrt{(\tfrac{1}{3})}\,\Phi^{\frac{3}{2}}_{-\frac{1}{2}} - \sqrt{(\tfrac{2}{3})}\,\Phi^{\frac{1}{2}}_{-\frac{1}{2}}, \quad (n^-) = \Phi^{\frac{3}{2}}_{-\frac{3}{2}}.$$

32*. We can expand the incident wave as follows:

$$\psi = \exp(ikz)\begin{pmatrix}1\\0\end{pmatrix}\delta(\pi-\pi_i)\,\delta(n-\tau_z)$$

$$= \sum_{l=0}^{\infty} i^l(2l+1)\,P_l(\cos\theta)\begin{pmatrix}1\\0\end{pmatrix}\delta(\pi-\pi_i)\,\delta(n-\tau_z)\frac{\sin(kr-\frac{1}{2}\pi l)}{kr}$$

$$= 2\sqrt{\pi}\sum_{l=0}^{\infty}\sum_{I} C^{I\tau_z}_{I_z}\Phi^{I}_{I_z}\, i^l\sqrt{(2l+1)}\,Y_{l0}\begin{pmatrix}1\\0\end{pmatrix}\frac{\sin(kr-\frac{1}{2}\pi l)}{kr}, \qquad (1)$$

where $C^{I\tau_z}_{I_z}$ are the Clebsch–Gordan coefficients which are given in the following table:

	$\tau_z = \frac{1}{2}$	$\tau_z = -\frac{1}{2}$
$C^{3/2}_{I_z}$	$\sqrt{(\frac{1}{2}+I_z/3)}$	$\sqrt{(\frac{1}{2}-I_z/3)}$
$C^{1/2}_{I_z}$	$-\sqrt{(\frac{1}{2}-I_z/3)}$	$\sqrt{(\frac{1}{2}+I_z/3)}$

(2)

We have used here (see preceding problem)

$$\delta(\pi-\pi_i)\,\delta(n-\tau_z) = \sum_I C^{I\tau_z}_{I_z}\Phi^{I}_{I_z}.$$

We introduce the Pauli functions for $m_j = \frac{1}{2}$ (see problem 29 of section 9):

$$Y_l^+ = \begin{pmatrix}\sqrt{\left(\dfrac{l+1}{2l+1}\right)}\,Y_{l0}\\[2mm] \sqrt{\left(\dfrac{l}{2l+1}\right)}\,Y_{l1}\end{pmatrix}; \quad Y_l^- = \begin{pmatrix}\sqrt{\left(\dfrac{l}{2l+1}\right)}\,Y_{l0}\\[2mm] -\sqrt{\left(\dfrac{l+1}{2l+1}\right)}\,Y_{l1}\end{pmatrix}$$

and expand $Y_{l0}\begin{pmatrix}1\\0\end{pmatrix}$ in terms of them:

$$Y_{l0}\begin{pmatrix}1\\0\end{pmatrix} = \frac{1}{\sqrt{(2l+1)}}\left[\sqrt{(l+1)}\,Y_l^+ + \sqrt{(l)}\,Y_l^-\right]. \qquad (3)$$

Substituting expression (3) into equation (1) we get

$$\psi = \frac{2\sqrt{\pi}}{kr}\sum_{l=0}^{\infty}\sum_{I} i^l[\sqrt{(l+1)}\,Y_l^+ + \sqrt{(l)}\,Y_l^-]\sin(kr-\tfrac{1}{2}\pi l)\,C_{I_z}^{I\tau_z}\Phi_{I_z}^{I}$$

$$= \frac{\sqrt{\pi}}{ikr}\sum_{l=0}^{\infty}\sum_{I}[\sqrt{(l+1)}\,Y_l^+ + \sqrt{(l)}\,Y_l^-]\,C_{I_z}^{I\tau_z}\Phi_{I_z}^{I}[\exp(ikr)-(-1)^l\exp(-ikr)]$$

$$= \frac{\sqrt{\pi}}{ik}\sum_{l=0}^{\infty}\sum_{I} C_{I_z}^{I\tau_z}\Phi_{I_z}^{I}[\sqrt{(l+1)}\,Y_l^+ + \sqrt{(l)}\,Y_l^-]\exp(ikr)/r$$

$$-\frac{\sqrt{\pi}}{ik}\sum_{l=0}^{\infty}\sum_{I} C_{I_z}^{I\tau_z}\Phi_{I_z}^{I}[\sqrt{(l+1)}\,Y_l^+ + \sqrt{(l)}\,Y_l^-](-1)^l\exp(-ikr)/r. \qquad (4)$$

34*. We shall use the expansion of the incident wave in terms of the eigenfunctions of the operators which are conserved (see equation (4) of the problem 32 of section 9).

Each term of this sum, which corresponds to well-defined values of l, J, and I, will be scattered independently of the others.

The number of particles corresponding to given values of l, J, and I will thus not be changed during the elastic scattering process, so that the influence of the scattering centre will be to introduce some phase factor $\exp(2i\delta_{l\pm}^I)$. The quantity $\delta_{l\pm}^I = \delta_{l,\,J=l\pm\frac{1}{2}}^I$ depends on l, J, and I but it is independent of I_z because of the hypothesis of isotopic invariance.

We must note that the scattering centre does not influence the incoming wave but only the outgoing one. The wave function of the system can thus be written, if we take scattering into account, in the form

$$\psi = \frac{\sqrt{\pi}}{ik}\sum_{I}\sum_{l=0}^{\infty} C_{I_z}^{I\tau_z}\Phi_{I_z}^{I}[\sqrt{(l+1)}\,Y_l^+\exp(2i\delta_{l+}^I) + \sqrt{(l)}\,Y^-\exp(2i\delta_{l-}^I)]\frac{\exp(ikr)}{r}$$

$$-\frac{\sqrt{\pi}}{ik}\sum_{I}\sum_{l=0}^{\infty} C_{I_z}^{I\tau_z}\Phi_{I_z}^{I}[\sqrt{(l+1)}\,Y_l^+ + \sqrt{(l)}\,Y_l^-]\frac{\exp(-ikr)}{r}. \qquad (1)$$

At large distances from the scattering centre we can write it in the form

$$\psi = \psi_{\text{inc}} + f\frac{\exp(ikr)}{r}.$$

The quantity f is the scattering amplitude. If we take from equation (1) the expression for ψ_{inc} [see equation (4) of the problem 32 of section 9] we get

$$f = \frac{\sqrt{\pi}}{ik}\sum_{I}\sum_{l=0}^{\infty} C_{I_z}^{I\tau_z}\Phi_{I_z}^{I}\{\sqrt{(l+1)}\,Y_l^+[\exp(2i\delta_{l+}^I)-1]$$

$$+\sqrt{(l)}\,Y_l^-[\exp(2i\delta_{l-}^I)-1]\}. \qquad (2)$$

The functions $\Phi^I_{I_z}$ can be expanded in terms of the eigenfunctions of the operator I_z

$$\Phi^I_{I_z} = C^{I\tau_z}_{I_z}\,\delta(\pi-\pi_i)\,\delta(n-\tau_z)+C^{I,-\tau_z}_{I_z}\,\delta(\pi-\pi_k)\,\delta(n+\tau_z), \qquad (3)$$

where the Clebsch–Gordan coefficients $C^{I\tau_z}_{I_z}$ are given in table (2) of the preceding problem and where $\pi_k = \pi_i+2\tau_z$. Substituting equation (3) into equation (2) we get

$$f = \frac{\sqrt{\pi}}{ik}\sum_I\sum_{l=0}^{\infty}\,[G^{I\tau_z}_{I\tau_z}\,\delta(\pi-\pi_i)\,\delta(n-\tau_z)+G^{I\tau_z}_{I_z,-\tau_z}\,\delta(\pi-\pi_k)\,\delta(n+\tau_z)]$$
$$\times\{\sqrt{(l+1)}\,Y_l^+[\exp(2i\delta^I_{l+})-1]+\sqrt{(l)}\,Y_l^-[\exp(2i\delta^I_{l-})-1]\},$$

where

$$G^{I\tau_z}_{I_z\tau_z'} = C^{I\tau_z}_{I_z}\,C^{I\tau_z'}_{I_z}$$

and τ_z' denotes the final state of the nucleon.

If we now put

$$f = f^{\tau_z}_{\tau_z}\,\delta(\pi-\pi_i)\,\delta(n-\tau_z)+f^{\tau_z}_{\tau_z}\,\delta(\pi-\pi_k)\,\delta(n+\tau_z),$$

we get

$$f^{\tau_z}_{\tau_z'} = \frac{\sqrt{\pi}}{ik}\sum_I\sum_{l=0}^{\infty} G^{I\tau_z}_{I_z\tau_z'}\{\sqrt{(l+1)}\,Y_l^+[\exp(2i\delta^I_{l+})-1]$$
$$+\sqrt{(l)}\,Y^-[\exp(2i\delta^I_{l-})-1]\}, \qquad (4)$$

where $G^{I\tau_z}_{I_z\tau_z'}$ is given in the following table for all reactions necessary for our problem:

Reaction	$p^+\to p^+$	$p^-\to p^-$	$p^-\to n^0$
$G^{\frac{3}{2}\tau_z}_{I_z\tau'}$	1	$\frac{1}{3}$	$\sqrt{(2)}/3$
$G^{\frac{1}{2}\tau_z}_{I_z\tau_z'}$	0	$\frac{2}{3}$	$-\sqrt{(2)}/3$

If we substitute these values of $G^{I\tau}_{I_z\tau_z'}$ into equation (4) we get finally

$$\left.\begin{aligned}
f(p^+,p^+) &= \frac{\sqrt{\pi}}{ik}\sum_{l=0}^{\infty}\{\sqrt{(l+1)}\,Y_l^+[\exp(2i\delta^{\frac{3}{2}}_{l+})-1]\\
&\qquad+\sqrt{(l)}\,Y_l^-[\exp(2i\delta^{\frac{3}{2}}_{l-})-1]\},\\
f(p^-,p^-) &= \frac{\sqrt{\pi}}{3ik}\sum_{l=0}^{\infty}\{\sqrt{(l+1)}\,Y_l^+[\exp(2i\delta^{\frac{3}{2}}_{l+})+2\exp(2i\delta^{\frac{1}{2}}_{l+})-3]\\
&\qquad+\sqrt{(l)}\,Y_l^-[\exp(2i\delta^{\frac{3}{2}}_{l-})+2\exp(2i\delta^{\frac{1}{2}}_{l-})-3]\},\\
f(p^-,n_0) &= \frac{\sqrt{(2\pi)}}{3ik}\sum_{l=0}^{\infty}\{\sqrt{(l+1)}\,Y_l^+[\exp(2i\delta^{\frac{3}{2}}_{+})-\exp(2i\delta^{\frac{1}{2}}_{l+})]\\
&\qquad+\sqrt{(l)}\,Y_l^-[\exp(2i\delta^{\frac{3}{2}}_{l-}-\exp 2i\delta^{\frac{1}{2}}_{l-})]\}.
\end{aligned}\right\} \qquad (5)$$

35*. The table of the coefficients $G^{I\tau_z}_{I_z\tau_z'}$ for all possible reactions of the pions with nucleons is of the form

No.	Reaction	I_z	$G^{\frac{3}{2}\tau_z}_{I_z\tau_z'}$	$G^{\frac{1}{2}\tau_z}_{I_z\tau_z'}$
1	$p^+ \to p^+$	$\frac{3}{2}$	1	0
2	$p^0 \to p^0$	$\frac{1}{2}$	$\frac{2}{3}$	$\frac{1}{3}$
3	$p^0 \to n^+$	$\frac{1}{2}$	$\sqrt{(2)}/3$	$-\sqrt{(2)}/3$
4	$p^- \to n^0$	$-\frac{1}{2}$	$\sqrt{(2)}/3$	$-\sqrt{(2)}/3$
5	$p^- \to p^-$	$-\frac{1}{2}$	$\frac{1}{3}$	$\frac{2}{3}$
6	$n^+ \to n^+$	$\frac{1}{2}$	$\frac{1}{3}$	$\frac{2}{3}$
7	$n^0 \to p^-$	$-\frac{1}{2}$	$\sqrt{(2)}/3$	$-\sqrt{(2)}/3$
8	$n^+ \to p^0$	$\frac{1}{2}$	$\sqrt{(2)}/3$	$-\sqrt{(2)}/3$
9	$n^0 \to n^0$	$-\frac{1}{2}$	$\frac{2}{3}$	$\frac{1}{3}$
10	$n^- \to n^-$	$-\frac{3}{2}$	1	0

(1)

Since the phases are independent of I_z because of our hypothesis of isotopic invariance it follows immediately from equation (4) of the preceding problem and table (1) that

(1) $f(p^+, p^+) = f(n^-, n^-)$,

(2) $f(p^-, p^-) = f(n^+, n^+)$,

(3) $f(p^0, n^+) = f(n^+, p^0) = f(p^-, n^0) = f(n^0, p^-)$,

(4) $f(p^0, p^0) = f(n^0, n^0)$.

The expressions for the first three amplitudes were given in the preceding problem. From table (1) one sees easily that

$$f(p^0, p^0) = \tfrac{1}{2}[f(p^+, p^+) + f(p^-, p^-)].$$

Using table (1) we get

(1) $f(p^+, p^+) = f(n^-, n^-) = f^{\frac{3}{2}}$,

(2) $f(p^-, p^-) = f(n^+, n^+) = \frac{1}{3}(f^{\frac{3}{2}} + 2f^{\frac{1}{2}})$,

(3) $f(p^0, n^+) = f(n^+, p^0) = f(p^-, n^0) = f(n^0, p^-) = \dfrac{\sqrt{2}}{3}(f^{\frac{3}{2}} - f^{\frac{1}{2}})$,

(4) $f(p^0, p^0) = f(n^0, n^0) = \frac{1}{3}(2f^{\frac{3}{2}} + f^{\frac{1}{2}})$.

36*. The differential scattering cross-section is equal to

$$\frac{d\sigma}{d\Omega} = |f|^2, \quad \text{where} \quad d\Omega = \sin\theta \, d\theta \, d\varphi.$$

The total scattering cross-section is equal to

$$\sigma = \int_0^{2\pi}\int_0^{\pi} |f|^2 \sin\theta \, d\theta \, d\varphi.$$

If we substitute here the scattering amplitudes for the reactions considered, which are given in problem 34 of section 9, and use the orthonormality of the Pauli functions

$$\iint (Y_l^+)^+ (Y_{l'}^+) \, d\Omega = \iint (Y_l^-)^+ (Y_{l'}^-) \, d\Omega = \delta_{ll'},$$

$$\iint (Y_l^+)^+ (Y_{l'}^-) \, d\Omega = \iint (Y_l^-)^+ (Y_{l'}^+) \, d\Omega = 0,$$

we get

$$\sigma(p^+, p^+) = \frac{4\pi}{k^2} \sum_{l=0}^{\infty} [(l+1)\sin^2 \delta_{l+}^{\frac{3}{2}} + l \sin^2 \delta_{l-}^{\frac{3}{2}}],$$

$$\sigma(p^-, p^-) = \frac{4\pi}{3k^2} \sum_{l=0}^{\infty} \{(l+1)[\sin^2 \delta_{l+}^{\frac{3}{2}} + 2\sin^2 \delta_{l+}^{\frac{1}{2}} - \tfrac{2}{3}\sin^2(\delta_{l+}^{\frac{3}{2}} - \delta_{l+}^{\frac{1}{2}})]$$
$$+ l[\sin^2 \delta_{l-}^{\frac{3}{2}} + 2\sin^2 \delta_{l-}^{\frac{1}{2}} - \tfrac{2}{3}\sin^2(\delta_{l-}^{\frac{3}{2}} - \delta_{l-}^{\frac{1}{2}})]\},$$

$$\sigma(p^-, n^0) = \frac{8\pi}{9k^2} \sum_{l=0}^{\infty} [(l+1)\sin^2(\delta_{l+}^{\frac{3}{2}} - \delta_{l+}^{\frac{1}{2}}) + l \sin^2(\delta_{l-}^{\frac{3}{2}} - \delta_{l-}^{\frac{1}{2}})].$$

37*. We shall give the detailed solution for the reaction (p^+, p^+). From problem 34 of section 9, we get for the scattering amplitude for S- and P-waves

$$f(p^+, p^+) = \frac{\sqrt{\pi}}{ik} [\alpha_0 Y_0^+ + \sqrt{(2)}\, \alpha_1 Y_1^+ + \beta_1 Y_1^-],$$

where

$$\alpha_0 = \exp(2i\delta_0^{\frac{3}{2}}) - 1; \quad \alpha_1 = \exp(2i\delta_{1+}^{\frac{3}{2}}) - 1; \quad \beta_1 = \exp(2i\delta_{1-}^{\frac{3}{2}}) - 1.$$

Using the explicit expression for the Pauli functions of problem 32 of section 9, we get for the differential scattering cross-section for S- and P-waves

$$\frac{k^2}{\pi}\frac{d\sigma}{d\Omega} = |Y_{00}|^2 |\alpha_0|^2 + \frac{1}{\sqrt{3}} |Y_{00}||Y_{10}| [2(\alpha_0 \alpha_1^* + \alpha_0^* \alpha_1) + (\alpha_0 \beta_1^* + \alpha_0^* \beta_1)]$$
$$+ \tfrac{1}{3} |Y_{10}|^2 [4|\alpha_1|^2 + |\beta_1|^2 + 2(\alpha_1 \beta_1^* + \alpha_1^* \beta_1)]$$
$$+ \tfrac{2}{3} |Y_{11}|^2 [|\alpha_1|^2 + |\beta_1|^2 - (\alpha_1 \beta_1^* + \alpha_1^* \beta_1)].$$

Since the spherical harmonics Y_{lm} are equal to

$$Y_{00} = \frac{1}{\sqrt{(4\pi)}}; \quad Y_{10} = \sqrt{\left(\frac{3}{4\pi}\right)} \cos\theta; \quad Y_{11} = \sqrt{\left(\frac{3}{8\pi}\right)} \sin\theta \exp(i\varphi),$$

we can write the differential cross-section in the form,

$$k^2 \frac{d\sigma}{d\Omega} = A + B\cos\theta + C\cos^2\theta, \tag{1}$$

where the coefficients A, B, and C are given by

$$\begin{aligned} A &= \tfrac{1}{4}[|\alpha_0|^2 + |\alpha_1|^2 + |\beta_1|^2 - (\alpha_1 \beta_1^* + \alpha_1^* \beta_1)]; \\ B &= \tfrac{1}{4}[2(\alpha_0 \alpha_1^* + \alpha_0^* \alpha_1) + (\alpha_0 \beta_1^* + \alpha_0^* \beta_1)]; \\ C &= \tfrac{3}{4}[|\alpha_1|^2 + (\alpha_1 \beta_1^* + \alpha_1^* \beta_1)]. \end{aligned}$$

To express the coefficients A, B, and C in terms of the phase shifts we use the identities

$$|\exp(2ix) - 1|^2 = 4\sin^2 x;$$

$$[\exp(2ix) - 1][\exp(-2iy) - 1] + [\exp(-2ix) - 1][\exp(2iy) - 1] = 4[\sin^2 x + \sin^2 y - \sin^2(x-y)].$$

We get then

$$\left.\begin{aligned} A(p^+,p^+) &= \sin^2 \delta_0^{3/2} + \sin^2(\delta_{1+}^{3/2} - \delta_{1-}^{3/2}); \\ B(p^+,p^+) &= 3\sin^2 \delta_0^{3/2} + 2\sin^2 \delta_{1+}^{3/2} + \sin^2 \delta_{1-}^{3/2} \\ &\quad - 2\sin^2(\delta_0^{3/2} - \delta_{1+}^{3/2}) - \sin^2(\delta_0^{3/2} - \delta_{1-}^{3/2}); \\ C(p^+,p^+) &= 3[2\sin^2 \delta_{1+}^{3/2} + \sin^2 \delta_{1-}^{3/2} - \sin^2(\delta_{1+}^{3/2} - \delta_{1-}^{3/2})]. \end{aligned}\right\} \tag{2}$$

We can easily extend our method to find the coefficients A, B, C for the reactions (p^-,p^-) and (p^-,n^0). We shall only give the final results:

$$\begin{aligned} A(p^-,p^-) &= \tfrac{1}{3}\sin^2 \delta_0^{3/2} + \tfrac{2}{3}\sin^2 \delta_0^{1/2} - \tfrac{2}{9}\sin^2(\delta_0^{3/2} - \delta_0^{1/2}) + \tfrac{1}{9}\sin^2(\delta_{1+}^{3/2} - \delta_{1-}^{3/2}) \\ &\quad - \tfrac{2}{9}\sin^2(\delta_{1-}^{3/2} - \delta_{1-}^{1/2}) + \tfrac{2}{9}\sin^2(\delta_{1-}^{3/2} - \delta_{1+}^{1/2}) + \tfrac{2}{9}\sin^2(\delta_{1+}^{3/2} - \delta_{1-}^{1/2}) \\ &\quad + \tfrac{4}{9}\sin^2(\delta_{1-}^{1/2} - \delta_{1+}^{1/2}) - \tfrac{2}{9}\sin^2(\delta_{1+}^{3/2} - \delta_{1+}^{1/2}); \end{aligned}$$

$$\begin{aligned} B(p^-,p^-) &= \sin^2 \delta_0^{3/2} + 2\sin^2 \delta_0^{1/2} + \tfrac{2}{3}\sin^2 \delta_{1+}^{3/2} + \tfrac{4}{3}\sin^2 \delta_{1+}^{1/2} + \tfrac{1}{3}\sin^2 \delta_{1-}^{3/2} \\ &\quad + \tfrac{2}{3}\sin^2 \delta_{1-}^{1/2} - \tfrac{2}{9}\sin^2(\delta_0^{3/2} - \delta_{1+}^{3/2}) - \tfrac{4}{9}\sin^2(\delta_0^{3/2} - \delta_{1+}^{1/2}) \\ &\quad - \tfrac{1}{9}\sin^2(\delta_0^{3/2} - \delta_{1-}^{3/2}) - \tfrac{2}{9}\sin^2(\delta_0^{3/2} - \delta_{1-}^{1/2}) - \tfrac{4}{9}\sin^2(\delta_0^{1/2} - \delta_{1+}^{3/2}) \\ &\quad - \tfrac{8}{9}\sin^2(\delta_0^{1/2} - \delta_{1+}^{1/2}) - \tfrac{2}{9}\sin^2(\delta_0^{1/2} - \delta_{1-}^{3/2}) - \tfrac{4}{9}\sin^2(\delta_0^{1/2} - \delta_{1-}^{1/2}); \end{aligned}$$

$$\begin{aligned} C(p^-,p^-) &= 2\sin^2 \delta_{1+}^{3/2} + 4\sin^2 \delta_{1+}^{1/2} + \sin^2 \delta_{1-}^{3/2} + 2\sin^2 \delta_{1-}^{1/2} \\ &\quad - \tfrac{2}{3}\sin^2(\delta_{1+}^{3/2} - \delta_{1+}^{1/2}) - \tfrac{1}{3}\sin^2(\delta_{1+}^{3/2} - \delta_{1-}^{3/2}) - \tfrac{2}{3}\sin^2(\delta_{1+}^{3/2} - \delta_{1-}^{1/2}) \\ &\quad - \tfrac{2}{3}\sin^2(\delta_{1+}^{1/2} - \delta_{1-}^{3/2}) - \tfrac{2}{3}\sin^2(\delta_{1+}^{1/2} - \delta_{1-}^{1/2}); \end{aligned} \tag{3}$$

$$\begin{aligned}A(p^-,n^0) = \tfrac{2}{9}[&\sin^2(\delta_0^{\frac{3}{2}}-\delta_0^{\frac{1}{2}})+\sin^2(\delta_{1+}^{\frac{3}{2}}-\delta_{1+}^{\frac{1}{2}})+\sin^2(\delta_{1+}^{\frac{3}{2}}-\delta_{1-}^{\frac{3}{2}})\\ &-\sin^2(\delta_{1+}^{\frac{3}{2}}-\delta_{1-}^{\frac{1}{2}})-\sin^2(\delta_{1+}^{\frac{1}{2}}-\delta_{1-}^{\frac{3}{2}})+\sin^2(\delta_{1+}^{\frac{1}{2}}-\delta_{1-}^{\frac{1}{2}})\\ &+\sin^2(\delta_{1-}^{\frac{3}{2}}-\delta_{1-}^{\frac{1}{2}})];\end{aligned}$$

$$\begin{aligned}B(p^-,n^0) = \tfrac{2}{9}[&-2\sin^2(\delta_0^{\frac{3}{2}}-\delta_{1+}^{\frac{3}{2}})+2\sin^2(\delta_0^{\frac{3}{2}}-\delta_{1+}^{\frac{1}{2}})-\sin^2(\delta_0^{\frac{3}{2}}-\delta_{1-}^{\frac{3}{2}})\\ &+\sin^2(\delta_0^{\frac{3}{2}}-\delta_{1-}^{\frac{1}{2}})+2\sin^2(\delta_0^{\frac{1}{2}}-\delta_{1+}^{\frac{3}{2}})-2\sin^2(\delta_0^{\frac{1}{2}}-\delta_{1+}^{\frac{1}{2}})\\ &+\sin^2(\delta_0^{\frac{1}{2}}-\delta_{1-}^{\frac{3}{2}})-\sin^2(\delta_0^{\frac{1}{2}}-\delta_{1-}^{\frac{1}{2}})];\end{aligned}$$

$$\begin{aligned}C(p^-,n^0) = \tfrac{2}{3}[&\sin^2(\delta_{1+}^{\frac{3}{2}}-\delta_{1+}^{\frac{1}{2}})+\sin^2(\delta_{1+}^{\frac{3}{2}}-\delta_{1-}^{\frac{1}{2}})\\ &-\sin^2(\delta_{1+}^{\frac{3}{2}}-\delta_{1-}^{\frac{3}{2}})+\sin^2(\delta_{1+}^{\frac{1}{2}}-\delta_{1-}^{\frac{3}{2}})-\sin^2(\delta_{1+}^{\frac{1}{2}}-\delta_{1-}^{\frac{1}{2}})]. \qquad (4)\end{aligned}$$

The results obtained are of great importance for the evaluation of the angular distribution for the scattering of pions by protons. We can verify experimentally equation (1) and determine the coefficients A, B, and C for the reactions (p^+,p^+), (p^-,p^-), and (p^-,n^0). We can then determine the six unknown phase shifts $\delta_0^{\frac{3}{2}}$, $\delta_0^{\frac{1}{2}}$, $\delta_{1+}^{\frac{3}{2}}$, $\delta_{1-}^{\frac{3}{2}}$, $\delta_{1+}^{\frac{1}{2}}$, $\delta_{1-}^{\frac{1}{2}}$ from the equations (2), (3), and (4).

However, the evaluation of the phase shifts possesses an ambiguity: first of all because of the fact that in equations (2), (3), and (4) the square of the sines of the phases and their differences enter so that we cannot determine their sign; secondly, there are several different sets of phases satisfying the experimental data. Out of them, the Fermi solution, in which the largest contribution to the scattering comes from the phase $\delta_{1+}^{\frac{3}{2}}$, gives the best agreement with experiments. This phase corresponds to a scattering over 90° for a pion energy $E \approx 195$ MeV in the laboratory system; the other phases $\delta_{1-}^{\frac{3}{2}}$, $\delta_{1+}^{\frac{1}{2}}$, $\delta_{1-}^{\frac{1}{2}}$ are small.

There are a number of other criteria which can be used to get rid of this ambiguity. The signs of the phases can be determined from considerations based on the causality principle, and also from experiments which take the Coulomb interaction especially into account. In choosing the correct solution one might be helped by an experiment about the polarisation of the recoil nucleons but this experiment has not as yet been performed. The expected values for the polarisation in the reactions (p^+,p^+), (p^-,p^-), and (p^-,n^0) are determined in the following problem.

38*. We consider in detail the reaction (p^+,p^+)

$$\alpha = \begin{pmatrix}1\\0\end{pmatrix} \quad \text{if} \quad s_z = \tfrac{1}{2}; \qquad \beta = \begin{pmatrix}0\\1\end{pmatrix} \quad \text{if} \quad s_z = -\tfrac{1}{2}.$$

If originally we had $s_z = \frac{1}{2}$ for the proton, we can write the scattering amplitude in the form

$$f_{\frac{1}{2}} = f_{\alpha\alpha}\,\alpha + f_{\alpha\beta}\,\beta. \qquad (1)$$

While if initially $s_z = -\frac{1}{2}$ we have

$$f_{-\frac{1}{2}} = f_{\beta\alpha}\,\alpha + f_{\beta\beta}\,\beta. \tag{2}$$

In these equations $f_{\alpha\alpha}$ and $f_{\beta\beta}$ are the scattering amplitudes without spin re-orientation and $f_{\alpha\beta}$ and $f_{\beta\alpha}$ those with spin re-orientation. The amplitudes $f_{\alpha\alpha}$ and $f_{\alpha\beta}$ can be determined from equation (5) of problem 34 of section 9, as the coefficients of the columns $\begin{pmatrix}1\\0\end{pmatrix}$ and $\begin{pmatrix}0\\1\end{pmatrix}$.

If we take only S- and P-waves into account we get

$$f_{\alpha\alpha} = \frac{\sqrt{\pi}}{ik}\left[\alpha_0 Y_{00} + \frac{1}{\sqrt{3}}(2\alpha_1+\beta_1)Y_{10}\right], \tag{3}$$

$$f_{\alpha\beta} = \frac{\sqrt{\pi}}{ik}\sqrt{(\tfrac{2}{3})}(\beta_1-\alpha_1)Y_{11}, \tag{4}$$

where α_0, α_1, β_1 were given in the preceding problem. If we use the expressions for the Pauli functions with $m_j = \frac{1}{2}$, we find easily

$$f_{\beta\alpha} = \frac{\sqrt{\pi}}{ik}\sqrt{(\tfrac{2}{3})}(\alpha_1-\beta_1)Y_{1,-1}; \quad f_{\beta\beta} = f_{\alpha\alpha}.$$

If we assume that the pions are scattered in the xz-plane, the polar angle $\varphi = 0$ and we have thus

$$f_{\alpha\beta} = -f_{\beta\alpha}.$$

We get then for the scattering amplitude for a proton with $s_z = -\frac{1}{2}$ [see equation (2)]

$$f_{-\frac{1}{2}} = -f_{\beta\alpha}\,\alpha + f_{\alpha\alpha}\,\beta. \tag{5}$$

Since the protons were originally unpolarised it follows that after the scattering they are still unpolarised along the z-axis.

One can show that after the scattering there is no polarisation in the xz-plane. Indeed the functions of the proton γ_θ and δ_θ, which correspond to $+\frac{1}{2}$ and $-\frac{1}{2}$ components of the spin along a z'-axis which in the xz-plane makes an angle θ with the z-axis, will be of the form (see problem 8 of section 4)

$$\begin{pmatrix}\gamma_\theta\\ \delta_\theta\end{pmatrix} = \exp(-i\hat{s}_y\,\theta)\begin{pmatrix}\alpha\\ \beta\end{pmatrix} = \begin{pmatrix}\cos\frac{1}{2}\theta & -\sin\frac{1}{2}\theta\\ \sin\frac{1}{2}\theta & \cos\frac{1}{2}\theta\end{pmatrix}\begin{pmatrix}\alpha\\ \beta\end{pmatrix}.$$

We get thus

$$f_{\frac{1}{2}} = f_{\alpha\alpha}\,\alpha + f_{\alpha\beta}\,\beta = (f_{\alpha\alpha}\cos\tfrac{1}{2}\theta - f_{\alpha\beta}\sin\tfrac{1}{2}\theta)\,\gamma_\theta + (f_{\alpha\alpha}\sin\tfrac{1}{2}\theta + f_{\alpha\beta}\cos\tfrac{1}{2}\theta)\,\delta_\theta;$$

$$f_{-\frac{1}{2}} = -f_{\alpha\beta}\,\alpha + f_{\alpha\alpha}\,\beta = -(f_{\alpha\alpha}\sin\tfrac{1}{2}\theta + f_{\alpha\beta}\cos\tfrac{1}{2}\theta)\,\gamma_\theta + (f_{\alpha\alpha}\cos\tfrac{1}{2}\theta - f_{\alpha\beta}\sin\tfrac{1}{2}\theta)\,\delta_\theta,$$

which means that there is no polarisation in any direction in the xz-plane.

The protons will, however, be polarised along the y-axis which is perpendicular to the plane of scattering. To find the magnitude of the polarisation, we express $f_{\frac{1}{2}}$ and $f_{-\frac{1}{2}}$ in terms of the spin eigenfunctions

$$\gamma = \frac{\alpha + i\beta}{\sqrt{2}}; \quad \delta = \frac{\alpha - i\beta}{\sqrt{2}},$$

which correspond to a spin direction parallel or anti-parallel to the y-axis. We get then

$$f_{\frac{1}{2}} = \frac{1}{\sqrt{2}}(f_{\alpha\alpha} - if_{\alpha\beta})\gamma + \frac{1}{\sqrt{2}}(f_{\alpha\alpha} + if_{\alpha\beta})\,\delta,$$

$$f_{-\frac{1}{2}} = \frac{-i}{\sqrt{2}}(f_{\alpha\alpha} - if_{\alpha\beta})\gamma + \frac{i}{\sqrt{2}}(f_{\alpha\alpha} + if_{\alpha\beta})\,\delta.$$

It follows that

$$W_+ \sim |f_{\alpha\alpha} - if_{\alpha\beta}|^2; \quad W_- \sim |f_{\alpha\alpha} + if_{\alpha\beta}|^2, \tag{6}$$

where W_+ and W_- are the probabilities that the spin after the scattering will be parallel or anti-parallel to the y-axis. We note that equation (6) is valid independent of the initial values s_z of the proton. If we substitute into equation (6) expressions (3) and (4) for $f_{\alpha\alpha}$ and $f_{\alpha\beta}$, we get

$$W_\pm \sim |\alpha_0 + (2\alpha_1 + \beta_1)\cos\theta \mp i(\beta_1 - \alpha_1)\sin\theta|^2$$

or

$$W_\pm \sim |[\exp(2i\delta_0^{\frac{3}{2}}) - 1] + [2\exp(2i\delta_{1+}^{\frac{3}{2}}) - 3 + \exp(2i\delta_{1-}^{\frac{3}{2}})]\cos\theta$$
$$\pm i[\exp(2i\delta_{1+}^{\frac{3}{2}}) - \exp(2i\delta_{1-}^{\frac{3}{2}})]\sin\theta|^2.$$

We find similarly for the reactions (p^-, p^-) and (p^-, n^0):

$$W_\pm(p^-, p^-) \sim |[\exp(2i\delta_0^{\frac{3}{2}}) - 3 + 2\exp(2i\delta_0^{\frac{1}{2}})] + [2\exp(2i\delta_{1+}^{\frac{3}{2}}) - 9$$
$$+ 4\exp(2i\delta_{1+}^{\frac{1}{2}}) + \exp(2i\delta_{1-}^{\frac{3}{2}}) + 2\exp(2i\delta_{1-}^{\frac{1}{2}})]\cos\theta$$
$$\pm i[\exp(2i\delta_{1+}^{\frac{3}{2}}) + 2\exp(2i\delta_{1+}^{\frac{1}{2}}) - \exp(2i\delta_{1-}^{\frac{3}{2}})$$
$$- 2\exp(2i\delta_{1-}^{\frac{1}{2}})]\cos\theta|^2;$$

$$W_\pm(p^-, n^0) \sim |[\exp(2i\delta_0^{\frac{3}{2}}) - \exp(2i\delta_0^{\frac{1}{2}})] + [2\exp(2i\delta_{1+}^{\frac{3}{2}}) - 2\exp(2i\delta_{1+}^{\frac{1}{2}})$$
$$+ \exp(2i\delta_{1-}^{\frac{3}{2}}) - \exp(2i\delta_{1-}^{\frac{1}{2}})]\cos\theta \pm i[\exp(2i\delta_{1+}^{\frac{3}{2}})$$
$$- \exp(2i\delta_{1+}^{\frac{1}{2}}) - \exp(2i\delta_{1-}^{\frac{3}{2}}) + \exp(2i\delta_{1-}^{\frac{1}{2}})]\sin\theta|^2.$$

39. The two nuclear reactions are each other's inverse, or

$$\underset{(A)}{n + p} \rightleftarrows \underset{(B)}{d + \gamma}. \tag{1}$$

The principle of detailed balancing for the reactions (1) can be written as

$$\frac{\bar{\sigma}_{A \to B}}{\bar{\sigma}_{B \to A}} = \frac{g_B p_B^2}{g_A p_A^2}. \tag{2}$$

In equation (2) $\bar{\sigma}_{A \to B}$, $\bar{\sigma}_{B \to A}$ are the cross-sections for the corresponding transitions, integrated over all directions of the velocities and summed over all spin directions of the final state, and averaged over those directions in the initial state: p_A and p_B are the momenta of the relative motion of the particles in the states A and B and g_A, g_B the spin (and polarisation) statistical weights of these states.

In our case we have

$$\bar{\sigma}_{A \to B} = \sigma_{\text{capt}} \text{ (capture cross-section)},$$

$$\bar{\sigma}_{B \to A} = \sigma_{\text{ph diss}} \text{ (photo-dissociation cross-section)}.$$

Since the neutron and proton spins are $\frac{1}{2}$ and the deuteron spin 1, while the photon has two states of polarisation, we have

$$g_A = (2.\tfrac{1}{2}+1)^2 = 4, \quad g_B = (2.1+1).2 = 6.$$

We shall take our system of reference to be fixed in the deuteron. We have then $p_B = p_\gamma = \hbar\omega/c$ where ω is the frequency of the photon. One can easily verify that our system of reference is practically the same as the centre of mass system of the states A and B. Indeed, in the exact centre of mass system we have by definition

$$\boldsymbol{p}_d + \boldsymbol{p}_\gamma = 0, \quad \text{or} \quad p_d = p_\gamma = \hbar\omega/c,$$

so that the deuteron velocity in this system of reference (and thus, conversely, the velocity of this system of reference relative to the deuteron) is equal to $v_d = p_d/2M = \hbar\omega/2Mc$. On the other hand, if the photon energy, $\hbar\omega$, is small compared to Mc^2, and not too close to the deuteron binding energy ϵ (we shall assume that these conditions are satisfied), the velocities of the separating nucleons will be of the order of

$$v_n \approx v_p \sim \sqrt{\left(\frac{\hbar\omega - \epsilon}{M}\right)} \sim \sqrt{\left(\frac{\hbar\omega}{M}\right)},$$

so that

$$\frac{v_d}{v_n} \sim \sqrt{\left(\frac{\hbar\omega}{Mc^2}\right)} \ll 1.$$

The relative velocity of the two systems of reference is thus, indeed, small compared with the velocity of the products of the reaction and we may assume that both systems of reference are approximately the same.

We have thus $p_A^2 \approx 2\mu E_A$ with $\mu = \frac{1}{2}M$, where μ is the reduced mass of the n–p system, M the nucleon mass, and E_A the energy of their relative motion (or the energy in the centre of mass system). From the energy conservation law we have $E_A = E_B$, and clearly

$$E_B = \hbar\omega + (-\epsilon) = \hbar\omega - \epsilon.$$

We have thus $p_A^2 = M(\hbar\omega - \epsilon)$.

Collecting all our results we get finally

$$\frac{\sigma_{\text{capt}}}{\sigma_{\text{ph diss}}} = \frac{3}{2}\frac{\hbar\omega}{Mc^2}\frac{\hbar\omega}{\hbar\omega - \epsilon}. \tag{3}$$

We emphasise that this result is independent of any definite assumption made about the mechanism of the reaction, but is based solely on the reversibility of quantum mechanics (that is, on the symmetry of its equation with respect to time reversal).

40. Let us note first of all that since the rest energy of a charged meson is larger than the rest mass of a neutral meson by about 5 MeV, while the maximum possible binding energy of the $(\pi^+ + \pi^-)$-system is only about 2 keV

$$\left(E = -\frac{\mu e^4}{2\hbar^2} = -\frac{\mu}{\mu_{\text{el}}}\cdot\frac{\mu_{\text{el}}e^4}{2\hbar^2} \approx -\tfrac{1}{2}.270.(13{\cdot}5 \text{ eV}) \approx -1800 \text{ eV},\right.$$

we do not consider possible here bound states produced by non-Coulomb interactions of the pions), the process under consideration is energetically possible, irrespective of the value of l.

The required selection rule arises because the two π^0-mesons, occurring in the final state, are identical. Indeed, since the pions have integer spin ($= 0$) the $(\pi^0 + \pi^0)$ system must be described by a wave function, symmetric in the two mesons. Since there is no spin–wave function (or, rather, since the spin–wave function $\equiv 1$), we must thus have a wave function which is symmetric in the coordinates of the two π^0-mesons. One can easily verify that the parity of the wave function of two particles 1 and 2 is the same as the symmetry of the coordinate wave function (the origin lies in the middle of the line 1–2); the wave function of the $\pi^0 + \pi^0$ system can thus only be even ($I = +1$), and hence using the relation $I = (-1)^l$ for the parity of the orbital angular momentum wave function, we find that the orbital angular momentum of the relative motion of two π^0-mesons can also be only even.

If we now take into consideration the conservation of the total (in our case this is simply the orbital) angular momentum, we conclude that the reaction is possible for an initial state with even l but impossible for an initial state with odd l.

We note further that we cannot conclude from our considerations that the pion possesses negative intrinsic parity since both in the initial and in the final state there are two mesons, so that the total intrinsic parity has the trivial value $+1$.

41. We shall apply to this process the conservation laws for angular momentum and parity and the Pauli principle. The total angular momentum J of the deuteron is equal to unity. If we take into account the assumed values of the spin and angular momentum of the π^--meson the total angular momentum of the original $(\pi^- + d)$-system is also equal to unity, and so is the total angular momentum of the final $(n+n)$-system. If we denote the orbital angular momentum and spin of the two-neutron system by L and S we see that the state with the required total angular momentum $J = 1$ $(\boldsymbol{J} = \boldsymbol{L} + \boldsymbol{S})$ can be realised in the following four ways:

$$(1)\ L = 0,\quad S = 1; \qquad (2)\ L = 1,\quad S = 0;$$

$$(3)\ L = 1,\quad S = 1; \qquad (4)\ L = 2,\quad S = 1.$$

The spin functions corresponding to $S = 1$ and $S = 0$ are respectively symmetric and antisymmetric in the two neutron spins. The symmetry of the coordinate functions of given L is the same as the parity $(-1)^L$ of these functions, since the reflection $\boldsymbol{r}_1 \to -\boldsymbol{r}_1$, $\boldsymbol{r}_2 \to -\boldsymbol{r}_2$ corresponds to a permutation of the coordinates of the two particles and in the system of reference with its origin at $(\boldsymbol{r}_1 + \boldsymbol{r}_2)/2$ even L corresponds to a symmetric and odd L to an antisymmetric function of the neutron coordinates.

It follows thus immediately that the combinations 1, 2, and 4 are forbidden by the Pauli principle, since the coordinates and the spin functions have the same symmetry so that the total wave function is symmetric in the two neutrons.

Reference to the Pauli principle shows thus that only state 3 which corresponds to an antisymmetric total wave function is possible; in fact there are three such states, since $J = 1$. However, a transition to such a state violates the parity conservation law since the final state is odd $[(-1)^L = (-1)^1 = -1]$, while the initial $(\pi^- + d)$-state was even, as follows from our assumptions and the addition rule for parity.

The fact that in actual fact the process considered takes place with appreciable probability is one proof that the π^--meson is pseudoscalar, that is, has negative intrinsic parity.

42. The least difference in rest mass energy of the π^- and π^0 meson for which the process under consideration is energetically possible is clearly equal to the sum of the deuteron binding energy ($\approx 2{\cdot}2$ MeV) and the difference in rest mass energy of the neutron and the proton ($\approx 1{\cdot}25$ MeV), that is, about $3{\cdot}5$ MeV. (We take into account that the

π^- meson is captured in a bound state and possesses thus not only kinetic energy but also "collects" some energy from the reaction, though only a very small amount of the order of some keV.) The actual rest mass energy difference is approximately 4·5 MeV. The reaction is thus energetically possible, but the particles produced in the reaction will possess small momenta—which is essential in the following argument.

We shall apply to the process considered the conservation laws of angular momentum and parity and the Pauli principle (for neutrons). The total angular momentum J of the final system $(n+n+\pi^0)$ must be equal to the total angular momentum of the initial (π^-+d)-system, that is, equal to unity (see the preceding problem). Let L and S denote the orbital angular momentum and spin of the subsystem of two neutrons and let l denote the orbital angular momentum of the π^0-meson with respect to this subsystem—or rather with respect to its centre of mass, which is approximately the same as the centre of mass of the system of all three particles.

We note first of all that apart from the strict selection rules imposed by the conservation laws and the Pauli principle, there is an additional fact which strongly decreases the probability of transitions which satisfy the selection rules, namely, the smallness of the momenta of the neutrons and the π^0-meson—from the conservation of angular momentum it follows that all three momenta, p, are of the same order of magnitude. Indeed, the matrix element for any possible transition contains under the integral sign a meson–nucleon interaction operator, which differs appreciably from zero only in a small region of the order of the range of nuclear forces, that is, of the order of the Compton wavelength $\hbar/\mu c$ of the pion. This operator acts on the wave functions of the meson and one of the nucleons. However, these wave functions vary in the region of interest near the origin of order $a \sim \hbar/\mu c$ as $(pr_\pi/\hbar)^l$ and $(pr_n/\hbar)^L$, so that the probability for an allowed transition contains clearly a factor $(pa/\hbar)^{2(L+l)}$.

It is convenient to estimate the value of $(pa/\hbar)^2$ as follows. In the most important case (where, as Migdal has shown, the two neutrons fly away at a small relative angle and with a small relative kinetic energy, thanks to the interaction of the two slow neutrons in the final state, so that the π^0-meson flies away with a momentum which is roughly twice that of each neutron) the π^0-meson carries away practically all "excess" kinetic energy $\Delta E \approx 4{\cdot}5\ \mathrm{MeV} - 3{\cdot}5\ \mathrm{MeV} = 1\ \mathrm{MeV} \sim \frac{1}{2}\epsilon$ where ϵ is the deuteron-binding energy, since $\mu \ll M$ so that $p^2 \sim \mu\epsilon$, and hence

$$\left(\frac{pa}{\hbar}\right)^2 \sim \frac{\mu\epsilon}{\hbar^2}\left(\frac{\hbar}{\mu c}\right)^2 = \frac{\epsilon}{\mu c^2} \sim \frac{1}{60} \ll 1.$$

The presence of three particles in the final state increases considerably—as compared to the preceding problem—the number of ways in which it can be realised with total angular momentum $J = 1$ and satisfying the Pauli principle and the parity conservation law. Since, however, $(pa/\hbar)^2 \ll 1$ we can henceforth consider only those allowed transitions for which the sum $L+l$ is minimum—and the probability thus maximum.

We consider two possibilities.

1. The charged and neutral pions have the same intrinsic parity. In this case the parity conservation law allows only even values of $L+l$, since the total orbital parity of the $n+n+\pi^0$ system, $(-1)^{L+l}$, must be equal to $+1$. If we take all combinations of L, S, and l which make $(L+l) = 0$, or 2, and $J = |\boldsymbol{L}+\boldsymbol{S}+\boldsymbol{l}| = 1$, we get the following states:

$$(1)\ L = 0,\quad S = 1,\quad l = 0; \qquad (2)\ L = 1,\quad S = 0,\quad l = 1;$$

$$(3)\ L = 1,\quad S = 1,\quad l = 1; \qquad (4)\ L = 2,\quad S = 1,\quad l = 0;$$

$$(5)\ L = 0,\quad S = 1,\quad l = 2.$$

From these five combinations only one, namely (3), satisfies the Pauli principle (see preceding problem). In this case $L+l = 2$, so that the probability for the reaction considered is proportional to $(pa/\hbar)^4$ if the π^-- and n^0-mesons have the same parity.

2. The charged and neutral mesons have opposite parity.

In that case $(L+l)$ must be odd, since $(-1)^{L+l} = -1$.

If we restrict ourselves to the most probable transitions, that is, if we take $L+l = 1$, we get for $J = 1$ the following states:

$$(1)\ L = 1,\quad S = 0,\quad l = 0; \qquad (2)\ L = 1,\quad S = 1,\quad l = 0;$$

$$(3)\ L = 0,\quad S = 0,\quad l = 1; \qquad (4)\ L = 0,\quad S = 1,\quad l = 1.$$

From these states (2) and (3) are allowed by the Pauli principle.

Since $L+l = 1$, the probability for the reaction considered is proportional to $(pa/\hbar)^2$ if the π^0- and π^--mesons have opposite parity; in other words, the reaction is much less forbidden than for case 1.

Experiments about the capture of π^--mesons by deuterons show that it proceeds through the reactions

$$\pi^- + d \to n + n,$$

or

$$\pi^- + d \to n + n + \gamma,$$

without involving a π^0-meson. This means that the reaction

$$\pi^- + d \rightarrow n + n + \pi^0$$

must be strongly prohibited.

From this we conclude that case 1 holds, that is, that the charged and the neutral pions have the same parity. It may be added that all experimental facts point to an intrinsic parity -1 for the pions.

43. We shall apply to the reaction

$$p + p \rightarrow p + p + \pi^0 \tag{1}$$

the laws of conservation of total angular momentum and parity and also the Pauli principle for the protons.

The fact that the π^0-meson is pseudoscalar means by definition that its spin is equal to zero, and its intrinsic parity -1. The final state of the reaction (1) is in the case considered by us a state of orbital angular momentum $L = 1$ and thus of orbital parity $(-1)^L = -1$. The total parity of the final state is thus $(-1).(-1) = +1$.

The parity conservation law chooses from the wave functions of the relative motion of the protons in the initial state only the spherical harmonics with even L [thus of parity $(-1)^L = +1$] because of the one-to-one connection between orbital parity and orbital angular momentum L.

As we have said several times, the orbital parity of a system of two identical particles is the same as the symmetry of their wave function with respect to the coordinates: in other words, the initial wave function of the $(p+p)$-system must be symmetric in the coordinates and thus antisymmetric in the proton spins, because of the Pauli principle. This wave function must thus correspond to a total spin $S = 0$, and the total angular momentum $(\boldsymbol{J} = \boldsymbol{S} + \boldsymbol{L})$ must thus be even.

On the other hand, the Pauli principle admits in the final state only a state of the protons which is antisymmetric in the spins $S = 0$, because we have assumed that $L = 0$. From the rule of addition of angular momenta we get for the final state the odd value $J = 1$ which violates the conservation law for total angular momentum.

This concludes our proof.

44. The nucleon isotopic spin $\tau = \frac{1}{2}$. Hence, in complete analogy with the rule for adding angular momenta, the isotopic spin T of a system of two nucleons must be either unity or zero. The first of these eigenvalues is three-fold degenerate with possible values $+1$, 0, and -1 for T_3: each of the three T, T_3 pairs corresponds to an isotopic function symmetric in the two nucleons. The eigenvalue $T = 0$ corresponds to one—antisymmetric—isotopic function.

We note that the whole problem is completely analogous to the problem of combining two spins $s = \frac{1}{2}$. We can thus compose the following table, which is immediately clear without further explanations, of the normalised isotopic functions and the corresponding values of T and T_3, where we have indicated the physical meaning of each state, and where 1 and 2 denote the isotopic variable τ_3 of the two nucleons.

T	T_3	Isotopic spin function	Symmetry	Interpretation
	+1	$\Psi^1_1 = \psi^{\frac{1}{2}}_{\frac{1}{2}}(1)\,\psi^{\frac{1}{2}}_{\frac{1}{2}}(2)$	symmetric	2 protons
1	0	$\Psi^1_0 = 1/\sqrt{(2)}\,[\psi^{\frac{1}{2}}_{\frac{1}{2}}(1)\,\psi^{\frac{1}{2}}_{-\frac{1}{2}}(2)+\psi^{\frac{1}{2}}_{\frac{1}{2}}(2)\,\psi^{\frac{1}{2}}_{-\frac{1}{2}}(1)]$	symmetric	proton+neutron
	−1	$\Psi^1_{-1} = \psi^{\frac{1}{2}}_{-\frac{1}{2}}(1)\,\psi^{\frac{1}{2}}_{-\frac{1}{2}}(2)$	symmetric	2 neutrons
0	0	$\Psi^0_0 = 1/\sqrt{(2)}\,[\psi^{\frac{1}{2}}_{\frac{1}{2}}(1)\,\psi^{\frac{1}{2}}_{-\frac{1}{2}}(2)-\psi^{\frac{1}{2}}_{\frac{1}{2}}(2)\,\psi^{\frac{1}{2}}_{-\frac{1}{2}}(1)]$	antisymmetric	proton+neutron

From the table it is clear, in particular, that the deuteron (whose coordinate-spin state is ${}^3S+{}^3D$) has $T = 0$. Indeed, its wave function corresponds to the values $L = 0$ and $L = 2$, that is, it is even; its spin is 1, and it is thus symmetric also in the spins of the two nucleons. The generalised Pauli principle requires, since the total wave function is a product, that it is antisymmetric in the isotopic variables of the nucleons; this requirement is satisfied only for $T = 0$.

One can say that the isotopic function of the deuteron is an "isotopic scalar", that is, simply a number, as far as its transformation properties are concerned, while the isotopic function of a nucleon is a two-component vector—that is, a spinor of the first rank—and the function of a pion, for instance, an ordinary vector with three components.

45. Since the deuteron is present both in the initial and in the final state of both reactions and since for the deuteron $T = T_3 = 0$, it does not enter into the addition of the isotopic spins. We can thus simply forget about it on both sides of the reactions. In other words, the deuteron plays the role of "catalyst" for the "dissociation" process of a proton p into a nucleon and a pion,

$$p \to \text{nucleon } (T = \tfrac{1}{2}) + \text{pion } (T = 1). \tag{1}$$

The left-hand side of the reaction (1) has $T = T_3 = \frac{1}{2}$. Since the total isotopic spin and its third component (that is, charge) are conserved, the right-hand side must have the same values of T and T_3. Expanding the isotopic wave function $\Psi^{\frac{1}{2}}_{\frac{1}{2}}$ in terms of the isotopic functions of the "subsystems", nucleon and pion, we have

$$\Psi^{\frac{1}{2}}_{\frac{1}{2}} = -\sqrt{(\tfrac{1}{3})}\,\psi^{\frac{1}{2}}_{\frac{1}{2}}\psi^{1}_{0} + \sqrt{(\tfrac{2}{3})}\,\psi^{\frac{1}{2}}_{-\frac{1}{2}}\psi^{1}_{1}.$$

The squares of the coefficients of this superposition ($\frac{1}{3}$ and $\frac{2}{3}$) give us clearly the required relative probability for the "dissociation" processes

$$p \rightarrow p + \pi^0 \quad \text{and} \quad p \rightarrow n + \pi^+,$$

and thus also the required ratio of the cross-sections.

46. Since for the deuteron $T = T_3 = 0$, the left-hand side of the reaction (1) corresponds to $T = 1$, $T_3 = 1$ and the left-hand side of the reaction (2) to $T = 1$, $T_3 = 0$.

The charge conservation law (T_3 = const) is clearly satisfied identically. The final states of both reactions (1) and (2) are characterised by well-defined values of the total isotopic spin T and are described respectively by the isotopic functions Ψ^1_1 and Ψ^1_0. The isotopic functions of the initial states are the products of the isotopic functions of the "subsystems", that is, the nucleons, and can be written (using the formulae for the addition of angular momenta) as the superposition of functions of two nucleon systems with a well-defined value of T: reaction (1):

$$\psi^{\frac{1}{2}}_{\frac{1}{2}}\psi^{\frac{1}{2}}_{\frac{1}{2}} = \Psi^1_1,$$

reaction (2):

$$\psi^{\frac{1}{2}}_{-\frac{1}{2}}\psi^{\frac{1}{2}}_{\frac{1}{2}} = -\sqrt{(\tfrac{1}{2})}\,\Psi^0_0 + \sqrt{(\tfrac{1}{2})}\,\Psi^1_0.$$

Two nucleons in the $(p+p)$-state give thus exactly the state of the "isotopic triplet" $T = 1$ which is necessary for reaction (1) because of the conservation laws for T and T_3, while in the $(n+p)$-state the triplet state $T = 1$ ($T_3 = 0$) which is required for reaction (2) only is present with a probability $\frac{1}{2}$. Since the other factors in the expression for the cross-section were assumed to be identical, the ratio of the two cross-sections is equal to the ratio of the probabilities of the necessary isotopic triplet state in the initial states, which is equal to two, as had to be proved.

47. Compare preceding two problems. The branching ratio is 2.

48. The required equality of the cross-sections expresses the charge symmetry of the nucleon–nucleon and nucleon–pion interactions, that is, the invariance of these interactions, the corresponding transition probabilities, ..., under a simultaneous replacement: $n \rightarrow p$, $p \rightarrow n$, $\pi^+ \rightarrow \pi^-$, $\pi^- \rightarrow \pi^+$. The equality of the cross-sections follows immediately from the transformation rule of the isotopic functions when isotopic spins are added, taking into account the generalised Pauli principle.

Indeed, since the wave function of two protons or two neutrons must be antisymmetric in their spins and coordinates, it must be symmetric in their isotopic spin variables, that is, it must correspond to a total isotopic spin $T = 1$. (This follows also very simply from the fact that $T \geqslant |T_3|$, where $T_3 = +1$ for $p+p$ and $T_3 = -1$ for $n+n$.) In this case the isotopic function of the $(p+p)$-system is the same as the function for π^+ (both are of the form Ψ_1^1), while the isotopic function of $n+n$ is the same as that of π^- (both are (Ψ_{-1}^1)). The right-hand sides of (1) and (2) correspond thus to the same isotopic spin function $\Psi_1^1 \Psi_{-1}^1$, and the matrix elements of the transitions from the initial isotopic state $\psi_{-\frac{1}{2}}^{\frac{1}{2}} \psi_{\frac{1}{2}}^{\frac{1}{2}}$ to the final state, and also the cross-sections (which are proportional to absolute square of these matrix elements) are the same.

Creation and annihilation operators; density matrix

2. $$|\Psi\rangle = \int ... \int d1 \,...\, dN\,|i_1, ..., i_N\rangle\langle i_1, ..., i_N|\Psi\rangle\,,$$

$$\hat{\Omega} = \int ... \int d1 \,...\, dN\, d1' \,...\, dN'\,|i_1, ..., i_N\rangle\langle i_1, ..., i_N|\hat{\Omega}|i_1', ..., i_N'\rangle\langle i_1', ..., i_N'|\,,$$

where $\int dj$ indicates integration over the spatial coordinates of the jth particle (and if necessary summation over its spin variables).

3. $$|i_1, ..., i_N\rangle = \frac{1}{\sqrt{N!}}\hat{a}^+(i_1)\hat{a}^+(i_2) \,...\, \hat{a}^+(i_N)|0\rangle\,.$$

7. $$\hat{\Omega} = \int di\,di'\langle i|\hat{\Omega}^{(1)}|i'\rangle\hat{a}^+(i)\hat{a}(i') + \tfrac{1}{2}\int di\,dj\,di'\,dj'\langle i, j|\hat{\Omega}^{(2)}|i', j'\rangle\hat{a}^+(i)\hat{a}^+(j)\hat{a}(j')\hat{a}(i')\,.$$

8. $$\hat{\Omega} = \sum_k \frac{\hbar^2 k^2}{2\mu}\hat{a}_k^+\hat{a}_k + \frac{1}{v}\sum_{k,\,q} V(q)\hat{a}_k^+\hat{a}_{k-q} + \frac{1}{2v}\sum_{k,\,k',\,q} U(q)\hat{a}_k^+\hat{a}_{k'}^+\hat{a}_{k'+q}\hat{a}_{k-q}\,, \tag{3}$$

where

$$V(q) = \int d^3r e^{-i(q\,\cdot\,r)}\,V(r)\,, \qquad U(q) = \int d^3r e^{-i(q\,\cdot\,r)}\,U(r)\,. \tag{4}$$

9. $\hat{a}^+(i)\hat{a}(i)|i_1, ..., i_N\rangle = N(i)|i_1, ..., i_N\rangle\,,$

where $N(i)$ is the number of times i occurs among the numbers $i_1, ..., i_N$.

The operator $\hat{n}(i)$ is thus an occupation number operator.

10. Use the fact that, if $[\hat{A}, \hat{B}]_-|\alpha\rangle = \epsilon\hat{B}|\alpha\rangle$, $\hat{B}|\alpha\rangle$ is an eigenfunction of $\hat{A}$, if $|\alpha\rangle$ is, and the corresponding eigenvalues differ by ϵ.

11. $\hat{\Psi}^+$ ($\hat{\Psi}$) is an operator which creates (annihilates) a particle at the point $\boldsymbol{r}$.

12. $$\hat{H} = \int\left\{\frac{\hbar^2}{2\mu}(\nabla\hat{\Psi}^+(\boldsymbol{r})\cdot\nabla\hat{\Psi}(\boldsymbol{r})) + V(\boldsymbol{r})\hat{\Psi}^+(\boldsymbol{r})\hat{\Psi}(\boldsymbol{r})\right\}d^3\boldsymbol{r} + \tfrac{1}{2}\int\int\hat{\Psi}^+(\boldsymbol{r})\hat{\Psi}^+(\boldsymbol{r}')U(\boldsymbol{r}, \boldsymbol{r}')\hat{\Psi}(\boldsymbol{r}')\hat{\Psi}(\boldsymbol{r})d^3\boldsymbol{r}\,d^3\boldsymbol{r}'\,.$$

13. The exact energy levels are found by making the transformation

$$\hat{a} \to \hat{\alpha} - \lambda\,, \quad \hat{a}^{+} \to \hat{\alpha}^{+} - \lambda\,,$$

which introduces new boson operators $\hat{\alpha}$ and $\hat{\alpha}^{+}$. The Hamiltonian then becomes

$$\hat{H} = \epsilon\hat{\alpha}^{+}\hat{\alpha} - \epsilon\lambda^2\,,$$

with eigenvalues

$$E_n = n\epsilon - \epsilon\lambda^2\,. \tag{1}$$

The same answer follows from perturbation theory. The unperturbed eigenvalues and eigenfunctions are (compare problem 3 of section 10)

$$E_n^{(0)} = n\epsilon\,, \quad |\psi_n\rangle = \frac{1}{\sqrt{n!}}(\hat{a}^{+})^n|0\rangle\,,$$

and the perturbations to the energy eigenvalues are

$$E_n^{(1)} = \langle\psi_n|\hat{H}_1|\psi_n\rangle\,, \quad E_n^{(2)} = \sum_m \frac{|\langle\psi_n|\hat{H}_1|\psi_m\rangle|^2}{E_n^{(0)} - E_m^{(0)}}$$

and the answer (1) follows immediately.

14*. Following the procedure of problem 8 of section 1 we can introduce the creation and annihilation operators of the two oscillators through the equations

$$\hat{a}_i = \frac{1}{\sqrt{2}}\left\{\sqrt{\left(\frac{\mu\omega}{\hbar}\right)}x_i + \sqrt{\left(\frac{\hbar}{\mu\omega}\right)}\frac{\partial}{\partial x_i}\right\}, \quad \hat{a}_i^{+} = \frac{1}{\sqrt{2}}\left\{\sqrt{\left(\frac{\mu\omega}{\hbar}\right)}x_i - \sqrt{\left(\frac{\hbar}{\mu\omega}\right)}\frac{\partial}{\partial x_i}\right\},$$

$$i = 1, 2\,. \tag{1}$$

Substituting this into the Hamiltonian gives

$$\hat{H} = \hbar\omega\{\hat{a}_1^{+}\hat{a}_1 + \hat{a}_2^{+}\hat{a}_2 + 1\} + \tfrac{1}{2}\lambda\hbar\omega\{\hat{a}_1\hat{a}_2 + \hat{a}_1^{+}\hat{a}_2^{+} + \hat{a}_1\hat{a}_2^{+} + \hat{a}_1^{+}\hat{a}_2\}\,. \tag{2}$$

This Hamiltonian is not diagonal in the number operators $\hat{n}_1$ ($\equiv \hat{a}_1^{+}\hat{a}_1$) and $\hat{n}_2$ ($\equiv \hat{a}_2^{+}\hat{a}_2$). We can proceed in two different ways. We can either diagonalise the original Hamiltonian, or diagonalise the Hamiltonian in the form (2).

In the first case, we introduce new variables y_1 and y_2 by rotating the axes:

$$x_1 = \alpha y_1 + \beta y_2\,, \quad x_2 = -\beta y_1 + \alpha y^2\,, \quad \alpha^2 + \beta^2 = 1\,. \tag{3}$$

If we take

$$\alpha = \beta = \frac{1}{\sqrt{2}}\,, \tag{4}$$

the Hamiltonian becomes

$$\hat{H} = -\frac{\hbar^2}{2\mu}\left(\frac{\partial^2}{\partial y_1^2}+\frac{\partial^2}{\partial y_2^2}\right)+\tfrac{1}{2}\mu\omega_1^2 y_1^2+\tfrac{1}{2}\mu\omega_2^2 y_2^2\,, \tag{5}$$

with

$$\omega_1^2 = \omega^2(1-\lambda)\,, \quad \omega_2^2 = \omega^2(1+\lambda)\,, \tag{6}$$

and introducing new annihilation and creation operators through the equations

$$\hat{b}_i = \frac{1}{\sqrt{2}}\left\{\sqrt{\left(\frac{\mu\omega_i}{\hbar}\right)}y_i+\sqrt{\left(\frac{\hbar}{\mu\omega_i}\right)}\frac{\partial}{\partial y_i}\right\}, \quad \hat{b}_i^+ = \frac{1}{\sqrt{2}}\left\{\sqrt{\left(\frac{\mu\omega_i}{\hbar}\right)}y_i-\sqrt{\left(\frac{\hbar}{\mu\omega_i}\right)}\frac{\partial}{\partial y_i}\right\},$$

$$i = 1, 2\,, \tag{7}$$

the Hamiltonian becomes

$$\hat{H} = \hbar\omega_1(\hat{b}_1^+\hat{b}_1+\tfrac{1}{2})+\hbar\omega_2(\hat{b}_2^+\hat{b}_2+\tfrac{1}{2})\,, \tag{8}$$

with energy levels

$$E_{n_1 n_2} = (n_1+\tfrac{1}{2})\hbar\omega_1+(n_2+\tfrac{1}{2})\hbar\omega_2\,. \tag{9}$$

A second way to obtain the result (9) is by introducing new annihilation and creation operators into equation (2). Writing

$$\hat{a}_1 = \alpha\hat{c}_1+\beta\hat{c}_2+\gamma\hat{c}_1^++\delta\hat{c}_2^+\,, \quad \hat{a}_2 = \xi\hat{c}_1+\eta\hat{c}_2+\zeta\hat{c}_1^++\vartheta\hat{c}_2^+\,, \tag{10}$$

with real coefficients $\alpha, \beta, \gamma, \delta, \xi, \eta, \zeta$, and ϑ, we find, first of all, from the requirement that the $\hat{c}_i$, $\hat{c}_i^+$ must satisfy the boson commutation relations (see problems 4 to 6 of section 10) that

$$\alpha^2+\beta^2-\gamma^2-\delta^2 = \xi^2+\eta^2-\zeta^2-\vartheta^2 = 1\,, \quad \gamma\xi+\delta\eta-\alpha\zeta-\beta\vartheta = 0\,. \tag{11}$$

Substituting expressions (10) into the Hamiltonian (2) gives

$$\begin{aligned}2\hat{H}/\lambda\hbar\omega = {} & \hat{c}_1^+\hat{c}_1\{(2/\lambda)(\alpha^2+\gamma^2+\xi^2+\zeta^2)+2(\alpha+\gamma)(\xi+\zeta)\}\\ & +\hat{c}_2^+\hat{c}_2\{(2/\lambda)(\beta^2+\delta^2+\eta^2+\vartheta^2)+2(\beta+\delta)(\eta+\vartheta)\}\\ & +(\hat{c}_1^{+2}+\hat{c}_1^2)\{(2/\lambda)(\alpha\gamma+\xi\zeta)+(\alpha+\gamma)(\xi+\zeta)\}\\ & +(\hat{c}_2^{+2}+\hat{c}_2^2)\{(2/\lambda)(\beta\delta+\eta\vartheta)+(\beta+\delta)(\eta+\vartheta)\}\\ & +(\hat{c}_1^+\hat{c}_2+\hat{c}_1\hat{c}_2^+)\{(2/\lambda)(\alpha\beta+\gamma\delta+\xi\eta+\zeta\vartheta)+(\alpha+\gamma)(\eta+\vartheta)\\ & \qquad +(\beta+\delta)(\xi+\zeta)\}\\ & +(\hat{c}_1^+\hat{c}_2^++\hat{c}_1\hat{c}_2)\{(2/\lambda)(\alpha\delta+\beta\gamma+\xi\vartheta+\eta\zeta)+(\alpha+\gamma)(\eta+\vartheta)\\ & \qquad +(\beta+\delta)(\xi+\zeta)\}\\ & +(2/\lambda)(\alpha^2+\beta^2+\xi^2+\eta^2+1)+(\alpha+\gamma)(\xi+\zeta)+(\beta+\delta)(\eta+\vartheta)\,,\end{aligned} \tag{12}$$

and we see that we can satisfy equations (11) and get rid of the terms which in $\hat{H}$ are not diagonal in the occupation number operators $\hat{c}_i^+\hat{c}_i$ by taking

$$\alpha = -\xi = \frac{1}{2\sqrt{2}}\{(1-\lambda)^{-1/4} - (1-\lambda)^{1/4}\},$$

$$\beta = \eta = \frac{1}{2\sqrt{2}}\{(1+\lambda)^{-1/4} - (1+\lambda)^{1/4}\}, \tag{13}$$

$$\gamma = -\zeta = \frac{1}{2\sqrt{2}}\{(1-\lambda)^{-1/4} - (1-\lambda)^{1/4}\},$$

$$\delta = \vartheta = \frac{1}{2\sqrt{2}}\{(1+\lambda)^{-1/4} - (1+\lambda)^{1/4}\}.$$

One obtains this result by noting, first of all, that one can satisfy two of the three equations (11) by putting $\alpha = -\xi$, $\beta = \eta$, $\gamma = -\zeta$, $\delta = \vartheta$. After that, by manipulating the remaining equations one finds

$$\frac{4}{\lambda}\beta\delta\sqrt{(1+\lambda)} + \frac{4}{\lambda}\alpha\gamma\sqrt{(1-\lambda)} = 1. \tag{14}$$

If one then puts each of the two terms on the left-hand side of equation (14) equal to $\frac{1}{2}$, one finds the results of equations (13).

Substituting the result (13) into the Hamiltonian (12) we find

$$\hat{H} = \hbar\omega_1\hat{c}_1^+\hat{c}_1 + \hbar\omega_2\hat{c}_2^+\hat{c}_2 + \tfrac{1}{2}\hbar(\omega_1+\omega_2),$$

which is the same as equation (8).

15. The conditions are

$$u_k^2 + v_k^2 = 1, \quad u_{-k}^2 + v_{-k}^2 = 1, \quad u_k v_{-k} + u_{-k} v_k = 0, \tag{2}$$

from which follows that one can characterise these coefficients by a single parameter ϑ_k, as follows:

$$\begin{aligned} u_k &= \cos\vartheta_k, \quad & v_k &= \sin\vartheta_k, \\ u_{-k} &= \cos\vartheta_k, \quad & v_{-k} &= -\sin\vartheta_k, \end{aligned} \tag{3}$$

where we have used the fact that the identity transformation should be part of the set of transformations (1).

$$\langle 0|\hat{a}_k^+\hat{a}_k|0\rangle = \frac{v_k^2}{u_k u_{-k} - v_k v_{-k}} = \sin^2\vartheta_k. \tag{4}$$

16. In this case we have instead of equations (2), (3), and (4) of the preceding problem

$$u_k^2 - v_k^2 = 1, \quad u_{-k}^2 - v_{-k}^2 = 1, \quad u_k v_{-k} - u_{-k} v_k = 0, \tag{2'}$$

and hence

$$u_k = \cosh\varphi_k\,, \quad v_k = \sinh\varphi_k\,, \quad u_{-k} = \cosh\varphi_k\,, \quad v_{-k} = \sinh\varphi_k\,. \quad (3')$$

$$\langle 0|\hat{a}_k^+\hat{a}_k|0\rangle = \sinh^2\varphi_k\,. \quad (4')$$

17*. If there were no interactions, we could use the operators $\hat{a}_{\boldsymbol{k}}$, $\hat{a}_{\boldsymbol{k}}^+$ of problem 8 of section 10 and the Hamiltonian would be, in terms of these operators (see problem 8 of section 10),

$$\hat{H}_0 = \sum_k \frac{\hbar^2 k^2}{2\mu}\hat{a}_k^+\hat{a}_k\,, \quad (3)$$

with eigenvalues

$$E = \sum_k \frac{\hbar^2 k^2}{2\mu} n_{\boldsymbol{k}}^{(0)}\,, \quad (4)$$

where the $n_{\boldsymbol{k}}^{(0)}$ would be integers, the eigenvalues of the occupation number operators $\hat{a}_{\boldsymbol{k}}^+\hat{a}_{\boldsymbol{k}}$.

The ground state is clearly the one where $n_0^{(0)} = N$, $n_{\boldsymbol{k}}^{(0)} = 0$ ($\boldsymbol{k} \neq 0$). If the interaction is weak, we would expect that for the low-lying states the number of particles in the $\boldsymbol{k} = 0$ (zero-momentum) state would still be large, or rather, that the expectation value of $\hat{a}_0^+\hat{a}_0$ would be of the order of the total number of particles N. As $\hat{a}_0\hat{a}_0^+ = \hat{a}_0^+\hat{a}_0 + 1$ and if we take N to be large compared to unity, as it will be for all physical systems, we see that $\hat{a}_0^+\hat{a}_0$ and $\hat{a}_0\hat{a}_0^+$ are practically the same so that to a very good approximation we can treat $\hat{a}_0$ and $\hat{a}_0^+$ as numbers and put them equal to $\sqrt{n_0}$, where n_0 is the expectation value of $\hat{a}_0^+\hat{a}_0$.

Let us now consider the Hamiltonian in terms of the $\hat{a}_{\boldsymbol{k}}$ and $\hat{a}_{\boldsymbol{k}}^+$ (see equation (3) of problem 8 of section 10):

$$\hat{H} = \hat{H}_0 + \hat{H}_1\,, \quad (5)$$

$$\hat{H}_1 = \frac{1}{2v}\sum_{\boldsymbol{k},\boldsymbol{k}',\boldsymbol{q}} U(q)\hat{a}_{\boldsymbol{k}}^+\hat{a}_{\boldsymbol{k}'}^+\hat{a}_{\boldsymbol{k}'+\boldsymbol{q}}\hat{a}_{\boldsymbol{k}-\boldsymbol{q}}\,. \quad (6)$$

In the operator $\hat{H}_1$ there is one term, for $\boldsymbol{k} = \boldsymbol{k}' = \boldsymbol{q} = 0$, with four operators $\hat{a}_0$ or $\hat{a}_0^+$, there are no terms with three such operators, and there are terms with two, one, or zero such operators. The terms with two such operators are

$$\frac{1}{2v}\sum_{\boldsymbol{q}}{}' U(q)\hat{a}_0^+\hat{a}_0^+\hat{a}_{\boldsymbol{q}}\hat{a}_{-\boldsymbol{q}} + \frac{1}{v}\sum_{\boldsymbol{q}}{}' U(0)\hat{a}_0^+\hat{a}_0\hat{a}_{\boldsymbol{q}}^+\hat{a}_{\boldsymbol{q}} + \frac{1}{v}\sum_{\boldsymbol{q}}{}' U(q)\hat{a}_0^+\hat{a}_0\hat{a}_{\boldsymbol{q}}^+\hat{a}_{\boldsymbol{q}}$$

$$+\frac{1}{2v}\sum_{\boldsymbol{q}}{}' U(q)\hat{a}_0\hat{a}_0\hat{a}_{\boldsymbol{q}}^+\hat{a}_{-\boldsymbol{q}}^+\,, \quad (7)$$

where the prime on the summation sign indicates that the term with $\boldsymbol{q} = 0$

is omitted, and where we have used the fact that $U(q)$ is a function of the absolute magnitude of $\boldsymbol{q}$. Putting now $\hat{a}_0 = \hat{a}_0^+ = \sqrt{n_0}$, and neglecting the terms with less than two zero-momentum operators, we find $\left(\sum_q \hat{a}_q^+ \hat{a}_q = n_0 + \sum_q' \hat{a}_q^+ \hat{a}_q = N\right.$ is the total number of particles, which is a constant for a Hamiltonian of the form (5)$\Big)$

$$\hat{H} = \sum_q \frac{\hbar^2 q^2}{2\mu} \hat{a}_q^+ \hat{a}_q + \frac{n_0^2}{2v} U(0) + \frac{n_0}{v} U(0) \sum_q' \hat{a}_q^+ \hat{a}_q$$
$$+ \frac{n_0}{2v} \sum_q' U(q) \{\hat{a}_q^+ \hat{a}_{-q}^+ + \hat{a}_q \hat{a}_{-q} + 2\hat{a}_q^+ \hat{a}_q\}, \qquad (8)$$

or

$$\hat{H} = \frac{N^2}{2v} U(0) + \sum_q \frac{\hbar^2 q^2}{2\mu} \hat{a}_q^+ \hat{a}_q + \frac{n_0}{2v} \sum_q' U(q) \{\hat{a}_q^+ \hat{a}_{-q}^+ + \hat{a}_q \hat{a}_{-q} + 2\hat{a}_q^+ \hat{a}_q\}. \qquad (9)$$

Let us now apply a transformation of the kind discussed in the preceding two problems:

$$\hat{b}_q = u_q \hat{a}_q + v_q \hat{a}_{-q}^+ . \qquad (10)$$

We saw in the preceding problem that this transformation—often called a *Bogolyubov transformation*, since it was first introduced by Bogolyubov—is a canonical one, that is, one which ensures that the $\hat{b}_q$, b_q^+ are again boson operators, if

$$u_q = u_{-q} = \cosh\varphi_q , \quad v_q = v_{-q} = \sinh\varphi_q . \qquad (11)$$

Substituting expressions (11) into the transformation (10) and performing the transformation we find for the Hamiltonian the expression

$$\hat{H} = \frac{N^2}{2v} U(0) + \sum_q' \hat{b}_q^+ \hat{b}_q \left\{ \frac{\hbar^2 q^2}{2\mu} (\cosh^2\varphi_q + \sinh^2\varphi_q) \right.$$
$$\left. + \frac{n_0}{v} U(q) (\cosh^2\varphi_q + \sinh^2\varphi_q - 2\cosh\varphi_q \sinh\varphi_q) \right\}$$
$$+ \sum_q' (\hat{b}_q \hat{b}_{-q} + \hat{b}_q^+ \hat{b}_{-q}^+) \left\{ -\frac{\hbar^2 q^2}{2\mu} \sinh\varphi_q \cosh\varphi_q \right.$$
$$\left. + \frac{n_0}{2v} U(q) (\cosh^2\varphi_q + \sinh^2\varphi_q - 2\sinh\varphi_q \cosh\varphi_q) \right\}$$
$$+ \sum_q' \frac{\hbar^2 q^2}{2\mu} \sinh^2\varphi_q + \frac{n_0}{v} \sum_q' U(q) (\sinh^2\varphi_q - \cosh\varphi_q \sinh\varphi_q). \qquad (12)$$

Therefore, if we put

$$\cosh^2\varphi_q + \sinh^2\varphi_q - 2\sinh\varphi_q \cosh\varphi_q = 2\alpha \sinh\varphi_q \cosh\varphi_q , \qquad (13)$$

with

$$\alpha = \alpha(q) \equiv \frac{v\hbar^2 q^2}{2\mu n_0 U(q)}, \tag{14}$$

or

$$\tanh\varphi_q = \alpha + 1 - \sqrt{(\alpha^2 + 2\alpha)}, \tag{15}$$

the terms with $\hat{b}_q \hat{b}_{-q}$ and $\hat{b}_q^+ \hat{b}_{-q}^+$ disappear, and substituting (15) into (12) we are left with

$$\hat{H} = E_0 + \sum_q{}' \epsilon_q \hat{b}_q^+ \hat{b}_q, \tag{16}$$

with

$$E_0 = \frac{N^2}{2v} U(0) + \tfrac{1}{2} \sum_q{}' \frac{n_0 U(q)}{v} \{-1 - \alpha + \sqrt{(\alpha^2 + 2\alpha)}\}, \tag{17}$$

and

$$\epsilon_q = \frac{\hbar^2 q^2}{2\mu} \left(\frac{\alpha + 2}{\alpha}\right)^{1/2}. \tag{18}$$

We note that when $\alpha \gg 1$, $\epsilon_q \approx \hbar^2 q^2 / 2\mu$, and the Hamiltonian (16) is essentially that of a system of free bosons (compare equation (3)). In the limit as $\alpha(q) \to 0$ we find that

$$\epsilon_q \to \hbar s q, \tag{19}$$

with

$$s = \left[\frac{n_0 U(0)}{\mu v}\right]^{1/2} \approx \left[\frac{N U(0)}{\mu v}\right]^{1/2}. \tag{20}$$

Equation (19) gives the energy of sound waves. Of course, in order that the ground state is stable it is necessary that $U(0) > 0$, or (if we use equation (4) of problem 8 of section 10)

$$\int U(r) d^3\mathbf{r} > 0, \tag{21}$$

which means that the interaction between the bosons must be basically a repulsive one.

The spectrum (18) of the quasi-particle excitations bears some resemblance to that of the elementary excitations in liquid helium.

18*. Introduce a set of matrices α_{ij} such that

$$\sum_i \alpha_{ij} \alpha_{ik}^* = \delta_{jk} = \sum_i \alpha_{ji}^* \alpha_{ki}. \tag{3}$$

In that case the transformation

$$\hat{a}_i = \sum_k \alpha_{ki} \hat{b}_k \tag{4}$$

will be canonical, as can be checked easily.

Substituting the transformation (4) into the Hamiltonian (1) we find

$$\hat{H} = \sum_{i,j,k,l} L_{ij} \alpha_{ki}^* \alpha_{lj} \hat{b}_k^+ \hat{b}_l , \tag{5}$$

and if we determine the α_{ij} from the requirement that

$$\sum_{i,j} L_{ij} \alpha_{ki}^* \alpha_{lj} = E_k \delta_{kl} , \tag{6}$$

we obtain expression (2). The problem is thus reduced to that of bringing a quadratic form onto diagonal form. From equation (6) it follows that the E_k are the eigenvalues of the secular equation

$$|L_{ij} - E\delta_{ij}| = 0 . \tag{7}$$

19*. The solution of this problem is analogous to that of the preceding problem. In this case one considers the transformation (compare the solution to problems 16 and 17 of section 10)

$$\hat{b}_i = \sum_j (\alpha_{ij} \hat{a}_j + \beta_{ij} \hat{a}_j^+) . \tag{4}$$

In order that the transformation (4) is a canonical one the α_{ij} and β_{ij} must satisfy the relations

$$\sum_j (\alpha_{ij} \alpha_{kj}^* - \beta_{ij} \beta_{kj}^*) = \delta_{ik} , \quad \sum_j (\alpha_{ij} \beta_{kj} - \beta_{ij} \alpha_{kj}) = 0 . \tag{5}$$

The inverse of transformation (4) is

$$\hat{a}_i = \sum_j (\alpha_{ji}^* \hat{b}_j - \beta_{ji}^* \hat{b}_j^+) , \tag{6}$$

and substitution into the Hamiltonian (1) leads to the Hamiltonian (3), provided the α_{ij}, β_{ij}, and E_i satisfy the equations

$$\begin{aligned} &\sum_k \{(E_i \delta_{jk} - L_{jk}^*) \alpha_{ik} + M_{kj}^* \beta_{ik}\} = 0 , \\ &\sum_k \{M_{kj} \alpha_{ik} + (E_i \delta_{jk} + L_{jk}) \beta_{ik}\} = 0 , \end{aligned} \tag{7}$$

so that the E_i are the eigenvalues of the secular equation

$$\begin{vmatrix} E\delta_{ij} - L_{ij}^* & M_{il}^* \\ M_{kj} & E\delta_{kl} + L_{kl} \end{vmatrix} = 0 . \tag{8}$$

The ground state energy E_0 is given by the equation

$$E_0 = -\sum_{k,l} E_k |\beta_{kl}|^2 . \tag{9}$$

We note that the problem reduces to the preceding one, if all $M_{ij} = 0$, in which case we can put all $\beta_{ij} = 0$, and equation (8) reduces to equation (7) of the preceding problem.

20. (i) The expectation value of A is given by the equation

$$\langle \hat{A} \rangle = \iint \Psi^*(q, x) \hat{A} \Psi(q, x) dq dx , \tag{2}$$

where dx and dq indicate integration over all coordinates of the smaller system and over the remainder of the larger system, respectively. If we use the definition (1) of $\hat{\rho}$, we get

$$\langle \hat{A} \rangle = \iint \langle x' | \hat{\rho} | x \rangle dx \langle x | \hat{A} | x' \rangle dx' = \mathrm{Tr} \hat{\rho} \hat{A} . \tag{3}$$

(ii) Using equation (3) with $\hat{A} \equiv 1$, we find

$$\mathrm{Tr} \hat{\rho} = \int \langle x | \hat{\rho} | x \rangle dx = 1 , \tag{4}$$

which follows also directly from the definition (1) and the normalisation of $\Psi(q, x)$.

(iii) Let $\varphi_n(x, t)$ be the complete orthonormal set of time-dependent eigenfunctions of $\hat{H}$, so that

$$i\hbar \frac{\partial \varphi_n}{\partial t} = \hat{H} \varphi_n . \tag{5}$$

We can now expand $\langle x | \hat{\rho} | x' \rangle$ in terms of the $\varphi_n(x, t)$ as follows

$$\langle x | \hat{\rho} | x' \rangle = \sum_{m, n} a_{mn} \varphi_m(x, t) \varphi_n^*(x', t) . \tag{6}$$

We note, first of all, that the a_{mn} give us the density matrix in the φ-representation.

Secondly, we can take the time-derivative of $\langle x | \hat{\rho} | x' \rangle$ and find

$$\begin{aligned} i\hbar \langle x | \dot{\hat{\rho}} | x' \rangle &= \sum_{m, n} a_{mn} i\hbar \{ \varphi_n^*(x', t) \dot{\varphi}_m(x, t) + \dot{\varphi}_n^*(x', t) \varphi_m(x, t) \} \\ &= \sum_{m, n} a_{mn} \int \{ \varphi_n^*(x', t) \langle x | \hat{H} | x'' \rangle dx'' \varphi_m(x'', t) \\ &\qquad - \varphi_m(x, t) \varphi_n^*(x'', t) dx'' \langle x'' | \hat{H}^* | x' \rangle \} \\ &= \int \{ \langle x | \hat{H} | x'' \rangle dx'' \langle x'' | \hat{\rho} | x' \rangle - \langle x | \hat{\rho} | x'' \rangle dx'' \langle x'' | \hat{H} | x' \rangle \} \\ &= \langle x | \hat{H} \hat{\rho} - \hat{\rho} \hat{H} | x' \rangle , \end{aligned} \tag{7}$$

or

$$i\hbar \dot{\hat{\rho}} = [\hat{H}, \hat{\rho}]_- . \tag{8}$$

21. The density matrix is now of the form (compare equation (1) of the preceding problem)

$$\langle x|\hat{\rho}|x'\rangle = \chi^*(x')\chi(x)\,, \tag{3}$$

and hence

$$\begin{aligned}\langle x|\hat{\rho}^2|x'\rangle &= \int\langle x|\hat{\rho}|x''\rangle dx''\langle x''|\hat{\rho}|x'\rangle \\ &= \int\chi^*(x'')\chi(x)dx''\chi^*(x')\chi(x'') = \chi^*(x')\chi(x) = \langle x|\hat{\rho}|x'\rangle\,. \end{aligned} \tag{4}$$

To prove sufficiency we proceed as follows. Let $\psi_n(x)$ be a complete orthonormal set and let us expand $\langle x|\hat{\rho}|x'\rangle$ as follows

$$\langle x|\hat{\rho}|x'\rangle = \sum_{m,\,n} c_{mn}\psi_m(x)\psi_n^*(x')\,. \tag{5}$$

Let us now assume that the set $\psi_n(x)$ is chosen such that it diagonalises $\hat{\rho}$ so that

$$c_{mn} = c_m\delta_{mn}\,. \tag{6}$$

If $\hat{\rho}$ is diagonal, so is $\hat{\rho}^2$ and equation (2) therefore means that

$$c_m^2 = c_m\,, \tag{7}$$

whence

$$\text{either } c_m = 1 \quad \text{or } c_m = 0\,. \tag{8}$$

From the normalisation of $\hat{\rho}$ (equation (4) of the preceding problem) it follows that

$$\sum c_m = 1\,. \tag{9}$$

It then follows from equations (8) and (9) that one of the c_m equals 1, while the others vanish,

$$c_{m_0} = 1\,, \quad c_m = 0\,, \quad m \neq m_0\,. \tag{10}$$

Hence

$$\langle x|\hat{\rho}|x'\rangle = \psi_{m_0}(0)\psi_{m_0}^*(x')\,, \tag{11}$$

which is exactly the form (3).

22. (i) Equation (1) follows easily from the fact that (a) the density matrix for the spin part of spin-$\frac{1}{2}$ particles must be a 2×2 matrix and (b) it must satisfy equation (4) of problem 20 of section 10,

$$\mathrm{Tr}\,\hat{\rho} = 1\,, \tag{2}$$

while

$$\mathrm{Tr}\,\hat{1} = 2\,, \quad \mathrm{Tr}\,\hat{\sigma}_i = 0\,, \quad i = x, y, z\,. \tag{3}$$

(ii) By evaluating $\langle\hat{\sigma}\rangle$ and using the fact that

$$\hat{\sigma}_x\hat{\sigma}_y = i\hat{\sigma}_z, \text{(and cyclic permutations)}, \hat{\sigma}_i^2 = \hat{1}, \tag{4}$$

and equations (3), it follows easily that

$$\boldsymbol{P} = \langle\hat{\sigma}\rangle \tag{5}$$

The vector $\boldsymbol{P}$ may therefore be called the (average) polarisation of the spin-$\frac{1}{2}$ particle.

(iii) We have

$$\frac{\partial \boldsymbol{P}}{\partial t} = \frac{\partial\langle\hat{\sigma}\rangle}{\partial t} = \left\langle\frac{\partial\hat{\sigma}}{\partial t}\right\rangle = -\frac{i}{\hbar}\langle[\hat{\sigma}, \hat{H}]_-\rangle, \tag{6}$$

where $\hat{H}$ is the (spin) Hamiltonian,

$$\hat{H} = -(\hat{\mu}\,.\,\mathcal{H}) = -\tfrac{1}{2}\gamma\hbar(\hat{\sigma}\,.\,\mathcal{H}), \tag{7}$$

with $\hat{\mu}$ and γ, respectively, the magnetic moment and the magnetogyric ratio,

$$\hat{\mu} = \gamma\hbar\hat{\sigma}\,. \tag{8}$$

Substituting (7) into (6) and using equations (4) we easily find that

$$\frac{\partial \boldsymbol{P}}{\partial t} = \tfrac{1}{2}i\gamma\langle[\hat{\sigma}, (\hat{\sigma}\,.\,\mathcal{H})]_-\rangle = -\gamma\langle[\mathcal{H}\wedge\hat{\sigma}]\rangle, \tag{9}$$

or

$$\frac{\partial \boldsymbol{P}}{\partial t} = -\gamma[\mathcal{H}\wedge\boldsymbol{P}]\,. \tag{10}$$

23. If the photon were in a pure state ψ_0,

$$\psi_0 = c_1^{(0)}\psi_1 + c_2^{(0)}\psi_2, \tag{5}$$

we would expect that

$$W = \left|\int\psi_0^*\psi^{\text{det}}\,d\tau\right|^2 = |c_1^{(0)*}c_1^{\text{det}} + c_2^{(0)*}c_2^{\text{det}}|^2\,. \tag{6}$$

If the photon state is described by a density matrix $\hat{\rho}$, we have

$$W = \langle\hat{\rho}^{\text{det}}\rangle = \text{Tr}\,\hat{\rho}\hat{\rho}^{\text{det}}\,. \tag{7}$$

One checks easily (i) that equation (7) leads to expression (6), if $\hat{\rho}$ corresponds to the pure state (5), in which case

$$\rho_{ij} = c_i^{(0)}c_j^{(0)*}, \tag{8}$$

and (ii) that if the photon is in a pure state such that $\psi_0 \equiv \psi^{\text{det}}$, $W = 1$, as should, of course, be the case.

24. To find the polarisation of photon A we must evaluate the expression

$$\boldsymbol{P}_{\mathrm{A}} = \mathrm{Tr}_{\mathrm{A}} \hat{\rho}_{\mathrm{A}} \hat{\boldsymbol{\omega}}_{\mathrm{A}} , \tag{2}$$

where $\hat{\rho}_{\mathrm{A}}$ is the density matrix referring to the state of photon A by itself. By analogy with equation (1) of problem 20 of section 10 we find

$$\hat{\rho}_{\mathrm{A}} = \mathrm{Tr}_{\mathrm{B}} \hat{\rho}_{\mathrm{AB}} , \tag{3}$$

where Tr_{B} (Tr_{A}) are traces over the degrees of freedom pertaining to photon B (A).

From equations (1) and (3) and the equations which are the analogues of equations (3) of problem 22 of section 10 it follows that

$$\hat{\rho}_{\mathrm{A}} = \tfrac{1}{2} \hat{1}_{\mathrm{A}} . \tag{4}$$

Comparing this with the results of problem 22 of section 10, or using equation (2), we find

$$\boldsymbol{P}_{\mathrm{A}} = 0 , \tag{5}$$

and, of course, similarly,

$$\boldsymbol{P}_{\mathrm{B}} = 0 : \tag{6}$$

both photons are unpolarised.

To find whether there is any correlation between the polarisations of the two photons we consider the case where we have two detectors which are set so as to respond 100 per cent, the first one to a polarisation $\boldsymbol{P}_{\mathrm{A}}^{\mathrm{det}}$ of photon A and the second one to a polarisation $\boldsymbol{P}_{\mathrm{B}}^{\mathrm{det}}$ of photon B. These detectors are characterised by the density matrices (compare equation (1) of problem 22 of section 10)

$$\hat{\rho}_{\mathrm{A}}^{\mathrm{det}} = \tfrac{1}{2}\{\hat{1}_{\mathrm{A}} + (\boldsymbol{P}_{\mathrm{A}}^{\mathrm{det}} . \hat{\boldsymbol{\omega}}_{\mathrm{A}})\}\hat{1}_{\mathrm{B}} , \quad \hat{\rho}_{\mathrm{B}}^{\mathrm{det}} = \tfrac{1}{2}\{\hat{1}_{\mathrm{B}} + (\boldsymbol{P}_{\mathrm{B}}^{\mathrm{det}} . \hat{\boldsymbol{\omega}}_{\mathrm{B}})\}\hat{1}_{\mathrm{A}} . \tag{7}$$

To find out whether there is a response at both counters we must evaluate the expression (compare equation (7) of the preceding problem)

$$W(\boldsymbol{P}_{\mathrm{A}}^{\mathrm{det}}, \boldsymbol{P}_{\mathrm{B}}^{\mathrm{det}}) = \mathrm{Tr}\, \hat{\rho}_{\mathrm{AB}} \hat{\rho}_{\mathrm{A}}^{\mathrm{det}} \hat{\rho}_{\mathrm{B}}^{\mathrm{det}} , \tag{8}$$

and the result is

$$W(\boldsymbol{P}_{\mathrm{A}}^{\mathrm{det}}, \boldsymbol{P}_{\mathrm{B}}^{\mathrm{det}}) = \tfrac{1}{4}\{1 - (\boldsymbol{P}_{\mathrm{A}}^{\mathrm{det}} \cdot \boldsymbol{P}_{\mathrm{B}}^{\mathrm{det}})\} . \tag{9}$$

We see that $W(\boldsymbol{P}_{\mathrm{A}}^{\mathrm{det}}, \boldsymbol{P}_{\mathrm{B}}^{\mathrm{det}}) = 0$, if $\boldsymbol{P}_{\mathrm{A}}^{\mathrm{det}}$ is parallel to $\boldsymbol{P}_{\mathrm{B}}^{\mathrm{det}}$, while it is a maximum when $\boldsymbol{P}_{\mathrm{A}}^{\mathrm{det}}$ and $\boldsymbol{P}_{\mathrm{B}}^{\mathrm{det}}$ are antiparallel: the two photons have opposite polarisations.

25. The density matrix $\hat{\rho}_{\mathrm{f}}$ after the scattering will be (compare equation (1) of problem 20 of section 10)

$$\hat{\rho}_{\mathrm{f}} = \hat{S} \hat{\rho}_{\mathrm{i}} \hat{S}^{\dagger} , \tag{3}$$

where $\hat{S}^\dagger$ is the Hermitean conjugate of $\hat{S}$ and where $\hat{\rho}_i$ is the density matrix before the scattering.

The polarisation after the scattering is given by the equation (see equation (5) of problem 22 of section 10)

$$\boldsymbol{P}_f = \langle\hat{\sigma}\rangle = \operatorname{Tr}\hat{\rho}_f\,\hat{\sigma} = \operatorname{Tr}\hat{\sigma}\hat{S}\hat{\rho}_i\hat{S}^\dagger \ . \tag{4}$$

As the polarisation before the scattering was zero, we have

$$\hat{\rho}_i = \tfrac{1}{2}\hat{1} \ , \tag{5}$$

and we then find easily, using equations (3) and (4) of problem 22 of section 10, that

$$\boldsymbol{P}_f = 2\mathrm{Re}(gh^*)\boldsymbol{n} \ . \tag{6}$$

Relativistic wave equations

1. For the case of a charged particle moving in an electromagnetic field j_μ should be given by the equation

$$j_\mu = \frac{e\hbar}{2mi}(\psi^* \partial_\mu \psi - \psi \partial_\mu \psi^*) - \frac{e^2 A_\mu}{mc}\psi^* \psi .$$

2. We write

$$\psi(\boldsymbol{r}, t) = \varphi(\boldsymbol{r}, t) \exp\left(-\frac{imc^2 t}{\hbar}\right), \tag{1}$$

and if E is the energy of the particle, we write

$$E = E' + mc^2 . \tag{2}$$

In the non-relativistic limit we have

$$E' \ll mc^2 , \tag{3}$$

and hence

$$\left| i\hbar \frac{\partial \varphi}{\partial t} \right| \sim E'\varphi \ll mc^2 \varphi . \tag{4}$$

Hence we get

$$\frac{\partial \psi}{\partial t} \approx -\frac{imc^2}{\hbar}\varphi \exp\left(-\frac{imc^2 t}{\hbar}\right), \tag{5}$$

$$\frac{\partial^2 \psi}{\partial t^2} \approx -\left\{\frac{2imc^2}{\hbar}\frac{\partial \varphi}{\partial t} + \frac{m^2 c^4}{\hbar^2}\varphi\right\} \exp\left(-\frac{imc^2 t}{\hbar}\right), \tag{6}$$

and hence, substituting into the Klein-Gordon equation [equation (4) of the preceding problem]

$$i\hbar \frac{\partial \varphi}{\partial t} = -\frac{\hbar^2}{2m}\nabla^2 \varphi . \tag{7}$$

Using equation (5) we find in the same limit

$$\boldsymbol{j} = \frac{e\hbar}{2mi}(\varphi^* \nabla \varphi - \varphi \nabla \varphi^*) , \qquad \rho = e\varphi^* \varphi . \tag{8}$$

Equations (7) and (8) are the free-particle Schrödinger equation and the corresponding expressions for $\boldsymbol{j}$ and ρ.

3. Putting

$$\psi = A \exp\left[\frac{i(\boldsymbol{p}.\boldsymbol{r})}{\hbar} - \frac{i\epsilon t}{\hbar}\right], \tag{1}$$

we find by substitution into the free-particle Klein-Gordon equation

$$\epsilon = \pm E_p , \qquad E_p = c\surd(\boldsymbol{p}^2 + m^2c^2) , \tag{2}$$

that is, there are two solutions,

$$\psi_+ = A_1 \exp \frac{i\{(\boldsymbol{p}.\boldsymbol{r}) - E_p t\}}{\hbar} , \tag{3}$$

$$\psi_- = A_2 \exp \frac{i\{(\boldsymbol{p}.\boldsymbol{r}) + E_p t\}}{\hbar} , \tag{4}$$

corresponding to

$$\rho_+ = \frac{eE_p}{mc^2} \psi_+^* \psi_+ , \tag{5}$$

$$\rho_- = -\frac{eE_p}{mc^2} \psi_-^* \psi_- . \tag{6}$$

We see here the appearance of an additional degree of freedom: the sign of the energy which can be positive or negative.

4. (i)

$$i\hbar \frac{\partial \varphi}{\partial t} = -\frac{\hbar^2}{2m} \nabla^2(\varphi + \chi) + mc^2 \varphi , \tag{2}$$

$$i\hbar \frac{\partial \chi}{\partial t} = \frac{\hbar^2}{2m} \nabla^2(\varphi + \chi) - mc^2 \chi .$$

(ii)

$$i\hbar \frac{\partial \Psi}{\partial t} = -\frac{\hbar^2}{2m} (\hat{\tau}_3 + i\hat{\tau}_2) \nabla^2 \Psi + mc^2 \hat{\tau}_3 \Psi , \tag{3}$$

where

$$\Psi = \begin{pmatrix} \varphi \\ \chi \end{pmatrix} . \tag{4}$$

(iii) $\rho = e(\varphi^* \varphi - \chi^* \chi) = e\Psi^\dagger \hat{\tau}_3 \Psi$. (5)

5. As in problem 3 of section 11 we find, of course, that the spinor

$$\Psi = \Phi e^{i\{(\boldsymbol{p}.\boldsymbol{r}) - \epsilon t\}/\hbar} , \quad \Phi = \begin{pmatrix} \varphi_0 \\ \chi_0 \end{pmatrix} , \tag{1}$$

is a solution of the wave equation, provided

$$\epsilon = \pm E_p , \quad E_p = c\surd(\boldsymbol{p}^2 + m^2c^2) . \tag{2}$$

If $\epsilon = \pm E_p$, we have

$$\varphi_0^{(\pm)} = \frac{mc^2 \pm E_p}{2\surd(mc^2 E_p)} , \quad \chi_0^{(\pm)} = \frac{mc^2 \mp E_p}{2\surd(mc^2 E_p)} , \tag{3}$$

where we have normalised the wavefunction to unit volume.

From equation (5) of the preceding problem it follows that

$$\rho_{\pm} = \pm e : \tag{4}$$

the solution with $\epsilon = +E_p$ ($\epsilon = -E_p$) corresponds to a positive (negative) "charge state".

In the non-relativistic limit we have $E_p \approx mc^2 + p^2/2m$, $p^2/2m \ll mc^2$, and from equation (3) we find

$$\varphi_0^{(+)} \sim 1\,, \quad |\chi_0^{(+)}| \sim \left(\frac{p}{2mc}\right)^2 \sim \left(\frac{v}{2c}\right)^2 \ll 1\,,$$

$$|\varphi_0^{(-)}| \sim \left(\frac{p}{2mc}\right)^2 \sim \left(\frac{v}{2c}\right)^2 \ll 1\,, \quad \chi_0^{(-)} \sim 1\,.$$

7. In this case we have the Klein–Gordon equation for a charged particle in an electromagnetic field described by the potentials

$$\boldsymbol{A} = 0\,, \quad \phi = -\frac{Ze}{r}\,. \tag{1}$$

As we are looking for a stationary solution, we put

$$\psi(\boldsymbol{r}, t) = \varphi(\boldsymbol{r})e^{-iEt/\hbar}\,, \tag{2}$$

and we look for solutions with $E > 0$. The equation for $\varphi(\boldsymbol{r})$ now becomes

$$\left\{\left(E + \frac{Ze^2}{r}\right)^2 - m^2c^4 + \hbar^2c^2\nabla^2\right\}\varphi(\boldsymbol{r}) = 0\,. \tag{3}$$

Changing to spherical polars and writing

$$\varphi(\boldsymbol{r}) = \frac{1}{r}R_l(r)Y_{lm}(\theta, \varphi)\,, \tag{4}$$

the radial function $R_l(r)$ satisfies the equation

$$\left\{\frac{d^2}{dr^2} - \frac{l(l+1) - Z^2\alpha^2}{r^2} + \frac{2Z\alpha E}{\hbar c r} - \frac{m^2c^4 - E^2}{\hbar^2c^2}\right\}R_l(r) = 0\,, \tag{5}$$

where α is the fine-structure constant,

$$\alpha = \frac{e^2}{\hbar c} \approx \frac{1}{137}\,. \tag{6}$$

If we are looking for bound solutions, $E < mc^2$. If we write

$$k(k+1) = l(l+1) - Z^2\alpha^2 \,, \tag{7}$$

$$\gamma = \frac{2Z\alpha E}{\hbar c\beta} \,, \tag{8}$$

$$\beta = \frac{2}{\hbar c}\sqrt{(m^2c^4 - E^2)} \,, \tag{9}$$

$$\rho = \beta r \,, \tag{10}$$

we can write equation (5) in the form

$$\left[\frac{d^2}{d\rho^2} + \frac{\gamma}{\rho} - \frac{k(k+1)}{\rho^2} - \tfrac{1}{4}\right] R_l(\rho) = 0 \,. \tag{11}$$

This equation has the same form as the analogous equation in the case of the non-relativistic hydrogen atom. It has as solutions confluent hypergeometric series multiplied by an exponential function and powers of ρ. In order that the wavefunction is normalisable, it is necessary that γ-k-1 is a non-negative integer, say, ν, or

$$\gamma = \nu + k + 1 \,, \quad \nu = 0, 1, 2, \ldots \,, \tag{12}$$

or, solving equation (7) for k,

$$\gamma = \nu + \tfrac{1}{2} + \sqrt{\{(l+\tfrac{1}{2})^2 - Z^2\alpha^2\}} \,, \quad \nu, l = 0, 1, \ldots \,. \tag{13}$$

From equations (8), (9), and (13) we now get for the energy eigenvalues, expanding in powers of $Z\alpha$

$$E = mc^2\left[1 - \frac{Z^2\alpha^2}{2n^2} - \frac{Z^4\alpha^4}{2n^4}\left(\frac{n}{l+\frac{1}{2}} - \tfrac{3}{4}\right) + \ldots\right] , \tag{14}$$

where $n = \nu + l + 1$ is the principal quantum number. The first term on the right-hand side of equation (14) is the rest-mass energy. The second term gives the non-relativistic energy levels of a particle in a Coulomb field. The third term gives a fine-structure splitting.

We have neglected here the finite size of the nucleus. As a pion is much heavier than an electron, the radius of its ground state orbit is about 270 times smaller than the Bohr radius, and this means that the correction for the finite nuclear size is an important one.

8. To find an equation of continuity we proceed as usual. We multiply equation (3) to the left by ψ_j^*, add to that equation the Hermitean conjugate of equation (3) multiplied to the right by ψ_j, and sum over j. The result is

$$\frac{1}{c}\frac{\partial}{\partial t}\Psi^\dagger\Psi + (\Psi^\dagger(\hat{\alpha}\,.\,\nabla)\Psi + (\nabla\Psi^\dagger)\hat{\alpha}^\dagger\Psi) + \frac{imc}{\hbar}(\Psi^\dagger\hat{\beta}\Psi - \Psi^\dagger\hat{\beta}^\dagger\Psi) = 0 \,. \tag{5}$$

As

$$\rho = e\Psi^{\dagger}\Psi , \tag{6}$$

we see that equation (5) reduces to the equation of continuity,

$$\frac{\partial \rho}{\partial t} + (\nabla \cdot j) = 0 , \tag{7}$$

if we take

$$j = ec\Psi^{\dagger}\hat{\alpha}\Psi , \tag{8}$$

provided

$$\hat{\alpha} = \hat{\alpha}^{\dagger} , \quad \hat{\beta} = \hat{\beta}^{\dagger} . \tag{9}$$

Let us now operate on equation (4) from the left with the operator

$$\frac{1}{c}\frac{\partial}{\partial t} - (\hat{\alpha} \cdot \nabla) - \frac{imc}{\hbar}\hat{\beta} . \tag{10}$$

The result is

$$\left\{\frac{1}{c^2}\frac{\partial^2}{\partial t^2} - \tfrac{1}{2}\sum_{k,\,l=1}^{3} (\hat{\alpha}_k\hat{\alpha}_l + \hat{\alpha}_l\hat{\alpha}_k)\partial_k\partial_l + \frac{m^2c^2}{\hbar^2}\hat{\beta}^2 - \frac{imc}{\hbar}\sum_{k=1}^{3} (\hat{\alpha}_k\hat{\beta} + \hat{\beta}\hat{\alpha}_k)\partial_k\right\} \Psi = 0 . \tag{11}$$

This equation reduces to the required equation,

$$\left\{\frac{1}{c^2}\frac{\partial^2}{\partial t^2} - \nabla^2 + \frac{m^2c^2}{\hbar^2}\right\}\Psi = 0 , \tag{12}$$

provided

$$\hat{\beta}^2 = \hat{1} , \quad [\hat{\alpha}, \hat{\beta}]_+ = 0 , \quad [\hat{\alpha}_k , \hat{\alpha}_l]_+ = 2\hat{1}\delta_{kl} , \tag{13}$$

where $\hat{1}$ is the unit matrix and $[\hat{A}, \hat{B}]_+ \equiv \hat{A}\hat{B} + \hat{B}\hat{A}$ is an anticommutator.

The first point to make is that $\hat{\alpha}$ and $\hat{\beta}$ must have at least four rows and four columns, if equations (13) and (9) are to be satisfied. Secondly, if we take 4×4 representations, there are several choices. We shall use in this section the following one:

$$\hat{\alpha} = \begin{pmatrix} 0 & \hat{\sigma} \\ \hat{\sigma} & 0 \end{pmatrix} , \quad \hat{\beta} = \begin{pmatrix} \hat{1} & 0 \\ 0 & -\hat{1} \end{pmatrix} , \tag{14}$$

where $\hat{\sigma}$ has as components the 2×2 Pauli matrices $\hat{\sigma}_x$, $\hat{\sigma}_y$, and $\hat{\sigma}_z$, and $\hat{1}$ is here the unit 2×2 matrix. We leave it as an exercise for the reader to prove that the matrices (14) satisfy equations (9) and (13).

9. As $\hat{\boldsymbol{\alpha}} = i\beta\hat{\boldsymbol{\gamma}}$, we find

$$j = iec\Psi^\dagger\hat{\beta}\hat{\boldsymbol{\gamma}}\Psi \,, \tag{4}$$

while

$$\rho = e\Psi^\dagger\hat{\beta}\hat{\beta}\Psi \,, \tag{5}$$

so that we can write

$$j_\mu = iec\Psi^\dagger\hat{\gamma}_4\hat{\gamma}_\mu\Psi \,, \tag{6}$$

or

$$j_\mu = iec\overline{\Psi}\hat{\gamma}_\mu\Psi \,, \tag{7}$$

where

$$\overline{\Psi} = \Psi^\dagger\hat{\gamma}_4 \tag{8}$$

is called the adjoint or Dirac conjugate of Ψ.

10. $$\sum_\mu \hat{\gamma}_\mu\hat{\gamma}_\mu = 4\,.\,\hat{1}\,, \quad \sum_\mu \hat{\gamma}_\mu\hat{\gamma}_\nu\hat{\gamma}_\mu = -2\hat{\gamma}_\nu\,,$$

$$\sum_\mu \hat{\gamma}_\mu\hat{\gamma}_\nu\hat{\gamma}_\rho\hat{\gamma}_\mu = 4\delta_{\nu\rho}\,\hat{1}\,, \quad \sum_\mu \hat{\gamma}_\mu\hat{\gamma}_\nu\hat{\gamma}_\rho\hat{\gamma}_\sigma\hat{\gamma}_\mu = -2\hat{\gamma}_\sigma\hat{\gamma}_\rho\hat{\gamma}_\nu\,.$$

12. As $p_4 = iE/c =$ constant, we can write

$$\Psi = \Phi(\boldsymbol{r})e^{-iEt/\hbar}\,, \tag{1}$$

and substitution into the Dirac equation for a free particle gives

$$E\Phi = \hat{H}_{\mathrm{D}}\Phi\,, \tag{2}$$

where

$$\hat{H}_{\mathrm{D}} \equiv c(\hat{\boldsymbol{\alpha}}\,.\,\hat{\boldsymbol{p}}) + mc^2\hat{\beta}\,. \tag{3}$$

If we consider the form of the matrices $\hat{\boldsymbol{\alpha}}$ and $\hat{\beta}$, it is clear that we can write them in the following form:

$$\hat{\boldsymbol{\alpha}} = \hat{\rho}_x\hat{\boldsymbol{\sigma}}\,, \quad \hat{\beta} = \hat{\rho}_z\hat{1}_\sigma\,. \tag{4}$$

To see how the $\hat{\rho}_k$ matrices operate we renumber the rows and columns of $\hat{\boldsymbol{\alpha}}, \hat{\beta}$ (and the components of the spinors Ψ and Φ) as follows:

$$1 \to 11\,, \quad 2 \to 12\,, \quad 3 \to 21\,, \quad 4 \to 22\,. \tag{5}$$

In that case the matrices $\hat{\boldsymbol{\sigma}}$ and $\hat{1}_\sigma$ in equations (4) are 2×2 matrices operating upon the second index and the $\hat{\rho}_k$ matrices (which have the same form as the Pauli matrices) operate on the first index. In terms of the $\hat{\rho}$ and $\hat{\boldsymbol{\sigma}}$ matrices equation (2) becomes

$$\hat{H}_{\mathrm{D}}\Phi \equiv \{c\hat{\rho}_x(\hat{\boldsymbol{\sigma}}\,.\hat{\boldsymbol{p}}) + \hat{\rho}_z\hat{1}_\sigma\}\Phi = E\Phi\,. \tag{6}$$

We are looking for states with a well-defined value of $\boldsymbol{p}$,

$$\Phi = \Phi_0 e^{i(\boldsymbol{p}\,.\,\boldsymbol{r})/\hbar} \,. \tag{7}$$

If we now introduce the spinors

$$\varphi \equiv \begin{pmatrix} \phi_1 \\ \phi_2 \end{pmatrix}, \quad \chi \equiv \begin{pmatrix} \phi_3 \\ \phi_4 \end{pmatrix}, \tag{8}$$

and use the fact that for the state described by the wavefunction (7) we have $\hat{\boldsymbol{p}}\Phi = \boldsymbol{p}\Phi$, we can write equation (6) in the form

$$\begin{aligned} (mc^2 - E)\varphi + c(\hat{\boldsymbol{\sigma}}\,.\,\boldsymbol{p})\chi &= 0\,, \\ c(\hat{\boldsymbol{\sigma}}\,.\,\boldsymbol{p})\varphi - (mc^2 + E)\chi &= 0\,. \end{aligned} \tag{9}$$

We note that these equations have a solution only, if

$$\begin{vmatrix} mc^2 - E & c(\hat{\boldsymbol{\sigma}}\,.\,\boldsymbol{p}) \\ -c(\hat{\boldsymbol{\sigma}}\,.\,\boldsymbol{p}) & mc^2 + E \end{vmatrix} = 0\,, \tag{10}$$

or, using the results of problem 15 of section 5

$$E = \pm E_p\,, \quad E_p = c\sqrt{(p^2 + m^2c^2)}\,. \tag{11}$$

If we introduce the sign operator $\hat{\Lambda}$ through the equation

$$\hat{\Lambda} = \frac{\hat{H}_{\mathrm{D}}}{\sqrt{\hat{H}_{\mathrm{D}}^2}} = \frac{c(\hat{\boldsymbol{\alpha}}\,.\,\hat{\boldsymbol{p}}) + \hat{\beta}mc^2}{c\sqrt{(p^2 + m^2c^2)}}\,, \tag{12}$$

we see, first of all, that it commutes with $\hat{H}_{\mathrm{D}}$ and that $\hat{\Lambda}^2 = \hat{1}$ so that its eigenvalues are ± 1; in fact, the eigenvalues are E/E_p. The eigenfunctions with eigenvalue $+1$ are called the positive energy eigenfunctions and those with eigenvalue -1 the negative energy eigenfunctions.

We have now found wavefunctions which are simultaneously eigenfunctions of $\hat{H}_{\mathrm{D}}$ (eigenvalue E), $\hat{\boldsymbol{p}}$ (eigenvalue $\boldsymbol{p}$), and $\hat{\Lambda}$ (eigenvalue E/E_p). From equations (9) we see that we can express one of the two two-component functions in terms of the other one. If we write

$$\varphi = Ae^{i(\boldsymbol{p}\,.\,\boldsymbol{r})/\hbar}u\,, \quad u \equiv \begin{pmatrix} u_1 \\ u_2 \end{pmatrix}, \tag{13}$$

u will be a spinor of the kind discussed in section 4, which can be acted upon by the Pauli spin matrices. From equation (9) it follows that if φ is given by equation (13), we have for χ

$$\chi = A\frac{c(\hat{\boldsymbol{\sigma}}\,.\,\boldsymbol{p})}{mc^2 + E}e^{i(\boldsymbol{p}\,.\,\boldsymbol{r})/\hbar}u\,, \tag{14}$$

and hence the wavefunction Ψ is given by the expression

$$\Psi = A\begin{pmatrix} u \\ \dfrac{c(\hat{\sigma}\,.\,\boldsymbol{p})}{mc^2+E}u \end{pmatrix} e^{i(\boldsymbol{p}\,.\,\boldsymbol{r})/\hbar - iEt/\hbar}, \tag{15}$$

where A is a normalisation constant. If the particle is confined to unit volume, we find

$$A^2 = \frac{E_p + \lambda mc^2}{(2\pi\hbar)^3 2E_p}, \quad \lambda = \frac{E}{E_p}, \tag{16}$$

provided u is normalised by the relation

$$u^\dagger u = 1\,. \tag{17}$$

Let us now consider the helicity operator $\hat{h}$ defined by the equation

$$\hat{h} = (\hat{\Sigma}\,.\,\boldsymbol{n}), \tag{18}$$

where $\boldsymbol{n} = \boldsymbol{p}/p$ and

$$\hat{\Sigma} = \hat{1}_\rho \hat{\sigma}, \tag{19}$$

where the unit operator $\hat{1}_\rho$ here operates on the same index as the $\hat{\rho}_k$. If we choose the z-axis along $\boldsymbol{n}$, $\hat{h}$ has the form

$$\hat{h} = \begin{pmatrix} 1 & 0 & 0 & 0 \\ 0 & -1 & 0 & 0 \\ 0 & 0 & 1 & 0 \\ 0 & 0 & 0 & -1 \end{pmatrix}. \tag{20}$$

The helicity operator commutes with $\hat{H}_\mathrm{D}$ so that we can find wavefunctions which are simultaneously eigenfunctions of $\hat{H}_\mathrm{D}, \hat{\boldsymbol{p}}, \hat{\Lambda}$, and $\hat{h}$. The eigenvalues of $\hat{h}$ are ± 1, and the eigenfunctions correspond to the spin of the particle being parallel or antiparallel to the momentum. The eigenvalue $+1$ of $\hat{h}$ corresponds to $u = \alpha \equiv \begin{pmatrix} 1 \\ 0 \end{pmatrix}$ and the eigenvalue -1 to $u = \beta \equiv \begin{pmatrix} 0 \\ 1 \end{pmatrix}$.

13. In the non-relativistic limit we can write (compare problem 2 of section 11)

$$E_p = mc^2 + E', \quad E' \approx p^2/2m \ll mc^2, \tag{1}$$

and hence in the case when $E = +E_p$ we find that

$$\chi = \frac{c(\hat{\sigma}\,.\,\boldsymbol{p})}{E' + 2mc^2}\varphi \approx \frac{(\hat{\sigma}\,.\,\boldsymbol{p})}{2mc}\varphi \ll \varphi, \tag{2}$$

while in the case when $E = -E_p$ we have

$$-\varphi = \frac{c(\hat{\sigma}.\boldsymbol{p})}{E' + 2mc^2}\chi \approx \frac{(\hat{\sigma}.\boldsymbol{p})}{2mc}\chi \ll \chi . \tag{3}$$

We see that in the non-relativistic limit we have two large and two small components; ψ_1 and ψ_2 are the large components in the positive energy case and ψ_3 and ψ_4 are the large components in the negative energy case.

In terms of the spinors φ and χ the charge and current densities have the form

$$\rho = e(\varphi^\dagger\varphi + \chi^\dagger\chi) , \quad \boldsymbol{j} = ec(\varphi^\dagger\hat{\sigma}\chi + \chi^\dagger\hat{\sigma}\varphi) . \tag{4}$$

In the non-relativistic limit we have for the positive energy case

$$\chi \approx \frac{(\hat{\sigma}.\hat{\boldsymbol{p}})}{2mc}\varphi = -i\hbar\frac{(\hat{\sigma}.\nabla)\varphi}{2mc} , \tag{5}$$

and hence

$$\rho = e(\varphi^\dagger\varphi + \chi^\dagger\chi) \approx e\varphi^\dagger\varphi - e\frac{\hbar^2}{4m^2c^2}\varphi^\dagger\nabla^2\varphi , \tag{6}$$

$$\boldsymbol{j} = ec(\varphi^\dagger\hat{\sigma}\chi + \chi^\dagger\hat{\sigma}\varphi) \approx \frac{e\hbar}{2mi}\{\varphi^\dagger\nabla\varphi - (\nabla\varphi^\dagger)\varphi\} + \frac{e\hbar}{2m}[\nabla \wedge \varphi^\dagger\hat{\sigma}\varphi] , \tag{7}$$

where we have used the relation

$$\hat{\sigma}(\hat{\sigma}.\nabla)\varphi = \nabla\varphi - i[\nabla \wedge \hat{\sigma}]\varphi . \tag{8}$$

The last term on the right-hand side of equation (7) is the current density due to the spin of the particle.

14. The procedure is similar to that in problem 1 of section 2. The Dirac equation is of the form

$$\hat{H}_{\mathrm{D}}\Phi = (E - e\phi)\Phi . \tag{3}$$

In the regions $z < 0$ and $z > 0$ we have, respectively, the solutions (compare equation (15) of problem 12 of section 11)

$$\Phi_{z<0} = A\begin{pmatrix} u \\ \dfrac{cp_1}{mc^2 + E_p}u \end{pmatrix}\exp(ip_1z/\hbar) + B\begin{pmatrix} u \\ \dfrac{-cp_1}{mc^2 + E_p}u \end{pmatrix}\exp(-ip_1z/\hbar) , \tag{4}$$

$$\Phi_{z>0} = C\begin{pmatrix} u \\ \dfrac{cp_2}{mc^2 + E_p - V_0}u \end{pmatrix}\exp(ip_2z/\hbar) , \tag{5}$$

where

$$E_p = c\sqrt{(p_1^2 + m^2c^2)}\,, \tag{6}$$

$$p_2^2c^2 + m^2c^4 = (E_p - V_0)^2\,. \tag{7}$$

The reflection coefficient R and the transmission coefficient T are given by the equations

$$R = \frac{|j_{\rm ref}|}{|j_{\rm inc}|}\,, \quad T = \frac{|j_{\rm tr}|}{|j_{\rm inc}|}\,, \tag{8}$$

where the current densities are the z-components of the expressions given by equation (8) of problem 8 of section 11; we find for them

$$j_{\rm ref} = -\frac{2c^2p_1}{mc^2+E_p}|B|^2\,, \tag{9}$$

$$j_{\rm inc} = \frac{2c^2p_1}{mc^2+E_p}|A|^2\,, \tag{10}$$

$$j_{\rm tr} = \frac{2c^2p_2}{mc^2+E_p-V_0}|C|^2\,, \tag{11}$$

while B and C can be expressed in terms of A by using the continuity of Φ and $\partial\Phi/\partial z$ at $z = 0$, whence we have

$$A + B = C\,, \tag{12}$$

$$A - B = \frac{p_2}{p_1}\frac{mc^2+E_p}{mc^2+E_p-V_0}C\,. \tag{13}$$

Combining equations (8) to (13) we find

$$R = \frac{(1-r)^2}{(1+r)^2}\,, \quad T = \frac{4r}{(1+r)^2}\,, \tag{14}$$

where

$$r = \frac{p_2}{p_1}\frac{mc^2+E_p}{mc^2+E_p-V_0}\,. \tag{15}$$

15. According to problem 27 of section 3 the velocity operator is given by the equation

$$\frac{d\hat{\boldsymbol{r}}}{dt} = \frac{1}{i\hbar}[\hat{\boldsymbol{r}}, \hat{H}_{\rm D}]_- = c\hat{\boldsymbol{\alpha}}\,. \tag{1}$$

as the eigenvalues of $\hat{\boldsymbol{\alpha}}$ are ± 1, we see that the particle moves with the velocity of light! However, $\hat{\boldsymbol{\alpha}}$ does not commute with $\hat{H}_{\rm D}$ so that any state of the particle will be a superposition of states with velocities in different directions. In order to integrate equation (1), we must find $d\hat{\alpha}/dt$ for which we find

$$\frac{d\hat{\alpha}}{dt} = \frac{1}{i\hbar}[\hat{\alpha}, \hat{H}_{\mathrm{D}}]_{-} = \frac{2i}{\hbar}(c\hat{\boldsymbol{p}} - \hat{\alpha}\hat{H}_{\mathrm{D}}) . \tag{2}$$

Looking at states which are eigenfunctions of $\hat{\boldsymbol{p}}$ and $\hat{H}_{\mathrm{D}}$ we can integrate equation (2):

$$\hat{\alpha}(t) = \frac{c\hat{\boldsymbol{p}}}{E} + \left\{\hat{\alpha}(0) - c\frac{\hat{\boldsymbol{p}}}{E}\right\} e^{-2iEt/\hbar} , \tag{3}$$

and hence

$$\hat{\boldsymbol{r}}(t) = \boldsymbol{r}(0) + \frac{c^2\boldsymbol{p}}{E}t + \frac{ic\hbar}{2E}\left\{\hat{\alpha}(0) - \frac{c\boldsymbol{p}}{E}\right\}\{e^{-2iEt/\hbar} - 1\} . \tag{4}$$

The first two terms on the right-hand side represent the normal classical motion. The last term on the right-hand side describes the so-called "Zitterbewegung", a very rapid, small amplitude motion. Its amplitude is less than the Compton wavelength $\hbar/mc$, which for an electron equals 4.10^{-11} cm, and its period less than $\hbar/mc^2$, which for an electron equals 10^{-21} s.

It is interesting to note that if we consider positive energy states only, which we can do by applying to the wavefunction a "projection" operator $\hat{\Pi}_{+}$ defined by the equation

$$\hat{\Pi}_{+} = \tfrac{1}{2}(\hat{1} + \hat{\Lambda}) , \tag{5}$$

the oscillating parts disappear. We leave the proof of this as an exercise to the reader. If we are considering positive energy states only, the eigenfunctions of the particle coordinate operator are no longer δ-functions, but are "smeared out" over a region with linear dimensions of the order of the Compton wavelength.

16. As the $\hat{\gamma}_\mu$ are numbers, they will not change under the transformation (1), while

$$\hat{p}'_\mu = \sum_\nu a_{\mu\nu}\hat{p}_\nu . \tag{3}$$

Hence, the Dirac equation

$$\left\{\sum_\mu \hat{\gamma}_\mu \hat{p}_\mu - imc\right\}\Psi(x_\mu) = 0 , \tag{4}$$

changes to

$$\left\{\sum_\mu \hat{\gamma}_\mu \hat{p}'_\mu - imc\right\}\Psi'(x'_\mu) = 0 , \tag{5}$$

where $\Psi'(x'_\mu)$ is a new function of the new independent variables x'_μ. We want to find a unitary transformation $\hat{S}$ such that if

$$\Psi'(x'_\mu) = \hat{S}\Psi , \tag{6}$$

equation (5) reduces to equation (4). Equation (5) can be written in the form

$$\left\{\sum_{\mu,\nu} \hat{\gamma}_\mu a_{\mu\nu}\hat{p}_\mu - imc\right\}\hat{S}\Psi = 0 , \tag{7}$$

or

$$\left\{\sum_{\mu,\nu} \hat{S}^{-1}\hat{\gamma}_\mu \hat{S} a_{\mu\nu}\hat{p}_\nu - imc\right\}\Psi = 0 , \tag{8}$$

and we see that $\hat{S}$ must satisfy the equation

$$\sum_\mu \hat{S}^{-1}\hat{\gamma}_\mu \hat{S} a_{\mu\nu} = \hat{\gamma}_\mu . \tag{9}$$

Once we have found $\hat{S}$ from equation (9), the transformed wavefunction follows from equation (6).

Let us now consider the special cases. The first two cases correspond to classes of continuous transformations so that we can confine ourselves to considering infinitesimal transformations, in which case

$$a_{\mu\nu} = \delta_{\mu\nu} + \epsilon_{\mu\nu} , \tag{10}$$

where $\epsilon_{\mu\nu}$ is an infinitesimal second-rank tensor. If the $a_{\mu\nu}$ are to satisfy condition (2), we have

$$\delta_{\nu\nu'} = \sum_\mu a_{\mu\nu}a_{\mu\nu'} = \delta_{\nu\nu'} + (\epsilon_{\nu\nu'} + \epsilon_{\nu'\nu}) , \tag{11}$$

which means that $\epsilon_{\mu\nu}$ must be an antisymmetric tensor.

(a) In the case of a transformation from one inertial frame to another moving with a velocity v in the x-direction we have (v/c is here assumed to be infinitesimal)

$$\epsilon_{14} = -\epsilon_{41} = iv/c ; \quad \epsilon_{\mu\nu} = 0 , \quad \mu, \nu \neq 1, 4 \text{ or } 4, 1 . \tag{12}$$

(b) In this case we have ($\delta\varphi$ infinitesimal)

$$\epsilon_{12} = -\epsilon_{21} = \delta\varphi ; \quad \epsilon_{\mu\nu} = 0 , \quad \mu, \nu \neq 1, 2 \text{ or } 2, 1 . \tag{13}$$

In general we must find $\hat{S}$ from equation (9) with $a_{\mu\nu}$ given by equation (10). We expect that for an infinitesimal transformation

$$\hat{S} = \hat{1} + \tfrac{1}{2}\sum_{\mu,\nu} \hat{C}_{\mu\nu}\epsilon_{\mu\nu} , \tag{14}$$

and substituting this expression into equation (9) we find that

$$\tfrac{1}{2}\sum_{\lambda,\nu} [\hat{\gamma}_\mu , \hat{C}_{\lambda\nu}]_- = \sum_\nu \epsilon_{\mu\nu}\hat{\gamma}_\nu . \tag{15}$$

Writing the right-hand side of this equation in the form

$$\sum_{\nu} \epsilon_{\mu\nu}\hat{\gamma}_{\nu} = \tfrac{1}{2}\sum_{\lambda,\nu} \epsilon_{\mu\nu}\{\delta_{\lambda\mu}\hat{\gamma}_{\nu} - \delta_{\nu\mu}\hat{\gamma}_{\lambda}\}, \tag{16}$$

equation (15) becomes

$$\sum_{\lambda,\nu}\{\hat{\gamma}_{\mu}\hat{C}_{\lambda\nu} - \hat{C}_{\lambda\nu}\hat{\gamma}_{\mu} - \delta_{\lambda\mu}\hat{\gamma}_{\nu} + \delta_{\nu\mu}\hat{\gamma}_{\lambda}\}\epsilon_{\lambda\nu} = 0\,, \tag{17}$$

which is satisfied by

$$\hat{C}_{\lambda\nu} = \tfrac{1}{2}\hat{\gamma}_{\lambda}\hat{\gamma}_{\nu}\,, \tag{18}$$

as can be seen on inspection. Hence we find from equation (14) for an infinitesimal transformation

$$\hat{S} = \hat{1} + \tfrac{1}{4}\sum_{\mu,\nu}\epsilon_{\mu\nu}\hat{\gamma}_{\mu}\hat{\gamma}_{\nu}\,. \tag{19}$$

For instance, for a rotation around the z-axis we find, using equations (13) and the fact that $\hat{\gamma}_1\hat{\gamma}_2 = -\hat{\gamma}_2\hat{\gamma}_1 = i\hat{1}_{\rho}\hat{\sigma}_z$,

$$\hat{S} = \hat{1} + \tfrac{1}{2}i\delta\varphi\hat{1}_{\rho}\hat{\sigma}_z\,, \tag{20}$$

and for a finite transformation

$$\hat{S} = \exp(\tfrac{1}{2}i\varphi\hat{1}_{\rho}\hat{\sigma}_z)\,. \tag{21}$$

Similarly we find for the change from one inertial system to another using equations (12) and the fact that $\hat{\gamma}_1\hat{\gamma}_4 = -\hat{\gamma}_4\hat{\gamma}_1 = i\hat{\rho}_x\hat{\sigma}_x$,

$$\hat{S} = \hat{1} - \tfrac{1}{2}\frac{v}{c}\hat{\rho}_x\hat{\sigma}_x\,. \tag{22}$$

For a finite transformation with $\boldsymbol{v}$ in an arbitrary direction we find in a similar way

$$\hat{S} = \exp\{-\tfrac{1}{2}\varphi(\boldsymbol{n}\,.\,\hat{\boldsymbol{\alpha}})\}\,, \tag{23}$$

where $\varphi = \operatorname{artanh}(v/c)$ and $\boldsymbol{n} = \boldsymbol{v}/v$.

(c) In the case of an inversion we do not have an infinitesimal transformation. In this case

$$a_{\mu\nu} = \delta_{\mu\nu}a_{\mu}\,, \quad a_{11} = a_{22} = a_{33} = -1\,, \quad a_{44} = 1\,. \tag{24}$$

(d) In the case of time reversal we also do not have an infinitesimal transformation and now

$$a_{\mu\nu} = \delta_{\mu\nu}a_{\mu}\,, \quad a_{11} = a_{22} = a_{33} = 1\,, \quad a_{44} = -1\,. \tag{25}$$

In case (c) equations (9) become

$$\hat{\gamma}_k\hat{S} = -\hat{S}\hat{\gamma}_k\,, \quad k = 1, 2, 3\,; \quad \hat{\gamma}_4\hat{S} = \hat{S}\hat{\gamma}_4\,, \tag{26}$$

or

$$\hat{S} = \lambda\hat{\gamma}_4 , \quad |\lambda| = 1 . \tag{27}$$

In case (d) equations (9) become

$$\hat{\gamma}_k\hat{S} = \hat{S}\hat{\gamma}_k , \quad k = 1, 2, 3 ; \quad \hat{\gamma}_4\hat{S} = -\hat{S}\hat{\gamma}_4 , \tag{28}$$

or, using equations (4) of problem 12 of section 11 and (2) of problem 9 of section 11, from which it follows that

$$\hat{\gamma} = \hat{\rho}_y\hat{\sigma} , \quad \hat{\gamma}_4 = \hat{\rho}_z\hat{1}_\sigma , \tag{29}$$

we have

$$\hat{S} = \mu\hat{\rho}_y\hat{1}_\sigma , \quad |\mu| = 1 . \tag{30}$$

17. The proof follows, if one uses equation (23) from the preceding problem and the fact that if the particle momentum is $\boldsymbol{p}$, the velocity of the rest frame with respect to the laboratory frame is given by the equation

$$v = -\boldsymbol{p}/E . \tag{1}$$

18. Before considering the bilinear forms we must first find out how $\overline{\Psi}$ transforms. We have seen that

$$\Psi'(x'_\mu) = \hat{S}\Psi(x_\mu) . \tag{1}$$

Taking the Hermitean conjugate of this equation we find

$$\Psi'^\dagger(x'_\mu) = \Psi^\dagger(x_\mu)\hat{S}^\dagger , \tag{2}$$

and from that equation we get, using the definition of the adjoint,

$$\overline{\Psi}'(x'_\mu) = \Psi'^\dagger(x'_\mu)\hat{\gamma}_4 = \overline{\Psi}(x_\mu)\hat{\gamma}_4\hat{S}^\dagger\hat{\gamma}_4 . \tag{3}$$

The operator $\hat{\gamma}_4\hat{S}^\dagger\hat{\gamma}_4$ can be rewritten. Consider equation (9) of problem 16 of section 11 and use equation (2) of the same problem. We get

$$\hat{S}^{-1}\hat{\gamma}_\mu\hat{S} = \sum_\nu a_{\mu\nu}\hat{\gamma}_\nu . \tag{4}$$

Taking the Hermitean conjugate of equation (4) for $\mu = 4$ and using the fact that the a_{4k} are purely imaginary for $k = 1, 2, 3$ and a_{44} is real, we find

$$(\hat{S}^{-1}\hat{\gamma}_4\hat{S})^\dagger = a_{44}\hat{\gamma}_4 - \sum_{k=1}^{3} a_{4k}\hat{\gamma}_k . \tag{5}$$

Multiplying equation (5) from the right by $\hat{\gamma}_4$ and using the fact that $[\hat{\gamma}_k , \hat{\gamma}_4]_+ = 2\delta_{k4}\hat{1}$, we find

$$(\hat{S}^{-1}\hat{\gamma}_4\hat{S})^\dagger\hat{\gamma}_4 = \sum_\mu \hat{\gamma}_4 a_{4\mu}\hat{\gamma}_\mu , \tag{6}$$

or, using equation (4)

$$(\hat{S}^{-1}\hat{\gamma}_4\hat{S})^\dagger\hat{\gamma}_4 = \hat{\gamma}_4\hat{S}^{-1}\hat{\gamma}_4\hat{S}\,, \tag{7}$$

or, using the fact that $\hat{\gamma}_4 = \hat{\gamma}_4^{-1}$,

$$(\hat{\gamma}_4\hat{S}^\dagger\hat{\gamma}_4)\hat{\gamma}_4(\hat{\gamma}_4\hat{S}^\dagger\hat{\gamma}_4)^{-1} = \hat{S}^{-1}\hat{\gamma}_4\hat{S}\,, \tag{8}$$

whence we find

$$\hat{\gamma}_4\hat{S}^\dagger\hat{\gamma}_4 = \lambda\hat{S}^{-1}\,, \quad \lambda = +1 \text{ or } -1\,. \tag{9}$$

Consider now the identity

$$\hat{S}^\dagger\hat{S} = \hat{S}^\dagger\hat{\gamma}_4\hat{\gamma}_4\hat{S} = \lambda\hat{\gamma}_4\hat{S}^{-1}\hat{\gamma}_4\hat{S}\,, \tag{10}$$

or

$$\hat{S}^\dagger\hat{S} = \lambda\left\{a_{44}\hat{1} + \sum_{k=1}^{3} a_{4k}\hat{\gamma}_k\right\}\,. \tag{11}$$

Taking the trace, and using the fact that $\mathrm{Tr}\hat{S}^\dagger\hat{S} > 0$, we have

$$\mathrm{Tr}\hat{S}^\dagger\hat{S} = \lambda a_{44} > 0\,, \tag{12}$$

and hence

$$\begin{aligned}\hat{\gamma}_4\hat{S}^\dagger\hat{\gamma}_4 &= \hat{S}^{-1}\,, \quad \text{if } a_{44} > 0\,,\\ &= -\hat{S}^{-1}, \quad \text{if } a_{44} < 0\,.\end{aligned} \tag{13}$$

We can now study the transformation properties of the bilinear forms $\overline{\Psi}\hat{\Gamma}_j\Psi$, and we shall consider them for the infinitesimal transformations for which (see problem 16 of section 11)

$$\hat{S} = \hat{1} + \tfrac{1}{4}\sum_{\mu,\nu}\epsilon_{\mu\nu}\hat{\gamma}_\mu\hat{\gamma}_\nu \tag{14}$$

and

$$\hat{S}^{-1} = \hat{1} - \tfrac{1}{4}\sum_{\mu,\nu}\epsilon_{\mu\nu}\hat{\gamma}_\mu\hat{\gamma}_\nu\,, \tag{15}$$

or for inversions for which

$$\hat{S} = \lambda\hat{\gamma}_4\,, \quad |\lambda| = 1\,, \tag{16}$$

and hence

$$\hat{S}^{-1} = \lambda^{-1}\hat{\gamma}_4\,. \tag{17}$$

(a) G_1: We have

$$G_1' = \overline{\Psi}'\Psi = \overline{\Psi}\hat{S}^{-1}\hat{S}\Psi = G_1\,: \tag{18}$$

G_1 is a scalar.

(b) G_2 to G_5: Using equation (4), we have

$$G'_\nu = \overline{\Psi}'\hat{\gamma}_\nu\Psi' = \overline{\Psi}\hat{S}^{-1}\hat{\gamma}_\nu\hat{S}\Psi = \sum_\mu a_{\nu\mu}\overline{\Psi}\hat{\gamma}_\mu\Psi = \sum_\mu a_{\nu\mu}G_\mu : \qquad (19)$$

the G_ν transform as a four-vector.

(c) G_6 to G_{11}: We consider the symmetric and anti-symmetric combinations $G_{\mu\nu} \pm G_{\nu\mu}$ ($G_{\mu\nu} \equiv \overline{\Psi}\hat{\gamma}_\mu\hat{\gamma}_\nu\Psi$). As $\hat{\gamma}_\mu\hat{\gamma}_\nu + \hat{\gamma}_\nu\hat{\gamma}_\mu = 2\delta_{\nu\mu}\hat{1}$, the symmetric combination reduces to G_1. For the anti-symmetric combination we find

$$\begin{aligned}(G_{\mu\nu} - G_{\nu\mu})' &= \overline{\Psi}'[\hat{\gamma}_\mu, \hat{\gamma}_\nu]_-\Psi' = \overline{\Psi}\hat{S}^{-1}[\hat{\gamma}_\mu, \hat{\gamma}_\nu]_-\hat{S}\Psi \\ &= \overline{\Psi}[\hat{S}^{-1}\hat{\gamma}_\mu\hat{S}, \hat{S}^{-1}\hat{\gamma}_\nu\hat{S}]_-\Psi \\ &= \sum_{\rho,\sigma} a_{\mu\rho}a_{\nu\sigma}\overline{\Psi}[\hat{\gamma}_\rho, \hat{\gamma}_\sigma]_-\Psi = \sum_{\rho,\sigma} a_{\mu\rho}a_{\nu\sigma}(G_{\rho\sigma} - G_{\sigma\rho}) : \qquad (20)\end{aligned}$$

the anti-symmetric combinations transform as a second-rank tensor.

(d) G_{12} to G_{15}: If we introduce the matrix $\hat{\gamma}_5$ and use the fact that $\hat{\gamma}_\nu^{-1} = \hat{\gamma}_\nu$, we can write

$$G_{\mu\nu\rho} = \pm\overline{\Psi}\hat{\gamma}_5\hat{\gamma}_\sigma\Psi , \qquad (21)$$

where $\hat{\gamma}_\sigma$ is the one $\hat{\gamma}$-matrix which was not included in the product $\hat{\gamma}_\mu\hat{\gamma}_\nu\hat{\gamma}_\rho$ and where the sign depends on the order of the $\hat{\gamma}$-matrices. In fact, if we introduce the four-component quantity $\widetilde{G}_\mu$ with the following components

$$\begin{aligned}\widetilde{G}_1 &= \overline{\Psi}\hat{\gamma}_2\hat{\gamma}_3\hat{\gamma}_4\Psi , \\ \widetilde{G}_2 &= \overline{\Psi}\hat{\gamma}_4\hat{\gamma}_3\hat{\gamma}_1\Psi , \\ \widetilde{G}_3 &= \overline{\Psi}\hat{\gamma}_1\hat{\gamma}_2\hat{\gamma}_4\Psi , \\ \widetilde{G}_4 &= \overline{\Psi}\hat{\gamma}_3\hat{\gamma}_2\hat{\gamma}_1\Psi ,\end{aligned} \qquad (22)$$

we see that

$$\widetilde{G}_\mu = \overline{\Psi}\hat{\gamma}_\mu\hat{\gamma}_5\Psi . \qquad (23)$$

We then have

$$\widetilde{G}'_\mu = \overline{\Psi}'\hat{\gamma}_\mu\hat{\gamma}_5\Psi' = \overline{\Psi}\hat{S}^{-1}\hat{\gamma}_\mu\hat{\gamma}_5\hat{S}\Psi . \qquad (24)$$

From the definition of $\hat{\gamma}_5$ it follows that, if $\hat{S}$ is given by equation (14),

$$\hat{\gamma}_5\hat{S} = \hat{S}\hat{\gamma}_5 , \qquad (25)$$

and hence

$$\begin{aligned}\widetilde{G}'_\mu &= \overline{\Psi}\hat{S}^{-1}\hat{\gamma}_\mu\hat{S}\hat{\gamma}_5\Psi \\ &= \sum_\nu a_{\mu\nu}\overline{\Psi}\hat{\gamma}_\nu\hat{\gamma}_5\Psi = \sum_\nu a_{\mu\nu}\widetilde{G}_\nu : \qquad (26)\end{aligned}$$

under the transformation (14) the G_μ thus behave like a four-vector. However, if we consider the transformation (16), we have

$$\widetilde{G}'_\mu = \overline{\Psi}'\hat{\gamma}_\mu\hat{\gamma}_5\Psi' = \overline{\Psi}\hat{S}^{-1}\hat{\gamma}_\mu\hat{\gamma}_5\hat{S}\Psi = \overline{\Psi}\lambda^{-1}\hat{\gamma}_4\hat{\gamma}_5\hat{\gamma}_\mu\lambda\hat{\gamma}_4\Psi \, . \qquad (27)$$

From the definition of $\hat{\gamma}_5$, equation (29) of problem 16 of section 11 and the result of problem 15 of section 4 it follows that

$$\hat{\gamma}_5 = \hat{\gamma}_1\hat{\gamma}_2\hat{\gamma}_3\hat{\gamma}_4 = \hat{\rho}_y\hat{\sigma}_x\hat{\rho}_y\hat{\sigma}_y\hat{\rho}_y\hat{\sigma}_z\hat{\rho}_z\hat{1}_\sigma = -\hat{\rho}_x\hat{1}_\sigma \, , \qquad (28)$$

and hence we get from equation (27)

$$\widetilde{G}'_k = \widetilde{G}_k \, , \quad k = 1, 2, 3 \, ; \quad \widetilde{G}'_4 = -\widetilde{G}_4 : \qquad (29)$$

the $\widetilde{G}_\mu$ behave like a pseudo-vector.

(e) G_{16}: We now have

$$G'_{16} = \overline{\Psi}'\hat{\gamma}_5\Psi' = \overline{\Psi}\hat{S}^{-1}\hat{\gamma}_5\hat{S}\Psi \, . \qquad (30)$$

Under the transformation (14) we find, using equation (25) that

$$G'_{16} = \overline{\Psi}\hat{\gamma}_5\Psi = G_{16} \, , \qquad (31)$$

while under the transformation (16) we find

$$G'_{16} = \overline{\Psi}\hat{\gamma}_4\hat{\gamma}_5\hat{\gamma}_4\Psi = -G_{16} : \qquad (32)$$

G_{16} behaves like a pseudo-scalar.

Finally we note that under time-reversal, when $\hat{S}$ is given by equation (30) of problem 16 of section 11, the $\widetilde{G}_\mu$ and G_{16} transform as follows:

$$\widetilde{G}'_k = -\widetilde{G}_k \, , \quad k = 1, 2, 3 \, ; \quad \widetilde{G}'_4 = \widetilde{G}_4 \, , \qquad (33)$$

$$G'_{16} = -G_{16} \, . \qquad (34)$$

19. Let $\hat{S}_c$ be the operator which transforms Ψ^* into Ψ_c:

$$\Psi_c = \hat{S}_c\Psi^* \, . \qquad (3)$$

If we take the complex conjugate of equation (1) and bear in mind that $\hat{p}_4$ and $\boldsymbol{A}$ are real and $\hat{\boldsymbol{p}}$ and A_4 purely imaginary, we find that

$$\left[\left(\hat{\boldsymbol{\gamma}}^* \cdot \left\{\hat{\boldsymbol{p}} + \frac{e}{c}\boldsymbol{A}\right\}\right) - \hat{\gamma}_4^*\left(\hat{p}_4 + \frac{e}{c}A_4\right) - imc\right]\Psi^* = 0 \, . \qquad (4)$$

Comparing this with equation (2) we see that $\hat{S}_c$ must satisfy the equations

$$\hat{\boldsymbol{\gamma}} = \hat{S}_c^{-1}\hat{\boldsymbol{\gamma}}^*\hat{S}_c \, , \quad \hat{\gamma}_4 = -\hat{S}_c^{-1}\hat{\gamma}_4^*\hat{S}_c \, , \qquad (5)$$

or, bearing in mind that $\hat{\gamma}_2$ and $\hat{\gamma}_4$ are real and $\hat{\gamma}_1$ and $\hat{\gamma}_3$ are purely imaginary,

$$[\hat{S}_c, \hat{\gamma}_1]_+ = [\hat{S}_c, \hat{\gamma}_3]_+ = [\hat{S}_c, \hat{\gamma}_4]_+ = 0\,, \quad [\hat{S}_c, \hat{\gamma}_2]_- = 0\,, \tag{6}$$

whence we find that

$$\hat{S}_c = \hat{\gamma}_2\,. \tag{7}$$

20. One can show that the total angular momentum $\hat{\boldsymbol{J}}$, given by the equation

$$\hat{\boldsymbol{J}} = \hat{\boldsymbol{L}} + \hat{\boldsymbol{S}}\,, \tag{1}$$

with $\hat{\boldsymbol{L}}$ the orbital angular momentum,

$$\hat{\boldsymbol{L}} = [\boldsymbol{r} \wedge \hat{\boldsymbol{p}}]\,, \tag{2}$$

and $\hat{\boldsymbol{S}}$ the "spin angular momentum",

$$\hat{\boldsymbol{S}} = \tfrac{1}{2}\hbar\hat{\boldsymbol{\sigma}}\,, \tag{3}$$

commutes with $\hat{H}_D$. This means that a particle satisfying the Dirac equation has apart from its orbital angular momentum also an intrinsic angular momentum or spin—it is a spin-$\tfrac{1}{2}$ particle.

21. We consider the Dirac equation in the form similar to equations (9) of problem 12 of section 11:

$$(E - e\phi - mc^2)\varphi - c\left(\hat{\boldsymbol{\sigma}}\,.\,\left\{\hat{\boldsymbol{p}} - \frac{e}{c}\boldsymbol{A}\right\}\right)\chi = 0\,, \tag{1a}$$

$$c\left(\hat{\boldsymbol{\sigma}}\,.\,\left\{\hat{\boldsymbol{p}} - \frac{e}{c}\boldsymbol{A}\right\}\right)\varphi - (E - e\phi + mc^2)\chi = 0\,. \tag{1b}$$

Under the assumptions of the problem, and writing

$$E = E' + mc^2\,, \tag{2}$$

we have

$$|E' - e\phi| \ll mc^2\,. \tag{3}$$

From equation (1b) we then get

$$\chi = \frac{c\left(\hat{\boldsymbol{\sigma}}\,.\,\left\{\hat{\boldsymbol{p}} - \frac{e}{c}\boldsymbol{A}\right\}\right)}{E' + 2mc^2 - e\phi}\varphi \approx \frac{1}{2mc}\left(\hat{\boldsymbol{\sigma}}\,.\,\left\{\hat{\boldsymbol{p}} - \frac{e}{c}\boldsymbol{A}\right\}\right)\varphi\,, \tag{4}$$

or, after substituting into equation (1a)

$$E'\varphi = \left[\frac{\left(\hat{\boldsymbol{\sigma}}\,.\,\left\{\hat{\boldsymbol{p}} - \frac{e}{c}\boldsymbol{A}\right\}\right)^2}{2m} + e\phi\right]\varphi\,. \tag{5}$$

Using the result of problem 15 of section 4 we find

$$\left(\hat{\sigma}\,.\left\{\hat{p}-\frac{e}{c}A\right\}\right)^2 = \left(\hat{p}-\frac{e}{c}A\right)^2-\frac{e\hbar}{c}(\hat{\sigma}\,.[\nabla\wedge A])\,, \tag{6}$$

and using the fact that $\mathcal{H} = [\nabla\wedge A]$, we find from equation (5)

$$E'\varphi = \left[\frac{\left(\hat{p}-\frac{e}{c}A\right)^2}{2m}+e\phi-\frac{e\hbar}{2mc}(\hat{\sigma}\,.\mathcal{H})\right]\varphi\,. \tag{7}$$

We find an extra term:

$$-(\hat{\mu}\,.\mathcal{H}) = \mu_{\mathrm{B}}(\hat{\sigma}\,.\mathcal{H})\,, \tag{8}$$

where

$$\mu_{\mathrm{B}} = \frac{e\hbar}{2mc} = \text{Bohr magneton}\,. \tag{9}$$

The magneto-gyric ratio is clearly e/mc. Equation (7) was first suggested by Pauli and is called the Pauli equation.

22. We now put in equations (1) of the preceding problem $A = 0$, $e\phi = V(r)$, and using equation (2) of that problem we have

$$\{E'-V(r)\}\varphi - c(\hat{\sigma}\,.\hat{p})\chi = 0\,, \tag{1a}$$

$$c(\hat{\sigma}\,.\hat{p})\varphi - \{E'+2mc^2-e\phi\}\chi = 0\,. \tag{1b}$$

Eliminating χ we find

$$(E'-V)\varphi = \frac{(\hat{\sigma}\,.\hat{p})}{2m}\left(1-\frac{E'-V}{2mc^2}\right)(\hat{\sigma}\,.\hat{p})\varphi\,. \tag{2}$$

This equation can be written in the form

$$E'\varphi = \hat{H}'\varphi\,. \tag{3}$$

Before determining $\hat{H}'$ we draw attention to the fact that, if the original four-component wavefunction Φ were normalised, the two-component wavefunction φ will not be normalised. In order to see immediately from the form of $\hat{H}'$ what are the corrections as compared to the non-relativistic case, it is convenient to introduce a transformation which changes φ into a normalised function $\tilde{\varphi}$. This can be done as follows. To first order we have $\chi = (\hat{\sigma}\,.\hat{p})\varphi/2mc$, and thus we get, up to order $(v/c)^2$ from the normalisation condition

$$\int \rho d^3r = e\int(\varphi^\dagger\varphi+\chi^\dagger\chi)d^3r = e \tag{4}$$

the relation

$$\int \varphi^\dagger \left(1+\frac{\hat{\boldsymbol{p}}^2}{4m^2c^2}\right)\varphi d^3\boldsymbol{r} = 1\,. \tag{5}$$

As we want $\widetilde{\varphi}$ to satisfy the relation

$$\int \widetilde{\varphi}^\dagger \widetilde{\varphi} d^3\boldsymbol{r} = 1\,, \tag{6}$$

we find that up to order $(v/c)^2$

$$\widetilde{\varphi} = \left(1+\frac{\hat{\boldsymbol{p}}^2}{8m^2c^2}\right)\varphi\,. \tag{7}$$

Transforming equation (3) we then find

$$E'\widetilde{\varphi} = \hat{H}\widetilde{\varphi}\,, \tag{8}$$

with

$$\hat{H} = \left(1+\frac{\hat{p}^2}{8m^2c^2}\right)\hat{H}'\left(1-\frac{\hat{p}^2}{8m^2c^2}\right), \tag{9}$$

or

$$\hat{H} = \left(1+\frac{\hat{p}^2}{8m^2c^2}\right)\left\{\frac{(\hat{\sigma}\,.\hat{\boldsymbol{p}})}{2m}\left(1-\frac{E'-V}{2mc^2}\right)(\hat{\sigma}\,.\hat{\boldsymbol{p}})+V\right\}\left(1-\frac{\hat{p}^2}{8m^2c^2}\right). \tag{10}$$

As

$$(\hat{\sigma}\,.\hat{\boldsymbol{p}})f(\boldsymbol{r})(\hat{\sigma}\,.\hat{\boldsymbol{p}}) = f(\boldsymbol{r})\hat{\boldsymbol{p}}^2 - i\hbar\{(\nabla f\,.\hat{\boldsymbol{p}})+i(\hat{\sigma}\,.[\nabla f \wedge \hat{\boldsymbol{p}}])\}\,, \tag{11}$$

we find

$$\hat{H} = \frac{\hat{\boldsymbol{p}}^2}{2m}+V(\boldsymbol{r})-\frac{(E'-V(\boldsymbol{r}))^2}{2mc^2}-\frac{e\hbar(\hat{\sigma}\,.[\mathcal{E} \wedge \hat{\boldsymbol{p}}])}{4m^2c^2}-\frac{e\hbar^2}{8m^2c^2}(\nabla\,.\,\mathcal{E})\,, \tag{12}$$

where

$$\mathcal{E} = -e\nabla V \tag{13}$$

is the electrostatic field.

The first two terms on the right-hand side of equation (12) are the non-relativistic terms. The third term can be written to a first approximation as $p^4/8m^3c^2$ which comes from the relativistic mass correction. The fourth term can, for a central force, when $\nabla V = (\boldsymbol{r}/r)(dV/dr)$, be written in the form

$$\Delta\hat{H}_2 = \frac{dV}{dr}\frac{(\hat{\boldsymbol{S}}\,.\hat{\boldsymbol{L}})}{2m^2c^2r}\,, \tag{14}$$

where $\hat{S}$ and $\hat{L}$ are the spin- and orbital angular momentum operators. This term is called the spin-orbit interaction. The last term comes into play only at those positions where there are charges; for instance, in the case of a Coulomb field, $V \propto 1/r$ and $\Delta\hat{H}_2 \propto \delta(\boldsymbol{r})$.

23. Instead of starting from the unperturbed wavefunctions with well-determined values of l_z and s_z and later solving the secular equation, it is convenient to choose as initial functions the eigenfunctions with well-defined values of $\boldsymbol{l}^2$ and $\boldsymbol{j}^2$, where $\hat{\boldsymbol{j}} = \hat{\boldsymbol{l}} + \hat{\boldsymbol{s}}$ is the total angular momentum which can easily be shown to commute with $\Delta\hat{H}_2$ of the preceding problem. If we take into account that for these functions the following relation holds

$$\hat{\boldsymbol{j}}^2 = j(j+1) = l(l+1) + s(s+1) + 2(\hat{\boldsymbol{l}}\,.\,\hat{\boldsymbol{s}})\,, \tag{1}$$

we find that

$$\Delta E_2 = \overline{\Delta\hat{H}_2} = \frac{j(j+1)-l(l+1)-s(s+1)}{4m^2c^2}\hbar^2\,\overline{\frac{1}{r}\frac{dV}{dr}}\,. \tag{2}$$

For the hydrogen atom $V = -e^2/r$ and since (see problem 17 of section 5)

$$\overline{\frac{1}{r^3}} = \frac{1}{n^3(l+1)(l+\frac{1}{2})l}\left(\frac{me^2}{\hbar^2}\right)^3, \tag{3}$$

we find finally

$$\Delta E_2 = \frac{me^4}{\hbar^2}\left(\frac{e^2}{\hbar c}\right)^2\frac{j(j+1)-l(l+1)-s(s+1)}{4n^3 l(l+\frac{1}{2})(l+1)}\,. \tag{4}$$

This formula can be written more succinctly, as in our case $s = \frac{1}{2}$ and we have either $j = l-\frac{1}{2}$ or $j = l+\frac{1}{2}$. One can easily show that

$$2(\boldsymbol{l}\,.\,\boldsymbol{s}) = j(j+1)-l(l+1)-s(s+1) = \begin{cases} l\,, & \text{if } j = l+\frac{1}{2}\,,\\ -l-1\,, & \text{if } j = l-\frac{1}{2}\,, \end{cases} \tag{5}$$

so that for all values of j and l

$$\Delta E_2 = \frac{me^4}{\hbar^2}\left(\frac{e^2}{\hbar c}\right)^2\frac{1}{2n^3}\left[-\frac{1}{j+\frac{1}{2}}+\frac{1}{l+\frac{1}{2}}\right]. \tag{6}$$

Combining ΔE_2 with the correction which describes the relativistic correction to the mass, due to the particle's velocity (see problem 5 of section 7) and, for the $l = 0$ state, the correction due to the last term on the right-hand side of equation (12) of the preceding problem, we find

$$\Delta E = \frac{me^2}{\hbar^2}\left(\frac{e^2}{\hbar c}\right)^2\frac{1}{n^3}\left[\frac{3}{8n}-\frac{1}{2j+1}\right]. \tag{7}$$

This expression is independent of l, which means that two states with the same value of j but different values of l correspond to the same (degenerate) energy.

24. We start again from equations (1) of problem 21 of section 11, and now we can put

$$\phi = A_x = A_z = 0\,, \quad A_y = \mathcal{H}x\,, \tag{1}$$

if we take the z-axis along $\mathcal{H}$.

Eliminating χ we find

$$\{c^2\hat{\boldsymbol{p}}^2 + e^2\mathcal{H}^2x^2 - ec\mathcal{H}(\hbar\hat{\sigma}_z + 2x\hat{p}_y)\}\varphi = (E^2 - m^2c^4)\varphi\,. \tag{2}$$

We look for solutions which are eigenfunctions of $\hat{p}_y, \hat{p}_z$, and $\hat{\sigma}_z$ with eigenvalues p_y, p_z, and σ, and we write therefore

$$\varphi = \exp\{i(p_y y + p_z z)/\hbar\}\begin{pmatrix}1\\0\end{pmatrix}\Psi \quad \text{or} \quad \exp\{i(p_y y + p_z z)/\hbar\}\begin{pmatrix}0\\1\end{pmatrix}\Psi\,. \tag{3}$$

The function Ψ must satisfy the equation

$$\{c^2\hat{p}_x^2 + e^2\mathcal{H}^2x^2 - ec\mathcal{H}(\hbar\sigma + 2xp_y)\}\Psi = (E^2 - m^2c^4 - p_y^2c^2 - p_z^2c^2)\Psi\,, \tag{4}$$

or

$$\{c^2\hat{p}_x^2 + (e\mathcal{H}x + cp_y)^2\}\Psi = (E^2 - m^2c^4 - p_z^2c^2 + ec\mathcal{H}\hbar\sigma)\Psi\,. \tag{5}$$

The left-hand side operator is the Hamiltonian of a one-dimensional harmonic oscillator (compare the similar situation in the non-relativistic case; see problem 4 of section 6). The energy levels thus follow from the equation

$$E^2 - m^2c^4 - p_z^2c^2 + ec\hbar\mathcal{H}\sigma = |e|\mathcal{H}(2n+1)\hbar\,, \quad n = 0, 1, 2, \ldots\,. \tag{6}$$

One can easily check that in the non-relativistic limit the result is the same as that obtained in problem 3 of section 6.

25. (i) In this case we must satisfy instead of the four equations (13) of problem 8 of section 11 the equations

$$[\hat{\alpha}_k, \hat{\alpha}_l]_+ = 2\,.\,\hat{1}\,.\,\delta_{kl}\,, \tag{2}$$

and this can be done by three 2×2 matrices, taking

$$\hat{\boldsymbol{\alpha}} = c\hat{\boldsymbol{\sigma}}\,. \tag{3}$$

(ii) The discussion proceeds as in problem 20 of section 11.

(iii) In an energy eigenstate we have

$$i\hbar\frac{\partial\Psi}{\partial t} = E\Psi\,, \tag{4}$$

and as

$$c(\hat{\sigma}\,.\hat{\boldsymbol{p}})\Psi = i\hbar\frac{\partial\Psi}{\partial t}\,, \tag{5}$$

we find

$$cp\,\cos\gamma = E\,, \tag{6}$$

where γ is the angle between spin and momentum. As $E^2 = p^2c^2$ or $E = \pm pc$, the proof follows.

Index

The numbers refer to the problems; for instance, 5.14 is problem 14 of section 5.